Towards a Thermodynamic Theory for Ecological Systems

Towards a Thermodynamic Theory for Ecological Systems

Sven Erik Jørgensen
DFU, Environmental Chemistry
Universitetsparken 2
2100 Copenhagen
Denmark

Yuri M. Svirezhev
Potsdam Institute for Climate Impact Research
PO Box 601203
14412 Potsdam
Germany

2004

ELSEVIER

Amsterdam – Boston – Heidelberg – London – New York – Oxford
Paris – San Diego – San Francisco – Singapore – Sydney – Tokyo

ELSEVIER B.V.
Sara Burgerhartstraat 25
P.O. Box 211, 1000 AE
Amsterdam, The Netherlands

ELSEVIER Inc.
525 B Street, Suite 1900
San Diego, CA 92101-4495
USA

ELSEVIER Ltd
The Boulevard, Langford Lane
Kidlington, Oxford OX5 1GB
UK

ELSEVIER Ltd
84 Theobalds Road
London WC1X 8RR
UK

First edition 2004

Library of Congress Cataloging in Publication Data
A catalog record is available from the Library of Congress.

British Library Cataloguing in Publication Data
A catalogue record is available from the British Library.

ISBN: 0 08 044166 1 (hardbound)
ISBN: 0 08 044167 X (paperback)

♾ The paper used in this publication meets the requirements of ANSI/NISO Z39.48-1992 (Permanence of Paper). Printed in The Netherlands.

"Believe nothing, no matter where you read it,
or who said it,
no matter if I have said it,
unless it agrees with your own reason
and your own common sense."

Buddha

"Beware of Mathematicians
and those that make hollow prophesy.
There is a danger that they made a deal with Devil
In order to disconcert Souls and bring
the entire Humankind to Hell."

St. Augustine of Hippo

CONTENTS

PREFACE

This volume has two first authors because it is a result of very intensive teamwork between the two authors. We have had three brainstorm meetings, each of approximately one week's duration. There have been numerous ping-pong games (questions–answers–new proposals etc.) on the Internet. All the chapters have major contributions from both of us. We hope that the volume therefore demonstrates a synergistic effect, reflecting the positive teamwork that is behind the volume. The teamwork has been particularly fruitful because we have different scientific backgrounds, but still have ecological modelling and thermodynamics as a common platform. Sven Erik Jørgensen has, in addition to modelling and thermodynamics, a background in chemistry and biology (mainly system ecology), while Yuri M. Svirezhev is a passionate mathematician, who has used his mathematics during almost his entire career on biological–ecological problems. He has been able to present many of Sven Erik Jørgensen's previously published ideas with the right mathematical elegance, but there are also a lot of new ideas that are a result of the teamwork and the brain storming meetings. In turn, many concepts of mathematical ecology developed by Yuri Svirezhev are considered in the book from the new, thermodynamic point of view.

The application of thermodynamics on biological systems far from thermodynamic equilibrium is not new. It is possible to find numerous references on this topic in the past with Ostwald's, Bauer's and Prigogine's contributions as maybe the most important. Over the last three decades, many new and original contributions have been added to the previous theory, and we believe that today we have a solid and applicable theory of ecological systems far from thermodynamic equilibrium that is sufficiently developed to explain ecological observations (see the final chapter). We have built the presented theory very much on our own work using exergy as a core thermodynamic variable. We have, however, also touched on other approaches presenting the research from the last three decades in system ecology (for instance, the work of H.T. Odum, B.C. Patten, R. Ulanowicz, E. Tiezzi and F. Müller). Ecosystems are extremely complex systems, and it is therefore not surprising that the various approaches are to a large extent complementary. Presentation of a consistent and comprehensive theory is, however, facilitated by the application of one approach with which you are familiar. Therefore, the application of exergy to explain the ecosystem reactions and processes is the core theme. In addition, we believe that a consistent and comprehensive theory cannot be developed, at least not today, without thermodynamics. We hope, however, whatever background you may have as reader of this book, and whatever ecosystem approach you prefer, that you do agree: we have an ecosystem theory and we should use it much more widely in ecology. The application of thermodynamics on ecosystems requires a heavy use of mathematics; but to emphasise the application of the theory to understand ecosystems and to explain ecological observations, we have included an ecosystem theoretical summary in most chapters. We have here presented the implications of the theory of particular interest in system ecology.

Behind the presented theory are many important contributions from other scientists with whom we have cooperated for a shorter or longer time. They have inspired us by their thoughts, not only the thoughts resulting in joint publications, but also the many thoughts that have been "hanging in the air" at a brain storming. We would like to express our appreciation to all of them: Vyacheslav Alexeev, Brian Fath, Niels Ladegaard, Joao Marques, Henning Mejer, Felix Muller, Søren Nors Nielsen, Bernard C. Patten, Vladimir Petukhov, Vicente Santiago, Wolf Steinborn, Alexey Voinov, Maciej Zalewski, Nikolai Zavalishin and J. Zhang.

We are also grateful to Valentina Krysanova, Valery Pomaz, Alison Schlums, Stephen Sitch and Anastasia Svirejeva-Hopkins for their help in the preparation and editing of our manuscript.

Finally, we are very grateful to H.-J. Schellnhuber, the Director of the Potsdam Institute for Climate Impact Research, who has provided Potsdam's player in our team with the perfect conditions for working on this book.

Sven Erik Jørgensen and Yuri M. Svirezhev,
Copenhagen and Potsdam, September 2003.

Andiam. Incominciate!
***Leoncavallo* "Pagliacci"**

Chapter 1
Thermodynamics as a method: a problem of statistical description

Thermodynamics is full of highly scientific and charming terms and concepts, giving an impression of philosophical and scientific profundity. *Entropy, thermal death of the Universe, ergodicity, statistical ensemble*—all these words sound very impressive posed in any order. But, placed in the appropriate order, they can help us to find the solution of urgent practical problems. The problem is how to find this order...
(from table talks in Copenhagen and Potsdam).

1.1. Literary introduction

In the beginning, thermodynamics was an experimental science, and it was only after the work of Gibbs and Boltzmann that an understanding of the statistical basement of all thermodynamic relations appeared. Nevertheless, it is necessary to note that, despite all these discussions about determinism and randomness, metasystemic properties of large systems and the macroscopic description of ensembles consisting of the large number of "similar" microscopic units that are considered as historical facts today, a full understanding has still not been achieved. This is especially so in relation to sciences that differ from physics and chemistry, such as biology and social sciences, where we also deal with ensembles of many interacting individuals (particles, "molecules", etc.?), and where the idea of applying thermodynamics formalism is very attractive. But "before to discuss the problem, let us come to an agreement about definitions" (N. Timofeev-Resovsky).

Over many years of one of us delivering a course of lectures under the title "Mathematical biology" at the Moscow State University for *mathematicians*, the problem arose of the meaning of such terms as statistical ensemble, stochasticity or randomness, stochastic processes, how to pass from microscopic description to macroscopic one, etc.? We should like to avoid a superfluous "bourbakism" in these definitions and descriptions. A lot of different books were examined with an unexpected result. The best description of the nature of randomness, the relation between microscopic and macroscopic variables, and, as a special application, the role of stochasiticity and determinism in human history was given by Leo Tolstoy in his great novel "War and Peace". Let us cite these pages.

Towards a Thermodynamic Theory for Ecological Systems, pp. 1–11

"From the close of the year 1811 intensified arming and concentrating of the forces of Western Europe began, and in 1812 these forces—millions of men, reckoning those transporting and feeding the army—moved from the west eastwards to the Russian frontier, toward which since 1811 Russian forces had been similarly drawn. On the twelfth of June 1812, the forces of Western Europe crossed the Russian frontier and war began, that is, an event took place opposed to human reason and to human nature. Millions of men perpetrated against one another such innumerable crimes, frauds, treacheries, thefts, forgeries, issues of false money, burglaries, incendiarisms, and murders as in whole centuries are not recorded in the annals of all the law courts of the world, but which those who committed them did not at the time regard as being crimes.

What produced this extraordinary occurrence? What were its causes? The historians tell us with naive assurance that its causes were the wrongs inflicted on the Duke of Oldenburg, the non-observance of the Continental System, the ambition of Napoleon, the firmness of Alexander, the mistakes of the diplomatists, and so on.

Consequently, it would only have been necessary for Metternich, Rumyantsev, or Talleyrand, between a levee and an evening party, to have taken proper pains and written a more adroit note, or for Napoleon to have written to Alexander: "My respected Brother, I consent to restore the duchy to the Duke of Oldenburg"—and there would have been no war.

We can understand that the matter seemed like that to contemporaries. It naturally seemed to Napoleon that the war was caused by England's intrigues (as in fact he said on the island of St. Helena). It naturally seemed to members of the English Parliament that the cause of the war was Napoleon's ambition; to the Duke of Oldenburg, that the cause of the war was the violence done to him; to businessmen that the cause of the war was the Continental System which was ruining Europe; to the generals and old soldiers that the chief reason for the war was the necessity of giving them employment; to the legitimists of that day that it was the need of re-establishing *les bons principes*, and to the diplomatists of that time that it all resulted from the fact that the alliance between Russia and Austria in 1809 had not been sufficiently well concealed from Napoleon, and from the awkward wording of Memorandum No. 178. It is natural that these and a countless and infinite quantity of other reasons, the number depending on the endless diversity of points of view, presented themselves to the men of that day; but to us, to posterity who view the thing that happened in all its magnitude and perceive its plain and terrible meaning, these causes seem insufficient. To us it is incomprehensible that millions of Christian men killed and tortured each other either because Napoleon was ambitious or Alexander was firm, or because England's policy was astute or the Duke of Oldenburg wronged. We cannot grasp what connection such circumstances have with the actual fact of slaughter and violence: why because the Duke was wronged, thousands of men from the other side of Europe killed and ruined the people of Smolensk and Moscow and were killed by them.

To us, their descendants, who are not historians and are not carried away by the process of research and can therefore regard the event with unclouded common sense, an incalculable number of causes present themselves. The deeper we delve in search of these causes the more of them we find; and each separate cause or whole series of causes appears to us equally valid in itself and equally false by its insignificance compared to the magnitude of the events, and by its impotence—apart from the cooperation of all the other coincident causes—to occasion the event. To us, the wish or objection of this or that French corporal to serve a second term appears as much a cause as Napoleon's refusal to withdraw his troops beyond the Vistula and to restore the duchy of Oldenburg; for had he not wished to serve, and had a second, a third, and a thousandth corporal and private also refused, there would have been so many less men in Napoleon's army and the war could not have occurred.

If Napoleon had not taken offence at the demand that he should withdraw beyond the Vistula, and not ordered his troops to advance, there would have been no war; but had all his sergeants objected to serving a second term then also there could have been no war. Nor could there have been a war had there been no English intrigues and no Duke of Oldenburg, and had Alexander not felt insulted, and had there not been an autocratic government in Russia, or a Revolution in France and a subsequent dictatorship and Empire, or all the things that produced the French Revolution, and so on. Without each of these causes nothing could have happened. So all these causes—myriads of causes—coincided to bring it about. And so there was no cause for that occurrence, but it had to occur because it had to. Millions of men, renouncing their human feelings and reason, had to go from west to east to slay their fellows, just as some centuries previously hordes of men had come from the east to the west, slaying their fellows.

The actions of Napoleon and Alexander, on whose words the event seemed to hang, were as little voluntary as the actions of any soldier who was drawn into the campaign by lot or by conscription. This could not be otherwise, for in order that the will of Napoleon and Alexander (on whom the event seemed to depend) should be carried out, the concurrence of innumerable circumstances was needed without any one of which the event could not have taken place. It was necessary that millions of men in whose hands lay the real power—the soldiers who fired, or transported provisions and guns—should consent to carry out the will of these weak individuals, and should have been induced to do so by an infinite number of diverse and complex causes.

We are forced to fall back on fatalism as an explanation of irrational events (that is to say, events the reasonableness of which we do not understand). The more we try to explain such events in history reasonably, the more unreasonable and incomprehensible do they become to us.

Each man lives for himself, using his freedom to attain his personal aims, and feels with his whole being that he can now do or abstain from doing this or that action; but as soon as he has done it, that action performed at a certain moment in time becomes irrevocable and belongs to history, in which it has not a free but a predestined significance.

There are two sides to the life of every man, his individual life, which is the more free the more abstract its interests, and his elemental hive life in which he inevitably obeys laws laid down for him.

Man lives consciously for himself, but is an unconscious instrument in the attainment of the historic, universal, aims of humanity. A deed done is irrevocable, and its result coinciding in time with the actions of millions of other men assumes an historic significance. The higher a man stands on the social ladder, the more people he is connected with and the more power he has over others, the more evident is the predestination and inevitability of his every action.

The king's heart is in the hands of the Lord.

A king is history's slave.

History, that is, the unconscious, general, hive life of mankind, uses every moment of the life of kings as a tool for its own purposes.

Though Napoleon at that time, in 1812, was more convinced than ever that it depended on him, *verser (ou ne pas verser) le sang de ses peoples*—as Alexander expressed it in the last letter he wrote him—he had never been so much in the grip of inevitable laws, which compelled him, while thinking that he was acting on his own volition, to perform for the hive life—that is to say, for history—whatever had to be performed. "To shed (or not to shed) the blood of his peoples."

The people of the west moved eastwards to slay their fellow men, and by the law of coincidence thousands of minute causes fitted in and co-ordinated to produce that movement and war: reproaches for the non-observance of the Continental System, the Duke of Oldenburg's wrongs, the movement of troops into Prussia—undertaken (as it seemed to Napoleon) only for the purpose of securing an armed peace, the French Emperor's love and habit of war coinciding with his people's inclinations, allurement by the grandeur of the preparations, and the expenditure on those preparations and the need of obtaining advantages to compensate for that expenditure, the intoxicating honours he received in Dresden, the diplomatic negotiations which, in the opinion of contemporaries, were carried on with a sincere desire to attain peace, but which only wounded the self-love of both sides, and millions and millions of other causes that adapted themselves to the event that was happening or coincided with it.

...Nothing is the cause. All this is only the coincidence of conditions in which all vital organic and elemental events occur. In historic events the so-called great men are labels giving names to events, and like labels they have but the smallest connection with the event itself.

Every act of theirs, which appears to them an act of their own will, is in an historical sense involuntary and is related to the whole course of history and predestined from eternity." (*Volume III, Book I, Chapter I*)

"...The movement of humanity, arising as it does from innumerable arbitrary human wills, is continuous.

To understand the laws of this continuous movement is the aim of history. But to arrive at these laws, resulting from the sum of all those human wills, man's mind postulates arbitrary and disconnected units. The first method of history is to take an arbitrarily selected series of continuous events and examine it apart from others, though there is and can be no beginning to any event, for one event always flows uninterruptedly from another.

The second method is to consider the actions of some one man—a king or a commander—as equivalent to the sum of many individual wills; whereas the sum of individual wills is never expressed by the activity of a single historic personage.

Historical science in its endeavour to draw nearer to truth continually takes smaller and smaller units for examination. But however small the units it takes, we feel that to take any unit disconnected from others, or to assume a beginning of any phenomenon, or to say that the will of many men is expressed by the actions of any one historic personage, is itself false.

It needs no critical exertion to reduce utterly to dust any deductions drawn from history. It is merely necessary to select some larger or smaller unit as the subject of observation—as criticism has every right to do, seeing that whatever unit history observes must always be arbitrarily selected.

Only by taking infinitesimally small units for observation (the differential of history, that is, the individual tendencies of men) and attaining to the art of integrating them (that is, finding the sum of these infinitesimals) can we hope to arrive at the laws of history.

...To study the laws of history we must completely change the subject of our observation, must leave aside kings, ministers, and generals, and the common, infinitesimally small elements by which the masses are moved." (*Volume III, Book III, Chapter I*).

1.2. Ontic openness

One of the key questions in natural science in the XXth century was: is the world deterministic—in the sense that, if we would know the initial conditions in all details, could we also predict in all details how a system would develop—or is the world ontic open?

We cannot, and will probably never be able to, answer these two questions, but the world is under all circumstances too complex to enable us to determine the initial conditions. The uncertainty relations similar to Heisenberg's uncertainty relations in quantum mechanics are also valid in ecology. This idea has been presented in Jørgensen (1988, 1992c, 1997) but will be summarised below because the discussion in the next chapters is dependent on this uncertainty in our description of nature. The world may be ontic open = non-deterministic because the Universe has been created that way, or it may be ontic open = non-deterministic because nature is too complex to allow us to know a reasonable fraction of the initial conditions even for a subsystem of an ecosystem. We shall probably never be able to determine which of the two possibilities will prevail, but it is not of importance because we have anyhow to accept ontic openness in our description of nature.

A single method to get an information about a system is to observe it (from this point of view any experiment is an active observation). Let an ecosystem consist of n components so that it could be described by n variables, and a single act of observation is the determination of its state in the n-dimensional space. However, firstly, we do not know the value of n, i.e. a dimensionality of the state space, and secondly, we know nothing about a structure of the system, i.e. about relations between its variables, which determine the system structure. Note that a single observation with randomly chosen n does not give us any information about the structure and dimensionality. How many observations do we need in order to get this information? How do we organise the process of observation? We shall do it by a *recursive* method.

If the system is really one-dimensional then the single observation is enough for identification of its state. But if this hypothesis is wrong then we have to extend the space dimensionality by considering the case $n = 2$. This is a first step of our recursion. Let the two variables be x and y then the simplest non-linear relation between them is $y = a + bx + cx^2$ where a, b and c are constant. To determine their values, we need three observations. A second step is the introduction of the third variable, z. Then the simplest non-linear description of the ecosystem will be $y = a(z) + b(z)x + c(z)x^2$ where we assume again that the functions $a(z)$, $b(z)$ and $c(z)$ are parabols:

$$a(z) = a_1 + a_2 z + a_3 z^2, \quad b(z) = b_1 + b_2 z + b_3 z^2, \text{ and } c(z) = c_1 + c_2 z + c_3 z^2.$$

In order to determine all these coefficients we need nine observations. Continuing the process we obtain for $(n - 1)$th step, i.e. for the nth dimensionality, that the necessary number of observations will be equal to $N^n_{\text{obs}} = 3^{n-1}$. For instance, if $n = 20$ then $N^{20}_{\text{obs}} = 3^{19} \approx 10^9$, i.e. one billion observations!

Costanza and Sklar (1985) talk about the choice between the two extremes: knowing "everything" about "nothing" or "nothing" about "everything". The first refers to the use of all the observations on one relation to obtain a high accuracy and certainty, while the latter refers to the use of all observations on as many relations as possible in an ecosystem.

But, of course, the possibility that the practical number of observations may be increased in the future cannot be excluded. Ever more automatic analytical equipment is emerging on the market. This means that the number of observations that can be invested in one project may be one, two, three or even several magnitudes larger in one or more decades. However, a theoretical uncertainty relation can be developed. If we go to the limits given by quantum mechanics, the number of variables will still be low compared to the number of components in an ecosystem.

The Heisenberg uncertainty relations, $\Delta E \times \Delta t \geq h/2\pi$, where $h = 6.625 \times 10^{-34}$ J s is Planck's constant, where Δt is the uncertainty in time and ΔE in energy, may now be used to give the upper limit of the number of observations. Indeed, if we use all the energy that Earth has received during its lifetime of 4.5 billion years we get:

$$(1.73 \times 10^{17} \text{ W})(4.5 \times 10^9 \times 365.3 \times 24 \times 3600 \text{ s}) = 2.5 \times 10^{34} \text{ J},$$

where 1.73×10^{17} W is the energy flow of solar radiation. The value of Δt would, therefore, be in the order of 4×10^{-69} s. Consequently, an observation will take 4×10^{-69} s, even if we use all the energy that has been available on Earth as ΔE, which must be considered the most extreme case. The hypothetical number of observations possible during the lifetime of Earth would therefore be:

$$4.5 \times 10^9 \times 365.3 \times 3600/4 \times 10^{-69} \approx 1.5 \times 10^{84}.$$

This implies that, if to substitute this value into the formula related the number of variables in the ecosystem, n, and number of observations, N^n_{obs}, then we get:

$$n \approx 180.$$

From these very theoretical considerations, we can clearly conclude that we shall never be able to obtain a sufficient number of observations to describe even one ecosystem in all its details. These results are completely in harmony with Niels Bohr's complementarity theory. He expressed it as follows: "It is not possible to make one unambiguous picture (model or map) of reality, as uncertainty limits our knowledge." The uncertainty in nuclear physics is caused by the inevitable influence of the observer on the nuclear particles; in ecology the uncertainty is caused by the enormous complexity and variability.

No map of reality is completely correct. There are many maps (models) of the same piece of nature, and the various maps or models reflect different viewpoints. Accordingly, one model (map) does not give all the information and far from all the details of an ecosystem. In other words, the theory of complementarity is also valid in ecology.

The use of maps in geography is a good parallel to the use of models in ecology (Jørgensen and Bendoricchio, 2001). As we have road maps, aeroplane maps, geological maps, maps in different scales for different purposes, we have in ecology many models of the same ecosystems and we need them all if we want to get a comprehensive view of ecosystems. A map cannot, furthermore, give a complete picture. We can always make the scale larger and larger and include more details, but we cannot get all the details…for instance where all the cars of an area are situated just now, and if we could the picture would be invalid a few seconds later because we want to map too many dynamic details at the same time. An ecosystem also consists of too many dynamic components to enable us to model all the components simultaneously and, even if we could, the model would be invalid a few seconds later, where the dynamics of the system has changed the "picture."

In nuclear physics, we need to use many different pictures of the same phenomena to be able to describe our observations. We say that we need a pluralistic view to cover our observations completely. Our observations of light, for instance, require that we consider light as waves as well as particles. The situation in ecology is similar. Because of the immense complexity, we need a pluralistic view to cover a description of the ecosystems according to our observations. We need many models covering different viewpoints.

In addition to physical openness, there is also an epistemological openness inherent in the formal lenses through which humans view reality. Gödel's Theorem, which was published in January 1931, introduces an epistemic openness in a very strong way. The theorem requires that mathematical and logical systems (i.e. purely epistemic, as opposed to ontic) cannot be shown to be self-consistent within their own frameworks but only from outside. A logical system cannot itself (from inside) decide on whether it is false or true. This requires an observer from outside the system, and this means that even epistemic systems must be open.

We can distinguish between ordered and random systems. Many ordered systems have emergent properties defined as properties that a system possesses in addition to the sum of properties of the components—the system is more than the sum of its components. Wolfram (1984a,b) calls these *irreducible systems* because their properties cannot be revealed by a reduction to some observations of the behaviour of the components. It is necessary to observe the entire system to capture its behaviour because everything in the system is dependent on everything else due to direct and indirect linkages. The presence of irreducible systems is consistent with Gödel's Theorem, according to which it will never be possible to give a detailed, comprehensive, complete and comprehensible description of the world. Most natural systems are irreducible, which places profound restrictions on the inherent reductionism of science.

In accordance with Gödel's Theorem, the properties of order and emergence cannot be observed and acknowledged from within the system, but only by an outside observer. It is consistent with the proverb: "You cannot see the wood for the trees", meaning that if you only see the trees as independent details inside the wood you are unable to observe the system, the wood as a cooperative unit of trees. This implies that the natural sciences, aiming toward a description or ordering of the systems of nature, have meaning only for open systems. A scientific description of an isolated system, i.e. the presentation of an algorithm describing the observed, ordering principles valid for the system, is impossible. In addition, sooner or later an isolated ontic system will reach thermodynamic equilibrium, implying that there are no ordering principles, but only randomness. We can infer from this that an isolated epistemic system will always ultimately collapse inward on itself if it is not opened to cross fertilisation from outside. Thomas Kuhn's account of the structure of scientific revolutions would seem to proceed from such an epistemological analogy of the Second Law.

This does not imply (Jørgensen et al., 1999) that we can describe *all* open systems in *all* details. On the contrary, the only complete, detailed and consistent description of a system is the system itself. We can furthermore never know if a random system or subsystem is ordered or random because we have not found the algorithm describing the order. We can never know if it exists or we may find it later by additional effort. This is what modelling and model-making in accordance with our definition of life (Patten et al., 1997) is all about. A model is always a simplified or homomorphic description of some features of a system, but no model can give a complete or isomorphic description. Therefore, one might conclude that it will always require an infinite number of different models to realise a complete, detailed, comprehensive and consistent description of any entire system. In addition, it is also not possible to compute or totally explain our thoughts and conceptions of our limited, but useful description of open natural systems. Our perception of nature goes, in other words, beyond what can be explained and computed, which makes it possible for us to conceive irreducible (open) systems, though we cannot explain all the details of the system. This explains the applicability and usefulness of models in the adaptations of living things ("subjects", Patten et al., 1997) to their environment. It also underlines that the models in the best case will only be able to cover one or a few out of many views of considered systems. If we apply the definition of life proposed in Patten et al. (1997)—Life is things that make models—this implies that all organisms and species must make their way in the world based on only partial representations, limited by the perceptual and cognitive apparatus of each, and the special epistemologies or models that arise therefrom. The models are always incomplete but sufficient to guarantee survival and continuance, or else extinction is the price a failed model pays.

Following from Gödel's Theorem, a scientific description can only be given from outside open systems. Natural science cannot be applied to isolated systems at all (the Universe is considered open due to the expansion). A complete, detailed, comprehensive and consistent description of an open system can never be obtained. Only a partial, though useful, description (model) covering one or a few out of many views can be achieved.

Due to the enormous complexity of ecosystems we cannot, as already stressed, know all the details of ecosystems. When we cannot know all the details, we are not able to describe fully the initial stage and the processes that determine the development of the ecosystems—as expressed above, ecosystems are therefore irreducible. Ecosystems are not deterministic because we cannot provide all the observations that are needed to give a full deterministic description. Or, as expressed by Tiezzi: *ecosystems do play dice* (Tiezzi, 2003). This implies that our description of ecosystem developments must be open to a wide spectrum of possibilities. It is consistent with the application of chaos and catastrophe theory; see, for instance, Jørgensen (1992a,c, 1994, 1995a, 2002b). Ulanowicz (1997) makes a major issue of the necessity for systems to be causally open in order to be living—the open possibilities may create new pathways for development which may be crucial for survival and further evolution in a non-deterministic world. He goes so far as to contend that a mature insight into the evolutionary process is impossible without a revision of our contemporary notions on causality. Ulanowicz (1997) uses the concept of propensity to get around the problem of causality. On the one side, we are able to relate the development with the changing internal and external factors of ecosystems. On the other side, due to the uncertainty in our predictions of development caused by our lack of knowledge about all details, we are not able to give deterministic descriptions of the development, but we can only indicate which propensities will be governing.

To conclude: *Ecosystems have ontic openness. They are irreducible and, due to their enormous complexity which prohibits us from knowing all details, we will only be able to indicate the propensities of their development. Ecosystems are not deterministic systems.*

1.3. The scope of this volume

Science does not make sense without a theory. Without a theory our observations become only a beautiful pattern of impressions. All our knowledge in a scientific discipline has to be coherent to be able to apply the underlying theory to explain our observations. Ecology has for a long time only partially been able to condense the systematic collection of observations and knowledge about ecosystems into testable laws and principles. The authors of this volume are convinced that an ecological theory is now available as a tool in ecology due to the contributions of many system ecologists during the last decades, mainly through the application of thermodynamics to explain the reactions of ecosystems. It has been difficult and has taken a long time to construct the theoretical building of system ecology, but nature has not been created to be easily understood by human beings. It has been necessary to break with the long reductionistic tradition in science and use thermodynamics in a new holistic approach to understand ecosystems. Reductionistic science has had a continuous chain of successes since Descartes and Newton.

Lately, there has, however, been an increasing understanding for the need of syntheses of knowledge into a holistic image to be able to grasp the sense of complex systems such as ecosystems and social systems. It is today considered by many scientists the greatest challenge of science in the XXIst century to put together our many observations of complex systems into a completely understandable holistic picture.

A number of ecosystem theories have been published during the last three decades. They are all attempts to capture the features and characteristics of ecosystems, their processes and their reactions to changed conditions, i.e. changed forcing functions. The different theories look at first glance not to be consistent, but when we examine the different theories more carefully, it becomes clear that they represent different angles and view points. It was asserted in the first edition of S.E. Jørgensen's book "Integration of Ecosystem Theories: A Pattern" (1992) that the various theories actually form a pattern, and the later editions (second edition 1997 and third edition 2002) have only enhanced the perception that the theories form a pattern and that to a large extent they are consistent. During 2000, there have been several meetings where the fathers of the theories met and discussed the pattern. It is clear from these discussions that we today have an ecosystem theory which is rooted in a consensus of the pattern of ecosystem theories. It is the intention of this volume to present this ecosystem theory as it has taken form in the beginning of the XXIst century, but with particular emphasis on the thermodynamic interpretation of this ecosystem theory. It does not mean that the network interpretation by Ulanowitz and Patten or the green accounting using emergy by H.T. Odum are less important. They are just other angles to, in principle, the same ecosystem theory, as the quantum mechanic theory has been approached differently by Heisenberg's uncertainty relationships and by Schrödinger's wave functions.

The thermodynamic interpretation of an ecosystem theory by use of the concept of exergy has been chosen as the main focus of this volume. Exergy may be applied as a core concept in a thermodynamic edition of an ecosystem theory, as will be shown many times throughout the volume. The various approaches have, however, different advantages in different situations. When an ecosystem problem is best solved by use of an approach

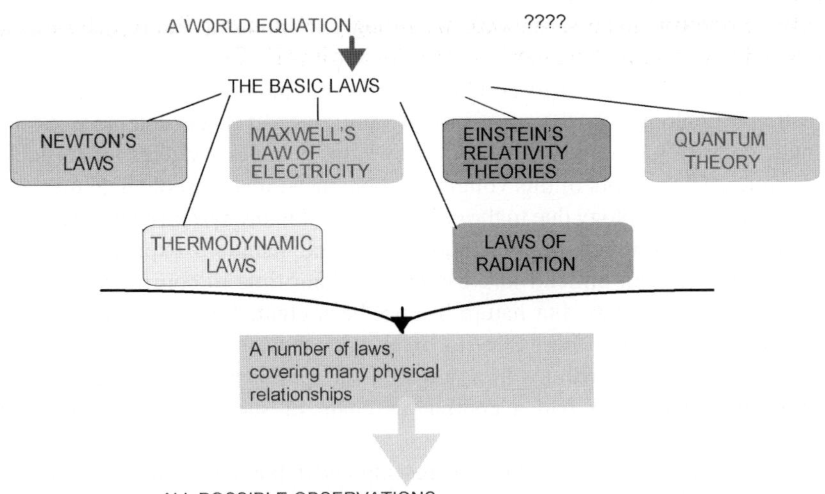

Fig. 1.1. The theoretical network of physics consists of a few fundamental laws, for instance the thermodynamic laws, from which other laws can be derived. All (or almost all) observations can be explained by a fundamental law or a derived law.

based on energy and exergy, these concepts should be applied, but when the processes and reactions concern the network, the use of a network theoretical approach may give clear advantages. The relationship between the different approaches will therefore be mentioned to emphasise the importance of a pluralistic view to describe an ecosystem. When a simple physical phenomenon such as light needs two descriptions, it is not a surprise that a very complex ecosystem needs many different complementary descriptions (Jørgensen, 1992c, 2001b, 2002b).

It is a very important step forward in ecology and system ecology that we now have a theory that several system ecologists can agree upon, as this is the prerequisite for further progress in system ecology. Furthermore, it makes it feasible to construct a network of laws, rules and observations as we know from physics, where a few fundamental laws can be applied to derive other laws which can be used to explain, if not all, then almost all physical observations (see Fig. 1.1). We do not know yet to what extent this is possible in ecology, but, assuming that it is the right time to start to build such a theoretical network in ecology, it should be possible at least to propose a promising direction for our thought and create some fragments of the network.

Our book demonstrates that it is possible with the present ecosystem theory in hand to start to build such a theoretical network in ecology and shows to what extent such a theoretical network has been established today. It may be concluded that we do have sufficient knowledge about the behaviour of ecosystems to be able to explain many observations, rules and regressions on the basis of an ecosystem theory.

The advantages of having an ecosystem theory is, of course, that it allows us to understand nature better, including the behaviour of ecosystems and their reactions to different perturbations. An ecosystem theory is, however, also applicable in environmental management, because it allows us to predict how ecosystems will react to various sets of man-controlled forcing functions.

Finally, in addition to our own book, we strongly recommend "Entropy for Biologists: an introduction to thermodynamics" by H.J. Morowitz (1970).

Chapter 2
The laws of classical thermodynamics and their application to ecology

When I had been a student, I have read with pleasure F. Wald's small book under the title "The Queen of the World and her Shadow". Energy and entropy were kept in mind. Now, when I understand these concepts deeper, I think that their positions should be interchanged. In the giant factory of natural processes the entropy law is a director who controls and manages all the business, while the energy conservation law is only an accountant who is keeping a balance between debit and credit.

Robert Emden, 1938.

2.1. Introduction

Following a classic biosphere and ecological tradition (Vernadsky, 1926; Lindeman, 1942), we can say that the structure, functioning and evolution of the biosphere, as a whole, and ecosystems in particular, are mainly determined by the flows of matter, heat and radiation. A physical study of their interaction is possible in the framework of thermodynamics. The subject of classic thermodynamics is the consideration of conformities in energy transformations when they are transferred between different physical bodies (which we shall call "systems") in the form of heat and work. Thermodynamics methods are applicable to so-called *macroscopic* systems consisting of a great number of particles. In physics, the particles are molecules.

Thermodynamics is now one of the most complete and elegant chapters of theoretical physics. However, the domain of the applications of its general concept and methods is much larger than purely physical and chemical systems. Generally speaking, a thermodynamics approach allows us to describe general (*macroscopic*, *systemic*) properties of the systems consisting of the large number of interacting elementary (*microscopic*) objects (*particles*). Therefore, when we talk about a thermodynamic approach in such a semi-empirical and phenomenological natural science as ecology, we hope implicitly that these methods help us to overcome an ecological "perdition of dimension". In other words, we could reduce a huge number of the individual descriptions of specimens, populations, their interactions, concentrations of different chemical substances, etc. to a few *macroscopic* variables and parameters, which will determine some generalised states of the ecosystem.

Towards a Thermodynamic Theory for Ecological Systems, pp. 13–40

Before applying the thermodynamic concepts and methods to ecological systems we have to tell our readers about them. But before that, we shall try to answer the question: "What, strictly speaking, do we understand by the notions heat, energy and entropy?" We also have to define: "What is meant by a 'system'?" The latter is a main word in the thermodynamic thesaurus. When we talk about a system, naturally we imply that we know what a "system border" and "system environment" are. The system exchanges matter and energy with its environment across its borders. A type of exchange generates the systems classification.

In an *isolated* system neither energy nor matter exchanges with its environment. In a *closed* system only energy exchanges with its environment, but it cannot exchange matter. An *open* system is the system which exchanges energy and matter with its environment. From a thermodynamic point of view, any ecosystem (except the biosphere, which is the biggest ecosystem of our planet) is an open system, which exchanges energy (incoming solar radiation and outgoing heat irradiation) and matter (water, carbon dioxide, nutrients, organic matter, etc.) with the environment. The biosphere is a typical closed system, since it exchanges only energy with space (matter exchange can be neglected with a good approximation). Certainly, all these classes are theoretical models of real systems. In reality, there are not purely isolated, closed and open systems. The same system can be isolated during some time intervals, but be open during the others. For instance, the biosphere is a closed system with respect to thousands of years, and it is an open system with respect to geological times, when we cannot neglect the matter exchange between the core and the mantle. Generally speaking, the successful setting of a system gives a more than 50% guarantee of solving the problem.

The main bodies in thermodynamics are the *system* and its *environment.* As a rule, this pair forms an absolutely isolated *supersystem*, while the system and its environment are exchanged by matter and energy through a border of the system (Fig. 2.1). There are two sorts of equilibriums in thermodynamics. The first, thermodynamic equilibrium, takes place when there is no exchange of matter and energy between the system and its environment, and its state does not change. The second, dynamic equilibrium (steady state), is maintained by non-zero flows of energy, matter (and entropy) across borders of the system.

The world of the classic Newton's mechanics is isotropic, i.e. there are not some special preferable directions with respect to both the space and the time. For instance, all equations of Newton's mechanics do not change (they are invariant) when the positive time is replaced by the negative one. Another situation applies in thermodynamics, when the time is strictly oriented from the past towards the future. What could generate this anisotropy of time? In standard textbooks on thermodynamics, the reason is shyly hidden, but for us it is very important, since one of the main concepts of our book, namely, the concept of exergy (see Chapter 5), is based on this reason.

The basic hypothesis is very simple: *the environment is much more than the system*, so that the latter is a very small part of the first. A state of the system depends on a state of the environment, and any change in the environment causes some change in the system. However, the inverse is not true: the system does not influence its environment. Any processes can take place within the system but they cause no effects on the state of the environment.

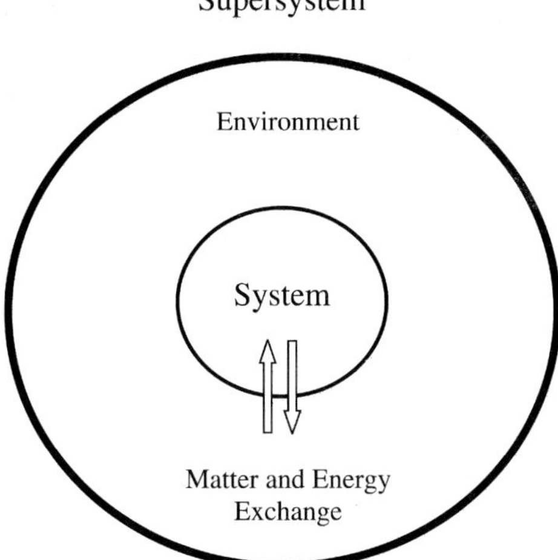

Fig. 2.1. System, its environment and their superposition into a supersystem.

It is obvious that acceptance of the hypothesis immediately provides the anisotropy of action: the action is not equal to the counteraction. In other words, we have to depart from the Third Newton's Law: action is equal to counteraction. But this immediately allows us to depart from the temporal symmetry of Newton's mechanics. Note that, generally speaking, thermodynamics does not need this mechanics.

A state of the system is determined by a collection of thermodynamic macroscopic parameters (variables) that in turn are divided between extensive and intensive. Each extensive parameter for the system is equal to the sum of the corresponding extensive parameters of the macroscopic subsystems, which are components of the original system. It is easy to see that the energy, mass, number of particles and volume are typical extensive parameters. In other words, the *extensivity* of parameters reflects such a fundamental property of the system as its *additivity*. But such parameters as the temperature, pressure and density are intensive ones. They are some mean characteristics of the system. In order to clarify the concept of *intensitivity*, we consider the following example. Let us have two volumes V_1 and V_2, the number of particles in which are equal to N_1 and N_2, respectively. Then the densities will be equal to $n_1 = N_1/V_1$ and $n_2 = N_2/V_2$. By joining these subsystems into a single system we have the following expression for density of particles in this joined system:

$$n = \frac{N_1 + N_2}{V_1 + V_2} = \frac{n_1 V_1 + n_2 V_2}{V_1 + V_2} = n_1 p_1 + n_2 p_2,$$

where $p_1 = V_1/(V_1 + V_2), p_2 = V_2/(V_1 + V_2)$ are the volume fractions of each subsystem in the total system. We can see that the total density is a *mean* of subsystem densities. Generally speaking, we can say that the system state (in relation to intensive parameters)

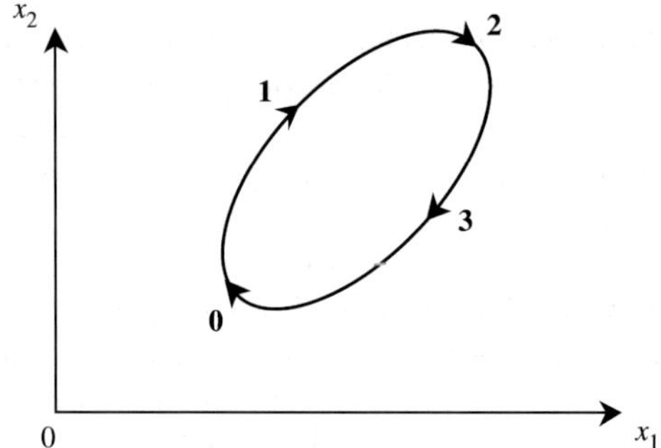

Fig. 2.2. Cyclic process: x_1 and x_2 are state variables, "0" = "4" are the initial and final states, "1", "2" and "3" are the intermediate states.

is described by mean parameters that are obtained by averaging corresponding parameters on the entire ensemble of subsystems.

Processes which take place in the system may be either equilibrium (*reversible*) or non-equilibrium (*irreversible*). Changes in the system caused by reversible processes can be neglected as a result of inverse consequence of local transitions without any changes in the environment. On the contrary, when the system moves inversely to its initial state in an irreversible process then the movement is accompanied by residual changes in the environment. Note that mainly reversible equilibrium processes are considered in classic thermodynamics.

We introduce also one important auxiliary concept, viz. the concept of *cyclic process.* Let the system start from some initial state, which is described by state variables. The system moves along the sequence of intermediate states and finishes where the movement started (Fig. 2.2). This is a *cyclic* process.

2.2. Matter and energy in mechanics and thermodynamics. Energy conservation as the First Law of Thermodynamics. Fundamental Gibbs equation

The conservation law of matter had probably already been accepted in ancient Greece and by the alchemists in medieval times, although it has also been presumed violated when the reproduction of microorganisms was observed. Lavoisier was the first to formulate the mass conservation principle when he stated, in 1785, that the total mass in a system remained constant after both chemical and physical processes.

The mass conservation principle is applied widely in chemical stoichiometric calculations, as the elements of the reactants are equal to the elements of the products. The mass conservation principle may also be formulated in everyday language: *nothing comes from nothing*. If we keep in mind the definition of an isolated system, then we can formulate the mass conservation principle in the following form: in the isolated system the total amount of matter is constant.

Along with matter, the ancient Grecians and medieval alchemists also considered a "very special matter" such as "fire" or "flogiston", i.e. *heat*. Now, we know that heat is not matter (in the proper sense) but it is one of the forms of energy. Heat can be transformed into other forms (mechanical energy, chemical, electrical, etc.), and the other forms of energy can be transformed into heat. *Energy cannot be created or destroyed.* This is the energy conservation principle called the First Law of Thermodynamics. It is obvious that in the isolated system the total amount of energy is constant. If the system is open then both matter and energy may be either accumulated or spent. This means that for both matter and energy the following equation, that could be called the "book keeping equation" since it can be also used in book keeping, is valid:

$$\text{Change of mass or energy} = \text{inputs} - \text{outputs}. \tag{2.1}$$

The change of energy $\mathrm{d}E$ can be represented as the sum of two items: the item which corresponds to internal processes within the system (i), and the item which corresponds to processes of exchange between the system and its environment (e):

$$\mathrm{d}E = \mathrm{d_i}E + \mathrm{d_e}E. \tag{2.2}$$

Since the energy cannot be created or destroyed within the system then $\mathrm{d_i}E = 0$.

The term $\mathrm{d_e}E$ can be represented in the form of the sum of work done on the system by its environment, δA, and heat δQ gained by the system from its environment:

$$\mathrm{d_e}E = \delta A + \delta Q. \tag{2.3}$$

This is a quantitative form of the First Law, initiated with James Joule's work (published in 1843 and 1849). He has shown that mechanical and electrical energy could be converted into heat, and demonstrated that the same amount of heat was always produced from a given amount of mechanical or electrical work, and that both the amount of heat and work as its equivalent are measured in the same units, calories or joules (1 cal = 4.18 J). Note that here we encounter the new concept of *work*, which is closely connected with transformations of energy and accompanies them.

An external force applied to a body, i.e. an action of the environment on a system, does work on it. In accordance with the general laws of mechanics, the work is the product of the force and the shift distance produced by it. The work can be used to change the kinetic and potential energy of the system, to change its chemical composition, etc. However, here and later on we shall exclude all the processes that lead to the changes in the kinetic and potential energy of the system. Then the total energy can be represented as $U = E - E_{\text{kinetic}} - E_{\text{potential}}$, where U is the so-called *internal energy* of the system.

It is necessary to distinguish the work δA done by the environment on the system, which counts as positive, and the work $\delta A'$ done by the system on its environment (and, certainly, changes it), which is negative. Analogously, we shall count the heat received by the system from its environment as positive, and vice versa.

Let us consider the following example: there is a cylindrical tube with piston of cross-section area σ filled by some "ideal" gas (system). The piston is densely adjacent to the interior lateral area of the tube. We press on the piston with force $p\sigma$, where p is the pressure, and compress the gas. The piston is shifted the distance dx. We consider this as a transition from the initial point "0" to the final point "1". Since in mechanics the work is defined as the product of force by distance then we (i.e. the environment) do the work $\delta A = -p\sigma\, dx = -p\, dV$, where the change of volume $dV < 0$, since gas is compressed. Then $\delta A > 0$. In addition, we can heat the gas, gaining as a result some amount of heat δQ. As a result we also increase the internal energy of the gas so that the change of internal energy $dU_{01} = \delta Q + \delta A$. After this the system is *capable* to do some work (on the environment), but does not do it until we (the environment) set free the piston, and the system starts to move backwards towards its initial state, i.e. "free" equilibrium. In the process of the movement it does the *useful* work $\delta A'$, gives back the environment some heat $\delta_e Q'$ and produces some heat $\delta_i Q'$ within the system. The latter is a consequence of irreversible processes within the system, for instance the friction of piston on the walls of the tube. As a result, the change of internal energy will be equal to $dU_{10} = \delta Q' + \delta A'$, where $\delta Q' = \delta_e Q' + \delta_i Q'$. In accordance with the energy conservation law applied to the sum of these two transitions (this is a typical cyclic process) $dU_{010} = dU_{01} + dU_{10} = 0$, but it is not necessary that $\delta A = -\delta A'$ and $\delta Q = -\delta Q'$. Therefore, the infinitesimal change in internal energy is the full differential but the changes in heat and work are not the full differentials. In other words, U is a state variable but A and Q are not the same, and the system energy cannot be split into heat energy and, for instance, mechanical or chemical energy.

The work done and the capability to do work are not equivalent. Even for a very simple system (such as an ideal gas in the tube with a piston), the system is capable to do the work only after the environment has done some work on the system. Although the two events are subdivided by a temporal interval in the dynamic case, this difference disappears in the static case. Thus, the fact that $\delta A'$ and δA become equivalent (but not equal!) gives birth to a lot of misunderstanding. But let us not forget about their original non-equivalence!

Both conservation laws (matter and energy) are independent of each other, but according to Einstein's famous equation $E = mc^2$, matter can be transferred to matter and energy to matter and the equivalent amount of mass and energy can be found by using this equation. E symbolises the energy, m the mass and c the light velocity ($c = 3 \times 10^8$ m/s). Therefore, these laws are not independent but, since the transfers from matter to energy and vice versa have interest for nuclear and plasma processes, in biology we can consider them as independent. For practical use of the conservation laws in ecosystems, it can be stated that matter and energy are conserved. However, Einstein's equation is very important in this theory since it provides the energy positiveness ($E_0 = m_0c^2 > 0$ where m_0 is the "rest-mass").

The energy conservation law is a fundamental of mechanics (and electrodynamics), but it is insufficient for the complete dynamical description of a mechanical system. We need also some relation connecting three main mechanical variables, mass m, force F and velocity v. The second Newton law gives the relation: $d(mv)/dt = F$, while the other two laws define the structure of "mechanical" space and time. However, in ecology (and, for instance, in chemical kinetics also) there are not any analogies of Newton's equations; using for the deduction of Lotka–Volterra equations the so-called *principle of collisions* is rather a phenomenological one.

Strictly speaking, the laws of mechanics (and electrodynamics) are already sufficient in order to acquire the statement about the conservation of energy, but there is not any kind of general theoretical proof of the First Law: this is an empirical generalisation of our experience. In particular, despite the huge number of attempts to create *perpetuum mobile*, up to now it has not been constructed.

Let us consider a macroscopic system that can be described by such extensive variables as the internal energy U, a presently unknown variable, entropy, S, volume V and numbers of different virtual particles $n_1, \ldots, n_m, \ldots$, and also such intensive variables as the temperature T, pressure P, etc. As some unit of quantity of matter, either a single particle or 1 mol will be used. Note that this is not important since in our general formulas the choice of unit is not essential. So, by having done the work δA on the system and gaining to it the external heat δQ, we transfer it (quasi-steadily, without the disturbance of equilibrium) from the initial equilibrium "0" to some new state "1". This is a *forced* transition, as a result of which the internal energy increases in the value of $dU_{01} = dU > 0$.

In accordance with the First Law, $d_eE = dU = \delta Q + \delta A$. Since energy cannot be created or destroyed within the system, then the latter changes only by means of exchange processes or external work. The work in turn changes the state of the system, i.e. changes its extensive variables (only small, infinitesimal differentials are considered). However, in reality, some part of the work is spent in order to overcome different "resistances" (such as friction, resistance of conductor by electrical current, etc.). Therefore, work can be represented as $\delta A = \delta A_{us} + \delta A_{irrev}$ where the item δA_{us} is "useful" work and the item δA_{irrev} is a contribution of real irreversible processes into the increase of internal energy. If the "0" state is a thermodynamic equilibrium, then in the course of inverse spontaneous transition to the equilibrium the internal energy dU is spent in the processes of performance of the different work and released in the form of heat. The main Gibbs idea was that dU can be represented as a bilinear form of intensive variables and differentials of conjugated extensive variables, so that $\delta A = \sum X_k\, dx_k + \sum N_m\, d_e n_m$ and $\delta Q + \delta A_{irrev} - \sum N_m\, d_i n_m = T\, dS$. Here, X_k and x_k correspondingly form the pair $\{X_k, x_k\}$ of conjugate (by them) variables of work. For instance, the pair corresponding to the work done for compression is the pressure $-p$ and the volume V, so that the corresponding bilinear term is $-p\, dV$. If n_m corresponds to the number of the mth sort of virtual particles then $dn_m = d_e n_m + d_i n_m$ are the sum of change of their numbers as a result of matter exchange between the system and its environment, and their birth and death within the system (for instance, as a result of biological "birth and death" processes or irreversible chemical reactions). Then the expression for the full differential of internal energy is written as

$$\begin{aligned} dU &= \sum X_k\, dx_k + \sum N_m\, d_e n_m + \sum N_m\, d_i n_m + T\, dS \\ &= \sum X_k\, dx_k + \sum N_m\, dn_m + T\, dS. \end{aligned} \tag{2.4}$$

This is a general formal representation of the famous Gibbs equation.

Let $d_i n_m = 0$, then the values of N_m are potentials that maintain the exchange of matter. In thermodynamics, the system is separated from its environment by *potential barriers*. In order for any virtual particle to penetrate from the environment into the system it has to overcome these barriers. For this, it has to perform certain work moving

Table 2.1
Different forms of energy and their intensive and extensive variables (potential and kinetic energies are denoted as mechanical energy)

Energy form	Extensive variable	Intensive variable
Heat	Entropy (J/K)	Temperature (K)
Expansion	Volume (m^3)	Pressure (Pa = kg/s^2 m)
Chemical	Moles (M)	Chemical potential (J/mol)
Electrical	Charge (A s)	Voltage (V)
Potential	Mass (kg)	(Gravity)(height) (m^2/s^2)
Kinetic	Mass (kg)	0.5(velocity)^2 (m^2/s^2)

in a field of the thermodynamic potential. A value of the specific work (per single particle) is equal to N_m. In the case when the particles are chemical molecules, then $N_m = \mu_m$ are *chemical* potentials. The value δA_{irrev} is a contribution of real irreversible processes (such as friction, heating of conductor by electrical current, etc.) into the increase of internal energy. Note that δA_{irrev} can also be represented as a product of an intensive variable and differential of a conjugated extensive variable. For instance, in the case of friction, $\delta A_{\text{irrev}} = f_{\text{fr}}\, \mathrm{d}s$ where f_{fr} denotes the force of friction and $\mathrm{d}s$ is the infinitesimal distance. In the case of electrical current, $\delta A_{\text{irrev}} = U\, de$ where U denotes the voltage and de is the change of charge. It is interesting that the first δA_{irrev} can be represented as $\delta A_{\text{irrev}} = f_{\text{fr}} v\, \mathrm{d}t$, and the second as $\delta A_{\text{irrev}} = UI\, \mathrm{d}t$, where v is the velocity of movement, I is the amperage and $\mathrm{d}t$ is the time differential. In both the cases the products $f_{\text{fr}}v$ and UI are powers.

We are keeping in mind that in thermodynamics there is such an intensive variable as the temperature T. What kind of conjugate extensive variable corresponds to it? Gibbs has postulated that this extensive variable S is *entropy*.

Finally, if we select the term connected with work of expansion or compression and consider only "chemical" particles, then the expression for the full differential of internal energy is represented in the form:

$$\mathrm{d}U = T\, \mathrm{d}S - p\, \mathrm{d}V + \sum_m \mu_m\, \mathrm{d}n_m. \tag{2.5}$$

This is a standard form of the fundamental Gibbs equation. Certainly, along with "thermal", "mechanical" and "chemical" products $T\, \mathrm{d}s$, $-p\, \mathrm{d}V$ and $\mu_m\, \mathrm{d}n_m$, we can also consider other pairs of intensive and extensive variables connected with two other forms of energy (see Table 2.1) and other sorts of virtual particles. Potential and kinetic energy is denoted here as mechanical energy.

2.3. Entropy and the Second Law of Thermodynamics. Nernst's theorem

All the laws of Newton's mechanics are symmetrical with respect to the direction of time: the Past and the Future do not differ from each other in Newton's world. However, in our world heat transfers only from a hot to a cold body, and never vice versa; the "arrow of time" does really exist, and there are always experiments allowing the "Observer" to distinguish the future from its past... Therefore, by falling outside the framework of "normal physics", and in order to ensure the link of our world to reality, we formulate the following statement: *in an isolated system some physical value determined by the value of energy is always increasing or remaining constant, but it is never decreasing.* This value is *entropy*. By the same token we have formulated the so-called *Second Law of Thermodynamics*. It is interesting that an enormous number of attempts ("*their name is legion*") were being undertaken and still are being undertaken to find such special conditions under which the Second Law could be violated... but unsuccessfully. On the other hand, the Second Law can only be confirmed experimentally, not theoretically.

The equations of classical physics and even quantum mechanics, as well as the general relativity theory, presume that time is reversible, which is not the case according to the Second Law of Thermodynamics. Prigogine and Stengers (1979) write that although quantum mechanics and the general relativity theory are revolutionary, as far as the concept of time is concerned they are direct descendants of classical dynamics and carry radical negation of the irreversibility of time. Einstein claimed that the irreversibility of time was an illusion. The Second Law of Thermodynamics clearly breaks the symmetry of time. Prigogine introduces a natural time ordering of dynamic states. For instance, a drop of blue ink in a glass of water will after some time inevitably colour all the water light blue; but the opposite process, that the light blue colour in a glass of water forms a single drop of concentrated blue ink, has an almost infinitely little probability to occur. The example illustrates the "arrow of time", which has ecological implications: the evolution.

Keep in mind that the energy conservation law can be proved theoretically if the basic equations, so-called "motion equations" that fully determine the dynamics of all particles participating in the process, are known. But all attempts to prove the law of increasing (non-decreasing) entropy were unsuccessful. Moreover, Henry Poincaré has proved that it is impossible to calculate entropy, even if co-ordinates, velocities, masses and momenta of all particles are known. Generally speaking, there is no theoretical proof of correctness of the First and Second Laws, so that all the proofs are experimental; they are a consequence of our proper experience in our world. The problem of induction does exist: maybe such extreme conditions exist that the First and Second Laws become non-applicable, either partly or wholly. In other words, maybe such virtual worlds with characteristic sizes and times that differ from ours exist, and in these worlds energy can be created or destroyed, and entropy of an isolated system decreases. Fortunately, we are living in the "right" world, where the biosphere and ecology obey the First and Second Laws.

Apparently, some "strangeness" of the Second Law has brought to its formulation a number of various equivalent forms:

- It is impossible to take heat from a reservoir and convert it into work without at the same time transferring heat from a hot to a cold reservoir.
- It is impossible to transfer heat from a cold to a hot reservoir without converting a certain amount of work into heat in the same process.
- Entropy of an isolated system always increases during irreversible processes.
- All real processes produce entropy.
- Time has only one direction—which is formulated in everyday language as "do not cry over spilt milk". This may be called the "arrow of time".

- All real processes are irreversible.
- All real processes result in a partial transfer of one of the other energy forms to heat, that unfortunately cannot be fully utilised to do work because of the difficulties in providing a reservoir at absolute zero of temperature and because it is released at the temperature of the environment.

It seems to us that such a diversity of definitions will much more easily allow us to find suitable "ecological" interpretations. However, one natural question arises: is entropy a real existing physical value (like energy), which can be measured, or is it some theoretical fiction, which is needed only for representing the observation in more or less elegant form? The positive answer is given by the famous Boltzmann formula:

$$S = k \ln W. \tag{3.1}$$

Hewed on the Boltzmann memorial at the Vienna cemetery, this formula is to live in the endless sky over the grave of the great Boltzmann. This is one of the greatest formulas, making bridges between thermodynamics and other sciences: probability theory, theory of information, dynamical systems theory, etc. It connects the entropy, S, and the "thermodynamic" probability of state, W; the coefficient of proportionality k is the so-called Boltzmann constant. The thermodynamic probability is equal to the number of possible states in which the system can be found. The Boltzmann constant is $k = 1.37 \times 10^{-23}$ J/K; the entropy is usually evaluated in these units. It is obvious that entropy is a measure of *order* (or, most likely, *disorder*) in the system. Entropy increases with increase in temperature and decreases with its fall. It is obvious that the temperature exists when the system can be only at a single state, i.e. $W = 1$, then $S = 0$. This temperature $T = 0$ is called *absolute zero*, and it is equal to $t^0 = -273.15$°C. The statement that *entropy is equal to zero at the absolute zero of temperature* is called *Nernst's theorem*. It is obvious that $S > 0$ at $T > 0$. We see that the symmetry is again violated, since the temperature cannot be negative. It is interesting that formally the temperature could be negative, but any body would strive to split spontaneously into scattering parts. In other words, for $T < 0$, bodies in equilibrium cannot exist (Landau and Lifshitz, 1995).

Entropy can only be created, it cannot be destroyed. Let us represent (as we did above for energy and matter) the entropy differential as the sum of two items: $dS = d_eS + d_iS$ where d_eS corresponds to the entropy exchange between the system and its environment, and d_iS describes the entropy production within the system. In accordance with the Second Law of Thermodynamics:

$$d_iS \geq 0. \tag{3.2}$$

Concerning the exchange term d_eS, Clausius has suggested that

$$d_eS = \delta Q/T, \tag{3.3}$$

where Q is the heat input into the system from its environment (gained heat). In an isolated system, when $d_eS = 0$ then $dS = d_iS \geq 0$. In a closed system $d_eS = \delta Q/T$, since $d_iS \geq 0$, then $dS = d_eS + d_iS \geq \delta Q/T$. The latter inequality means that if heat is brought into the system, then its entropy and, correspondingly, the disorder increase. On the contrary, if heat is withdrawn, then entropy decreases and the order increases. In a thermally (not completely) isolated system, when $\delta Q = 0$, $dS \geq 0$ too. Since $dS = Q + \delta A_{irrev} - \sum N_m \, d_i n_m$ and $d_e S = \delta Q/T$ then

$$d_i S = \delta A_{irrev} - \sum N_m d_i n_m. \tag{3.4}$$

From this follows that in a thermally isolated system ($\delta Q = 0$), within which there are neither chemical nor biological transformations (all $d_i n_m = 0$) and all processes are reversible ($\delta A_{irrev} = 0$), $dS \equiv 0$, i.e. the entropy is constant.

A system is said to be in equilibrium when it has no further tendency to change its properties. As previously mentioned, a dynamic equilibrium (steady state) can be maintained by equal process rates in opposite directions. The entropy of a thermally isolated system will increase until no further spontaneous changes can occur—all gradients have been eliminated—the thermodynamic equilibrium has been reached. The entropy therefore reaches its maximum at thermodynamic equilibrium. The criterion for thermodynamic equilibrium is, in other words, that entropy is at maximum.

2.4. Maximal work which the system can perform on its environment. Characteristic functions or thermodynamic potentials

Consider a thermally isolated system consisting of different particles, which are neither in thermal nor chemical equilibrium. Let the energy of the system at some initial time be E_0, and the energy in equilibrium (i.e. at the final state) as a function of the supersystem entropy in this state be $E(S)$. Since the supersystem is closed, then $E_0 = |A| + E(S)$ where $|A|$ is the absolute value of work. Differentiating $|A|$ with respect to the entropy at the final state we have $\partial |A|/\partial S = -(\partial E/\partial S)_V = -T$ where T is the temperature at this state. Since this derivative is negative then $|A|$ increases as S decreases. On the other hand, the entropy of a closed system cannot decrease. Therefore, the value of $|A|$ will be maximal if only the total entropy S remains constant in the course of the entire transition, i.e. when the transition to an equilibrium state is a reversible one. For any irreversible process $|A| < |A|_{max}$.

Since the full differential of internal energy is known (see Eq. (2.5)), then the internal energy is written as the following function of variables S, V and n_m:

$$U(S, V, n_m) = TS - pV + \sum_m \mu_m n_m. \tag{4.1}$$

This is the first *thermodynamic identity*, and the internal energy is the first characteristic function (or thermodynamic potential). (Sometimes we shall omit the term $\sum_m \mu_m \, dn_m$ describing the effect of chemical reaction.) If to differentiate formally the function U, assuming that it depends on the variables T, S, p and V, then we obtain

$$dU = T\,dS + S\,dT - p\,dV - V\,dp + \sum_m \mu_m \, dn_m + \sum_m n_m \, d\mu_m \quad \text{or}$$

$$dU + V\,dp - S\,dT = T\,dS - p\,dV. \tag{4.2}$$

In order to get the Gibbs equation we have to assume that the temperature, pressure and chemical potentials remain constant during the transition. But this means that the process of transition has to be quasi-stationary. This has already been required when we deduced the Gibbs equation. Since the right-hand side of equality (4.2) is a full differential (as the right-hand side of the Gibbs equation), then the left-hand side can be written as the full differential of the function Φ, so that

$$d\Phi = dU + V\,dp - S\,dT = d(U + pV - TS). \tag{4.3}$$

If the volume and temperature of the system remain constant during the transition then p and S have to be the variables. From Eq. (4.3) for $dT = 0$ we obtain

$$d\Phi = dU + V\,dp = d(U + pV). \tag{4.4}$$

The function

$$H(S, p) = U + pV \tag{4.5}$$

is called the *heat function* or *enthalpy* of the system. Its change during a process with a constant pressure is equal to the heat gained by the system. By taking into account the Gibbs equation, its full differential can be rewritten as

$$dH = T\,dS + V\,dp. \tag{4.6}$$

Since the transformation of the first identity into the other identities, defined with respect to the other characteristic functions (thermodynamic potentials), for instance enthalpy, does not affect the variables n_m (moreover, they could be not only molecules but any virtual particles, for instance biological individuals), it is clear that terms which are proportional to the differentials dn_m may be added to any thermodynamic identities. Then

$$dH = T\,dS + V\,dp + \sum_m \mu_m\,dn_m. \tag{4.7}$$

Let some chemical reaction occur in the system with a constant temperature. As a result, the initial mixture of the reacting substance transforms into some final products. This reaction could go through many stages, some of which either do not go to completion or are wholly unobserved. Since in this case the heat is a function of state, then the total heat released or absorbed in the process of chemical reaction (the heat of reaction) does not depend on paths from the initial mixture of reactants to the final products; it is only determined by the differences in their enthalpies. This is the Hess law, which is the main law of thermodynamic chemistry and a special corollary of the First Law of Thermodynamics.

The heat of reaction has biochemical significance. If heat is produced, the process is named exothermic. Decomposition of the organic matter in food is an exothermic process, which provides heterotrophic organisms with the energy that is needed for the maintenance of life. Endothermic processes require the addition of heat. A typical endothermic process is the formation of adenosine triphosphate (ATP) from phosphate and adenosine diphosphate (ADT), expressed in biochemistry as the following equation:

$$\text{ADP} + \text{P} + 42\ \text{kJ/mol} = \text{ATP}.$$

The opposite process, formation of P and ADP from ATP, similarly yields 42 kJ/mol. ATP is applied by the organisms as a suitable unit of energy wherever it is needed to carry out the necessary biochemical processes.

The heat of reaction may be either at a constant pressure or constant volume, which implies that the heat of reaction becomes, respectively, the change in enthalpy and the change in internal energy. In reactions involving only liquids, or when solid ΔV is negligible, there is practically no difference between the two heats of reaction. A convenient standard state for a substance may be taken to be the state when it is stable at 25°C and atmospheric pressure. The standard enthalpy of any compound is the heat of reaction, by which it is formed from its elements, reactants and products all being in the standard state. A superscript of zero indicates that the standard heat of formation with reactants and products is at a pressure of one atmosphere.

Before we introduce new characteristic functions, we shall define some special transition processes, so-called *adiabatic* ones. Strictly speaking, a process is named *adiabatic* if during the process the system entropy is conserved, i.e. $\mathrm{d}S = 0$. Any adiabatic process is reversible. The process can be realised if the system is thermally isolated, and its environment changes rather slowly.

If the pressure remains constant during the adiabatic process then T and V have to be the variables. From Eq. (4.3) for $\mathrm{d}p = 0$ we obtain

$$\mathrm{d}\Phi = \mathrm{d}U - S\,\mathrm{d}T = \mathrm{d}(U - TS). \tag{4.8}$$

The function

$$F(T, V) = U - TS \tag{4.9}$$

is called the *free energy* or *Helmholtz's free energy* of the system. Its change during an adiabatic process at constant pressure is equal to the work done on the system. The full differential of F (with regard to the differentials $\mathrm{d}n_m$) is equal to

$$\mathrm{d}F = -S\,\mathrm{d}T - p\,\mathrm{d}V + \sum_m \mu_m\,\mathrm{d}n_m. \tag{4.10}$$

This is the third thermodynamic identity.

At last, we have to get the characteristic function related to the variables T and p, i.e. for an adiabatic process at constant volume. In this case, as follows from Eq. (4.3), the function, which is called the *thermodynamic potential* (in the narrow sense of the word) or the *Gibbs free energy* or the *Gibbs potential* is defined as

$$G(T, p) = \Phi = U + pV - TS = F + pV = H - TS \tag{4.11}$$

and its full differential is equal to

$$\mathrm{d}G = -S\,\mathrm{d}T + V\,\mathrm{d}p + \sum_m \mu_m\,\mathrm{d}n_m. \tag{4.12}$$

This is the fourth thermodynamic identity.

Note that, if only the numbers of virtual particles change slightly, then the changes of characteristic functions that are also small will always be equal to one another if every one of them is considered with the appropriate pair of quantities constant, i.e.

$$(\mathrm{d}U)_{S,V} = (\mathrm{d}H)_{S,p} = (\mathrm{d}F)_{T,V} = (\mathrm{d}G)_{T,p} = \sum_m \mu_m \,\mathrm{d}n_m. \tag{4.13}$$

Under these conditions, chemical potentials depend only on the composition of the mixture consisting of $n_1, n_2, \ldots$ particles, i.e. on their frequencies, and they do not depend on their total number.

2.5. Chemical equilibrium, chemical affinity and standard energies of biochemical reactions. Function of dissipation

The classic thermodynamics has introduced two more energy-based functions, named the work function, A, and the Gibbs free energy, G. They have been introduced because it is desirable to obtain criteria for thermodynamic equilibrium under practical conditions, which means that the temperature is approximately constant in addition to either constant volume in bomb calorimeters or constant pressure in chemostates.

Most chemical processes in the laboratory or in organisms are carried out at constant volume, pressure and also temperature. As follows from the results of Section 2.4 under these conditions, the Helmholtz and Gibbs free energies, F and G, play the most important role; moreover, they practically coincide, $\mathrm{d}F \approx \mathrm{d}G$. This implies that $\mathrm{d}F \approx \mathrm{d}G = -\delta A$. At thermodynamic equilibrium, all gradients are eliminated according to the definition. This means that no work can be performed. This implies that at constant temperature and pressure $(\mathrm{d}G)_{T,p} = \sum_m \mu_m \,\mathrm{d}n_m = 0$, which is an important consequence of the energy conservation law. A general dynamic equilibrium, but not a thermodynamic equilibrium, is possible by equalising process rates in opposite directions to ensure that the steady state is maintained.

Consider the system consisting of one ith sort of particles distributed between two phases, a and b. The transfer of $\mathrm{d}n_i$ number of particles from a into b is accompanied by the change of Gibbs potential (or free energy), $\mathrm{d}G$. Since $-\mathrm{d}n_i^{\mathrm{a}} = \mathrm{d}n_i^{\mathrm{b}}$ then at $p, T = \text{const}$ $(\mathrm{d}G)_{p,T} = \mu_i^{\mathrm{a}}\,\mathrm{d}n_i^{\mathrm{a}} + \mu_i^{\mathrm{b}}\,\mathrm{d}n_i^{\mathrm{b}} = \mathrm{d}n_i^{\mathrm{a}}(\mu_i^{\mathrm{a}} - \mu_i^{\mathrm{b}})$, and in the equilibrium $(\mathrm{d}G)_{p,T} = \mathrm{d}n_i^{\mathrm{a}}(\mu_i^{\mathrm{a}} - \mu_i^{\mathrm{b}}) = 0$ so that $\mu_i^{\mathrm{a}} = \mu_i^{\mathrm{b}}$, i.e. the equality of chemical potentials is the condition of equilibrium for the process of transfer of ith sort of particles between two phases. Moving to the thermodynamic equilibrium, the system is always moving toward the state with lower chemical potential.

It was shown in standard text-books on chemical thermodynamics that the chemical potential:

$$\mu_i = \mu_i^{(0)} + RT\ln[c_i], \tag{5.1}$$

where R is the gas constant, T is the temperature, $[c_i]$ is the concentration (or frequency) of ith particles, $\mu_i^{(0)}$ is a constant depending on the temperature and the origin of matter (substance).

Standard free energies (or thermodynamic potentials) of formation of chemical compounds are very important for calculation of chemical equilibriums, since their knowledge allows us to estimate the "useful" work, various energetic effects and possible directions of the chemical evolution of the system.

Let there be a chemical reaction:

$$\nu_1 A_1 + \nu_2 A_2 + \cdots \leftrightarrow \nu'_1 A'_1 + \nu'_2 A'_2 + \cdots, \tag{5.2}$$

where $A_1, A_2, \ldots, A'_1, A'_2, \ldots$ are reacting substances, and $\nu_1, \nu_2, \ldots, \nu'_1, \nu'_2, \ldots$ are their stoichiometric coefficients. At the thermodynamic equilibrium:

$$(\mathrm{d}G)_{p,T} = \sum \mu_i \nu_i = \sum \mu_i^{(0)} \nu_i + RT \sum \nu_i \ln c_i^{\mathrm{eq}} = 0, \tag{5.3}$$

where c_i^{eq} are equilibrium concentrations and the summation is produced over all substances, both products and reactants. After simple transformation, we obtain from Eq. (5.3):

$$-\sum \mu_i^{(0)} \nu_i = RT \ln\left[\frac{(c'_1)^{\nu'_1}(c'_2)^{\nu'_2}\cdots}{(c_1)^{\nu_1}(c_2)^{\nu_2}\cdots}\right]_{\mathrm{eq}} = RT \ln K(T), \tag{5.4}$$

where $K(T)$ is the so-called *equilibrium constant* depending only on the temperature.

Using these formulas, we can define a so-called *standard value of the change of thermodynamic (Gibbs) potential* ΔG_0, which corresponds to the temperature 25°C and the conditions when the concentrations (activities) of all products and reactants are equal to 1. Indeed, after denoting $\Delta G_0 = \sum \mu_i^{(0)} \nu_i = -RT \ln K(T)$, we have

$$\Delta G = \Delta G_0 + RT \sum \nu_i \ln c_i = -RT \ln K(T) + RT \sum \nu_i \ln c_i.$$

It is obvious that if all $c_i = 1$, then $\Delta G = \Delta G_0$.

There are different methods for the determination of ΔG_0 in biochemical reactions. For instance, if we know the equilibrium constant then $\Delta G_0 = -RT \ln K(T)$. The other method is based on the Hess law when we sum the partial changes of Gibbs potential corresponding to all intermediate stages of reaction:

$$\Delta G_0 = \sum \Delta G_0(\text{products}) - \sum \Delta G_0(\text{reactants}). \tag{5.5}$$

The thermodynamic potential describes the chemical affinity under conditions of constant temperature and pressure: $\Delta G = G(\text{products}) - G(\text{reactants})$.

When the potential is zero there is no net work obtainable by any change or reaction at constant temperature and pressure. The system is in a state of thermodynamic equilibrium. When the change of Gibbs potential is positive for a proposed process, net work must be put into the system to effect the reaction, otherwise it cannot take place. When it is negative, the reaction can proceed spontaneously by providing useful net work.

As an example we consider the reaction of glucose oxidation in the process of respiration: $C_6H_{12}O_6 + 6O_2 = 6CO_2 + 6H_2O$. The value of ΔG_0 was found by the Hess method: $\Delta G_0 = -2840$ kJ/mol. It is known that the general direction of photosynthesis is opposite to the respiration; therefore, the photosynthetic process of formation of one molecule of glucose from water and carbon dioxide demands an increase of thermodynamic potential by the value of $\Delta G_0 = 2840$ kJ/mol > 0. Therefore, it cannot occur spontaneously and demands an additional energy inflow. The latter is provided by solar photons.

Note that for real systems, when the interaction between particles is rather strong, it is convenient to introduce a new function for the considered substances, called *fugacity*, *f*. Fugacity is defined from the following equation:

$$\mu_i = \mu_i^{(0)} + RT \ln[f_i/f_i^{(0)}]. \tag{5.6}$$

A standard state may, however, be defined as the state of unit fugacity, as the standard state for ideal gases was the state of unit pressure. It is now possible to set up an expression for the equilibrium constant which is true in general not only for real (non-ideal) gases but for substances in any state of aggregation. It can be shown that the fugacity can be replaced by concentrations (pressures for gases) in many calculations with a good approximation. For solutions, it is possible to find the fugacity by multiplying the concentration with an activity coefficient that can be found by empirical equations. For an aquatic solution, the fugacity coefficient is close to 1.00 with a total concentration of dissolved matter of less than 1 g/l.

It is emphasised in this context that these thermodynamic calculations of equilibrium constant and standard heat and free energy are also valid for biochemical processes that are the reactions of interest in an ecosystem.

We distinguish different forms of energy. All forms can be described as some product of "quantitative" (extensive) and "qualitative" (intensive) variables. These pairs of extensive and intensive variables for different energy forms are summarised in Table 2.1. Work can be performed when the extensive variable is changed from one level of the intensive variable to another, and the work = the extensive variable × the difference of the intensive variable between the two levels. As energy is conserved, work implies that one energy form is transferred to another energy form. The energy of a considered system is often defined as "the ability to do work". It is presumed that the system is transferred to the level where the intensive variable is zero, whereby the work performed becomes equal to the energy content of the system.

The change in the number of moles of any chemical substance caused by a chemical reaction within the system is proportional to its stoichiometric coefficient: $\mathrm{d_i}n_m = \nu_m\,\mathrm{d}\xi$, where ξ is called the *co-ordinate of reaction* or the *degree of its completeness*. From this formula it follows that

$$\frac{\mathrm{d_i}n_1}{-\nu_1} = \frac{\mathrm{d_i}n_2}{-\nu_2} = \cdots = \frac{\mathrm{d_i}n_1'}{\nu_1'} = \frac{\mathrm{d_i}n_2'}{\nu_2'} = \cdots, \tag{5.7}$$

where the sign " − " corresponds to reactants and " + " to products, $\mathrm{d}n_1 < 0, \mathrm{d}n_2 < 0, \ldots, \mathrm{d}n_1' > 0, \mathrm{d}n_2' > 0, \ldots$. Let $V = \text{const}$ be the volume of the considered homogenous system. Then the rate of reaction is defined as

$$v^{\mathrm{ch}} = \frac{1}{V\nu_m}\frac{\mathrm{d_i}n_m}{\mathrm{d}t} = \frac{1}{V}\frac{\mathrm{d}\xi}{\mathrm{d}t}. \tag{5.8}$$

If the mth substance simultaneously participates in several reactions, then the total change in the number of moles, $d_i n_m$, is

$$\frac{1}{V}\frac{d_i n_m}{dt} = \sum_{k=1}^{K} \nu_{mk} v_k^{ch}, \tag{5.9}$$

where k is the index of reaction and K is the total number of simultaneously resulting reactions.

Since the system is homogenous, we can take into consideration such functions of *density* as the density of entropy production and the specific concentrations (or, simply, concentrations) of substance:

$$\sigma = \frac{1}{V}\frac{d_i S}{dt}, \qquad c_m = \frac{n_m}{N}.$$

Analogously, we define the local function of dissipation:

$$\Psi = \frac{1}{V} T \frac{d_i S}{dt} = T\sigma.$$

If to take into account the matter exchange between the system and its environment, then

$$\frac{dc_m}{dt} = \frac{d_e c_m}{dt} + \sum_{k=1}^{K} \nu_{mk} v_k^{ch}. \tag{5.10}$$

The main hypothesis of the theory of irreversible processes (Glansdorff and Prigogine, 1971) is that the fundamental Gibbs equation (see Section 2.2) is also valid at any local time. Since Eq. (3.4) is one form of fundamental equation then its local form is written as (we set $N_m = \mu_m$ and $(\delta A_{irrev}/\delta t) = (d' A_{irrev}/dt)$)

$$\frac{1}{V}\frac{d_i S}{dt} = \frac{1}{VT}\left(\frac{d' A_{irrev}}{dt} - \sum_m \mu_m \frac{d_i n_m}{dt}\right) \text{ or } \sigma = \frac{1}{VT}\frac{d' A_{irrev}}{dt} - \frac{1}{T}\sum_m \mu_m \frac{d_i c_m}{dt}. \tag{5.11}$$

The chemical affinity of kth reaction is defined as

$$A_k = -\sum_m \nu_{mk}\mu_k. \tag{5.12}$$

Then

$$\sigma = \frac{1}{VT}\frac{d' A_{irrev}}{dt} + \sum_{k=1}^{K} \frac{A_k}{T} v_k^{ch}. \tag{5.13}$$

This is a very important relation in the thermodynamics of irreversible processes, which will be used in Chapter 3.

Finally, we would like to clarify the concept of affinity. Consider as an example a simple reaction of the transformation of mth substance into $(m+1)$th one: $c_m \rightarrow c_{m+1}$. The stoichiometric coefficients and chemical potentials are ν_m, ν_{m+1} and μ_m, μ_{m+1}, respectively.

In accordance with Eq. (5.12) $A = -(-\nu_m\mu_m + \nu_{m+1}\mu_{m+1}) = \nu_m\mu_m - \nu_{m+1}\mu_{m+1}$. In the special case, $\nu_m = \nu_{m+1} = 1$, $A = \mu_m - \mu_{m+1}$, i.e. the value of affinity is equal to the difference of chemical potentials.

2.6. Illustrations of thermodynamics in ecology

If the law of matter (mass) conservation is used for chemical compounds that can be transformed into other chemical compounds, Eq. (2.1) must be changed to

$$V\frac{dc_i}{dt} = \text{input} - \text{output} + \text{formation} - \text{transformation}, \tag{6.1}$$

where V is a volume and c_i is a concentration of ith substance.

This principle is widely used in the class of ecological models called biogeochemical models. The equation is set up for the relevant elements, e.g. for eutrophication models for C, P, N and perhaps Si (see, for instance, Jørgensen, 1976, 1992a; Jørgensen and Mejer, 1977).

For terrestrial ecosystems, mass per unit of area is often applied in the mass conservation equation, so that

$$A\frac{dm}{dt} = \text{input} - \text{output} + \text{formation} - \text{transformation}, \tag{6.2}$$

where A is an area and m is a mass per unit of area.

The transformation of solar energy to chemical energy by plants conforms with the First Law (see also Fig. 2.3), which is represented here as

$$\begin{aligned}&\text{Solar energy assimilated by plants}\\&\quad = \text{chemical energy of plant tissue growth} + \text{heat energy of respiration}.\end{aligned} \tag{6.3}$$

For the next level in the trophic chains, the herbivorous animals, the energy balance can also be set up:

$$F = \text{As} + \text{UD} = \text{Gr} + \text{Re} + \text{UD}, \tag{6.4}$$

where F is the food intake converted to energy (J), As is the energy assimilated by the animals, UD is the undigested food or the chemical energy of faeces, Gr is the chemical energy of animal growth and Rc is the heat energy of respiration. These considerations

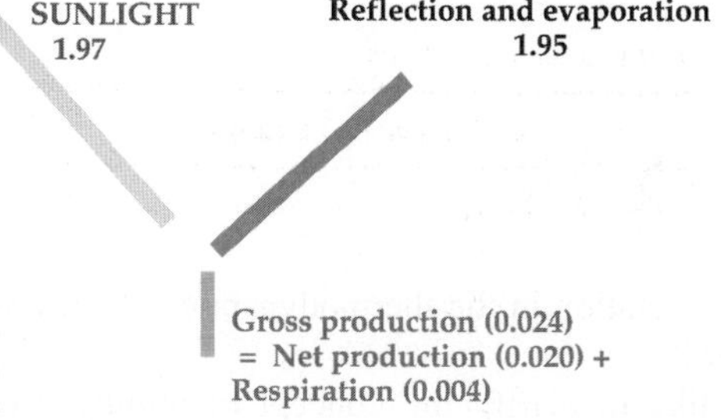

Fig. 2.3. Fate of solar energy incident upon the perennial grass–herb vegetation of an old field community in Michigan. All values in J/m^2 year.

Table 2.2a
Combustion heat of animal material

Organism	Species	Heat of combustion (kcal/ash-free gram)
Ciliate	*Tetrahymena pyriformis*	−5.938
Hydra	*Hydra littoralis*	−6.034
Green hydra	*Chlorohydra viridissima*	−5.729
Flatworm	*Dugesia tigrina*	−6.286
Terrestrial flatworm	*Bipalium kewense*	−5.684
Aquatic snail	*Succinea ovalis*	−5.415
Brachiipode	*Gottidia pyramidata*	−4.397
Brinc shrimp	*Artemia* sp. (nauplii)	−6.737
Cladocera	*Leptodora kindtii*	−5.605
Copepode	*Calanus helgolandicus*	−5.400
Copepode	*Trigriopus californicus*	−5.515
Caddis fly	*Pycnopsyche lepido*	−5.687
Caddis fly	*Pycnopsyche guttifer*	−5.706
Spit bug	*Philenus leucopthalmus*	−6.962
Mite	*Tyroglyphus lintneri*	−5.808
Beetle	*Tenebrio molitor*	−6.314
Guppie	*Lebistes reticulatus*	−5.823

Table 2.2b
Energy values in an *Andropogus virginicus* "old-field" community in Georgia

Component	Energy value (kcal/ash-free gram)
Green grass	−4.373
Standing dead vegetation	−4.290
Litter	−4.139
Roots	−4.167
Green herbs	−4.288
Mean	−4.251

Table 2.2c
Combustion heat of migratory and non-migratory birds

Sample	Ash-free material (kcal/g)	Fat ratio (% dry weight as fat)
Fall birds	−8.08	71.7
Spring birds	−7.04	44.1
Non-migrants	−6.26	21.2
Extracted bird fat	−9.03	100.0
Fat extracted: fall birds	−5.47	0.0
Fat extracted: spring birds	−5.41	0.0
Fat extracted: non-migrants	−5.44	0.0

Table 2.2d
Combustion heat of components of biomass

Material	ΔH protein (kcal/g)	ΔH fat (kcal/g)	ΔH carbohydrate (kcal/g)
Eggs	−5.75	−9.50	−3.75
Gelatin	−5.27	−9.50	–
Glycogen	–	–	−4.19
Meat, fish	−5.65	−9.50	–
Milk	−5.65	−9.25	−3.95
Fruits	−5.20	−9.30	−4.00
Grain	−5.80	−9.30	−4.20
Sucrose	–	–	−3.95
Glucose	–	–	−3.75
Mushroom	−5.00	−9.30	−4.10
Yeast	−5.00	−9.30	−4.20

Source: Morowitz (1968).

pursue the same lines as those mentioned in context with Eq. (6.3), where the mass conservation principle was applied. The conversion of biomass to chemical energy is illustrated in Table 2.2. The energy content per 1 g ash-free organic material is surprisingly uniform, as illustrated in Table 2.2 (see also the discussion in Chapter 7). Notice that the energy content in the table is the energy of "dead matter" applied as fuel due to its elementary composition. Table 2.2d indicates ΔH, which symbolises the increase in enthalpy, defined (see Section 2.4) as $H = U - pV$. The energy is originated from the following chemical process: organic matter + oxygen = carbon dioxide + water + other inorganic compounds as nitrate, sulphate and phosphate.

Biomass can therefore be translated into energy, and this is also true of transformations through trophic chains. This implies that the short trophic chains of "grain to human" should be preferred to the longer and more wasteful "grain to domestic animal to human". The problem of food shortage cannot, however, be solved so simply, since animals produce proteins with a more favourable amino acid composition for human food (lysine is missing in plant proteins) and eat plants that cannot all be used as human food today. But food production can, to a certain extent, be increased by making the trophic chains as short as possible.

These relationships (see, for instance, Jørgensen, 2001b) can also be illustrated by means of so-called ecological pyramids that can either represent the number of individuals, the biomass (or energy content) or the energy flows on each level in the trophic chains or trophic networks. Only the energy flow forms a true pyramid due to the loss of heat by respiration. The pyramids based on numbers are affected by variation in size and the biomass pyramids by the metabolic rates of individuals.

However, as will be shown in the next chapters, energy in ecosystems cycles like matter, if we consider the chemical energy carried by biomass. It will make the interpretation of trophic levels more complicated.

Ecological efficiency should also be mentioned here; see Table 2.3 (Jørgensen, 2002b), where some useful definitions are listed, and Table 2.4, where the efficiency values are

Table 2.3
Ecological efficiency

Concept	Definition
Lindeman's efficiency	Ratio of energy intake level n to $n-1$: F_n/F_{n-1}
Trophic level assimilation efficiency	$\mathrm{As}_n/\mathrm{As}_{n-1}$
Trophic level production efficiency	P_n/P_{n-1}
Tissue growth efficiency	P_n/As_n
Ecological growth efficiency	P_n/F_n
Assimilation efficiency	As_n/F_n
Utilisation efficiency	F_n/UD_{n-1}

Variables: F, trophic input; UD, undigested food; As, assimilated food; R, respiration; P, net production; E, excretion; G, growth; n, trophic level. $F = \mathrm{As} + \mathrm{UD}$, $\mathrm{As} = P + R$ and $P = G + E$.

exemplified (Table 2.4). All these definitions use the energy balance of the ecosystem or organisms as its active component, i.e. they are based on the First Law. The items of the balance are similar to the ones used in equality (6.4): the trophic input $F = \mathrm{As} + \mathrm{UD}$ where As is an assimilated fraction and UD is an undigested food.

There is a close relationship between energy flow rates and organism size (denoted as the allometric principle) and some of the most useful of these relationships are illustrated in Jørgensen (1988, 1990, 1994) (see also Peters, 1983). Since many rate parameters are closely related to the rate of energy exchange, it is possible to find unknown parameters for various organisms on the basis of knowledge about the same parameters for other organisms, providing that the sizes of the organisms are known. This is illustrated in the four references mentioned above. The size or what also may be called the *openness*, which is a ratio of the surface area relative to the volume, will be further touched on in the next chapter.

Any self-sustaining ecosystem will contain a wide spectrum of organisms ranging in size from tiny microbes to large animals and plants. The small organisms account in most cases for most of the respiration (energy turnover), whereas the larger organisms comprise most of the biomass. It is therefore important for the ecosystem to maintain both small and large organisms as it will mean that both the energy turnover rate and the energy storage in the form of biomass are maintained.

The developments and reactions of ecosystems in general are not only a question of the energy flow, as will be touched upon many times in this volume. Matter and information also play a major role. So, matter, energy and information form the "triangle of freemasons". No transfer of energy is possible without matter and information and no matter can be transferred without energy and information. The relationships between energy and information will be discussed in more detail in Chapters 4 and 5. The higher the levels of information are, the higher is the utilisation of matter and energy for the development of ecosystems farther from the thermodynamic equilibrium; see also Chapter 5.

E.P. Odum has described the development of ecosystems from the initial stage to the mature stage as a result of continuous use of the self-design ability (Odum, 1969, 1971a,b). See the significant differences between the two stages of systems listed in Table 2.5 and notice that the major differences are on the level of information. The content of information

Table 2.4a
Assimilation efficiency (As/F) for selected organisms (after various authors)

Taxa	As/F value
Internal parasites	
Entomophagous Hymenoptera *Ichneumon* sp.	0.90
Carnivores	
Amphibian (*Nectophrynoides occidentalis*)	0.83
Lizard (*Mabuya buettneri*)	0.80
Praying mantis	0.80
Spiders	0.80–0.90
Warm- and cold-blooded herbivores	
Deer (*Odocoileus* sp.)	0.80
Vole (*Microtus* sp.)	0.70
Foraging termite (*Trinervitermes* sp.)	0.70
Impala antelope	0.60
Domestic cattle	0.44
Elephant (*Loxodonta*)	0.30
Pulmonate mollusc (*Cepaea* sp.)	0.33
Tropical cricket (*Orthochtha brachycnemis*)	0.20
Detritus eaters	
Termite (*Macrotermes* sp.)	0.30
Wood louse (*Philoscia muscorum*)	0.19

Table 2.4b
Tissue growth efficiency (P/As) for selected organisms (after various authors)

Taxa	P/As value
Immobile, cold-blooded internal parasites	
Ichneumon sp.	0.65
Cold-blooded, herbivorous and detritus-eating organisms	
Tropical cricket (*Orthochtha brachycnemis*)	0.42
Other crickets	0.16
Pulmonate mollusc (*Cepaea* sp.)	0.35
Termite (*Macrotermes* sp.)	0.30
Termite (*Trinervitermes* sp.)	0.20
Wood louse (*Philoscia muscorum*)	0.16
Cold-blooded, carnivorous vertebrates and invertebrates	
Amphibian (*Nectophrynoides occidentalis*)	0.21
Lizard (*Mabuya buettneri*)	0.14
Spiders	0.40
Warm-blooded birds and mammals	
Domestic cattle	0.057
Impala antelope	0.039
Vole (*Microtus* sp.)	0.028
Elephant (*Loxodonta*)	0.015
Deer (*Odocoileus* sp.)	0.014
Savanna sparrow (*Passerculus* sp.)	0.011
Shrews	Even lower values

Table 2.4c
Ecological growth efficiency ($P/F = (\mathrm{As}/F)(P/\mathrm{As})$) for selected organisms (after various authors)

Taxa	P/F value
Herbivorous mammals	
Domestic cattle	0.026 (0.44 × 0.057)
Impala antelope	0.022 (0.59 × 0.039)
Vole (*Microtus* sp.)	0.020 (0.70 × 0.285)
Deer (*Odocoileus* sp.)	0.012 (0.80 × 0.014)
Elephant (*Loxodonta*)	0.005 (0.30 × 0.015)
Birds	
Savanna sparrow (*Passerculus* sp.)	0.010 (0.90 × 0.011)
Herbivorous invertebrates	
Termite (*Trinervitermes* sp.)	0.140 (0.70 × 0.20)
Tropical cricket (*Orthochtha brachycnemis*)	0.085 (0.20 × 0.42)
Other crickets (New Zealand taxa)	0.050 (0.31 × 0.16)
Pulmonate mollusc (*Cepaea* sp.)	0.130 (0.33 × 0.30)
Detritus-eating and soil-eating invertebrates	
Termite (*Macrotermes* sp.)	0.090 (0.30 × 0.30)
Wood louse (*Philoscia muscorum*)	0.030 (0.19 × 0.16)
Tropical earthworm (*Millsonia anomala*)	0.005 (0.076 × 0.06)
Carnivorous vertebrates	
Lizard (*Mabuya* sp.)	0.100 (0.80 × 0.14)
Amphibian (*Nectophrynoides occidentalis*)	0.180 (0.83 × 0.21)
Carnivorous invertebrate	
Spiders	0.350 (0.85 × 0.42)
Internal parasites	
Ichneumon sp.	0.580 (0.90 × 0.65)

increases in the course of an ecological development because an ecosystem encompasses an integration of all the modifications that are imposed on the environment.

Thus, it is in the background of genetic information that systems develop which allows interaction of information with the environment. Herein lies the importance in the feedback organism–environment, which means that an organism can only evolve in an evolving environment. The differences between the two stages include entropy and exergy. This latter concept will be discussed later in Chapters 4 and 5.

The conservation laws of energy and matter set limits to the further development of "pure" energy and matter, while information may be amplified (almost) without limit. These limitations lead to the concept of limiting factors that is playing a significant role in ecology, including systems ecology. Patten et al. (1997) have speculated how a world without the conservation principles would look. Things would behave erratically. Something could arise from nothing. Mathematical counting would be meaningless. They conclude that, if there is a scientific law more fundamental than the rest, it is probably the conservation principles of matter, energy, momentum and electrical charge.

A major design principle observed in natural systems is the feedback of energy from storages to stimulate the inflow pathways as a reward from receiver storage to the inflow source (Odum, 1971a,b). By this feature, the flow values developed reinforce the processes

Table 2.5
Differences between initial stage and mature stage are indicated

Properties	Early stage	Late or mature stage
Energetic		
P/R	>> 1	Close to 1
P/B	High	Low
Yield	High	Low
Specific entropy	High	Low
Entropy production per unit of time	Low	High
Exergy	Low	High
Information	Low	High
Structure		
Total biomass	Small	Large
Inorganic nutrients	Extrabiotic	Intrabiotic
Diversity, ecological	Low	High
Diversity, biological	Low	High
Patterns	Poorly organised	Well organised
Niche specialisation	Broad	Narrow
Size of organisms	Small	Large
Life cycles	Simple	Complex
Mineral cycles	Open	Closed
Nutrient exchange rate	Fast	Slow
Life span	Short	Long
Selection and homeostatis		
Internal symbiosis	Undeveloped	Developed
Stability (resistance to external perturbations)	Poor	Good
Ecological buffer capacity	Low	High
Feedback control	Poor	Good
Growth form	Fast growth	Feedback controlled growth
Types	r-strategists	K-strategists

that are doing useful work. In other words, feedback allows the circuit to learn. A wider use of the cosystem's self-organisation ability in environmental, or rather ecological, management has been proposed by Odum (1988).

2.7. Ecosystem as a biochemical reactor

The fuel of ecosystems is organic matter, detritus. It is therefore relevant to calculate the free energy of dead organic matter. The chemical potential of dead organic matter, indexed $i = 1$, can be expressed in classic thermodynamics (e.g. Russel and Adebiyi, 1993) as

$$\mu_1 = \mu_1^0 + RT \ln(c_1/c_1^0). \tag{7.1}$$

The difference $\mu_1 - \mu_1^0$ is known for detrital organic matter, which is a mixture of carbohydrates, fats and proteins. The enthalpy values in Table 2.2 could be applied. Approximately 18.7 kJ/g are applied for the free energy content of average detritus. Obviously, the value is higher for detritus originated from birds, as they on average have higher fat contents (see Table 2.2c). Coal has a free energy content of about 30 kJ/g and mineral oil of 42 kJ/g. Both coal and mineral oil are a concentrated form of detritus from previous periods of the Earth. c_1 is the concentration of the detritus in the considered ecosystem and c_1^0 is the concentration of detritus in the same ecosystem but at a thermodynamic equilibrium.

Generally speaking, if we use the probabilistic interpretation of chemical kinetics, then the probability of finding the *i*th component at a thermodynamic equilibrium is equal to

$$\mathrm{Pr}_i^0 = c_i^0 / \sum_{i=0}^{n} c_i^0. \tag{7.2}$$

Since the inorganic component, c_0, is very dominant at thermodynamic equilibrium, then $c_0^0 >> \sum_{i=1}^{n} c_i^0$, and expression (7.2) can be approximated as

$$\mathrm{Pr}_i^0 \approx c_i^0 / c_0^0. \tag{7.3}$$

As a result of the combination of Eqs. (7.1) and (7.3), we get

$$\mathrm{Pr}_1^0 \approx (c_1/c_0^0)\exp\left[-\frac{\mu_1 - \mu_1^0}{RT}\right]. \tag{7.4}$$

The equilibrium constant for the process describing the aerobic (presence of oxygen) decomposition of detritus at 300 K can be found based upon the above-mentioned values. We could presume that the detritus molecular weight is about 100,000 (more accurately 104,400; Morowitz, 1968) and its typical composition is 3500 carbon, 6000 hydrogen, 3000 oxygen and 600 nitrogen:

$$C_{3500}H_{6000}O_{3000}N_{600} + 4350O_2 \rightarrow 3500CO_2 + 2700H_2O + 600NO_3^- + 600H^+. \tag{7.5}$$

Then the equilibrium constant

$$K = \frac{[CO_2]^{3500}[NO_3^-]^{600}[H^+]^{600}}{[C_{3500}H_{6000}O_{3000}N_{600}][O_2]^{4350}}, \tag{7.6}$$

since water is omitted from the expression of K. Then we have $\Delta G_0 = -RT \ln K(T)$, so that $-\Delta G_0 = 18.7$ kJ/g × 104,400 g/mol = 1952 MJ/mol = 8.2 J/mol × 300 ln K, which implies that ln K = 793,496 or K is about 10,344,998.

In other words, the equilibrium constant is enormous. The spontaneous formation of detritus in the form of a compound with the molecular weight of about 100,000 has therefore a very small probability. Even if we consider detritus with a low molecular weight corresponding to detritus partially decomposed, the K value is still very high. If we presume a 100 times smaller molecular weight, the exponent is 100 times smaller or is about 3500—still a very high K value. It is therefore understandable that detritus is decomposed spontaneously and thereby yields energy to the heterotrophic organisms. The opposite process corresponds to what may be the result of the photosynthesis, the conversion of solar radiation (energy) into chemical energy.

Fig. 2.4 shows the resulting biochemical reactions of an ecosystem, i.e. how the ecosystem works as a biochemical reactor. The biologically important elements are cycling

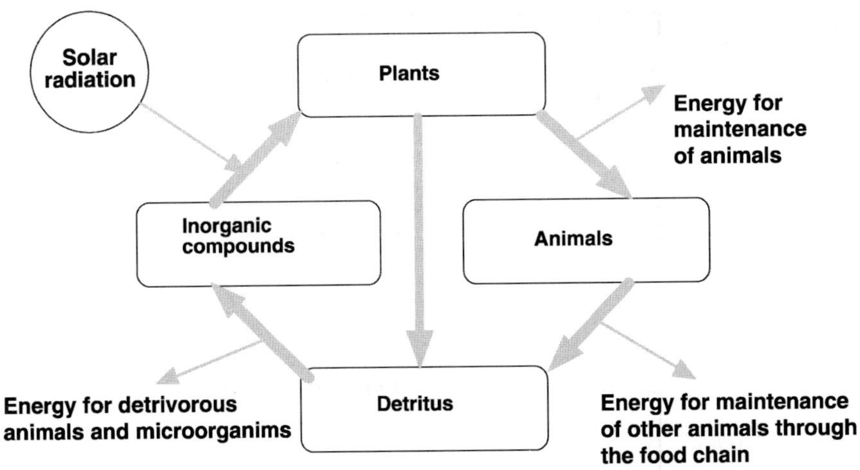

Fig. 2.4. An ecosystem is a biochemical reactor. The input of energy comes from the solar radiation. The biologically important elements cycle and carry the energy which is utilised by heterotrophic organisms to support the life processes.

and used again and again to build up biochemically important compounds, as for instance proteins, lipids and carbohydrates. These compounds are carrying the energy of the solar radiation and thereby supporting the maintenance of life and the cycling processes. The cycle may be compared with a Carnot cycle (given in detail in Chapter 5). The hot reservoir (the Sun) delivers the energy, which is utilised to do work. The heat energy is delivered to the cold reservoir at the ambient temperature (the temperature of the Earth). The work, after it

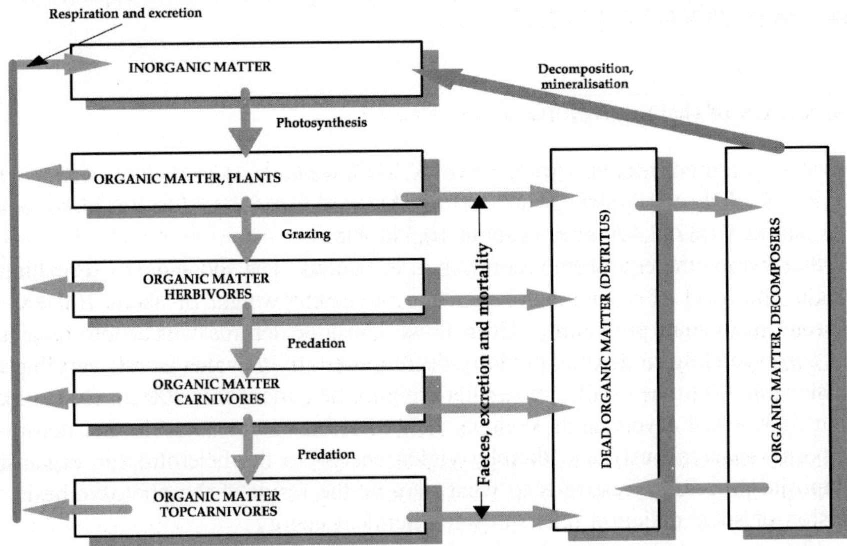

Fig. 2.5. The biochemical cycling of matter in ecosystems (from Jørgensen, 2000b).

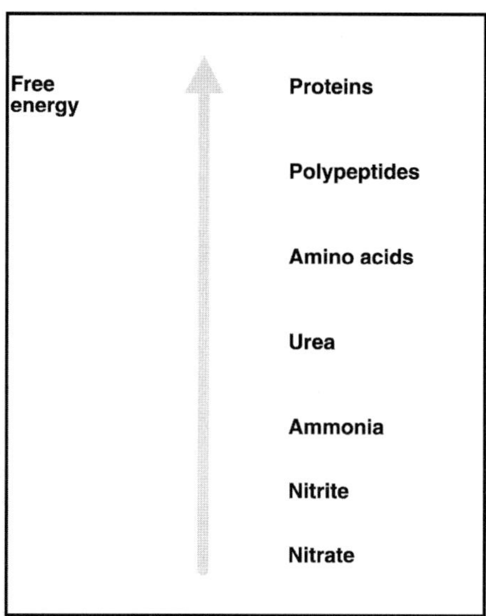

Fig. 2.6. The level of free energy for some biologically important nitrogen compounds.

has been performed, is transformed into heat, which is also delivered to the environment. Fig. 2.5 shows the cycling of matter in more detail.

Fig. 2.6 illustrates the hierarchy in free energy of various ecologically important nitrogen compounds. Proteins with the highest free energy are the result of the photosynthesis or of the biochemical synthesis in heterotrophic organisms. They are important food components for the heterotrophic organism as the suppliers of the important blocks, amino acids and of energy.

2.8. Summary of the important ecological issues

It is the objective of this volume to demonstrate that we do have an ecosystem theory, mainly based on thermodynamics for ecosystems. The theory should be applied as any other theory to explain our observations. The fundament of any ecosystem theory will inevitably be the basic thermodynamics laws: the First, the Second and the Third Law of Thermodynamics. The introduction of the concept of exergy will later enable us to apply these three fundamental laws more directly in an ecological context, because exergy is a thermodynamic variable that can be applied far from thermodynamic equilibrium. However, after presentation of the three basic laws of classic thermodynamics, we can already glimpse the enormous importance the laws have for ecology.

The conservation laws are widely applied in all ecological models, which are based on a "book-keeping" of matter and energy, i.e. the use of Eq. (2.1). The holistic description of the ecosystem embodied in the model thereby becomes the consequence of the thermodynamic laws or, expressed slightly differently: the description of an ecosystem by a model reflects the constraints of the thermodynamic laws on the ecosystem.

The concept of an ecosystem, widely used in ecology, makes it possible to distinguish the system and the environment in a thermodynamic sense. System ecology is concerned with the exchange of mass and energy between the system and the environment, and the influence of these exchange processes on the ecosystem and its processes.

Thermodynamics is furthermore applied in biochemistry to understand the relationships between chains of biochemical processes and the corresponding energy budget. The catabolic processes are important as the supplier of energy for maintenance of the life processes. When the energy demand is covered by the catabolic processes, additional energy may be used to build up biomass—to cover the anabolism. The organisms need, in other words, to transport energy between cells—from cells where the energy is produced to cells where the energy is needed. The organisms use small packages of ATP, which are easily transported and are able to release 41.8 kJ/mol by the following process:

$$\mathrm{ATP} \leftrightarrow \mathrm{ADP} + \mathrm{P}$$

to solve that problem. All these considerations are in accordance with the thermodynamic laws.

Also, the Second Law gives us a deep insight into the function of ecosystems. The function is based on entirely irreversible processes. Energy that can do work is lost as heat to the environment. It is the cost of keeping the ecological machine cycling and repeatedly recycling the matter, so it can be used to build continuously new biomass, which in the long run contains more and more information. The energy for maintenance of the cycling and recycling is provided by solar radiation that supplies the work to the ecosystem needed for recycling the matter. Heat is produced due to "friction" in this process caused by the life processes. Ecosystems therefore require an input source of useful energy to drive the life processes.

We cannot understand ecosystems by Newton's Laws, because the processes are irreversible and the time arrow is governing. The process is one way: solar radiation → plant biomass is built according to the photosynthesis → the biochemical energy in the plant biomass is utilised throughout the entire food web → new biomass is formed continuously, which implies that new "solutions" to life under the continuously changing conditions are formed. The result is the evolution, a distinct one-way process—no symmetry. The consequence is that we cannot describe ecosystems by Newton's Laws. Ecological models cannot be developed from Newton's mechanics but they must be based on the thermodynamics.

The calculations of the equilibrium constants for the decomposition of high molecular weight proteins show that the decomposition processes are spontaneous and provide a considerable amount of energy (that can do work). It is only possible to avoid the decomposition process by feeding energy to the living organisms to compensate for the inexorable production of entropy = inexorable reducing gradients to obtain a more probable situation. Living systems require an energy source that can provide the energy needed to maintain the system far from thermodynamic equilibrium. Without the energy source, the system will inevitably move toward thermodynamic equilibrium, where there are no gradients in space or time—the system is therefore dull and has no life. *An energy source is a necessary prerequisite for living systems.* Is it also a sufficient condition for creation of life? We will turn back to this question a few times in the coming chapters.

Chapter 3

Second and Third Law of Thermodynamics in open systems

...Only living systems are never in equilibrium and permanently performing work at the expense of its free energy against the equilibrium required by laws of physics and chemistry under current external conditions.
...A source of the work done by living systems is at the final account free energy peculiar to this molecular structure, to this state of molecules.
...This non-equilibrium state, this deformed molecular structure... is maintained or permanently restored at the expense of the energy of continuous processes of equalisation which is flowing past within a living matter.
E. Bauer, "Theoretical biology", 1935.

3.1. Open systems and their energy balance

Ecosystems are open systems in the sense that they are open for mass and energy transfer. Ecosystems receive energy from solar radiation and water from precipitation, dry deposition from the atmosphere, inputs by wind and flows of various types plus migration of species. A system that is closed for in- and outputs of energy and mass is called an isolated system, while a system that is closed to in- and outputs of mass, but open to energy transfers, is named a closed system. A non-isolated system is a closed or open system. If an ecosystem is isolated, it would inevitably move toward thermodynamic equilibrium and become a dead system with no gradients to do work—or, as expressed in Chapter 2, $\mathrm{d}G = 0$ and $\mathrm{d}S = 0$ at a maximum S value. The openness explains why an ecosystem can maintain life and stay far from thermodynamic equilibrium because maintenance of life requires input of energy, which of course is only possible if an ecosystem is at least non-isolated.

The use of the Second Law of Thermodynamics for open systems is crucial. At first glance, it looks like ecosystems violate the Second Law because they are moving away from thermodynamic equilibrium by formation of a biological structure. Ecosystems receive, however, energy as solar radiation, which can compensate for the steady transfer of work to heat. Several proposals on how to apply the Second Law of Thermodynamics on systems far from thermodynamic equilibrium have been given, as will be demonstrated in Section 3.3, but before we turn to this central issue

Towards a Thermodynamic Theory for Ecological Systems, pp. 41–67

for open systems, what openness implies for the properties of ecosystems and how it is possible to quantify openness will be discussed.

In ecosystem steady states, the formation of biological compounds (anabolism) is in approximate balance with their decomposition (catabolism). The energy captured by an ecosystem can, in principle, be any form of energy (electromagnetic, electrical, magnetic, chemical, mechanical, etc.) but for the ecosystems on Earth the short-wave energy of solar radiation (electromagnetic energy) plays the major role. One must keep in mind that there are only four types of energy in our physical world: gravitational, electromagnetic, and energies of strong and weak interactions (in microphysics). All biosphere processes use only electromagnetic energy. Until humans were a part of the biosphere and used only an electromagnetic form of energy, we should hope that all so-called "ecological crises" could be resolved. But as soon *Homo sapiens* began to use the two latter forms of energy (nuclear power plants) and, moreover, intends to use the thermonuclear synthesis, she enters into a *principal* contradiction with the biosphere. As a consequence, these new crises apparently become unresolved. However, this is a speculative hypothesis, too.

The following reaction chain summarises the consequences of energy openness (Jørgensen et al., 1999): *source*: solar radiation → *anabolism* (charge phase): incorporation of high quality energy, with entrained work capacity (and information), into complex bio-molecular structures, entailing anti-entropic system movement away from equilibrium → *catabolism* (discharge phase): deterioration of structure involving release of chemical bond energy and its degradation to lower states of usefulness for work (heat) → *sink*: dissipation of degraded (low work capacity and high entropy) energy as heat to the environment (and, from Earth, to deep space), involving entropy generation and return toward thermodynamic equilibrium.

This same chain can also be expressed in terms of matter: *source*: geochemical substrates relatively close to thermodynamic equilibrium → *anabolism*: inorganic chemicals are moulded into complex organic molecules (with low probability, it means that the equilibrium constant for the formation process is very low, low entropy, and large distance from thermodynamic equilibrium) → *catabolism*: synthesised organic matter is ultimately decomposed into simple inorganic molecules again; the distance from thermodynamic equilibrium decreases, and entropy increases → *cycling*: the inorganic molecules, returned to near-equilibrium states, become available in the nearly closed material ecosphere of Earth for repetition of the matter charge–discharge cycle.

Input environments of ecosystems serve as sources of high quality energy whose high contents of work and information and low entropy raise the organisational states of matter far from equilibrium. Output environments, in contrast, are sinks for energy and matter lower in work capacity, higher in entropy, and closer to equilibrium. Since, in the organisation of ecosystems, output environments feed back to become portions of input environments, living systems operating in the ecosphere, which is energetically non-isolated but materially nearly closed, must seek an adaptive balance between these two aspects of their environmental relations in order to sustain their continued existence.

3.2. The Second Law of Thermodynamics interpreted for open systems

Let us first expand on the conclusions that we have already made to give more detail on the difference between isolated and open systems and thereby understand better the application of the Second Law to open systems (Jørgensen et al., 1999).

If ecosystems were isolated, no energy or matter could be exchanged across their boundaries. The systems would spontaneously degrade their initially contained work capacity, i.e. exergy (for details see Chapter 5) and increase their entropy, corresponding to a loss of order and organisation, and increase in the randomness of their constituents and microscopic states. This dissipation process would cease at equilibrium, where no further motion or change would be possible. The physical manifestation would ultimately be a meltdown to the proverbial "inorganic soup" containing degradation products dispersed equally through the entire volume of the system. All gradients of all kinds would be eliminated, and the system would be frozen in time in a stable, fixed configuration. The high-energy chemical compounds of biological systems, faced suddenly with isolation, would decompose spontaneously to compounds with high entropy contents. The process would be progressive, to higher and higher entropy, and would, by the presence of oxygen, end with a mixture of inorganic residues—carbon dioxide and water, nitrates, phosphates, sulphates, etc. These simpler compounds could never be reconfigured into the complex molecules necessary to carry on life processes without the input of new low-entropy energy to be employed in biosynthesis. An isolated ecosystem could therefore, in the best case, sustain life for only a limited period of time, less than that required from the onset of isolation to reach thermodynamic equilibrium. This local situation is comparable to the "thermal death" of the Universe, seen by physicists of a century ago as the ultimate outcome of the Second Law of Thermodynamics. Thus, thermodynamic equilibrium is the global attractor for all physical processes isolated from their surroundings. Having reached it, no further changes are possible. In this "frozen" state, even time would have no meaning as its passage could not be verified by reference to any changes. Observations of properties could not be made, only inferred, because observation requires some kind of exchanges between the system and an observer. There would be no internal processes, because no gradients would exist to enable them. There would only be uninterrupted and uninterruptable stillness and sameness which would never change. The system would be completely static at the thermodynamic equilibrium. Thus, in a peculiar way, isolated systems can only be pure abstractions in reality, submitting neither to time passage, change, nor actual observation. They are the first "black holes" of physics, and the antithesis of our systems plus their environments which are the core model for systems ecology. No ecosystem could ever exist and be known to us as an isolated system.

The change in entropy for an *open* system, $\mathrm{d}S$, consists of an external, exogenous contribution from the environment, $\mathrm{d}_\mathrm{e}S = (q^S_\mathrm{in} - q^S_\mathrm{out})\mathrm{d}t$, where q^S_in and q^S_out are inflows and outflows of entropy, and an internal, endogenous contribution due to system state, $\mathrm{d}_\mathrm{i}S$, which should always be positive by the Second Law (Prigogine, 1955). Prigogine uses the concept of entropy and the Second Law of Thermodynamics far from thermodynamic equilibrium, which is outside the framework of classic thermodynamics, but he uses the concepts only locally.

There are three possibilities for the entropy balance:

$$\frac{dS}{dt} = \frac{d_eS}{dt} + \frac{d_iS}{dt} > 0, \tag{2.1}$$

$$\frac{dS}{dt} = \frac{d_eS}{dt} + \frac{d_iS}{dt} < 0, \tag{2.2}$$

$$\frac{dS}{dt} = \frac{d_eS}{dt} + \frac{d_iS}{dt} = 0. \tag{2.3}$$

The system loses order in the first case. Gaining order (case 2), is *only* possible if

$$-\frac{d_eS}{dt} > \frac{d_iS}{dt} > 0. \tag{2.4}$$

This means that, if order is to be created in a system ($dS/dt < 0$), d_eS/dt must be negative, and therefore $q_{in}^S < q_{out}^S$.

Creation of order in a system must be associated with a greater flux of entropy out of the system than into the system. This implies that the system must be open or at least non-isolated.

Case 3, Eq. (2.3), corresponds to a stationary situation, for which Ebeling et al. (1990) use the following two equations for the energy (*U*) balance and the entropy (*S*) balance:

$$\frac{dU}{dt} = 0 \quad \text{or} \quad \frac{d_eU}{dt} = -\frac{d_iU}{dt} \tag{2.5}$$

and

$$\frac{dS}{dt} = 0 \quad \text{or} \quad \frac{d_eS}{dt} = -\frac{d_iS}{dt}. \tag{2.6}$$

Usually, the thermodynamic processes are isotherm and isobar. This implies that we can interpret the third case (Eqs. (2.3) and (2.5), and (2.6)) by use of the Gibbs free energy:

$$\frac{d_eG}{dt} = T\frac{d_iS}{dt} > 0. \tag{2.7}$$

It means that a "*status quo*" situation for an ecosystem requires input of free energy to compensate for the loss of free energy and corresponding formation of heat due to maintenance processes, i.e. respiration and evapotranspiration. If the system is not receiving a sufficient amount of free energy, the entropy will increase. If the entropy of the system continues to increase, the system will approach thermodynamic equilibrium—the system will die. This is in accordance with Ostwald (1931): *life without the input of free energy is not possible.*

The entropy produced by the life processes can be exported by three processes: (1) transfer of heat to the environment, (2) exchange of material with the environment, and (3) biochemical processes in the system. The first process (heat transfer) is of particular importance.

An energy flow of about 10^{17} W by solar radiation ensures the maintenance of life on Earth. The surface temperature of the Sun is 5800 K and of the Earth on average about 280 K. This implies that the following export of entropy per unit of time takes place from Earth to the open space:

$$10^{17}\ \mathrm{W}(1/5800\ \mathrm{K} - 1/280\ \mathrm{K}) \approx 4 \times 10^{14}\ \mathrm{W/K} \tag{2.8}$$

corresponding to 1 $\mathrm{W/m^2\ K}$.

Ecosystems can maintain a certain concentration of low-entropy compounds against the second-law dissipation gradient because they are not isolated. Ecosystems receive a continuous supply of free energy or negentropy (potential entropy, not yet released (see Schrödinger, 1944)) from outside to compensate for the positive entropy produced internally as a consequence of the Second Law of Thermodynamics ($d_iS > 0$). On Earth, solar radiation is the main source of this input of free energy, negentropy or low-entropy energy. The incoming energy has low entropy, while the outgoing energy has higher entropy.

All ordered structures require low entropy for maintenance and, therefore, for a system to maintain structure or increase its internal order, it must receive input of low-entropy energy from external sources. *Structure*, in this context, is a spatial or temporal order, describable in terms of information theory (see Chapter 4). Prigogine uses the term *dissipative structure* to denote self-organising systems, thereby indicating that such systems dissipate energy (produce entropy) for the maintenance of their organisation (order). The following conclusions are appropriate:

All systems, because they are subject to the Second Law of Thermodynamics, are inherently dissipative structures. To offset the dissipative processes, they require inputs of low-entropy energy to maintain or produce more internal organised structure, measurable in terms of information content. Thus, all real systems must be open or, at least, non-isolated.

Ecosystems, in common with all real systems, have, as previously noted, a global attractor state, thermodynamic equilibrium. Through their openness, they avoid reaching this state by importing low entropy, or matter carrying information from their surroundings. This anabolism combats and compensates for the catabolic deterioration of structure; the two processes operate against one another. Note that the equilibrium "attractor" represents a resting or refractory state, one that is passively devolved if system openness or non-isolation is compromised (Jørgensen et al., 1999). The term is also commonly used to express the situation when a system is actively pushed or "forced" towards a *steady state*. Though widespread, we do not subscribe to this usage and make a distinction between steady states and equilibriums for two reasons:

1. The state–space system theory we outlined in the "conservation" chapter (Patten et al., 1997) precludes anything in system dynamics but a unique input–state–output relationship. Therefore, given an initial state, state–space theory asserts that there exists one and only one sequence of inputs that will put an open system in a given state at a specified final time. For this terminal state to be an "attractor", many input sequences would have to be able to place the system in it, and from many initial states—the attractor would be hard to avoid.

2. As observed above, a steady state is a forced (non-zero input) condition; there is nothing "attractive" about it. Without a proper forcing function, it will never be reached or maintained. A steady state that is constant may appear in equilibrium, but it is really far from equilibrium and maintained by a steady input of energy or matter. We regard equilibrium as a zero-input condition. What are often recognised as local attractors in mathematical models really have no counterparts in nature. Steady states are forced conditions, not to be confused with unforced equilibriums, which represent states to which systems settle when they are devoid of inputs. The only true natural attractor in reality, and it is global, is the unforced thermodynamic equilibrium. See Patten et al. (1997) for further clarification of the basis for these distinctions.

The overall results of these processes are consistent with the energy charge–discharge cycles introduced and discussed in detail in Patten et al. (1997). The consequences of the openness may be stated in three different ways to emphasise the different perspectives provided by, respectively, energy, entropy, and thermodynamic equilibrium (Jørgensen et al., 1999):

1. *Energy*. Energy, mainly as solar radiation but also contained in certain inorganic substances of geochemical origin, is transferred in anabolism to high molecular weight, high energy, organic compounds (charge phase). These compounds are decomposed in catabolism along step-wise molecular pathways to release high quality energy which powers biological processes (discharge phase). This energy is degraded in the performance of work to lower quality energy in the form of heat which leaves the system, necessitating subsequent anabolic recharge. The quantity of energy is conserved, but not the energy quality (measured by its ability to do work). The energy quality is degraded by this charge–discharge cycle.
2. *Entropy*. A certain quantity of negentropy is embodied in solar and geochemical energy. Some of this negentropy is built in anabolic processes into the organised structure of biomolecules (charge phase). Progressive decomposition of these molecules in catabolism releases the entropy as the low-entropy energy is converted to high-entropy heat (discharge phase). The entropy reduces the overall order and organisation of the system and its environment. Thus, organised (low-entropy) states of ecosystems and their organisms can only be achieved and maintained at the expense of their respective environments by, in effect, "pumping out disorder" as high-entropy released heat.
3. *Thermodynamic equilibrium*. The energy of solar radiation and inorganic compounds used in photosynthesis and chemosynthesis is high quality energy far from thermodynamic equilibrium (as reflected in their exergy and negentropy contents). Systems move further from equilibrium (and their energies and negentropies increase) to the extent that this energy becomes incorporated into their organised structure (charge phase). If a far-from-equilibrium system becomes isolated, or is otherwise severed from its energy sources, then it will spontaneously decay by irreversible processes toward thermodynamic equilibrium (discharge phase). Continual input of high quality energy as a forcing function is required for a system to achieve and maintain a steady state far from equilibrium.

In each of the above three descriptions of the same input–output phenomenon associated with system non-isolation and openness, the opposition of successive charge and discharge phases can be seen as an antagonism or "combat" between anabolic and catabolic processes. The first seeks to build up structural organisation against the gradient provided by the Second Law of Thermodynamics. The second inexorably tears it down toward thermodynamic equilibrium. Far from equilibrium, energy of high quality is degraded to heat, exergy is consumed in performing work, and negative entropy is converted to positive entropy as self-organising systems are spontaneously and irreversibly drawn toward the global attractor that is thermodynamic equilibrium. Openness or non-isolation is a necessary condition for avoidance of this state, for it is only by the exchange of energy or matter across system boundaries that the far-from-equilibrium condition can be reached and sustained. This is because the total entropy is not preserved but will steadily increase for any irreversible process. An ecosystem must, therefore, to repeat the previous arguments, be able to dissipate the generated entropy as heat to its environment. Otherwise, temperature would increase without bound and life in ecosystems as we know it could not be supported.

The following conclusion pertains to the two environments of every ecosystem:

Input environments of ecosystems serve as sources of high quality energy whose high contents of free energy and low entropy raise the organisational states of matter far from equilibrium. Output environments, in contrast, are sinks for energy higher in entropy, and closer to equilibrium. Since, in the organisation of ecosystems, output environments feed back to become portions of input environments, living systems operating in the ecosphere, which is energetically non-isolated but materially nearly closed, must seek an adaptive balance between these two aspects of their environmental relations in order to sustain their continued existence.

3.3. Prigogine's theorem and the evolutionary criterion by Glansdorff–Prigogine

Revenons à nos moutons and keep in mind Eq. (5.13) of Chapter 2 written for the density of entropy production, σ:

$$\sigma T = \frac{1}{V}\frac{\mathrm{d}'A_{\mathrm{irrev}}}{\mathrm{d}t} + \sum_{k=1}^{K} A_k v_k^{\mathrm{ch}}. \tag{3.1}$$

If the particles of mth sort under the action of internal force F_m move with the current density J_m, then, in accordance with Ohm's rule, $\mathrm{d}'A_{\mathrm{irrev}} = V\sum_m F_m J_m \mathrm{d}t$. By substituting it into Eq. (3.1), we get

$$\sigma T = \sum_{m=1}^{M} F_m J_m + \sum_{k=1}^{K} A_k v_k^{\mathrm{ch}}. \tag{3.2}$$

Formally, the right side of Eq. (3.2) can be written in a bilinear form with $(3M + K)$ items:

$$\sigma T = \sum_{\alpha=1}^{3M+K} X_\alpha J_\alpha, \tag{3.3}$$

where X_α and J_α are the generalised thermodynamic forces and fluxes defined as

$$X_\alpha = F^l_m,\ J_\alpha = J^l_m\ (l = 1, 2, 3;\ m = 1, \dots, M;\ \alpha = 1, \dots, 3M),$$
$$X_\alpha = A_k,\ J_\alpha = v^{\text{ch}}_k\ (k = 1, \dots, K;\ \alpha = 1, \dots, 3M + K). \tag{3.4}$$

The index $l = 1, 2, 3$ appears here since the forces and fluxes are three-dimensional vectors.

The product σT is a function of dissipation Ψ (see Section 2.5).

At thermodynamic equilibrium, all the thermodynamic forces and fluxes simultaneously vanish; therefore, in the vicinity of the equilibrium, the fluxes can be considered as linear functions of forces:

$$J_\alpha = \sum_{\beta=1}^{3M+K} L_{\alpha\beta} X_\beta, \qquad L_{\alpha\beta} = \left(\frac{\partial J_\alpha}{\partial X_\beta} \right)_0. \tag{3.5}$$

This is the so-called Onsager's linear relation, and the $L_{\alpha\beta}$ is named *Onsager's coefficients*. It is obvious that in this case the expression for $\Psi = \sigma T$ can be represented as a quadratic form:

$$\sigma T = \sum_{\alpha,\beta=1}^{3M+K} L_{\alpha\beta} X_\alpha X_\beta. \tag{3.6}$$

From the Second Law, it follows that $\sigma \geq 0$, where $\sigma = 0$ only for $X_\alpha = 0$. This means that the matrix $\|L_{\alpha\beta}\|$ must be positive definite that, in turn, imposes certain constraints on Onsager's coefficients. For instance, for $\alpha, \beta = 1.2$, these constraints have the form: $L_{11}, L_{22} > 0;\ 4L_{11}L_{22} > (L_{12} + L_{21})^2$.

Other constraints appear from temporal and spatial symmetries of the system. Since the equations of microphysics are symmetrical with respect to time, then $L_{\alpha\beta} = L_{\beta\alpha}$ (Onsager, 1931).

Let us consider some spatial–isotropic medium, where all processes are divided into two classes: scalar (chemical reactions) and vector (particles flows). By virtue of isotropy, scalar causes cannot induce anisotropic effects, and vice versa (Curie–Prigogine's principle). Thus, Onsager's coefficients corresponding to coupling between forces and fluxes, described by tensors of different ranges, are equal to zero.

Let us calculate the variations of entropy production, forces and fluxes at $T = \text{const}$ in the vicinity of some steady state (dynamic equilibrium) $\sigma^{(0)}, X^{(0)}_\alpha, J^{(0)}_\alpha$:

$$T\delta\sigma = \sum_{\alpha,\beta} \left[(L_{\alpha\beta} X^{(0)}_\alpha \delta X_\beta + L_{\alpha\beta} X^{(0)}_\beta \delta X_\alpha) + L_{\alpha\beta} \delta X_\alpha \delta X_\beta \right].$$

Since at the dynamic equilibrium, $\mathrm{d}\sigma/\mathrm{d}t = 0$, then the term within brackets also vanishes. Therefore

$$T\delta\sigma = \sum_{\alpha,\beta=1}^{3M+K} L_{\alpha\beta}\delta X_\alpha \delta X_\beta. \tag{3.7}$$

Since the matrix $\|L_{\alpha\beta}\|$ is positive definite then $\delta\sigma \geq 0$, $\delta(\mathrm{d_i}S/\mathrm{d}t) = V\delta\sigma \geq 0$, where $\delta\sigma = \delta(\mathrm{d_i}S/\mathrm{d}t) = 0$ only if $\delta X_\alpha = 0$. This is the main statement of Prigogine's theorem: *the entropy production of linear irreversible process reaches a maximum in a dynamic equilibrium*. When the system approaches the equilibrium, the forces and fluxes change in such a way that the entropy production is always decreasing (Prigogine, 1955).

The theorem can be considered as a criterion of the system evolution in a linear region; a criterion for non-linear domain has been formulated by Glansdorff and Prigogine (1971). We shall formulate the criterion for a partial case of chemical reactions, which takes place within a system. Since in this case, $\mathrm{d}'A_{\mathrm{irrev}}/\mathrm{d}t = 0$, then we have from Eq. (3.1):

$$\sigma = \sum_{k=1}^{K} \frac{A_k}{T} v_k^{\mathrm{ch}}. \tag{3.8}$$

By varying the chemical affinities (see Eq. (5.12) in Chapter 2), $\delta A_k = -\sum_m v_{mk}\delta\mu_k$, and substituting these variations into Eq. (3.8), we obtain

$$\delta_X\sigma = -\frac{1}{T}\sum_{k=1}^{K}\sum_m v_{mk} v_k^{\mathrm{ch}} \delta\mu_k. \tag{3.9}$$

Since (see Eq. (5.9) in Chapter 2)

$$\sum_{k=1}^{K} v_{mk} v_k^{\mathrm{ch}} = \frac{1}{V}\frac{\mathrm{d_i}n_m}{\mathrm{d}t} = \frac{\mathrm{d}c_m}{\mathrm{d}t} \tag{3.10}$$

and

$$\mathrm{d}\mu_m = \sum_l \frac{\partial\mu_m}{\partial c_l}\mathrm{d}c_l$$

then for isotherm processes we get finally from Eq. (3.9)

$$\frac{\mathrm{d_X}\sigma}{\mathrm{d}t} = -\frac{1}{T}\sum_{m,l}\left(\frac{\partial\mu_m}{\partial c_l}\right)\frac{\mathrm{d}c_m}{\mathrm{d}t}\frac{\mathrm{d}c_l}{\mathrm{d}t}. \tag{3.11}$$

Gibbs has shown the matrix $\|\partial\mu_m/\partial c_l\|$ to be positive definite, therefore

$$\frac{\mathrm{d_X}\sigma}{\mathrm{d}t} \leq 0 \quad \text{or} \quad \frac{\mathrm{d_X}}{\mathrm{d}t}\left(\frac{\mathrm{d_i}S}{\mathrm{d}t}\right) \leq 0. \tag{3.12}$$

This is the Glansdorff–Prigogine evolutionary criterion, or Glansdorff–Prigogine's theorem: the change of entropy production caused by variations of forces is either negative or equal to zero. As a partial case of the theorem for linear process, we get the following

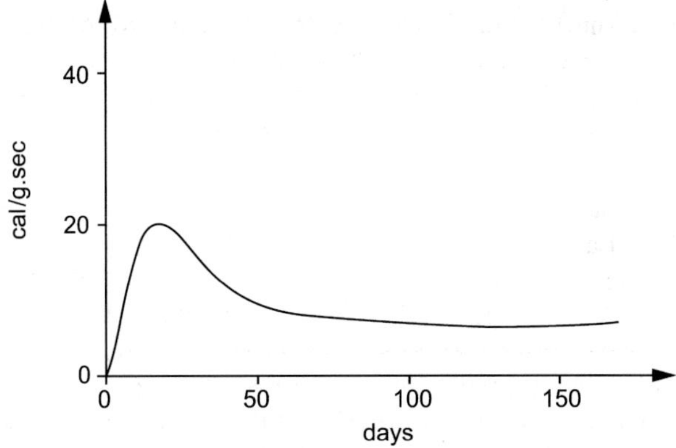

Fig. 3.1. A typical temporal pattern of the heat production by the fertilised amphibian egg.

form of Prigogine's criterion:

$$\frac{\mathrm{d}\sigma}{\mathrm{d}t} = 2\frac{\mathrm{d}_X\sigma}{\mathrm{d}t} \le 0 \left(\text{and correspondingly, } \frac{\mathrm{d}}{\mathrm{d}t}\left(\frac{\mathrm{d_i}S}{\mathrm{d}t}\right) \le 0\right). \tag{3.13}$$

There are a lot of investigations where the rate of entropy change in biological systems is estimated by means of experimental measurement of their heat production. These experiments could be correct if the entropy of reactants does not depend on the degree of completeness of the chemical reactions. In this case, the affinity of process is directly connected with its thermal effect, $A = -(\partial H/\partial \xi)_{T,p}$ where the H is enthalpy, and the entropy production must be proportional to the effect, so that

$$\frac{\mathrm{d_i}S}{\mathrm{d}t} \approx -\frac{1}{T}\left(\frac{\mathrm{d}Q}{\mathrm{d}t}\right)_{T,p}. \tag{3.14}$$

A typical temporal pattern of heat production by the fertilised amphibian egg is shown in Fig. 3.1 (Lurie and Wagensberg, 1979).

Fig. 3.1 is a good illustration to Prigogine's theorem: one can see that the entropy production for a rather large period of time is a monotonous decrease with time tending to a certain positive value.

3.4. The Third Law of Thermodynamics applied on open systems

The First Law of Thermodynamics is often applied to ecosystems, first of all when the energy balances of ecosystems are made. Also, the Second Law of Thermodynamics is applied to the ecosystem when we consider the entropy production of ecosystems as a consequence of the maintenance of the system far from thermodynamic equilibrium. This section is concerned with the application of the Third Law of Thermodynamics to ecosystems (see also Chapter 2).

The lesser-known Third Law of Thermodynamics states that the entropies, S_0, of pure chemical compounds are zero, and that entropy production, ΔS_0, by chemical reactions between pure crystalline compounds is zero at absolute temperature, 0 K. The Third Law implies, since both $S_0 = 0$ (absolute order) and $\Delta S_0 = 0$ (no disorder generation), that disorder does not exist and cannot be created at absolute zero temperature. But at temperatures higher than zero of Kelvin, disorder can exist ($S > 0$) and be generated ($\Delta S > 0$). The Third Law defines the relation between entropy production, ΔS, and the Kelvin temperature, T:

$$\Delta S = \int_0^T \Delta c_p d \ln T + \Delta S_0 \tag{4.1}$$

where Δc_p is the increase in heat capacity by the chemical reaction. Since order is absolute at absolute zero, its further creation is precluded there. At higher temperatures, however, order can be created.

Entropy production implies degradation of energy from a state of high utility (large T) to a state of low utility (small T); compare also Carnot's Cycle (see Chapter 5). Ecosystems have, in other words, a global attractor state, the thermodynamic equilibrium, but will never reach this state as long as they are not isolated and receive exergy (energy that can do work; see Chapter 5) from outside to combat the decomposition of their compounds. As ecosystems have an energy through-flow, the attractor becomes the steady state, where the formation of new biological compounds is in balance with the decomposition processes. As seen from these perspectives of the Second Law of Thermodynamics for open (non-isolated) systems, it is vital for ecosystems to be non-isolated.

It has been stated a few times that it is *necessary* for an ecosystem to transfer the generated heat (entropy) to the environment and to receive low-entropy energy (solar radiation) from the environment for formation of dissipative structure. The next obvious question would be: will energy source and sink also be *sufficient* to initiate formation of dissipative structure, which can be used as source for entropy combating processes?

The answer to this question is "Yes." It can be shown by the use of simple model systems and basic thermodynamics; see Morowitz (1968, 1978). He shows that a flow of energy from sources to sinks leads to an internal organisation of the system and to the establishment of element cycles. The type of organisation is, of course, dependent on a number of factors: the temperature, the elements present, the initial conditions of the system, and the time available for the development of organisation. It is characteristic for the system, as pointed out above, that the steady state of an open system does *not* involve chemical equilibrium.

An interesting illustration of the creation of organisation (dissipative structure) as a result of an energy flow through ecosystems concerns the possibilities to form organic matter from the inorganic components which were present in the primeval atmosphere. Since 1897, many simulation experiments have been performed to explain how the first organic matter was formed on Earth from inorganic matter. All of them point to the conclusion that energy interacts with a mixture of gases to form a large set of randomly synthesised organic compounds. Most interesting is perhaps the experiment performed by Stanley Miller and Harold Urey at the University of Chicago in 1953, because it showed that amino acids can be formed by sparking a mixture of CH_4, H_2O, NH_3 and H_2; corresponding approximately to the composition of the primeval atmosphere.

Prigogine and his colleagues have shown that open systems that are exposed to an energy through-flow exhibit coherent self-organisation behaviour and are known as dissipative structures. Formations of complex organic compounds from inorganic matter as mentioned above are typical examples of self-organisation. Such systems can remain in their organised state by exporting entropy outside the system, but are dependent on outside energy fluxes to maintain their organisation, as was already mentioned and emphasised above. Glansdorff and Prigogine (1971) have shown that the thermodynamic relationship of far from equilibrium dissipative structures is best represented by coupled non-linear relationships, i.e. autocatalytic positive feedback cycles.

Given this necessary condition, simple energy flow through a system provides a sufficient condition. Creation of order is inevitable. On Earth, the surface temperature difference between sun and planet guarantees this. Morowitz (1968, 1978) showed, as mentioned above, that energy through-flow is sufficient to produce cycling, a prerequisite for the ordering processes characteristic of living systems.

A system at 0 K, on the other hand, is without any creative potential, because no dissipation of energy can take place at this temperature. A temperature greater than 2.726 ± 0.01 K, where 2.726 K is the temperature of deep space, is therefore required before order can be created. At 0 K, the world is dead and still as the temperature is a measure of the velocity of atoms. The so-called Bose–Einstein condensate is formed, where all atoms are the same and behave like one single atom. This was predicted by Bose and Einstein in the 1920s, but has recently been shown experimentally at temperatures very close to 0 K.

The velocity is zero at 0 K by definition and therefore determined without uncertainty. This explains that the position is undetermined according to Heisenberg's uncertainty equation. At 0 K there is therefore no structure, no gradients, and no complexity. No entropy can be formed because all mass is everywhere and nowhere and without form and structure. There is no disorder to create and therefore no entropy to produce. The system is trapped between complete order because all the mass occupies all the space, and complete disorder because all the space is occupied by mass—a complete dissipation has taken place. At 0 K, no creativity is possible, no differences (gradients), no structure and even no physical activity, because all velocities are zero. Everything is dull and dead. Even the light has stopped. Time has no meaning because time is determined by the rate of changes.

These extreme conditions at 0 K elucidate the meaning of the concept entropy. Entropy is, on the one side, the price we have to pay for order, structure, organisation and creativity, but without entropy there would be no order, structure, organisation and creativity.

Moreover, it explains the meaning behind the Second Law. Because heat is an energy form which is generated by transformation of all other energy forms and because a 100% effective transformation of heat to work cannot take place because in the Carnot Cycle the cold reservoir can never be maintained at 0 K. Energy that can do work is inexorably lost to energy that cannot do work. This is the condition which is imposed on us: time and all reactions are irreversible.

3.5. Thermodynamics of living organisms

The processes of self-organisation and evolution of living organisms can be interpreted from the thermodynamics point of view. Namely, the problem was considered in the basic works of Ostwald (1931), Bauer (1935), von Bertallanffy (1942, 1952, 1956), Schrödinger (1944) and Prigogine and Wiame (1946). In this section, we shall follow the book of Ebeling et al. (1990).

To keep the organism at steady state, the energy balance must be maintained, $\mathrm{d}U/\mathrm{d}t = 0$. Nevertheless, it is not sufficient: the entropy balance must also be fulfilled, $-\mathrm{d_e}S/\mathrm{d}t = \mathrm{d_i}S/\mathrm{d}t > 0$. The function of metabolism in a living organism, from the thermodynamic point of view, is to "suck" the amount of entropy which is at least equal to entropy produced within the organism by means of the exchange of matter and heat with the environment.

Let $\mathrm{d}'Q$ be heat brought to the organism during time $\mathrm{d}t$, and $\mathrm{d_e}n_k$ be a number of moles or molecules of kth component brought to the organism during the same time. Then

$$\mathrm{d_e}S = \frac{\mathrm{d}'Q}{T} + \sum_k s_k \mathrm{d_e}n_k \tag{5.1}$$

where s_k is a specific entropy of kth component. The required export of entropy is provided by the following processes: (1) heat transfer, (2) matter exchange, and (3) transformation of matter within the system. The first (heat conductivity and heat irradiation) plays the main role. For this, a certain difference of temperatures has to be maintained between the organism and its environment, $T_{\mathrm{org}} - T_{\mathrm{env}} = \Delta T > 0$. There are two groups of organisms with different types of thermal control: poikilotherms with $\Delta T \approx \mathrm{const}$ and homeotherms with $T_{\mathrm{org}} \approx \mathrm{const}$. The latter can transfer a quantity of heat q_{org} in a unit time into the environment by means of heat conductivity. In addition, the organism can get the radiation energy q_{rad} in the course of the same time.

If we assume radiation incoming to the organism to be radiation from a "black body" with temperature T_{rad}, then it corresponds to the entropy inflow $q^S_{\mathrm{rad}} = 4q_{\mathrm{rad}}/3T_{\mathrm{rad}}$, where the factor (4/3) is Planck's *form-factor*, which takes into account some peculiarity of entropy transport by means of radiation (Landau and Lifshitz, 1995). Finally, also taking into account matter exchange, the entropy export is written as

$$-\frac{\mathrm{d_e}S}{\mathrm{d}t} \approx -\frac{4q_{\mathrm{rad}}}{T_{\mathrm{rad}}} + q_{\mathrm{org}}\left(\frac{1}{T_{\mathrm{env}}} - \frac{1}{T_{\mathrm{org}}}\right) - \sum_k s_k \frac{\mathrm{d_e}n_k}{\mathrm{d}t}. \tag{5.2}$$

A more exact balance has to take into account different types of radiation. For instance, if we consider the energy of solar radiation absorbed by plants during the process of photosynthesis, q_{phot}, then for a more correct description we must assume the existence of different thermal components q^i_{phot} with different radiative temperatures T^i_{phot} (see also Chapter 10). Summing all these contributions we get

$$-\frac{\mathrm{d_e}S}{\mathrm{d}t} \approx -\sum_i \frac{4q^i_{\mathrm{phot}}}{3T^i_{\mathrm{rad}}} + q_{\mathrm{org}}\left(\frac{1}{T_{\mathrm{env}}} - \frac{1}{T_{\mathrm{org}}}\right) + \frac{4q_{\mathrm{irrad}}}{3T_{\mathrm{org}}} - \frac{4q_{\mathrm{assim}}}{3T_{\mathrm{env}}} - \sum_k s_k \frac{\mathrm{d_e}n_k}{\mathrm{d}t}. \tag{5.3}$$

Here, q_{irrad} and q_{assim} are the energies of radiation emitted and assimilated by the organism into and from the environment. A necessary condition of functioning and developing for the living organism is a positive balance of the entropy export (see Eq. (2.4))

$$-\sum_i \frac{4q^i_{phot}}{3T^i_{rad}} + q_{org}\left(\frac{1}{T_{env}} - \frac{1}{T_{org}}\right) + \frac{4q_{irrad}}{3T_{org}} - \frac{4q_{assim}}{3T_{env}} - \sum_k s_k \frac{d_e n_k}{dt} > \frac{d_i S}{dt} > 0, \tag{5.4}$$

where

$$\frac{d_i S}{dt} = \frac{1}{T_{org}}\left(\frac{dA'_{diss}}{dt} - \sum_k \mu_k \frac{d_i n_k}{dt}\right), \tag{5.5}$$

and the term dA'_{diss}/dt is the power of dissipative forces.

The phenomenological theory of cell fission can be a good illustration of this feature (Volkenstein, 1988). The entropy balance in a cell considered as a sphere with radius R is described by the equation

$$\frac{dS}{dt} = \frac{d_i S}{dt} - \left|\frac{d_e S}{dt}\right| = \alpha\frac{4}{3}\pi R^3 - \beta 4\pi R^2. \tag{5.6}$$

We assume that the entropy production within the cell is proportional to its volume, and the entropy outflow is proportional to its surface (with α and β as the coefficients of proportionality). The cell grows until the stationary state at $dS/dt = 0$ with $R^* = 3\beta/2\alpha$ will be attained. If $R > R^*$, then the internal entropy production does not compensate its outflow, and the cell has to die. However, if the cell is divided, then the volume is conserved but the surface increases. As a result, dS/dt again becomes negative. Since at this moment $\Delta_i S = 2\alpha(4\pi(R')^3/3 = \alpha 4\pi R^3/3$, then the radius, R', of two new cells will be equal to $R' = R/\sqrt[3]{2}$. The entropy outflow for these cells is $\Delta_e S = 2\beta 4\pi R^2 = 8\beta\pi R^2/\sqrt[3]{4}$, and at $R = R^* = 3\beta/\alpha$ the negative increment of the total entropy is $\Delta S = 36\pi(\beta^3/\alpha^2)$ $(1 - \sqrt[3]{2}) \approx -29.4(\beta^3/\alpha^2)$.

The picture will be more complex for organisms which are more complex than a cell. Theoretically, it should be expected that two processes—the total growth of biomass and cells differentiation—most strongly influence entropy change. In order to separate these effects, we consider the change of *specific* entropy, i.e. entropy per unit of biomass, $s = S/B$, where B is the total biomass:

$$\frac{d}{dt}\left(\frac{S}{B}\right) = \frac{1}{B}\frac{dS}{dt} - \frac{1}{B}\left(\frac{S}{B}\right)\frac{dB}{dt} \quad \text{or} \quad \frac{dS}{dt} = B\frac{ds}{dt} + s\frac{dB}{dt}. \tag{5.7}$$

The process of differentiation leads to a decrease in the specific entropy, since system order increases, while biomass growth corresponds to the positiveness of the derivative dB/dt. Therefore, the change of entropy of the organism is determined by a combination of negative (differentiation) and positive (growth) terms, and a monotonous dependence of the entropy change on time should not be expected.

Returning to Eq. (5.6), we can rewrite it in a more general form. For this, we write the entropy as $S = \rho s V$ where ρ is the biomass density and V is the organism volume, so that $B = \rho V$. We assume again that $\mathrm{d_i}S/\mathrm{d}t = aV$ and $\mathrm{d_e}S/\mathrm{d}t = bF$, but now the values of a and b are not coefficients. Indeed, since $(1/V)\mathrm{d_i}S/\mathrm{d}t = a$ is nothing else than the specific entropy production, in accordance with Prigogine, $\mathrm{d}a/\mathrm{d}t \leq 0$, $a(t) \to a^*$ when $t \to \infty$. As before, b is assumed constant. We also assume that volume V and surface area F are connected by the relation $\Phi(F, V) = 0$. For standard, simply-connected geometric figures with central symmetry, where volume and surface area are determined by a single characteristic size, $F = fV^\gamma$ where $\gamma = 2/3$. For instance, for a sphere with radius R we have $V = (4/3)\pi R^3$ and $F = 4\pi R^2$ then $\gamma = 2/3$ and $f = (4\pi)^{1/3}(3)^{2/3} \approx 4.84$ and for a cube with side l we have $V = l^3$ and $F = 6l^2$ then $f = 6$ and $\gamma = 2/3$, too. By combining Eqs. (5.6) and (5.7), we get ($\rho = \text{const}$)

$$\frac{\mathrm{d}V}{\mathrm{d}t} = \frac{V}{\rho s(t)}(a(t) - bfV^{\gamma-1}) - \frac{V}{s}\frac{\mathrm{d}s}{\mathrm{d}t}. \tag{5.8}$$

We assume that there are the limit values of volume, V^*, specific entropy production, $a^* = \lim_{t\to\infty} a(t)$, and specific entropy, $s^* = \lim_{t\to\infty} s(t)$, so that (1) we can neglect by the term $(V/s)\mathrm{d}s/\mathrm{d}t$ for large t, (2) we can express the value of V^* as a function of parameters: $(V^*)^{1-\gamma} = bf/a^*$. By denoting a new variable, $z = V/V^*$, we can write the asymptotic analogue of Eq. (5.8) as follows

$$\frac{\mathrm{d}z}{\mathrm{d}t} = \left(\frac{a^*}{\rho s^*}\right)z\left(1 - \frac{1}{z^{1-\gamma}}\right). \tag{5.9}$$

It is easy to see that the derivative of z is negative within the interval (0,1); this implies that the equilibrium with $z^* = 1$ is unstable! This means that, in the process of growing old, organisms do not reach their final stationary state, where the constant rate of entropy production would be observed. The latter means death for living organisms. So, we have proved (thermodynamically) the impossibility of infinite life for any organism. It is interesting that if we use the Kostitzin–von Bertallanffy (Kostitzin, 1937; von Bertallanffy, 1956) growth equation based on the conservation law of energy, we would get another result.

Let the energy contained in an organism be equal to $E = \rho eV$ where e is a specific energy per unit of biomass. The energy balance is

$$\frac{\mathrm{d}E}{\mathrm{d}t} = Q_{\mathrm{in}}^{\mathrm{e}} - M. \tag{5.10}$$

The main assumptions in the model are the following: (1) input of energy is proportional to surface area of the organism, $Q_{\mathrm{in}}^{\mathrm{e}} = \varepsilon\rho F$, (2) metabolism M is proportional to its volume (or biomass), $M = r\rho V$, where r is a specific metabolism per unit of biomass. Then Eq. (5.10) can be represented as ($\rho = \text{const}$, $e = \text{const}$)

$$\frac{\mathrm{d}V}{\mathrm{d}t} = \frac{V^\gamma}{e}(\varepsilon f - rV^{1-\gamma}). \tag{5.11}$$

It is easy to see that if $r(t) \to r^*$, i.e. to the stationary level of basic metabolism, then Eq. (5.11) has the equilibrium $V^* = \sqrt[1-\gamma]{\varepsilon f/r^*}$, and it is stable! From this follows the so-called *allometric* principle:

$$r^* \sim (W^*)^{\gamma - 1}, \quad \gamma < 1, \tag{5.12}$$

where r^* is, for instance, the intensity of oxygen uptake per unit mass, and W^* is the total mass or weight of the organism. The principle is very popular in physiological ecology (see, for instance, Pianka, 1978).

Finally, we consider one curious example. Let there be two organisms which differ from each other only by the geometry of the surface across which energy transports into organisms: values of form-factors, f_1 and f_2, and such a geometric index as γ: γ_1 and γ_2. Note that $\gamma = 2/3$ for figures of simple shape. Let $V_2^* > V_1^*$ then the following inequality must hold: $f_2(1 - \gamma_1) > f_1(1 - \gamma_2)$. Assume that the organism strives to increase its size (mass, weight, volume) in the process of its evolution. As is seen from the equality, there are two strategies to realise the aim: (1) to increase f and (2) to increase γ (of course, a mixed strategy is also possible). In order to realise the first strategy, the organism has to increase its form-factor, for instance to transform from cubic to spherical. Aristotle could be right when he spoke about pre-humans as spherical figures. This could be a reason that we do not observe living organisms of cubic shape.

The second strategy is a principal complication of surface. For instance, if the surface is a fractal with dimension greater than 2 (but less than 3), then $\gamma > 2/3$. Let us look at lungs of superior animals: this is a typical fractal structure.

3.6. Quantification of openness and allometric principles

As we have already seen above, the surface area of a species is a fundamental property. The surface area indicates quantitatively the size of the boundary to the environment. Flow rates are often formulated in physics and chemistry as area times a gradient, which can be utilised to set up useful relationships between size and rate coefficients in ecology. Loss of heat to the environment must, for instance, be proportional to the surface area and to the temperature difference, according to the law of heat transfer. The rate of digestion, the lungs, hunting ground, etc. are, on the one hand, determinants for a number of parameters, and, on the other hand, they are all dependent on the size of the animal. It is therefore not surprising that many rate parameters for plants and animals are highly related to the size, which implies that it is possible to get very good first estimates for most parameters based only upon size. Naturally, the parameters are also dependent on several characteristic features of the species, but their influence is often minor compared to the size, and providing good estimates is valuable for many ecological models, at least as a starting value in the calibration phase. It is possible, however, to take these variations into account by the use of a form-factor. The form-factor may vary considerably among species.

The conclusion of these considerations must therefore be that there should be many parameters that might be related to simple properties, such as size of the organisms, and that such relationships are based upon fundamental biochemistry and thermodynamics.

Above all, there is a strong positive correlation between size and generation time, t_g, ranging from bacteria to the biggest mammals and trees (Bonner, 1965). The relationship

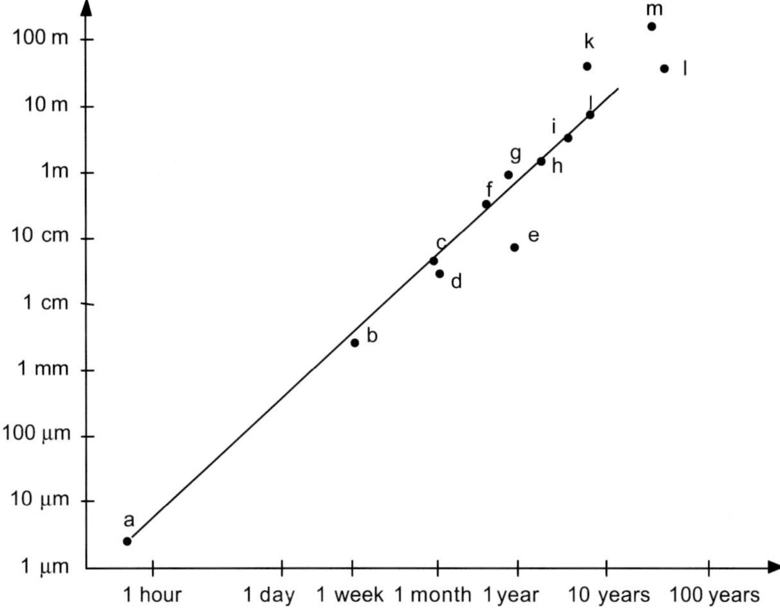

Fig. 3.2. Size and generation time plotted on log–log scale: (a) pseudomonas, (b) daphnia, (c) bee, (d) house fly, (e) snail, (f) mouse, (g) rat, (h) fox, (i) elk, (j) rhino, (k) whale, (l) birch, (m) fir.

is illustrated in Fig. 3.2. This relationship can be explained by use of the relationship between size (surface) and total metabolic action per unit of body weight mentioned above. It implies that the smaller the organism, the greater the metabolic activity. It is interesting that Bonner's relation could be obtained directly from the latter statement.

Indeed, taking into consideration the new variables: $z = V/V^*$, where $V^* = \sqrt[1-\gamma]{\varepsilon f/r^*}$, and $\tau = (r^*/e)t$, Eq. (5.11) can be written in a dimensionless form:

$$\frac{dz}{d\tau} = z^{\gamma}(1 - z^{1-\gamma}). \tag{6.1}$$

Since the right side of the equation does not depend on any parameters, then the dimensionless characteristic time of the system must be equal to 1, $\tau_{ch} \approx 1$. Since the length of generation can be considered as such a type of dimension characteristic time then $(r/e)t_g \approx 1$. From the expression for V^*, we have $r^* = \varepsilon f/(V^*)^{1-\gamma}$; substituting it into the previous expression, we obtain

$$t_g = \frac{e}{\varepsilon f}(V^*)^{1-\gamma}. \tag{6.2}$$

Finally, setting $\gamma = 2/3$ and bearing in mind that the characteristic size $L^* \sim (V^*)^{1/3}$, we get $t_g \sim L^*$, i.e. Bonner's relation.

Note that, since the generation time is a characteristic value both for organism and for population the relation between the organism description and population one, then Bonner's relation allows us to construct a bridge between these descriptions. Indeed, the most popular models in ecology are Malthusian ones, $dN/dt = \alpha N$, and logistic, $dN/dt = \alpha N(1 - (N/K))$, where N is a population size, and the intrinsic rate $\alpha \approx 1/t_g$. But since $t_g \sim L^*$ and the weight $W^* \sim V^* = L^{*3}$, then $\alpha = A/(W^*)^{1/3}$ where $A = \text{const}$. In the logarithmic scale, the relation will be written as $\log \alpha = \log A - (1/3)\log W^*$. This implies that α is related to the size of the organism but, as shown by Fenchel (1974), actually falls into three groups: unicellular, poikilo- and homeotherms (see Fig. 3.3).

The analogous allometric relations (principles) are expressed in the following equations giving the respiration, feed consumption and ammonia excretion for fish when the weight, W, is known:

$$\begin{aligned} \text{Respiration} &= A_R W^{0.80}, \\ \text{Feed consumption} &= A_{FC} W^{0.65}, \\ \text{Ammonia excretion} &= A_{AE} W^{0.72}. \end{aligned} \tag{6.3}$$

This is also expressed in Odum's equation (E.P. Odum, 1959, p. 56):

$$r = kW^{-1/3} \tag{6.4}$$

where k is roughly a constant for all species, equal to about 5.6 kJ/$g^{2/3}$ day, and r is the specific metabolism (metabolic rate) per unit weight.

Similar relationships exist for other animals. The constants in these equations might be slightly different due to differences in shape, but the equations are otherwise the same. All these examples illustrate the fundamental relationship in organisms between size (surface) and the biochemical activity. The surface quantitatively determines the contact with the environment, and through that the possibility of taking up food and excreting waste substances.

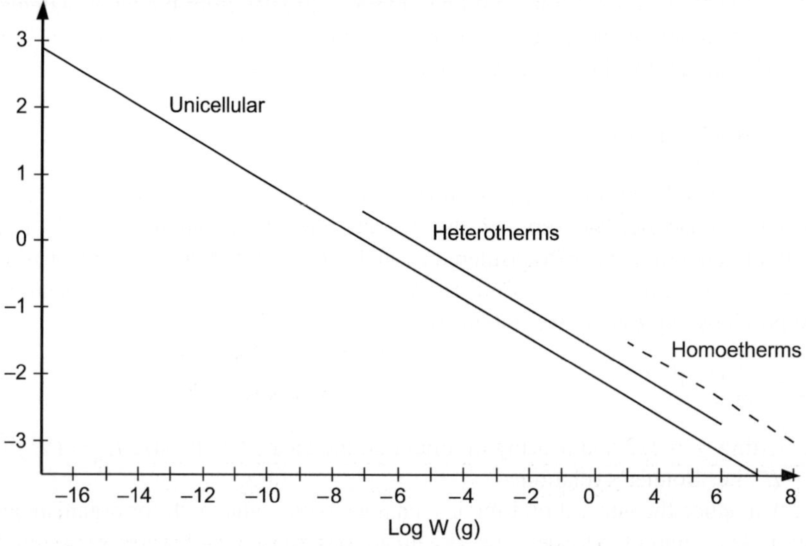

Fig. 3.3. Intrinsic rate of natural increase against weight for various animals.

The same relationships are shown in Figs. 3.4–3.6, where biochemical processes involving toxic substances are applied as illustrations. These figures are constructed from data in the literature and, as can be seen, the excretion rate and uptake rate (for aquatic organisms) follow the same trends as the metabolic rate. This is of course not surprising, as excretion is strongly dependent on metabolism and the direct uptake is dependent on the surface.

These considerations are based on allometric principles (Peters, 1983; Straskraba et al., 1997), which in other words can be used to assess the relationship between the size of the units in the various hierarchical levels and the process rates, determining the need for the rate of energy supply. All levels in the entire hierarchy of an ecosystem are, therefore, due to the hierarchical organisation, characterised by a rate that is ultimately determined by its size. Note that the degree of openness plays a significant role in the establishment of all these relations.

Intuitively, it is clear that the degree of openness must be proportional to the area available for exchange of energy and matter, relative to the volume, i.e. it must be inversely proportional to space co-ordinate L. It may also be expressed as:

the supply rate $= k \cdot$ gradient $\cdot$ area relative to the rate of needs,

which is proportional to the volume or mass. An ecosystem must, as previously mentioned, be open or at least non-isolated to be able to import the energy needed for its maintenance. Table 3.1 illustrates the relationship between hierarchical level, openness, and the four scale hierarchical properties presented in Simon (1973). The openness is expressed here as the ratio of area to volume. For the higher levels in the hierarchy, approximate values are used. As we move upwards in the hierarchy, the exchange of energy (and matter) becomes

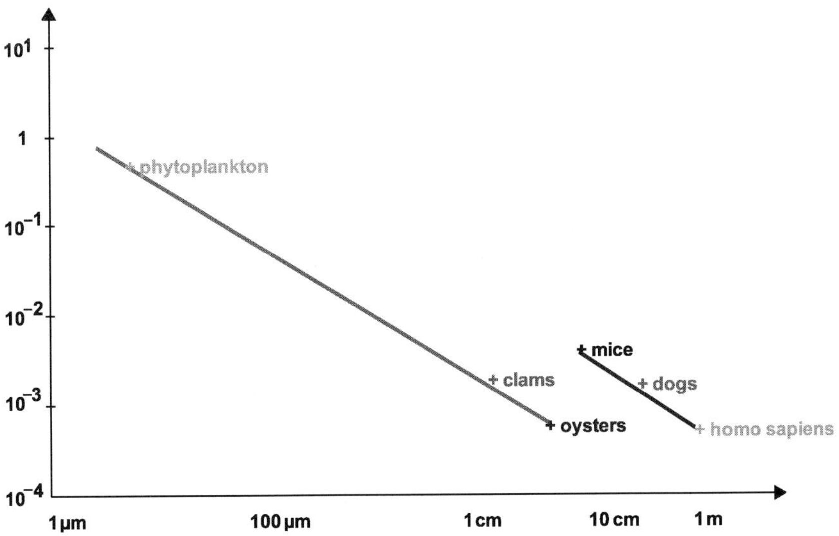

Fig. 3.4. Excretion rate of Cd (1/24 h) plotted against the size of various animals: (1) *Homo sapiens*, (2) mice, (3) dogs, (4) oysters, (5) clams, (6) phytoplankton.

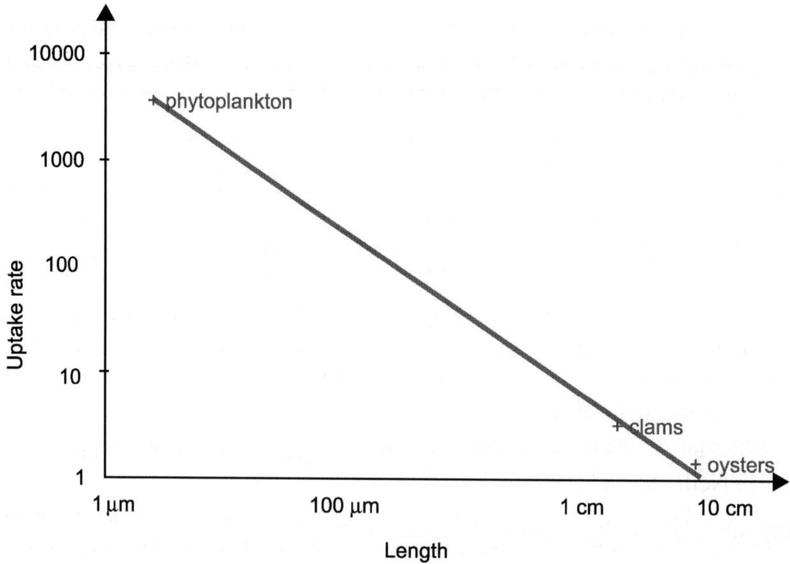

Fig. 3.5. Uptake rate (μg/g 24 h) plotted against the size of various animals (CD): (1) phytoplankton, (2) clams, (3) oysters (Jørgensen, 1984).

increasingly more difficult due to a decreasing openness. It becomes increasingly more difficult to cover needs, which explains why energy density, time scale and dynamics decrease according to the inverse space scale or openness or, expressed differently, as the rates are adjusted to make the possible supply of energy sufficient. These considerations

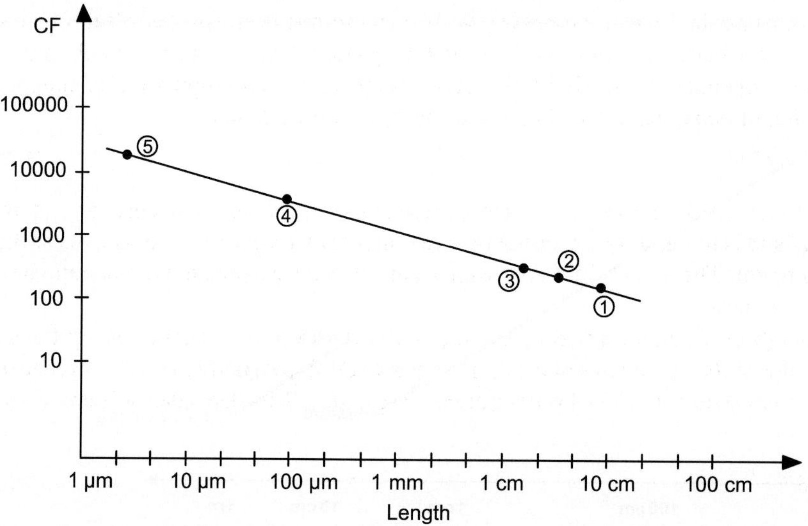

Fig. 3.6. BCF (Biological Concentration Factor) denoted CF for Cd versus size: (1) goldfish, (2) mussels, (3) shrimps, (4) zooplankton, (5) algae (brown-green).

Table 3.1
The relationship between hierarchical levels, the approximate magnitudes of their openness and approximate values of the typical four scale hierarchical properties (energy portions/volume, space scale, time scale, and behavioural frequencies)

Hierarchical level	Openness (m^{-1})	Energy (kJ/m^3)	Space scale (m)	Time scale (s)	Dynamics (g/m^3 s)
Molecules	10^9	10^9	10^{-9}	$<10^{-3}$	10^4–10^6
Cells	10^5	10^5	10^{-5}	10^1–10^3	1–10^2
Organ	10^2	10^2	10^{-2}	10^4–10^6	10^{-1}–10^{-3}
Organism	1	1	1	10^6–10^8	10^{-3}–10^{-5}
Populations	10^{-2}	10^{-2}	10^2	10^8–10^{10}	10^{-5}–10^{-7}
Ecosystems	10^{-4}	10^{-4}	10^4	10^{10}–10^{12}	10^{-7}–10^{-9}

are consistent with the relationship between size and time scale of levels in the hierarchy, as presented by O'Neill et al. (1986).

The energy received by ecosystems as solar radiation comes in small packages (quanta, $h\nu$, where h is the Planck constant and ν is the frequency), which makes only utilisation on the molecular level possible. The energy can be used on all hierarchical levels by an interactive coupling. The exchange of energy and matter on each level is dependent on openness, measured by the available area for exchange of energy and matter relative to the volume. Openness becomes the measure of the dynamics of the hierarchical level. Openness is inverse to hierarchical space scale.

Exchange of matter and information with the environment of open systems is not absolutely necessary, as energy input (non-isolation) is sufficient (the system is non-isolated) to ensure maintenance far from equilibrium. However, it often gives the ecosystem some additional advantages, for instance by input of chemical compounds needed for certain biological processes or by immigration of species offering new possibilities for a better ordered structure of the system. The importance of the latter consequence of openness is clearly illustrated in the general relationship between number of species, SD, of ecosystems on islands and the area of the islands, A:

$$SD = CA^z \tag{6.5}$$

where C and z are constants. The perimeter relative to the area of an island determines how "open" the island is to immigration or dissipative emigration from or to other islands or the adjacent continent. The unit (L^{-1}) is the same as the above-used area to volume ratio as a measure of openness.

Another type of allomeric relation at the population level was manifested by Peters (1983). His plot of the abundance of different species of different body mass shows a steep decline in abundance of species of progressively larger size. This dependence is described by the equation

$$N = 3W^{-1} \tag{6.6}$$

where N (ind/km^2) is the population density and W (kg) the mean weight. From this formula, a curious conclusion follows: the mean spatial density of biomass for any animals at the second trophic level is the same: $NW = 3\ \mathrm{kg/km^2}$ (see also Chapter 12).

Different species have very different types of energy use to maintain their biomass. For example, the blue whale uses most (97%) of the energy available for increasing the biomass for growth and only 3% for reproduction. Whales are what we call K-strategists, defined as species having a stable habitat with a very small ratio between generation time and the length of time the habitat remains favourable. It means that they will evolve toward maintaining their population at its equilibrium level, close to the carrying capacity. K-strategists are in contrast to r-strategists which are strongly influenced by any environmental factor. Due to their high growth rate, they can, however, utilise suddenly emergent favourable conditions and increase the population rapidly. Many fishes, insects, and other invertebrates are r-strategists. The adult female reproduces every season she is alive and the proportion going into reproduction can be over 50%.

3.7. The temperature range needed for life processes

The input of energy to ecosystems is in the form of the solar photon flux by small portions (quanta). This implies that the exergy (energy that can do work; see Chapter 5) at first can only be utilised at molecular (lowest) levels in the hierarchy. The appropriate atoms or molecules must be transported to the place where order is created. Diffusion processes through a solid are extremely slow, even at room temperature. The diffusion of molecules through a liquid is about three orders of magnitude faster than in a solid at the same temperature. Diffusion coefficients for gases are ordinarily four orders of magnitude greater than for liquids. This implies that the creation of order (and also the inverse process, disordering) is much more rapid in liquid and gaseous phases than in solids. The temperature required for a sufficiently rapid creation of order is consequently considerably above the lower limit mentioned above, 2.726 K. As far as diffusion processes in solids, liquids and gases are concerned, gaseous diffusion allows the most rapid mass transport. However, many molecules on Earth that are necessary for ordinary carbon-based life do not occur in a gaseous phase, and liquid diffusion, even though it occurs at a much slower rate, is of particular importance for biological ordering processes.

The diffusion coefficient increases significantly with temperature. For gases, the diffusion coefficient varies with temperature approximately as $T^{3/2}$ (Hirschfelder et al., 1954), where T is the absolute temperature. Thus, we should look for systems with the high-order characteristic of life at temperatures considerably higher than 2.726 K. The reaction rates for biochemical anabolic processes on the molecular level are highly temperature-dependent (see Straskraba et al., 1997). The influence of temperature may be reduced by the presence of reaction-specific enzymes, which are proteins formed by anabolic processes. The relationship between the absolute temperature, T, and the reaction rate coefficient, k, for a number of biochemical processes can be expressed by the following general equation (see any textbook in physical chemistry):

$$\ln k = b - A/RT, \tag{7.1}$$

where A is the so-called activation energy, b is a constant and R is the gas constant. Enzymes are able to reduce the activation energy (the energy the molecules require to perform the biochemical reaction). Similar dependence of temperature is known for a wide spectrum of biological processes, for instance growth and respiration. Biochemical and biological kinetics point, therefore, towards ecosystem temperatures considerably higher than 2.726 K.

The high efficiency in the use of low-entropy energy at the present "room temperature" on Earth works hand in hand with the chemical stability of the chemical species characteristic of life on Earth. Macromolecules are subject to thermal denaturation. Among the macromolecules, proteins are most sensitive to thermal effects, and the constant breakdown of proteins leads to a substantial turnover of amino acids in organisms. According to biochemistry, an adult man synthesises and degrades approximately 1 g of protein nitrogen per kilogram of body weight per day. This corresponds to a protein turnover of about 7.7% per day for a man with a normal body temperature. A too high temperature of the ecosystem (more than about 340 K) will therefore enhance the breakdown processes too much. A temperature range between 260 and 340 K seems, from these considerations, the most appropriate to create the carbon-based life that we know on Earth. An enzymatic reduction of the activation energy makes it possible to realise basic biochemical reactions in this temperature range, without a too high decomposition rate, which would be the case at a higher temperature. In this temperature range, anabolic and catabolic processes can, in other words, be in proper balance.

3.8. Natural conditions for life

The conditions for the creation of life-ordering processes out of disorder (or, more specifically, chemical order by formation of complex organic molecules and organisms from inorganic matter) can now be deduced from the First, Second and Third Laws of Thermodynamics:

1. It is necessary that the system be open (or at least non-isolated) to exchange energy (as well as mass) with its environment;
2. An influx of low-entropy energy that can do work is necessary;
3. An outflow of high-entropy energy (heat produced by transformation of work to heat) is necessary (this means that the temperature of the system must inevitably be greater than 2.726 K);
4. Entropy production accompanying the transformation of energy (work) to heat in the system is a necessary cost of maintaining the order; and
5. Mass transport processes at a not too low rate are necessary (a prerequisite).

This implies that the liquid or gaseous phase must be anticipated. A higher temperature will imply a better mass transfer, but also a higher reaction rate. An increased temperature also means a faster breakdown of macromolecules, and therefore a shift towards catabolism. A temperature approximately in the range of 260–340 K must therefore be anticipated for carbon-based life.

The rates of biochemical reactions on the molecular level are determined by the temperature of the system and the exergy (energy that can do work; see Chapter 5) supply to the system. Hierarchical organisation ensures that the reactions and the exergy available on the molecular level can be utilised on the next level, the cell level, and so on throughout the entire hierarchy: *molecules → cells → organs → organisms → populations → ecosystems*. The maintenance of each level is dependent on its openness to exchange energy and matter. The rates in the higher levels are dependent on the sum of many processes on the molecular level. They are furthermore dependent on the slowest processes in the chain: *supply of energy and matter to the unit → the metabolic processes → excretion of waste heat and waste material*. The first and last of these three steps limit the rates and are determined by the extent of openness, measured by the area available for exchange between the unit and its environment relative to the volume. These considerations are based on allometric principles (Section 3.6; Peters, 1983; Straskraba et al., 1997).

In addition to the five conditions given above, it is necessary to add a few biochemically determined conditions. The carbon-based life on Earth requires first of all an abundant presence of water to deliver the two important elements, hydrogen and oxygen, as solvent for compounds containing the other needed elements (see below), as a compound which is liquid at a suitable temperature with a suitable diffusion coefficient, a suitable specific heat capacity to buffer temperature fluctuations and a suitable vapour pressure to ensure a suitable cycling (purification) rate of these crucial chemical compounds.

Life on Earth is characterised by about 25 elements. Some of these elements are used by life processes in micro amounts, and it cannot be excluded that other elements could have replaced these elements on other planets somewhere else in the Universe. Several metal ions are, for instance, used as coenzymes and are often important parts of high molecular organic complexes. Other ions may be able to play similar roles for biochemical processes and complexes. It is, on the other hand, difficult to imagine carbon-based life without at least most of the elements used in macro amounts, such as nitrogen for amino acids (proteins—the enzymes) and amino bases, phosphorus for ATP and phosphorous esters in general and sulphur for formation of some of the essential amino acids.

The biochemically determined conditions can therefore be summarised in the following two points:

6. Abundant presence of the unique solvent water is a prerequisite for the formation of life forms similar to the life forms as we know from Earth.
7. The presence of nitrogen, phosphorus, and sulphur and some metal ions seems absolutely necessary for the formation of carbon-based life.

A last and eighth condition should be added: the seven other conditions should be maintained within reasonable ranges for a very long period of time. The genes may ensure that, if an advantageous property of an organism has been developed, the property can be hereditary and the following generations will be able to maintain the advantageous property. The probability to create (complex) life spontaneously is so low that even the time from 'the big bang' would not have been sufficient. It is therefore necessary that the development toward life is made step-wise with conservation of each achieved progress to allow further development to ride on the shoulders of the progress already made. Many mechanisms are probably involved in the emergence of a progressive property on the first hand, but indisputable random processes based on trial-and-error are also important in the emergence of progressive properties. This implies that carbon life is not formed overnight. The history of evolution on Earth shows that, after a suitable temperature was achieved and water was abundant, probably of the order of 10^8 years or more (Nielsen, 1999) were needed to form the first living cells with some type of primitive genes to ensure a continuous development (evolution) from inorganic components dissolved in water. Phytoplankton fossils have recently been found at Isua, Greenland by Minik Rosing (Nielsen, 1999). The age of the fossils was determined to be 3.8 billion years old, or about 100 million years after the termination of the massive bombardment of meteors that characterised the first 600–700 million of years after Earth was born. Numerous theories have been published to explain how this development may have happened, probably in many steps: inorganic matter formed organic molecules by a through-flow of low-entropy energy, organic molecules formed high molecular organic compounds, self-catalytic processes occurred, complex organic molecules were brought randomly in contact by adsorption on clay particles, and many other processes are mentioned in these presented theories. Which theory is right is not important in this context. The focal point is that the seven above-mentioned conditions must be fulfilled for a sufficiently long period of time, which leads to the eighth condition:

8. As the formation of life from inorganic matter requires a very long time, probably of the order of 10^8 years or more, the seven conditions have to be maintained in the right ranges for a very long time, which probably exceeds about 10^8 years.

After the Mars Pathfinder mission, it has been discussed whether Mars hosts or has hosted life. Clearly, conditions 1–7 are not met on Mars today. The climate is too harsh and water is far from being present in the amount needed for the planet to bear life. There are, however, many signs of a warmer and wetter climate at an earlier stage. It therefore looks as if the seven conditions may have been valid and the question is: "...have they also prevailed for a sufficient period of time?" If later missions to Mars will show that it is the case, the next obvious question is: "Will life inevitably be the result of self-organising processes if the eight conditions (the eighth condition about sufficient time should of course be included) are fulfilled? It should be expected that primitive life has been present on Mars at an early stage, provided that the warmer and wetter conditions have prevailed for sufficient time. The further evolution from unicellular organisms (maybe prokaryotic) to more and more complex organisms such as we know from Earth could not be realised on Mars because the climate changed and the water disappeared. Latest investigations of Mars-originated meteorites have made it almost certain that there has previously been microbiological life on Mars. The latest geological investigation has furthermore shown that there has previously been plenty of water on Mars, which also points toward a prior existence of life on Mars. It is, of course, still an open question how long a time this microbiological life has been present. The Mars Pathfinder mission will be able to answer this question.

Another possibility for life in our solar system exists on Europa, one of the moons of Jupiter (Sweinsdottir, 1997). Europa is characterised by a coverage of ice. It implies that there is plenty of water on Europa, which means that one of the important conditions for life is fulfilled. Some researchers (Sweinsdottir, 1997) suggest that the chance to find life on Europa is higher than on Mars. Europa has of course much less sunlight and the surface temperature is probably too low, but volcanic activity in the deeper parts of the oceans on Europa is very probable, and it could provide the needed low-entropy energy for the formation and maintenance of life.

The probable number of civilisations, N, which could be detected in our Galaxy is expressed by the Drakes equation:

$$N = n_e f_p f_l f_i f_c R L. \tag{8.1}$$

What do all these factors mean? n_e is the number of planets that have an appropriate distance from the star to ensure a suitable temperature, i.e. that water is mainly in the liquid form (see also Section 3.7). Astronomers believe that this number is about 1.0. The value of a relative number of stars with planets $f_p \approx 0.5$. The probability f_l that life is formed under these temperature conditions in accordance with our discussion above is estimated as 1.0. This is because life is a result of low-entropy energy flow and the right temperature and the presence of the elements that are important for life. The value of f_i is the probability that the right conditions have been maintained a sufficiently long time to allow the entire evolution from the most simple life form to man. It is hardly possible to estimate this number very accurately, but we may estimate that 1 out of 100 planets would be able to maintain approximate life conditions for a very long period of time. f_c is the probability that radio-telescopes are developed to receive signals from other intelligent lifeforms (it is estimated to be 0.1). R is the number of stars formed per year, and L is the lifetime for advanced civilisations. R is about 1–10 according to astronomers; let us say 1.0. These estimations of the factors give that $N = 0.0005L$. There should, in other words, be a reasonable probability to find other civilisations in our Galaxy.

The Drakes equation is maybe too simple and does not consider other factors that are important for the evolution of life. For instance, it cannot be excluded that the size and distance of our moon from Earth are of utmost importance for the rate of evolution. The changes on Earth caused by the gravity of the moon including the tide may be the disturbances that are needed to ensure a good mixing of the oceans and provoke challenges for the organisms to evolve. It cannot be excluded that the probability to have a moon of a suitable size and at a suitable distance, f_m, is as low as 0.02.

Another important factor for evolution could be the catastrophes that have probably occurred 3–4 times during the last 500 million years due to a collision between Earth and an asteroid or a comet. From time to time—not too frequent and not too rare—it may be beneficial to start from a radically different situation to give new emergent life forms a chance to take over and come up with new and better solutions to survival. This is probably what happened (see also Section 11.8) when the dinosaurs were replaced by the mammals about 65 million years ago, when probably an asteroid collided with Earth and created a very difficult situation for the survival of the various life forms present on Earth at that time. The probability that a planet has a suitable frequency of catastrophes is dependent on the formation of the planet system. We do not know to what extent other sun systems have asteroids and comets in approximately the same amount as our sun system, f_{ca}, to provide a suitable frequency of catastrophes. It could therefore be necessary to estimate that it has only a probability of 0.1, although this estimation is very uncertain.

With this expansion of the Drakes equation:

$$N = (n_e f_p f_l f_i f_c R L) f_m f_{ca} \tag{8.2}$$

we get $N = 10^{-6} L$, and the probability to find civilisations is reduced considerably, which may explain why we have still not got a radio signal from other civilisations. If $L = 10{,}000$ years to come up with a guess that is between very pessimistic and very optimistic guesses, the probability is as low as 10^{-2} to receive a radio signal—one that may even be from a civilisation very far away.

With this expansion of the Dirac's equation

Chapter 4

Entropy, probability and information

... nobody knows what entropy is in reality, that is why in the debate you will always have an advantage.

John von Neumann

4.1. Entropy and probability

The Boltzmann formula

$$S = k \ln W \tag{1.1}$$

connects entropy, S, and the "thermodynamic" probability, W; the thermodynamic probability is equal to the number of possible states in which the system can be found; Boltzmann's constant $k = 1.38 \times 10^{-23}$ J/K. It is in these units that the physical entropy is usually evaluated. Note that the expression "the number of possible states in which the system can be found" needs additional explanation.

Let the system contain N particles, which do not depend on each other and can be combined in different combinations. The number of these combinations, which we shall nominate "possible states", is equal to $W = N^N$. Then $S = k \ln(N^N) = kN \ln N$.

Now let these particles possess some properties: in classic thermodynamics this can be energy, specific membership in ecology, etc. and the distribution of particles in relation to these properties is known, described by the vector $\mathbf{N} = \{N_1, \ldots, N_n\}$, $\sum_{i=1}^{n} N_i = N$. Here the system is divided by n boxes, each of which corresponds to a certain combination of properties, for instance to a certain interval of energy or to certain biological species. In other words, if there is a space of parameters describing the properties, and the space is divided by a rather large number of cells, then these cells are our boxes. It is assumed that these boxes are isolated from each other. This means that a particle from one box cannot pass to others, and particles of ith sort cannot produce particles of another sort. This is certainly a very strict constraint if we deal with biological particles. The value of N_i is the number of particles in ith box, i.e. the number of particles possessing such values of the properties which define ith box. The total number of this kind of distribution is equal to

$$W = \frac{N!}{\prod_{i=11}^{n} N_i!}. \tag{1.2}$$

Towards a Thermodynamic Theory for Ecological Systems, pp. 69–93

For large $N > 100$ a good approximation is given by Stirling's formula: $N! \approx (N/e)^N$, then

$$S = k\left[\ln(N!) - \sum_{i=1}^{n} \ln(N_i!)\right] \approx k\left(N \ln N - \sum_{i=1}^{n} N_i \ln N_i\right) = -kN \sum_{i=1}^{n} p_i \ln p_i, \quad (1.3)$$

where $p_i = N_i/N$. The frequencies p_i can be considered as *probabilities* of a given particle to have ith set of properties. It is obvious that entropy S is maximal (at fixed N) when the probabilities are equal to each other, any particle can be detected at any point of the space of properties, or, in other words, the system is not ordered. Then $S_{max} = kN \ln n$. In the opposite case, when all p_i (with the exception of single $p_k = 1$) are equal to zero, i.e. all particles are concentrated in kth box, then $S_{min} = 0$. From n possible states of the system the single state was chosen, so that we can say that it is maximally ordered. Since Boltzmann's constant k is very small, in order for the entropy to have a reasonable value compared with the thermal effects in real physical systems, the value N must be enormous. This is true if our particles are molecules (for instance, the number of molecules in 1 mol, the so-called Avogadro number, $N_a \approx 6 \times 10^{23}$). Then, even if some substance has only two energy levels (really any substance has many more), 1 mol of the substance can have the entropy $S_{max} \approx 5.7$ J/K. This is a very reasonable value. However, when we deal with "biological" particles, for instance specimens, as a rule their number in such integrity as an ecosystem is not too large in order to get reasonable values corresponding to the ecosystem energetics. Certainly, we can increase the number by increasing the volume of the ecosystem, but we cannot do it to an unlimited degree without destroying its integrity. This contradiction forces us to think about the following: could we apply the entropy concept to an ecosystem directly? Perhaps we would have to use another coefficient of proportionality instead of Boltzmann's constant in the relation between entropy and the logarithm of possible states. We shall try to answer these questions in Section 4.2.

4.2. Entropy and information

At all levels of biological organisation there are information "machines", which produce, memorise, transmit, transform and perceive information. Here, as we did above when we introduced, for consideration, the pair "system and its environment", we shall consider the pair "system and its Observer" to speak about information. In the information approach, the Observer plays the role of the "environment", so that there is a relation of equivalency between these actors. When the Observer receives information about a real state of the system, then the initial uncertainty of his knowledge about the system decreases. The simplest model of the uncertainty is the assumption that the system can be at one of the W^0 equiprobable states with probability $p^0 = 1/W^0$. The received information allows us (the Observer) to conclude that really the number of possible states is equal to $(W^1 < W^0)$ with probability $p^1 = 1/W^1$. The increment of information is defined as

$$\delta I = I^0 - I^1 = (-\log_2 p^0) - (-\log_2 p^1) = \log_2 W^0 - \log_2 W^1. \quad (2.1)$$

Let us assume that in the simplest case we receive information that the system can be situated only at a unique state. Then $I^1 \equiv 0$, hence we can define the lower boundary (zero) for information. After this, as we are often doing in thermodynamics, we forget about the Observer and consider the result of the experiment as some internal immanent property of the system, i.e. we state that the information contained in the system with W possible equiprobable states is equal to

$$I = \log_2 W = -\log_2 p, \tag{2.2}$$

where $p = 1/W$. Information is measured in *bits*: 1 bit $= \log_2 2$, i.e. the system, which can be at two possible states, contains one bit of information. The expression $I = \log_2 W$—named Shannon's entropy—coincides (with an accuracy of the constant factor) with Boltzmann's entropy $S = k \ln W$, so that 1 bit is equal to 0.96×10^{-23} J/K, i.e. a very small thermodynamic value. The entropy in bits is equal to $S_{\text{bit}} = (1/k \ln 2)S_{\text{J/K}}$.

The formal similarity of entropy and information has a very deep meaning. Entropy is a deficiency of information for the full description of the system, or information is a deficiency of entropy, i.e. the difference between the maximal entropy of the given system and entropy, which the system really possesses. The latter is elucidated after receiving information about the system. The relation between entropy and information was established by Brillouin (1956). The equivalency of entropy and information is similar (in some sense) to Einstein's relation between mass and energy $E = mc^2$, or $m = (1/c^2)E$. The transferring factor between mass and energy, $1/c^2 \approx 10^{-21}$ s^2/cm^2, is very small, as is the same factor between information and entropy: 1 bit $= k \ln 2 = 0.96 \times 10^{-23}$ J/K. It is interesting that both in Einstein's case and in Brillouin's case the same gnoseologic principle is used: the pair "system and Observer". The equivalency has a real physical sense: the increase of entropy is a charge for gained information. The value $k \ln 2$ is the minimal cost of a bit. It is obvious that one bit of information is gained as the result of flipping a coin. However, the entropy release, when the coin strikes the floor, is much more than $k \ln 2$. How can the cost of one bit in units of work be evaluated? Let us consider the following simple model. There are N molecules of ideal gas at temperature T and pressure p, which fill the volume V. As a result of fluctuation the volume decreased to $V - \delta V$, so that the work of compression $\delta A = p\,\delta V$. If the probability to find a single molecule within the volume V is equal to 1, then it will be equal to $(V - \delta V)/V = 1 - (\delta V/V)$ for the reduced volume. For N molecules the probability is equal to $[1 - (\delta V/V)]^N$, and the increment of information

$$\delta I = \frac{1}{\ln 2} \ln\left\{\frac{1}{[1 - (\delta V/V)]^N}\right\} \approx \frac{1}{\ln 2} N(\delta V/V) \text{ bits.}$$

In accordance with the gas law $pV = kNT$, from which $p = kNT/V$. By substituting p into the expression for work we get $\delta A = kTN(\delta V/V)$. By comparing the expressions for δA and δI, we see that the work that is necessary to get the unit of information is proportional to the temperature at which the information is determined. If $T = 300$ K, then this work is equal to $kT \ln 2 \approx 2.9 \times 10^{-21}$. This is a lower estimation of the necessary work.

It is necessary to note that the Brillouin principle of equivalency between entropy and information is valid only for *micro-information*, i.e. for information about the realisation at a given time of possible *microscopic* states of the system. Micro-information cannot, in principle, be memorised and transmitted, since any microscopic state is very unstable and very quickly passes into another by means of thermal fluctuations. In biology (and in techniques too) the system perceives, memorises and transmits only *macro-information* (more about this later on), which is not connected with physical entropy by Brillouin's relation.

Blumenfeld (1977b) estimated (very roughly, of course) the amount of information contained in a human organism. He assumed that the main amount of information is determined by the fully ordered disposition of 3×10^{25} amino acid residues in proteins contained in 7 kg of human body nitrogen. This corresponds to $I_1 = (1/\ln 2) \times 3 \times 10^{25} \times \ln(3 \times 10^{25}) \approx 2.5 \times 10^{27}$ bits of information (we use formula (2.2) where $W = 3 \times 10^{25}$). The other contributions are appreciably lower. For instance, 150 g of DNA contain only 6×10^{23} bits. If we keep in mind that the human body consists of about 10^{13} individual cells, then the ordered structure of the human body (we assume that all these cells are unique, and they cannot, in principle, be replaced) contains $I_2 = (1/\ln 2) \times 10^{13} \ln \times 10^{13} \approx 4.3 \times 10^{14}$ bits of information. In every cell there are 10^8 ordered polymeric molecules that correspond to $I_{\text{cell}} = (1/\ln 2) \times 10^8 \ln 10^8 \approx 2.7 \times 10^9$ bits. Even if this number is multiplied by the total number of cells, we receive $I_3 = 2.7 \times 10^{22}$ bits. Thus, the maximal contribution is given by the protein information, but even it is very small in thermodynamic terms: $S_1 = 0.96 \times 10^{-23} I_1 \approx 2.4 \times 10^4$ J/K ≈ 6 kcal/K. This is the entropy of 1 kg of crystal NaCl. If the process of the creation of the information takes place at $T \approx 300$ K, then the corresponding work $\delta A = 7.2 \times 10^6$ J ≈ 1700 kcal, i.e. it is approximately equal to the human daily metabolism. If we make the fully reasonable assumption that the mean rate of protein denaturation is equal to 1/day, then we can say that the "supporting" metabolism compensates the entropy production (as a result of destructive processes) within the human body, and by the same token the human being.

Of course, Blumenfeld's approach could be considered as some "zero" approximation. Undoubtedly, life processes are possible because they are enzymatic (enzymes are proteins with a special structure), and the sequence of the amino acids is crucial for the processes. Therefore, the information in a living organism is a question not only of the number of amino acids we have but also of the sequence. In the mid-1990s we thought that we had 250,000 genes determining on average 700 amino acids in the right sequence (the right sequence is determining). Today the number of genes is more in the order of 40,000 but it has also been found that human genes may determine as much as 38,000 amino acids in the right sequence (on average about 5000 amino acids). Later on, we shall take into account in our "information" calculations these arguments using the amount of amino acids in the right sequence as a measure of the information content. Unfortunately, we do not know the number of amino acids in the right sequence but we can estimate it with the number of non-nonsense genes, which also are not known in all details and particularly not for many species. Then we have a long list of DNA which could also be used, but different species have different non-nonsense genes.

4.3. The system as a text and its information entropy

Ludwig von Wittgenstein said that any physical object or system could be represented as a text, written in a special language with a proper alphabet and grammar.

Let the text be a single word with length N. If the alphabet contains n symbols (for English $n = 29$: 26 letters, one blank, one comma and one full stop), then every symbol is repeated in the word $N_1, N_2, \ldots, N_n$ times ($\sum_{i=1}^{n} N_i = N$). The total number of different words from N symbols of this n-symbols language is equal to $W = N!/\prod_{i=11}^{n} N_i!$. Then the total information contained in the word (text) is equal to $I = -N\sum_{i=1}^{n} p_i \log_2 p_i$ (in bits) where $p_i = N_i/N$. Comparing this formula with Eq. (1.2) we see that they coincide (to within units of measurement). The specific information, i.e. the information per symbol, will be equal to

$$I_s = -\sum_{i=1}^{n} p_i \log_2 p_i. \tag{3.1}$$

This is the so-called Shannon measure of information, or the information entropy (Shannon and Weaver, 1963; Pierce, 1980). Since receiving information reduces uncertainty, the Shannon concept can be formulated as

Information per symbol = mean value of uncertainty per symbol.

We prove how changing the level of uncertainty leads to a gain in information. Let us pass from the distribution of probability $\mathbf{p}_0 = \{p_1^0, \ldots, p_n^0)$ to the distribution $\mathbf{p} = \{p_1, \ldots, p_n\}$. Since the distribution of probability changes, the uncertainty changes too. How can we estimate the change of information that results from this transition? If the probability of some event changes from p^0 to 1, the change of information is equal to $\delta I = \log_2(1/p^0) = -\log_2 p^0$, if $p^0 \to p$ then $\delta I = \log_2(p/p^0)$. If the entire distribution $\mathbf{p}_0 \to \mathbf{p}$, then the change of information is equal to the sum of partial changes $\delta I_i = \log_2(p_i/p_i^0)$ multiplied by the finite probabilities p_i:

$$\delta I = \sum_{i=1}^{n} p_i \log_2(p_i/p_i^0). \tag{3.2}$$

This value, also called Kullback's measure for the increment of information (Kullback, 1959), is always positive (it is equal to zero only if $\mathbf{p}_0 = \mathbf{p}$); therefore, in this case we can say about a *gain* in information. So, knowledge about the transition $\mathbf{p}_0 \to \mathbf{p}$ decreases uncertainty and gives the gain of information. Later on, Kullback's measure δI will be denominated as $K = K(\mathbf{p}_0, \mathbf{p})$.

In the Shannon concept symbols (letters) are considered as primary elements of a language. However, a text can consist of separate words, so that the words (not letters) could as such be considered primary elements. For instance, if the alphabet contains n letters then n^r words of r letters could be constructed under the alphabet. If $p_{\underbrace{ij\cdots k}_{r}}$ is the probability of formation of the r-words then the information entropy of rth order will be equal to (Yaglom and Yaglom, 1973)

$$I^{(r)} = -\sum_{i,j,\ldots,k=1}^{n} p_{\underbrace{ij\cdots k}_{r}} \log_2 p_{\underbrace{ij\cdots k}_{r}}, \quad r = 1, 2, \ldots. \tag{3.3}$$

It is obvious that $I^{(1)} = I$, i.e. Shannon's entropy. Under the assumption that the source of information is stationary and generates an ergodic Markov sequence,

Khinchin (1953, 1957) has proved that

$$I^{(r+1)} \leq \frac{r+1}{r} I^{(r)} \quad \text{and} \quad I = \lim_{r \to \infty} (I^{(r)}/r). \tag{3.4}$$

Let us consider a text which is written in the English language. At zero level of perception (or description) we know only the number of symbols ($n = 29$). Then the information per symbol $I^{(0)} = \log_2 29 \approx 4.85$ bits. At the first level of perception we take into account the frequencies of the symbols (letters); then $I^{(1)} \approx 4.03$ bits. At the next levels, when we take into account double, triple, etc. correlations, i.e. words of two letters, three letters, etc., we get the following values of information per symbol (Ebeling et al., 1990):

$$\frac{I^{(2)}}{2} \approx 3.32 \text{ bits}, \quad \frac{I^{(3)}}{3} \approx 3.10 \text{ bits}, \ldots.$$

We see that the amount of information decreases as it is transmitted from lower to higher levels. This implies that at each level there is *redundant* information. Certainly, only non-redundant information has a cost, but the repeating of information, its *redundancy*, provides the reliability of its transmission defending it from errors and destruction of the text by noise. The redundancy of information at rth level of perception (or description) can be defined as (Klix, 1974)

$$R^{(r)} = 1 - \frac{I^{(r)}}{\max I^{(r)}}, \quad r = 1, 2, \ldots, \tag{3.5}$$

where $\max I^{(r)} = r \log_2 n$. For $r = 0$ the redundancy is defined as $R^{(0)} = 0$. Then the corresponding redundancies in English will be equal to $R^{(0)} = 0$, $R^{(1)} = 0.15$, $R^{(2)} = 0.30$ and $R^{(3)} = 0.35$. The latter, for instance, implies that only 35% of letters are redundant at the third level, i.e. 65% of randomly distributed letters are sufficient for the understanding of the text. The cost of information can be defined as the degree of non-redundancy (Volkenstein, 1988):

$$C^{(r)} = \frac{1}{1 - R^{(r)}}. \tag{3.6}$$

Then for each level we have $C^{(0)} = 1$, $C^{(1)} = 1.18$, $C^{(2)} = 1.43$, $C^{(3)} = 1.54$.

We used here one of the simplest definitions of the *cost of information*. In fact, this problem "What is the cost of information?", in spite of continuing discussion, is still far from its completion. This discussion falls outside the framework of our book, but nevertheless we shall cite one example.

So, there is some aim. Let probabilities of its attainment before and after receiving information be equal to P_0 and P_1, respectively. Then the cost of information is equal to $C = \log_2(P_1/P_0)$ (Kharkevich, 1963). However, if the aim is unattained without information ($P_0 = 0$), then the cost of any finite information is equal to infinity. This is not properly understandable.

4.4. Diversity of biological communities

The concept of stability is one of the most important in ecology (for details see Chapter 6). It is evident that only a stable community can exist over the course of a rather long time. Clearly that is possible if the sizes of the populations constituting the community do not undergo any large fluctuations. This definition is closer to the thermodynamic (or more correctly, to the statistical physics) notion of system stability. In thermodynamics (statistical physics) a system is believed to be stable when large fluctuations that can take the system far from the equilibrium or even destroy it are unlikely (see, for instance, Landau and Lifshitz, 1995). Evidently, the general thermodynamic concepts (for instance, the stability principle associated in the case of closed systems with the Second Law and, in the case of open systems, with Prigogine's theorem) should be applicable to biological (and, in particular, ecological) systems. As an illustration of such a phenomenon, we consider a well-known problem of the relationship between the species diversity of a community and its stability.

Ecologists consider it almost as an axiom that communities which are more complex in structure and richer in comprising species are more stable. Any popular ecological textbook (for instance, E. Odum's book, Fundamentals of Ecology) would convince you of this. This is explained in the following way: different species adapt differently to environmental variations. Therefore, a variety of species may respond with more success to different environmental variations than a community composed of a small number of species, and hence the former will be more stable. In other words, the more diverse the community is, the more stable it is. Perhaps this motivates the fact that Shannon's information entropy was suggested as a measure of species *diversity* (Margalef, 1951, 1968; MacArthur, 1955):

$$D = -\sum_{i=1}^{n} p_i \ln p_i, \tag{4.1}$$

where $p_i = N_i / \sum_{i=1}^{n} N_i$, n is the number of species in the community and N_i is the population size of ith species. Comparing Eqs. (3.1) and (4.1) we see that they are fully identical, only the interpretation of notations differs. Margalef and MacArthur also suggested using the value of D as a measure of stability: the more the D is, the more stable is the community. Therefore, when a community moves to its climax, then its diversity increases. In accordance with this "logic", the community is the most stable if D is maximal. But, as could readily be shown, in this case the community structure is such that specimens of any species occur with the same frequency ($\max_{p_i} D$ is attained at $p_i^* = 1/n$). In other words, the diversity of a community is maximal when the distribution of species is uniform, or when there are no abundant or rare species, and no structures. However, observations in real communities show that this is never the case, and that there is always a hierarchical structure with a dominant species. What is the reason for such a contradiction? It probably lies in the formal application of models and concepts taken from physics and information theory to systems that do not suit this type of definition. Both Boltzmann's entropy in thermodynamics and statistical physics and Shannon's entropy in the theory of information make sense only for populations of weakly interacting particles. A typical example of such a system is the ideal gas: its macroscopic state is an additive function of the microscopic states of its molecules.

Let us remember the original formulation of Boltzmann's entropy: $S_B \sim \ln W$ where W is the probability of the state of the system. In the general formulation, Boltzmann's formula is applicable to any system, not only to systems with weak interactions. But as soon as we use the standard formula $S = -k\sum_{i=1}^{n} p_i \ln p_i$, we implicitly use the classic thermodynamic model of the ideal gas.

The use of the entropy measure to such objects as the ideal gas is well founded. Moreover, stability of the equilibrium—entropy is maximal in this state—is associated with the Second Law. However, the stable structure of a biological community is the consequence of interactions between populations rather than a function of the characteristics of individual species, i.e. the biological community is a typical system with strongly interacting elements. But as soon as we become concerned with such systems the entropy measure is no longer appropriate. There is one more argument against the use of diversity as a goal function relating to stability. The entropy increases (tending to a maximum) only in closed systems, but any biological system is an open system in a thermodynamic sense, so that its total entropy is changed in an arbitrary way. When the system is in equilibrium (we speak of a dynamic equilibrium) the rate of the entropy production inside a system is positive and minimal. This is Prigogine's theorem. In this case, in relation with stability, the goal function is the *rate* of entropy production, not entropy.

Notice, however, that in large numbers of competitive communities at initial stages of their successions, far from climax, an increase in diversity may be observed. It seems that in these cases diversity is a "good" goal function for stability. This is explained in the following way: in the initial stages, far from equilibrium, the competition is still weak, and the community may well be regarded as a system with weak interactions. Moreover, interestingly, for communities of aquatic organisms (especially, phyto- and zooplankton) diversity increases along the entire transition from the initial state until climax. The latter can be considered as some dynamic equilibrium. The reason is the same: aquatic communities are the systems with weak interactions.

By summarising all the arguments considered above, we can say that the causal link between diversity and stability is not evident and univalent, as it seemed earlier. Nevertheless, there are empirical facts, which bring us to think about some very special properties of diversity in application to biological communities and about "linguistic" analogies between "natural" alphabetic languages (English, Russian, etc.), social systems and biological communities. The values of D for many communities tend to concentrate within a fairly narrow interval with a supremum of about five bits per individual (Margalef, 1995). There is an impression that Nature avoids both very low and very high diversity. The same picture is seen in alphabetic languages where information per letter, as a rule, does not exceed five (Ebeling et al., 1990). If Shannon's entropies are estimated for the distribution of human population with respect to professional groups in developed countries, then their values also do not exceed this limit. However, the analogous estimation, made, for instance, on a beach, gives a much higher value (Margalef, 1995).

So, we can shortly summarise that the maximum diversity principle can be considered basically true, but with certain constraints, which define at least the non-uniform structure of the given biological community.

4.5. Simple statistical models of biological communities

If we look at the titles of many thermodynamics textbooks we can see that almost all of them are named "Thermodynamics and Statistical Physics". And this is not surprising since this pair represents two sides of one coin. Today both disciplines are considered as the analysis of complex systems properties from the physical point of view. However, the way that they look at the object of their study, a system, is different. Thermodynamics considers a system as some "black box", i.e. only inputs and outputs are known, and what happens within the box is neither interesting nor understandable. In contrast, statistical physics (mechanics) "penetrates" within the system in order to understand how input is transformed into output. The system is "transparent" for statistical mechanics. Naturally, the latter gives more information about the system than thermodynamics, but all the information cannot always be used (it is often not needed). For instance, if we consider the trajectory of a swarm of bees then the individual dynamics of different bees within the swarm are not interesting: the movement of the entire formation is, in general, important to us.

In spite of a certain "complementarity" of thermodynamic and statistical approaches they are inseparably connected with each other: thermodynamic values are averaged on the whole system means of physical values, which are considered in detail by statistical physics. In this sense the statistical approach would justify the thermodynamics from the mechanical viewpoint, and because of this, it is called its mechanical justification. On the other hand, the same thermodynamic values which are observed in physical reality can be considered as some kind of "guide" through the labyrinth of statistical theory, although the guide is "rather blind" (J.W. Gibbs). Apparently, it is insufficient to know the system in general, we need some representation about some "intimate" processes working at a microscopic level. In other words, we need some model of the microstructure of the system. Note that conclusions will be different for different models.

The simplest model of a biological community is a system (ensemble) of N virtual particles belonging to n different types, so that N_i, $i = 1, \dots, n$ is the number of particles of ith type. If the type is a biological species then N_i is the population size. Note that a single population can play a role of community. In this case N_i is the number of some groups (age cohorts, size groups, etc.). It is obvious that any microscopic state of the system is described by the vector $\mathbf{N} = \{N_1, \dots, N_n\}$. If we introduce into consideration the *frequencies* $p_i = N_i/N$, $N = \sum_{i=1}^{n} N_i$, then the system state can also be described by the distribution $\mathbf{p} = \{p_1, \dots, p_n\}$, which is called in ecology the *species abilities* vector, and the value of the total size of community, N. We shall also name the vector $\mathbf{p}$ as the *vector of community composition* (*structure*). Note that if different compositions can be considered as different microscopic states, then the total size N is a typical extensive macroscopic variable. From this point of view the frequencies p_i are intensive variables. It is evident that communities with different total sizes may have the same composition and vice versa.

We see that every particle already possesses one property: it belongs to one of the n types, which, following our biological orientation, are named species. Let us assume that, except for a "specimen marker", every individual possesses a set of m quantitative indicators, which are all measurable. For instance, each specimen of ith species has biomass m_i, characteristic size l_i, rate of metabolism R_i, mean lifespan τ_i, etc. We shall denominate the vector of these indicators for ith species as $\mathbf{x}^i = \{x_1^i, \dots, x_m^i\}$, where the component x_k^i is a value of one of these indicators. Scalar products

$$\hat{x}_k = \sum_{i=1}^{n} x_k^i p_i, \quad k = 1, \ldots, m \tag{5.1}$$

are the *mean values* of each indicator for the community. For instance, $\hat{m} = \sum_{i=1}^{n} m_i p_i$ is the mean individual biomass or the biomass of some "mean" individual in the community. It is obvious that the total community biomass will be equal to $M = \sum_{i=1}^{n} m_i N_i = \hat{m}N$. Analogously, the mean individual size (or the size of the "mean" individual) will be equal to $\hat{l} = \sum_{i=1}^{n} l_i p_i$. If r_i is the rate of energy loss for ith specimen caused by metabolic processes, then $\hat{r} = \sum_{i=1}^{n} r_i p_i$ is the rate of energy metabolic loss for the "mean" individual in the community, and $R = \hat{r}N$ is the rate of total energy loss as a result of the community metabolism. An average of τ_i over all individuals of the community gives us the value $\hat{\tau} = \sum_{i=1}^{n} \tau_i p_i$, which is the lifetime of the "mean" community individual. The value $\hat{\tau}$ can also be interpreted as the mean period of renewal of the community composition: during the interval of time $\hat{\tau}$, individuals of the current generation are fully replaced by individuals of the next generation. Keeping in mind that the thermodynamic macroscopic variables are the mean values of statistical microscopic variables, we can say that the community macroscopic state is described by the total community number N and the vector of mean individual characteristics $\hat{\mathbf{x}} = \{\hat{x}_1, \ldots, \hat{x}_m\}$. Note that the same mean characteristics as others is the information contents of the "mean" individual:

$$D = I_s = -\sum_{i=1}^{n} p_i \log_2 p_i. \tag{5.2}$$

This value is Shannon's information entropy. Since entropy or information is an extensive variable, the total contents of information in the community will be equal, $I = NI_s = ND$.

We assume that the community evolves to an equilibrium, in which the information (or the diversity) contained in the community reaches maximum. All other mean characteristics $\hat{x}_1, \ldots, \hat{x}_m$ are assumed to be constant; values of the corresponding constants $C_1, \ldots, C_m$ and C_N are determined only by the environment and do not depend on macroscopic states of the community. This gives us the constraints:

$$\sum_{i=1}^{n} x_k^i p_i = C_k = \hat{x}_k^*, \quad k = 1, \ldots, m. \tag{5.3}$$

The solution of this problem determines an equilibrium composition of the community, $\{p_1^*, \ldots, p_n^*\}$, which depends on the constants. Note that the constants can be dependent on one another. Indeed, if it is assumed that it is because of their thermodynamic interpretation (energy, entropy, etc.), then they could be coupled by thermodynamic laws and identities. For instance, the simplest energy balance of the community can be represented as

$$\frac{\mathrm{d}M}{\mathrm{d}t} = Q - \hat{r}N - \frac{\hat{m}N}{\hat{\tau}}, \tag{5.4}$$

where Q is the flow of free energy or enthalpy into the community, and $M = \hat{m}N$ is its total biomass expressed in the same energy units. If we assume that both $\hat{m}$ and the total size of community, N, are constant, and $(\mathrm{d}M/\mathrm{d}t) = 0$ then from Eq. (5.4) we get

$$\frac{Q}{N} = \hat{r} + \frac{\hat{m}}{\hat{\tau}}, \tag{5.5}$$

i.e. the relation connecting the mean values of individual biomass, metabolism and life span with the total number of individuals in the community and the free energy inflow maintaining its existence. So, in addition to constraints (5.3), the others can exist, in particular concerning thermodynamic laws and identities. Since neither criterion (5.2) nor constraints (5.3), nor thermodynamic identities depend explicitly on N, then neither does the final result depend on N. This will be true for any number of particles. But if even one constraint contains N then we must include a new constraint $N = N^* = \text{const}$.

We shall analyse one simple but sufficiently reasonable example, where there is a single constraint $\hat{m} = \sum_{i=1}^{n} m_i p_i = \text{const}$. Then the problem of maximisation of the diversity $D = -\sum_{i=1}^{n} p_i \log_2 p_i = -\gamma \sum_{i=1}^{n} p_i \ln p_i$, where $\gamma = 1/\ln 2 \approx 1.44$, under the constraint $\hat{m} = \text{const}$ and norming condition $\sum_{i=1}^{n} p_i = 1$ reduces to the standard problem of maximisation for the function $D' = -\gamma \sum_{i=1}^{n} p_i \ln p_i + \lambda_1 \sum_{i=1}^{n} m_i p_i + \lambda_2 \sum_{i=1}^{n} p_i$. The maximum necessary conditions will be

$$\frac{\partial D'}{\partial p_i} = -\gamma(\ln p_i + 1) - \lambda_1 m_i - \lambda_2 = 0, \quad i = 1, \ldots, n. \tag{5.6}$$

Here λ_1 and λ_2 are Lagrange multipliers. Solving Eq. (5.6) we get $p_i = \mathrm{e}^{-1-(\lambda_2/\gamma)} \times \mathrm{e}^{-(\lambda_1/\gamma)m_i}$. From the norming condition $\sum_{i=1}^{n} p_i = 1$ we have $p_i = \mathrm{e}^{-\beta m_i} / \sum_{i=1}^{n} \mathrm{e}^{-\beta m_i}$ where $\beta = \lambda_1/\gamma$. In order to find β we multiply both sides of Eq. (5.6) by m_i and sum from 1 to n. As a result we obtain

$$\hat{m} = \frac{\sum_{i=1}^{n} m_i \,\mathrm{e}^{-\beta m_i}}{\sum_{i=1}^{n} \mathrm{e}^{-\beta m_i}} \approx \frac{\int_0^\infty m\,\mathrm{e}^{-\beta m}\,\mathrm{d}m}{\int_0^\infty \mathrm{e}^{-\beta m}\,\mathrm{d}m} = \frac{1}{\beta}.$$

Finally, the distribution

$$p_i = \frac{1}{Z}\mathrm{e}^{-m_i/\hat{m}}, \quad Z = \sum_{i=1}^{n} \mathrm{e}^{-m_i/\hat{m}}, \quad i = 1, \ldots, n \tag{5.7}$$

provides a maximum for diversity. This exponential distribution is named Boltzmann's distribution. It is consistent with our intuitive concept that larger organisms are less probable than smaller.

An analogous method is used in statistical physics where entropy is maximised under the condition that the mean energy $\hat{E} = \sum_{i=1}^{n} E_i p_i = kT$ is a constant. This gives a well-known *canonical* distribution $p_i = (1/Z)\,\mathrm{e}^{-E_i/kT}$, $Z = \sum_{i=1}^{n} \mathrm{e}^{-E_i/kT}$, $i = 1, \ldots, n$. The same method is used in the theory of information. Optimal coding is attained when the probabilities of symbols are equal to $p_i = \mathrm{e}^{-1.44 c t_i}$, where c is the channel capacity and t_i is the transition time of ith symbol.

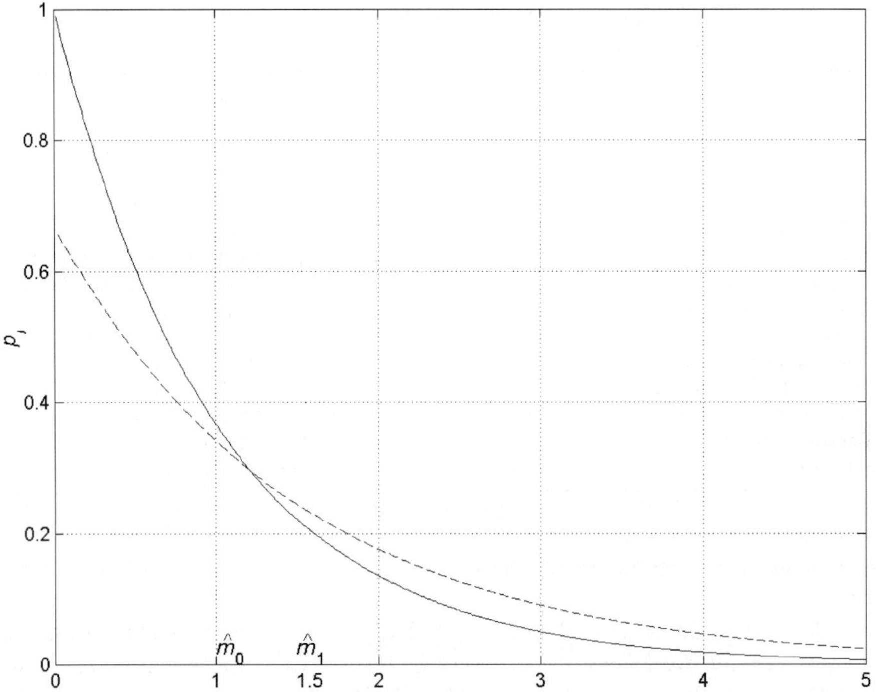

Fig. 4.1. Distributions of biomass for two cases: $\hat{m}_0 = 1$ (solid line) and $\hat{m}_1 = 1.5$ (dashed line).

Analysis of data about the distribution of individuals in fish populations with respect to their biomasses using fishery statistics (Pritz, 1974; Lurie and Wagensberg, 1984) showed a good conformity of these empirical distributions with theoretical ones.

Let us consider the following hypothetical situation: an environment became more favourable in relation to a community, and its total equilibrium biomass M_1 became bigger than its previous biomass M_0. The change brought about an increase both in the total number and in the mean biomass of the community, so that $\hat{m}_1 > \hat{m}_0$. In turn, this led to the change in distribution (5.7) (Fig. 4.1).

One can see that the growth of mean biomass increases the number of large individuals; the number of small individuals, on the contrary, decreases.

4.6. Information analysis of the global vegetation pattern

If we look at a standard botanical description of some territory, we can see that it contains, firstly, a list of species (types, forms, etc.) of plants represented in the territory, and secondly, the percentage of cover, p_i, i.e. the percentage of the total territory covered by ith species. This is a typical linguistic construction, in which the alphabet of the corresponding language is formed by the names of all the species contained in the list. It immediately appears useful to apply information methods to its analysis, in particular to

Table 4.1
Different types of vegetation (biomes)

1. Polar desert	16. Dry steppe
2. Tundra	17. Sub-boreal desert
3. Mountainous tundra	18. Sub-boreal saline desert
4. Forest tundra	19. Subtropical semi-desert
5. North taiga	20. Subtropical desert
6. Middle taiga	21. Mountainous desert
7. South taiga	22. Alpine and sub-alpine meadows
8. Temperate mixed forest	23. Evergreen tropical rain forest
9. Aspen–Birch lower taiga	24. Deciduous tropical forest
10. Deciduous forest	25. Tropical xerophytic woodland
11. Subtropical mixed forest	26. Tropical savannah
12. Xerophytic woods and shrubs	27. Tropical desert
13. Forest steppe	28. Mangrove forest
14. Temperate dry steppe	29. Saline land
15. Savannah	30. Subtropical and tropical woodland

estimate redundancy and the cost of information contained in the text at different levels of description.

Now let us consider a description of the global vegetation pattern (GVP). At the first level of description we have a list of biomes, or vegetation types. In accordance with Walter (1964, 1968) and Bazilevich (1973, 1993)—see also Svirezhev (2002)—the number of different biomes is equal to 30, and they are listed in Table 4.1. The corresponding biome maps can be found in the above-mentioned books.

If we know nothing about these biomes except for their denomination and number, then the natural assumption at the zero level of description (and perception) is that they are absolutely equivalent in the list. If we consider the list as some alphabet with 30 letters, then the information per letter is equal to $I^{(0)} = \log_2 30 = 4.9$ bits. However, this is not the case with biomes, since they are not entirely equivalent. Certain biomes occupy large areas, while the areas of others are negligibly small (for instance, biomes no. 9 and no. 28 in Table 4.1), and the production of some biomes is also much higher than the production of others, etc. In other words, in reality, the GVP is a hierarchical structure with respect to different characteristics, many of which are known (see Table 4.2). Therefore, at the next (first) level of description all these properties might be taken into account. For instance, using the data from Table 4.2 we can calculate the relative areas of biomes, $p_i = \sigma_i / \sum_{i=1}^{30} \sigma_i$, where σ_i is the total area of ith biome.

The information per letter at this level of description is equal to $I^{(1)} = -\sum_{i=1}^{30} p_i \log_2 p_i = 4.41$ bits. The redundancy of information at this level is equal to $R_1 = 1 - (4.41/4.9) = 0.1$, and its cost $C_1 = 1/(1 - 0.1) = 1.11$. These values are calculated by using the formulas of Section 4.3.

Up to this point we have not taken into account the spatial pattern of global vegetation, i.e. the spatial correlation between different pairs of biomes, which is a very important characteristic of the GVP. For instance, the correlation between tundra and north taiga biomes is very high, but in contrast the correlation between tundra and evergreen tropical

Table 4.2
Annual net primary production, P (kg C/(m^2 year), density of living biomass, B (kg C/m^2) and density of dead organic matter, D (kg C/m^2, in 1 m soil) for different types of vegetation (biomes); a, biome type; b, biome area ($\times 10^6$ km^2)

a	*b*	*P*	*B*	*D*	*a*	*b*	*P*	*B*	*D*
1	2.55	0.068	0.148	0.938	16	2.66	0.15	0.32	7.04
2	2.93	0.144	0.76	3.08	17	2.08	0.18	0.45	6.8
3	2.23	0.15	0.76	3.06	18	2.59	0.096	0.18	4.56
4	1.55	0.26	1.5	5.02	19	1.99	0.14	0.32	4.94
5	5.45	0.22	3.2	4.52	20	7.16	0.044	0.096	0.87
6	5.73	0.25	6.2	6.06	21	1.15	0.18	0.32	9.49
7	6.60	0.26	7.4	11.5	22	3.54	0.3	0.76	13.4
8	2.12	0.35	8.0	16.1	23	10.4	1.3	18.0	13.4
10	7.21	0.53	15.0	16.9	24	7.81	0.95	16.0	13.1
11	5.75	0.71	14.2	14.4	25	9.18	0.54	2.4	10.6
12	3.91	0.23	1.5	8.4	26	17.1	0.5	2.4	10.2
13	3.72	0.3	0.76	23.3	27	11.5	0.068	0.144	1.4
14	4.29	0.32	0.76	18.1	29	0.37	0.068	0.15	2.75
15	1.66	0.44	1.5	14.8	30	0.9	0.78	16.0	12.1

Since the areas of 9th and 28th biomes are negligibly small, they are omitted.

rain forest is very low. In order to estimate the spatial correlation between different pairs of biomes the Walter–Bazilevich biome map is used. For this we introduce a new concept of the border Γ_{ij} between ith and jth biomes. The border contains points belonging both to ith and jth biomes. The total length of the border, which, in a general case, can consist of several separate parts, is denoted by the same symbol. It is obvious that when the border is longer, the interaction between the bounding biomes is more intensive. Using the biome map the lengths of all borders are calculated. By defining $p_{ij} = \Gamma_{ij}/\sum_{i,j=1}^{30}\Gamma_{ij}$ we get the following expression for the amount of information per two-letter word at this level of description:

$$I^{(2)} = -\sum_{i,j=1}^{30} p_{ij}\log_2 p_{ij} \approx 7.2 \text{ bits.} \tag{6.1}$$

For the redundancy and the cost of information we get correspondingly: $R_2 = 0.265$ and $C_2 = 1.36$. Since at this level the elementary unit is a two-letter word, the information per letter will be equal to $I^{(2)}/2 \approx 3.6$ bits.

Note that in addition to their area, the biomes are also characterised by three values: annual productivity P_i, living biomass B_i, and dead organic matter D_i (see Table 4.2). It is obvious that we can define the frequencies p_i^{P}, p_i^{B} and p_i^{D} for each value as $p_i^{\mathrm{P}} = P_i\sigma_i/\sum_{i=1}^{30}P_i\sigma_i$, $p_i^{\mathrm{B}} = B_i\sigma_i/\sum_{i=1}^{30}B_i\sigma_i$ and $p_i^{\mathrm{D}} = D_i\sigma_i/\sum_{i=1}^{30}D_i\sigma_i$. The description in terms of productivity, living or dead organic matter can be considered as description at the first level with information per letter defined as

$$(I^{(1)})^{\mathrm{P,B,D}} = -\sum_{i=1}^{n} p_i^{\mathrm{P,B,D}}\log_2 p_i^{\mathrm{P,B,D}}. \tag{6.2}$$

Using the data from Table 4.2 we can calculate the corresponding value of information (per letter), its cost and redundancy:

1. For productivity: $(I^{(1)})^{\mathrm{P}} = 3.61$ bits, $C_2^{\mathrm{P}} = 1.32$, $R_2^{\mathrm{P}} = 0.24$.
2. For living biomass: $(I^{(1)})^{\mathrm{B}} = 3.27$ bits, $C_2^{\mathrm{B}} = 1.5$, $R_2^{\mathrm{B}} = 0.33$.
3. For dead organic matter: $(I^{(1)})^{\mathrm{D}} = 4.13$ bits, $C_2^{\mathrm{D}} = 1.17$, $R_2^{\mathrm{D}} = 0.16$.

These values show that information about the biome productivity, living biomass and dead organic matter is more valuable than the information about the distribution of the biome areas. Information about the distribution of living biomass has the maximal cost ($C_2^{\mathrm{B}} = 1.5$).

It is necessary to say a few words about information at the global and regional levels of scaling. If we keep in mind that the total number of species in the contemporary biosphere is equal to $n \approx 10^6$, then information per species at the zero level of description and the global scale is equal to $(I^{(0)})_{\mathrm{B}} = \log_2(10^6) \approx 19.9$ bits. At this level of description all specimens of a community differ from each other only by one indication, namely, by its membership of one or another species. However, as we saw above, if the biosphere is represented as a composition of higher taxonomic units (biomes), which in turn represent the composition of species, i.e. words are formed by groups of letters representing species, then the information per word is less than 19.9 bits. The same effect is observed when we estimate the amount of information per species for some taxonomic unit, which is lower than the biosphere.

The point is that when we describe a community belonging to some ecotype (for instance, either an aquatic community in a lake, or a plant community of some biome, etc.) then the alphabet may be significantly shorter. Then the amount of information per letter will be less than in the previous case. This seeming loss of information is a result of its redundancy at this "regional" or "local" level in comparison with the global level. In fact, the information about all the species present in the biosphere is absolutely redundant when we describe some regional or local community. Here we need only the information about those species, which are typical for some of the considered locality. Note that the "lost" information is not actually lost; it is usually used to extract the locality from the biosphere.

Let us consider the following example. The number of different plant species typical for the Russian dry steppe is equal to 150. Then the information per species contained in the botanical description of any steppe community (list of species) is equal to $(I^{(0)})_{\mathrm{R}} = \log_2 150 \approx 7.23$ bits.

Finally, we can give another interpretation of these results. Indeed, what is implied the value of information equal to 19.9 bits per species? Note that this is a global value. Since a considered system is the Globe then this statement is equivalent to the following: the probability to find even one (and more) individuals of *given* species at any arbitrary point on the Globe is equal to $2^{-19.9}$. The same probability, but calculated for the list of steppe species (150) and the region of the Russian steppe, will be equal to $2^{-7.23}$. An analogous interpretation occurs for biomes. At the zero level of description the probability to find one and more plants representing one (of 30) given biome at any point of land is equal to $2^{-I^{(0)}}$. At the first level, when we know the area of each biome, the same probability but calculated for one plant will be higher, $2^{-I^{(1)}}$, $I^{(1)} < I^{(0)}$. Hence, the greater is the area of a certain biome, the higher is the probability to meet its representative at any arbitrary point.

At the second level the probability to find a pair of plants representing a given pair of biomes in the close vicinity of any point of land is equal to $2^{-I^{(2)}/2}$. It is also higher than the previous probability.

In addition to area each biome can be characterised by either productivity, or living biomass or dead organic matter. Then the corresponding probabilities $\mathrm{Pr}_1^{\mathrm{P,B,D}} = 2^{-(I^{(1)})^{\mathrm{P,B,D}}}$ are the probabilities to find at any point of land a plant with characteristics which are typical for a given biome. Note that all these statements can be paraphrased as "…the probability to find a plant-representative of any arbitrary biome at a given point of land…"

4.7. Diversity of the biosphere

Life on Earth is represented by a huge number of diverse forms, and it is necessary to maintain all this diversity. This is one of the main thermodynamic roles of solar energy. Otherwise, from the point of view of classic thermodynamics, the film of life would have to be homogenous, and, moreover, life would not have occurred at all. Nevertheless, life exists and its diversity is very high. In order to resolve this contradiction we suggest a simple thermodynamic model (see also Svirezhev and Svirejeva-Hopkins, 1997).

Let the biota (the living matter of the biosphere) be subdivided by n sets. These sets could be interpreted as different hierarchical taxonomic units: biomes, ecosystems, communities and species. We assume that every ith set contains N_i virtual biosphere "particles", so that the total number of particles in the biota is equal to $N = \sum_{i=1}^{n} N_i$. Generally speaking, the particles must differ from each other, but, as a first approximation, we assume that they are equal. In particular, they all have a unique mass, so that the sets differ from each other only by mass.

Let us assume that at some initial moment the particles were mixed up in some "pre-biosphere" substance, and this "pre-biosphere" system had no structure. How could such an ordered structure as the biosphere have arisen? We think that it is the result of the work of a demon named *Ecodemon*, from the famous Maxwell family of demons (Fig. 4.2). He distributes the biosphere particles among the boxes, removing them from the "pre-biosphere" pool. As a result, each ith box contains N_i particles. In this way a new structure arises, which may be called *biota*. The transition from a fully mixed system ("pre-biosphere chaos") to the "structured" biosphere is accompanied by the entropy reduction $\delta S = -\sigma N \sum_{i=1}^{n} p_i \ln p_i$, where σ is the specific entropy of the mass unit of some "pre-biosphere" substance. In our case this substance is a mixture of chemical elements from which living matter can be constructed: 106 molecules of CO_2 + 90 molecules of H_2O + 16 molecules of NO_3 + 1 molecule of PO_4 + a few molecules of some mineral elements (Odum, 1971a,b). Note that when we speak about the "mixture", we do not mean a real physical mixture, but some likely sum of elements already in existence but—very importantly—still not reacting with each other so that "living matter" is still not formed. Since the specific molar entropies of CO_2, H_2O (vapour), NO_3 and PO_4 are equal to 214, 154, 256 and 301 (all the values are expressed in J/mol K), and their molar masses are 44,

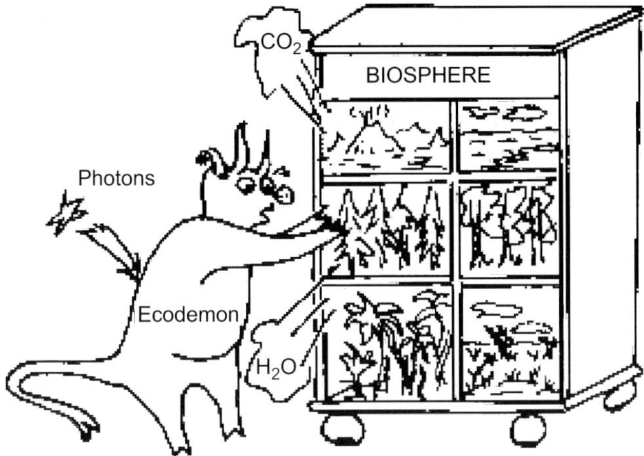

Fig. 4.2. Ecodemon is working…

18, 62 and 95 g, respectively, then

$$\sigma = \frac{214 \times 106 + 154 \times 90 + 256 \times 16 + 1 \times 301}{44 \times 106 + 18 \times 90 + 62 \times 16 + 1 \times 95} \approx 5.55 \text{ J/g K}. \tag{7.1}$$

This value can also be considered as entropy of 1 g of protoplasm, that is a primary non-structured living matter, from which *Ecodemon* has to construct the structured biosphere. One-gram portions of protoplasm are our biosphere particles.

Let us consider the following cyclic process: every year a certain amount of new biomass, δN, is created, and, since the biosphere is at steady state, the same amount of dead organic matter is decomposed. It is obvious that the value of δN is equal to the net annual production of the biosphere, NPP. The decomposition process is accompanied by a release of heat, which is equal to the enthalpy of δN. If δN is expressed in the units of dry biomass (d.w.), and the enthalpy of 1 g d.w. is equal to $h = 18.9 \times 10^3$ J/g, then NPP $=$ $h\,\delta N$ (J). If we assume the entropy production is equal to the thermal effect divided by temperature T, then $\delta S = (\text{NPP})/T$ where $T = 288$ K is the annual mean planetary temperature.

We assume (and this is our basic hypothesis) that this entropy production is balanced by the entropy decrease in the process of the creation of new biomass. The decrease has to be equal to information entropy. In other words, the value of NPP is the annual work performed by *Ecodemon*, which feeds on the enthalpy.

If we keep in mind that $I = -(1/\ln 2)\sum_{i=1}^{n} p_i \ln p_i$ is the information entropy or diversity of the system and NPP $= \sigma\,\delta N I \ln 2$ then

$$I = \frac{\text{NPP}}{\delta N \sigma T} = 1.44 \frac{h}{\sigma T}. \tag{7.2}$$

Since $h = 18.9 \times 10^3$ J/g and $\sigma \approx 5.55$ J/g then $I \approx 17$ bits.

Let us estimate the probability of spontaneous creation of the biosphere. It is equal to

$$\mathrm{Pr} = \mathrm{e}^{-I \ln 2} = \mathrm{e}^{-11.8} = 7.8 \times 10^{-6}, \tag{7.3}$$

i.e. is very small. But if the contemporary biosphere is a result of a sufficiently large number of attempts L, in accordance with a simple probabilistic model (Chernavsky and Chernavskaya, 1984), the probability of its creation will be equal to

$$\mathrm{Pr}_L = \frac{L\,\mathrm{Pr}}{1 + L\,\mathrm{Pr}}. \tag{7.4}$$

How to evaluate the number of attempts? The photosynthetic biosphere with vegetation existed during approximately the last 10^9 years (Rutten, 1971). The mean time of the biosphere renovation is equal to $\tau = B/\mathrm{NPP}$ where $\mathrm{NPP} = 1.2 \times 10^{17}$ g d.w./year is the annual net production and B is the total biomass of the biosphere equal to 1.8×10^{18} g d.w. (Svirezhev et al., 1985). Then $\tau = 15$ years, and if we assume that one attempt is nothing else than one cycle of the biosphere renovation, then $L = 10^9/15 \approx 6.7 \times 10^7 = 0.66 \times 10^8$, and

$$\mathrm{Pr}_L = \frac{6.7 \times 10^7 \times 7.8 \times 10^{-6}}{1 + 6.7 \times 10^7 \times 7.8 \times 10^{-6}} \approx 0.998. \tag{7.5}$$

Thus, the probability of the biosphere creation is close to 1. In other words, *there is nothing surprising in the existence of the contemporary biosphere*.

It is also very interesting that the probability Pr depends neither on the mass of the biosphere nor on its productivity (you can see it from the formula for I). The probability Pr depends only on two factors:

1. Work of the climatic machine, which determines the Earth temperature.
2. Composition of input components for construction of living matter; namely, the composition determines the value of σ.

Photosynthesis reaction uses two gases as basic substance for the formation of living matter: carbon dioxide and water vapour. Certainly, we can imagine other hypothetical reactions, which would use other elements and substances (for instance, silicon instead of carbon), but this would give other values of σ and, as a consequence, other values of diversity. As a result, the probabilities of existence of such virtual biospheres would differ from the similar probability for the really existing one.

Note, however, that the probability Pr_L nevertheless depends in an implicit way on the total mass of the biosphere and its productivity because the number of attempts (L) is defined by these values.

Knowing the value of I, the number of components (elements, elementary units, etc.) of the biosphere could be estimated. If these elements are relatively independent and they occur with almost the same frequencies, then $I = \log_2 n$ and the number of elements $n = 2^I = 2^{17} \approx 2 \times 10^5$. It is interesting that this number is close to the number of biological species on our planet. An impression appears that our *Ecodemon* uses species as boxes. However, if the elementary units within the biosphere are organised in "hard" structures (like trophic chains and trophic levels) with the exponential distributions of frequencies p_i, then $n \approx I$, i.e. the number of different sorts of these elementary units would be relatively small.

4.8. Information and evolutionary paradigm: selection of information

Note that some statements about the connection between evolution and information today seem truisms. In the course of evolution the amount of information increases. The information grows by means of either being "selected" from the noise, which masks it, or the creation of new information as a result of the memory of some random choice. For instance, the information increases in the process of crystallisation of liquid, but it is not new information. It is known as a priori information, i.e. which was earlier hidden by "entropy noise". In the course of evolution the appearance of new information is a result of some "memory" and genetic recombinations. New information is created when every individual appears in this world. New information is created also when new species and superior taxa appear. From the linguistic point of view, this means that in the course of evolution the length of the alphabet increases.

The genetic information is transmitted from one generation to another just after its transformation within an ecosystem, which consists of populations. In turn, populations are the elementary units of microevolution (Svirezhev, 1989). Still Timofeev-Resovsky has emphasised that for evolution the *quality* (not the quantity) or the *cost of information* is important (Timofeev-Resovsky, 1958, 1961b).

If we look in any detailed physical handbook we see that, for instance, entropy of 1 mol of CO_2 (this gas is very popular in ecology) at 25°C and 1 atm is equal to 214 J/K. Let us remember that the number of molecules in 1 mol is equal to $n_{\text{mol}} = R/k$, where R is the gas constant, $R \approx 8.314$ J/K mol, so that $n_{\text{mol}} \approx 6.07 \times 10^{23}$ molecules. Therefore, the entropy of a single molecule of CO_2 will be equal to $213.5/6.07 \times 10^{23} = 3.52 \times 10^{-22}$ J/K. From Eq. (1.1) we get $\ln W = 3.5 \times 10^{-22}/1.37 \times 10^{-23} \approx 25.7$ and $W = e^{25.7} \approx 1.4 \times 10^{11}$. The latter means that every CO_2 molecule may be situated at one of the 1.4×10^{11} different states.

Now we would like to tell the following story, which is absolutely heretical from a physical point of view. Since the CO_2 molecules differ from each other by the values of some parameter (in thermodynamics either the energy or temperature can be considered as the parameter), we can assume that there is a pool filled by $n_{\text{m}} = 1.4 \times 10^{11}$ different molecules of CO_2. We are two *Panglossists* (see Chapter 12), i.e. we believe that we are living in the best world; therefore, the CO_2 molecule in our world is the best of the possible number of 1.4×10^{11} molecules. Thus, our CO_2 is a result of sequential choice and we next test the chosen molecule in order to find the single "best" one among them. We again ask *Ecodemon* to help us. The probability that already the first attempt of *Ecodemon* would be successful is equal to $p = 1/n_{\text{m}} = 1/1.4 \times 10^{11} \approx 0.7 \times 10^{-11}$. But *Ecodemon* can solve the problem by exhaustion with n_{m} steps (maximum). If every step is about 10^{-2} s (mean time of carbon oxygenation), then in the most pessimistic scenario all the processes will be finished in the course of 45 years. This is only an instant in the geological time scale. It is obvious that similar arguments take place for the whole chemosphere of our Universe, which has already existed for 12×10^9 years. Therefore, we can say that all chemical elements and their (non-organic) compositions already existed at the initial stage of the evolution of the Universe.

Note that in these considerations we use a teleological paradigm (we assume that some "best" objects exist) and an evolutionary approach (i.e. we can find these "best" objects if we use some sequential evolutionary procedure and some selective principle). This method is very popular in the biological sciences where the selective principle is natural selection, but it is also used implicitly in theoretical physics. However, if in biology the entire course of evolution is interesting then in physics we are interested only in the initial and final states of the studied system; the ways that lead from the initial to the final state are not essential. This kind of method is called *thermodynamic*. From the evolutionary point of view, such a thermodynamic approach has a right to exist only if the system manages to test all intermediate states before attaining in the course of "reasonable" time some final state that may be considered as equilibrium. In other words, it is assumed that the velocity of the system's movement across intermediate states is sufficiently high. We saw that clearly with chemical systems. For biological systems the answer to the question "Is it possible for biological objects to attain some 'best' state in the process of biological evolution in a 'reasonable' time?" is not evident.

This is a very old problem but it has again become modern since we have known that genetic information is represented in the form of polynucleotide sequences, so that a single nucleotide base can be considered as one letter of a genetic text (genome). For instance, the human genome contains about 10^{10} nucleotide bases (indeed, the characteristic size of a nucleotide base pair is equal to 6×10^{-10} m; since the human genome is represented in the form of 2 m of "double-stranded" DNA, the total number of nucleotide base pairs will be $1/3 \times 10^{-10} \approx 0.35 \times 10^{10}$). Since the alphabet of nucleotides contains only four letters, the number of different genetic texts (genomes) will be equal to $(4)^{0.35 \times 10^{10}} \approx (10)^{2 \times 10^{9}}$. This number is monstrously immense; for instance, the number of all elementary particles in our Universe is 10^{100} and has been in existence for only 4×10^{17} s! Here we first meet the so-called "combinatorial" numbers, which are not comparable with "physical" ones (the number of particles is a typical "physical" number). However, combinatorial operations with physical numbers can generate combinatorial numbers.

Thus, it is an absolutely non-realistic idea to attain a single "best" combination in the course of "reasonable" time, if we have such a number of possible combinations. In other words, the sorting of combinations by natural selection is out of the question (Blumenfeld, 1977a). Therefore, evolutionary trajectories are random (non-deterministic) and they cannot be repeated. Therefore, Darwinism, which is based on the paradigm of natural selection, is not true. All of us are only a result of the "game of chance" on this planet Earth.

> *Again I saw that under the sun the race is not to the swift, nor the battle to the strong, nor bread to the wise, nor riches to the intelligent, nor favour to the men of skill; but time and chance happen to them all.*
>
> The Bible. Ecclesiastes, 9.

But the situation is not absolutely so pessimistic and there are some ways out of the contradiction. The first way is "thermodynamic", when the problem is assumed to be irrelevant since only the initial and final states are interesting for us; we assume only that the transition from the initial to the final state is possible without any detailed description in *which* way it is realised. There are also other possibilities.

If we look at the picture of evolution (as a whole), we see that evolution possesses a very long "memory": the genome of dinosaur cannot change in such a way that in the next evolutionary step a mammoth would occur. Timofeev-Resovsky said about this: "There cannot appear the mutation of the colour of the tail tip in the human genome". One of the main evolutionary problems, the appearance of new species, is always meeting the question of how the appearance can be explained, since it represents a jump in the continuous evolutionary movement under mutation pressure. This contradiction resolves the so-called hypothesis of "Meccano" (Timofeev-Resovsky, 1961b; Eigen, 1981; Ivanitzky et al., 1985). (We all know this children's construction set consisting of miniature metal or plastic parts from which mechanical models can be made.) We show the principal mechanism of the hypothesis by the following example (Shnol, 1989).

Let at the beginning five-letter short words be formed spontaneously from a four-letter alphabet: the possible number of these is $4^5 = 1024$. They can be sorted out by some criterion in the course of a relatively short time $t_1 \approx \tau \times 1024$, where τ is the time of one test. From these words the 20 most "perfect" words are selected. Then we begin to compose phrases from them. By assuming (for simplicity) that each of these phrases contains five words, their number is $(20)^5 = 3 \times 10^6$. From these phrases we select again the five best ones, and compose stanzas. From these stanzas we compose a poem, etc. "I process thousands of tons of verbal ore in order to find one unique word" (Mayakowsky, the Russian poet). Step by step the perfection of the text is reached. Even if something is insufficiently perfect at higher hierarchical levels then we shall change only individual blocks, but do not split all the text into single letters. Any time will be insufficient for a senseless work. Thus, we assume that evolution in general, and the evolution of genetic information in particular, is proceeding in accordance with the principle of "Meccano".

Let us try to formalise our preceding reasoning. At the first stage of aggregation we construct r_1-letter words in an R_0-letter alphabet. The number of possible words is equal to $(R_0)^{r_1}$, and then the time of total sorting will be equal to $t_1 = \tau_1(R_0)^{r_1}$ where τ_1 is the test time at this stage. Then we select from these words R_1 the best words, which will be "letters" at the next stage when we shall construct r_2-word phrases. The possible number of these is equal to $(R_1)^{r_2}$, and the time of total sorting will be equal to $t_2 = \tau_2(R_1)^{r_2}$, etc. until at the final fth stage we get a text of given length n. It is obvious that $n \approx r_1 \times r_2 \times \cdots \times r_f$. The total time of such type of evolution is equal to

$$t_c = \prod_{i=1}^{f} \tau_i (R_{i-1})^{r_i}. \tag{8.1}$$

One can see that the time of sorting by the "Meccano" principle t_c is essentially less than the time of total sorting, which is equal to $t_0 = \tau R_0^n$. This becomes especially visual, if we simplify Eq. (8.1) by setting $\tau_i = \tau$, $R_i = R$ and $r_i = r$. Then we get (Ivanitzky et al., 1985)

$$t_c = \tau \frac{\ln n}{\ln r} R^r. \tag{8.2}$$

For instance, for a virus genome the length of amino acids sequences is $n = 10^3$. Since the number of different amino acids is equal to 20, $t_0 = \tau \times 20^{1000} \approx \tau \times 10^{1333}$, and the evolutionary time is enormous. If the "Meccano" principle is working, and $R = 20$, $r = 5$, then the time is very small: $t_c = \tau(\ln 1000/\ln 5) \times 20^5 \approx \tau \times 5 \times 10^8$. For the human genome $n = 7 \times 10^9$, and the evolutionary time increases insignificantly: $t_c = \tau(\ln 7 \times 10^9/\ln 5) \times 20^5 \approx \tau \times 1.5 \times 10^9$. Therefore, the real differences between evolutionary times are explained by other reasons. The rate of evolutionary perfection is so high that it does not determine the direction of evolution. In this case the use of the "thermodynamics paradigm" is correct, and everything is determined only by the initial and final states of the evolving system. As a result, the evolution is strictly determined, i.e. the evolving system is moved from "molecules towards *Homo sapiens*" along deterministic, not random, trajectories. Note, however, that this is true only if the "constructor" algorithm exists in Nature.

In a certain sense a reality of the algorithm is evident: sexual recombination, crossover, transduction and other such processes are, generally speaking, a shuffling of ready-made blocks of genes. But biological evolution, i.e. properly, life, begins when molecules originate that are able to perform *matrix convariant reduplication*, i.e. to reproduce (Timofeev-Resovsky, 1961b). Unfortunately, we do not know even the chemical characters of these molecules. These could be spontaneously originating short sequences of RNA since RNA possesses catalytic properties. Recently, it was shown that there are short ($r \approx 5$) universal blocks in polynucleotides and polypeptides chains (Shnol, 1989).

4.9. Genetic information contained in an organism: hierarchy of information and its redundancy

The chemical alphabet of genetic information contains only four "letters": the nucleotides *adenosine*, *guanine*, *thiamine* and *cytosine* (*uracyle*). Then the information contained in the sequence of n nucleotides in the DNA chain is equal to $I^{(1)} \log_2(4^n) = 2n$ bits. This is the first level of reception of the information. At the next level of reception, the information that is coded in a protein chain (enzyme) synthesised on the sequence of n nucleotides has to be written. Indeed, each of 20 different amino acids is coded by three nucleotides (three-letter words), so that although the information per amino acid is equal to $I^{(2)} = \log_2(20)^n = 4.32n$ bits, the information per nucleotide (letter) will be equal to $I^{(2)}_{\text{nuc}} = I^{(2)}/3 = 1.44n$ bits, i.e. less than at the previous level. The decrease of information is a result of degeneration of the triplet code: the total number of codons ($4^3 = 64$) is higher than the number of amino acids (20). The redundancy of information at this level $R^{(2)} = 1 - (I^{(2)}/I^{(2)}_{\max}) = 1 - (\log_2(20)^N/\log_2(4)^{3N}) \approx 0.28$ and, correspondingly, its cost $C^{(2)} = 1/(1 - R^{(2)}) \approx 1.4$.

At the next level we take into account the so-called *neutral mutations*, when it is possible to replace one amino acid by others without the change of protein properties. Hence, the number of really irreplaceable amino acids decreases, and the amount of information decreases too. These mutations are very difficult to estimate, but it is known that their number connects to pair correlations between amino acids (two-letter words).

Note also that these correlations are significantly higher than the triplet, quadruplet, etc. correlations (Ebeling et al., 1990). Therefore, we can conclude that the genetic text (genome) contains, as a rule, six-letter words in a 20-letter alphabet. Then the possible number of n-length texts is equal to $w_n = (20)^{n/6}$, and the corresponding information per nucleotide $I_{\text{nuc}}^{(3)} = (n/6)\log_2(20)$ bits. The redundancy and the cost of information $R^{(3)} = 0.64$ and $C^{(3)} = 2.6$.

The main problem here is what is implied by the length of genetic text, n? The first idea, which was already used in Section 4.8, is to connect the value of n with the amount of DNA in cells of individuals at a certain taxonomic level. In general, the estimates of nuclear DNA contents are provided in picograms (pg, 1 pg $= 10^{-12}$ g) or in base pairs (bp) of double-stranded DNA. Each strand is a linear polynucleotide chain consisting of four nucleotides, two purines and two pyrimidines, and it is commonly accepted that the average molecular weight for each nucleotide is approximately 618 Da. The conversion factors are: 1 bp $= 1.02 \times 10^{-9}$ pg $= 618$ Da (Li and Grauer, 1991). For instance, the lowest amount of DNA (when only the non-repetitive DNA is taken into account) in cells of the group *Annelids* is equal to 0.07 pg (Fonseca et al., 2000). By converting to nucleotides (1 pg $= 0.98 \times 10^9$ bp) we get for one nucleotide chain: $n = (0.98 \times 10^9) \times 0.07 \times (1/2) = 3.43 \times 10^7$ nucleotides. Then the possible number of genetic texts (virtual genomes) consisting of six-letter words $w_n = (20)^{n/6} = (20)^{5.7\times10^6}$, and the corresponding information $I_{\text{nuc}}^{(3)} \approx 1.71 \times 10^7$ bits. Note that while estimating w_n Fonseca et al. stopped at the second level, when only a degeneration of the code is taken into account. Then $w_n = (20)^{n/3} = (20)^{1.14\times10^7}$ and $I_{\text{nuc}}^{(2)} = 3.42 \times 10^7$ bits. If we repeat all these calculations for *Mammals* (the lowest amount of DNA: 3 pg), then $w_n = (20)^{n/6} = (20)^{2.44\times10^8}$ and $I_{\text{nuc}}^{(3)} \approx 7.32 \times 10^8$ bits.

Another approach uses such values as the number of non-nonsense genes, g_i, and the number of amino acids in each gene's code (each gene is determined, on an average, by a sequence of about 700 amino acids; Li and Grauer, 1991). Then the length of nucleotide text will be equal to $n = 700g$. For the group *Annelids* $g = 10,500$, and $n = 7.35 \times 10^6$. Since we consider only non-nonsense genes, a partial ordering has already been taken into account; therefore, $w_n = (20)^n = (20)^{7.35\times10^6}$ and the corresponding amount of information (information content) $I_{\text{nuc}} = 2.2 \times 10^7$ bits. If we repeated all these calculations for *Mammals* ($g = 1.4 \times 10^5$), then $w_n = (20)^{0.98\times10^8}$ and $I_{\text{nuc}} \approx 2.93 \times 10^8$ bits.

Comparing these estimates with estimates made by the DNA-content method, we see that they are relatively close: $I_{\text{nuc}}^{(3)} \approx 1.71 \times 10^7$ and $I_{\text{nuc}} = 2.2 \times 10^7$ bits for *Annelids*, and $I_{\text{nuc}}^{(3)} \approx 7.32 \times 10^8$ and $I_{\text{nuc}} \approx 2.93 \times 10^8$ bits for *Mammals*. It is interesting that for *Annelids* the first estimation is slightly less than the second one, but for *Mammals* we see the opposite picture. Besides, if we take into account the accuracy of all these estimations, then there is a good agreement between them. And finally, we would like to emphasise that these results will be actively used in the next chapter.

4.10. Summary of the important ecological issues

It is important to understand that entropy measures the information needed to describe the system, while information that we have in the genes, for instance, represents negative entropy—because the information *is* available. As negative entropy does not exist, we have to introduce another concept. It is the concept of exergy, which will be introduced in the next chapter. The relationship between information and entropy, introduced with Boltzmann's famous equation, makes it understandable that Shannon's expression for biodiversity is strongly related to the entropy expression.

The relationship between biodiversity and stability is rather complex and not straightforward as previously believed in ecology. However, higher biodiversity means that a wider spectrum of properties is available for survival under changing conditions; therefore, higher biodiversity implies a wider spectrum of buffer capacities and type of resistances against changes.

Many distributions in ecology follow Boltzmann's distribution, which is consistent with the power law (Bak, 1996) for self-organising critical systems. It has been shown that several distributions in Nature—the deviation from an average situation, the spatial distribution of species—follow the power law (Jørgensen et al., 1998). It is interesting that the Boltzmann's distribution also explains why a more favourable situation implies that the number of large individuals increases and the number of small individuals decreases.

It is *shown* that the biosphere is a highly probable consequence of the right temperature, the presence of the elements needed for construction of living matter and sufficient time. Thermodynamically, it is not a puzzle that the biosphere exists, because Earth has the elements necessary for construction of life, has the right temperature, as also discussed in Chapter 3, and Earth has now existed for more than 4.5 billion years!

When the information hidden in the number of species or in the distribution of different biomes is analysed, it is possible to show that the observed patterns in Nature have a high probability. The high number of species (probably close to 10^7—it was shown that the number would probably be close to 3×10^6 at present; we know approximately this number of species, but it is presumed that there is the same number of still unknown species) or the number and distribution of biomes is a consequence of their properties, the long time and the time arrow.

The amount of information in the genes is astronomically big because not only the number of amino acids, but also the sequences of these amino acids determine the life processes. It is, in principle, not different from a book, where the sequence of the letters is important and carries the information—not just the number of letters. Earlier in the chapter, Blumenfeld's estimations are far too modest. In addition, the genes contain even more information than just about the amino acid sequence, namely, about the "management" of this information, which becomes increasingly more complicated with the increasing complexity of the organisms (Taft et al., submitted). This implies that organisms have a high content of information that can be translated to free energy and exergy, as we shall see in the next chapters. This makes it understandable that ecosystems cannot be easily replaced—ecosystem *conservation* is therefore urgently needed in environmental management context. When an ecosystem is lost, an enormous amount of entropy is produced due to the loss of the huge amount of stored information in ecosystems.

Additional References

Bak, P., 1996. How Nature Works. Springer, New York, 212 pp.

Jørgensen, S.E., Mejer, H., Nielsen, S.N., 1998. Ecosystem as self-organizing critical systems. Ecol. Model. 111, 261–268.

Taft, R.J., Mattrick, J.S., Andrew, P.S. Genome-wide increases in non-coding DNA positively correlate with increasing biological complexity (submitted for publication).

Additional References

Bak P. 1996. How Nature Works. Springer, New York, 212 pp.

Chapter 5

Work, exergy and information

And yet, to me, what is this quintessence of dust!
W. Shakespeare (Hamlet)

5.1. The work done by a system imbedded into an environment

One of the founders of modern thermodynamics, Ostwald, did not like the word "entropy". He generally tried to avoid the use of both the word and the concept of entropy. He tried to replace the latter by the concept of work. Since the main interaction in thermodynamics is the interaction between the system and its environment (keeping in mind that the interaction is non-symmetric), there are two sorts of work, namely, *work done by the system on its environment* and *work done by the environment on the system embedded within it* (see also Chapter 2).

This work can be used in order to bring the system into the state of macroscopic motion or to displace the chemical equilibrium of the system or to displace the state of a system in some external "field" defined by those "potentials" about which we were speaking above.

We shall count work A done by the external forces (which are associated with the environment) on the system as positive. Conversely, the negative work, $A < 0$ will be associated with the work $|A|$ done by the system on its environment. A supersystem "the system + its environment" can be considered as a closed system, but its components are not in thermodynamic equilibrium with each other. In the course of the equilibrium becoming established between the system and its environment the former can perform a work on the latter. The supersystem can arrive at a number of different final states (in particular as regards its energy and entropy), because the transition to equilibrium can also differ. Therefore, the total work, which can be performed by some non-equilibrium system, depends also on the way along which thermodynamic equilibrium is reached. Naturally, the following question arises: along which way will this work be maximal? Note that since we are interested in the work performed by a non-equilibrium system, we have not taken into account the work which might be performed during a general expansion (or compression) of a supersystem. The work could also be done by a system, which itself is in equilibrium. This means that neither the total volume nor the number of particles of each sort contained within the supersystem change, but remain constant. It is interesting that the conservation of the total number of particles of each sort is a necessary

Towards a Thermodynamic Theory for Ecological Systems, pp. 95–126

condition for the existence of the corresponding global biogeochemical cycle. Let us consider the global carbon cycle (GCC) as an example.

Let us have one local ecosystem as the system and another part of the biosphere as its environment. Naturally, the whole biosphere is our supersystem. Living matter is a combination of carbon atoms, which have been drawn from the surrounding biosphere (in particular, from atmospheric carbon dioxide) and "packed" into different "boxes" called genes, proteins, organisms, species, etc. upstairs in the hierarchy of the organisation of living matter. A certain amount of carbon atoms are continually released into the atmosphere while at the same time other atoms are returned in the process of the decomposition of living matter. If these amounts are permanently changed, tending to constant when the ecosystem tends to equilibrium with another part of the biosphere (all this is, namely, the GCC), then the total amount of carbon (and, as a consequence, the total number of atoms of carbon) remain constant (Fig. 5.1).

We shall now consider the problem of maximal work in a special situation. We follow here, in general, the Landau and Lifshitz book (1995). Suppose the system is in an external environment whose temperature T_0, pressure p_0 and (chemical) potentials $\mu^0 = \{\mu_1^0, ..., \mu_n^0\}$ differ from those of the system (T, p and $\mu = \{\mu_1, ..., \mu_1\}$). The internal energy U_0, volume V_0 and number of particles $\mathbf{N}^0 = \{N_1^0, ..., N_n^0\}$ of the environment are so large that changes in them due to processes taking place in the system do not produce any significant change in the temperature, pressure or potentials of the environment; these may, therefore, be considered to be constant.

Let at its initial state the system be in thermodynamic equilibrium with its environment. In order to change the initial state "0" and to "move" the system to another state, "1", in which it would not be in thermodynamic equilibrium (although always in a *dynamic* equilibrium), i.e. to realise the transition $0 \rightarrow 1$, the environment has to do a certain work δA_{01}. If the system is a thermally isolated object then the work done by the environment on the system for a given change in its state (i.e. from the given initial to the final state) would be a uniquely defined quantity equal to the change in the internal energy of the system. The coupling of two systems, when one of them becomes the environment of another and the environment also is involved in the process, makes the result no longer unique, and the question arises as to what is the minimal work an environment can do for a given change in state.

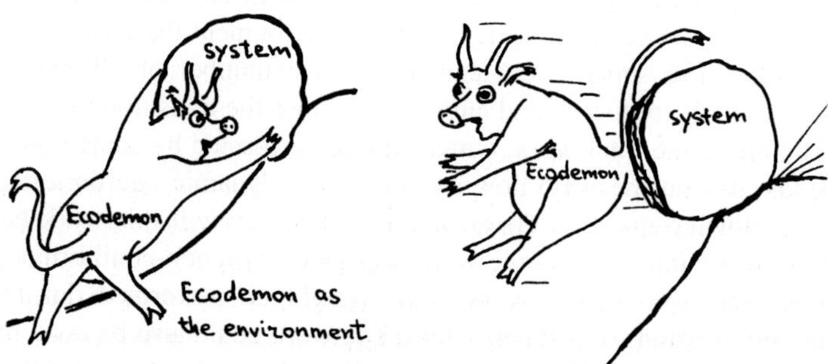

Fig. 5.1. The system and its environment. Ecodemon is working on the system (left), and the system is working on the ecodemon (right). Ecodemon is working as Sisyphus.

If, in a transition from one state to another, the environment does work on the system, then in an inverse transition from the state "1" to the initial state "0", the system must do work on the environment. To a direct transition, which requires the external source to supply the minimum work $\min(\delta A)_{01} = \delta A_{\min}$, and the transition is *forced* for the system, there is a corresponding inverse transition, in which the system does the maximum work $\max|(\delta A)_{10}| = |\delta A_{\max}|$. The latter transition is started when the action which supports it stops and the system begins spontaneously to move towards its thermodynamic equilibrium. It is evident that the amounts $\delta A_{\min}$ and $|\delta A_{\max}|$ are identical. During these transitions, the system can exchange heat, matter and work with the environment.

Thus, the total change ΔU in the internal energy of the system for some (not necessarily small) change of state is made up of three parts: the "useful" work δA done on the body by the environment, the work of compression done by the environment and the heat received from the environment. As was pointed out, owing to the great size of the environment, its temperature, pressure and chemical potentials may be considered to be constant; hence the work done by it on the system is $p_0 \Delta V_0$, the work done by environmental particles is $\sum_{i=1}^{n} \mu_i^0 \Delta N_i^0$, and the heat given out by it is $-T_0 \Delta S_0$ (letters with sub- and superscripts refer to the environment and those without them to the system). Thus we have

$$\Delta U = \delta A + p_0 \, \mathrm{d}V_0 + \sum_{i=1}^{n} \mu_i^0 \, \mathrm{d}N_i^0 - T_0 \, \mathrm{d}S_0. \tag{1.1}$$

Since the volume of the environment and the system together remains constant $\Delta V_0 = -\Delta V$. From the conditions of conserving the number of particles and the invariability of chemical potentials we have for each sort of particle $\Delta N_i^0 = -\Delta N_i;\ i = 1, \ldots, n$. Furthermore, in accordance with the law of increase of entropy we have $\Delta S + \Delta S_0 > 0$; thus $\Delta S_0 > -\Delta S$. Hence, from Eq. (1.1) we obtain:

$$\delta A \geq \Delta U - T_0 \Delta S + p_0 \Delta V - \sum_{i=1}^{n} \mu_i^0 \Delta N_i. \tag{1.2}$$

Equality is attained for a reversible process. Thus we again conclude that the transition occurs with the minimum expenditure of work (and hence the inverse transition with the maximum work done) if it is reversible. The value of the minimum work is given by the formula

$$\delta A_{\min} = \Delta(U - T_0 S + p_0 V - \sum_{i=1}^{n} \mu_i^0 N_i). \tag{1.3}$$

(T_0, P_0 and μ_i^0 being constant, can be taken under the operator Δ), i.e. this work is equal to the change in the quantity $U - T_0 S + p_0 V - \sum_{i=1}^{n} \mu_i^0 N_i$. For maximum work the formula must obviously be rewritten with the opposite sign, $\delta A_{\max} = -\delta A_{\min}$, since the initial and final states are interchanged. Here it is necessary to note that the operator Δ is the difference between values at the first and zero states, and not the opposite. For instance, $\delta A_{\min} = \Phi_1 - \Phi_0 = \Delta\Phi$ (here Φ embraces all terms that are within brackets in Eq. (1.3)). Then $\delta A_{\max} = \Phi_0 - \Phi_1 = -\Delta\Phi$.

If, at each instant during the process, the system is in a dynamic equilibrium (but not, of course, in thermodynamic equilibrium with the environment), then for an infinitesimal change of state, Eq. (1.3) may be written in a different form. Substituting the expression for the full differential of internal energy (Gibbs' equation: $dU = T\,dS - p\,dV + \sum_{i=1}^{n} \mu_i\,dN_i$) into the differential form of Eq. (1.3), $dA_{min} = dU - T_0\,dS + p_0 V - \sum_{i=1}^{n} \mu_i^0\,dN_i$, we obtain:

$$dA_{min} = |dA_{max}| = (T - T_0)dS - (p - p_0)dV + \sum_{i=1}^{n} (\mu_i - \mu_i^0)dN_i. \qquad (1.4)$$

There are two special cases. If the volume and the temperature of the system remain constant and the latter is equal to that of the environment, then from Eq. (1.3) we have

$$\delta A_{max} = -\Delta\left(U - TS + \sum_{i=1}^{n} \mu_i N_i\right) = -\Delta F, \qquad (1.5)$$

i.e. the maximal work which can be performed by the system tending to an equilibrium with its environment is equal to the decrease in the free energy of the system. If the temperature and pressure are constant, and $T = T_0, p = p_0$ we have

$$\delta A_{max} = -\Delta G, \qquad (1.6)$$

i.e. the system doing the work on the environment decreases its corresponding thermodynamic potential. Note that both conditions are typical for biochemical reactions and the ecological system.

Since $\delta A_{max} \leq 0$, $\Delta(U - T_0 S + p_0 V - \sum_{i=1}^{n} \mu_i^0 N_i) \leq 0$. This means that as a result of processes undergone by the system the quantity $U - T_0 S + p_0 V - \sum_{i=1}^{n} \mu_i^0 N_i$ will decrease so that at equilibrium it will be a minimum.

In particular, during spontaneous transition "1"→"0" at constant temperature $T = T_0$ and constant pressure $p = p_0$ the thermodynamic potential G of the system decreases, and during transitions at constant temperature $T = T_0$ and volume, its free energy F also decreases. Note that the deduction given here does not assume that the temperature and the volume (or pressure) of the system remain constant during the whole of the transition: it can be asserted that the thermodynamic potential (or free energy) of the system decreases in every process in which the temperature and pressure (or volume) are the same (and are equal to those of the environment) at the beginning and the end, even if they change at some stage of the process. For this it is sufficient that the process (transition) is quasi-stationary (as discussed above).

Yet, another thermodynamic meaning can be given to the maximum work. Let S_t be the total entropy of the supersystem (the system together with its environment). If the system is in equilibrium with the environment then S_t is a function of their total internal energy U_t: $S_t = S_t(U_t)$. If the system is not in equilibrium with the environment then their total entropy (for the same value of their total energy U_t) differs from the value $S_t = S_t(U_t)$ by some quantity $\delta S_t < 0$.

In Fig. 5.2 the solid line represents the function $S_t = S_t(U_t)$ and the vertical line *ab* the quantity $-\delta S_t$. The horizontal line *bc* is the change in the total energy during the transition from the state of thermodynamic equilibrium with the environment to the state represented by the point *b*. In other words, this line represents the minimum work an environment must do to bring the system from a state of equilibrium with the environment to the given state.

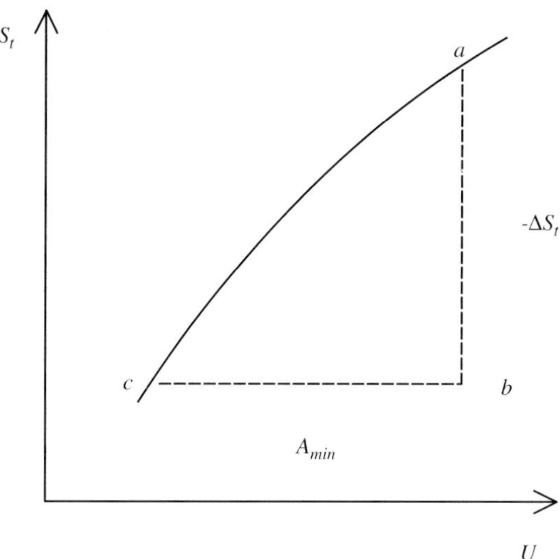

Fig. 5.2. Entropy versus internal energy.

The state of equilibrium we are referring to (the point c in Fig. 5.2) obviously does not coincide with the state of equilibrium corresponding to the given value, U_t (the point a).

Since the system represents only a very small part of the whole supersystem, the processes taking place in it lead to relatively insignificant changes in the total energy and entropy. From Fig. 5.2 it thus follows that $\Delta S_t \approx -\{dS_t(U_t)/dU_t\}\delta A_{min} = \{dS_t(U_t)/dU_t\}\delta A_{max}$, where the derivative dU_t/dS_t is the equilibrium temperature of the supersystem, i.e. the temperature T_0 of the environment. Thus,

$$\delta A_{max} = T_0 \Delta S_t = T_0(S_t - S_t^{eq}) = \Delta U - T_0 \Delta S + p_0 \Delta V - \sum_{i=1}^{n} \mu_i^0 \Delta N_i. \tag{1.7}$$

This equality expresses how much the entropy of the closed supersystem differs from its maximum possible value if the system is not in equilibrium with its environment, where ΔU, ΔS, ΔV and ΔN_i are, respectively, the differences between the energy, entropy, volume, the number of particles in the system and their values in the state of total thermodynamic equilibrium.

We can see that when we want to describe the behaviour and properties of an open system which is connected very closely with its surrounding environment (this is a typical definition of ecosystem), then the concept of maximum work considered above plays a very important role. For this reason this work has been honoured with a special term: *exergy*. Exergy (Ex) is defined as the amount of work (= entropy-free energy) a system can perform when it is brought into thermodynamic equilibrium with its environment, i.e. $|A_{max}| = \text{Ex}$ (Jørgensen et al., 1999). The state of the environment is usually named as the *reference state*. Note that exergy is, therefore, not a state variable. For instance, as a reference state we can select the same system but at thermodynamic equilibrium, i.e. such that all components are inorganic and at the highest oxidation state if sufficient oxygen is present (nitrogen as nitrate, sulphur as sulphate and so on). The reference state in this case will correspond to the ecosystem without life forms and with all chemical energy utilised or as an "inorganic soup". This usually implies that we consider $T = T_0$ and $p = p_0$, which means that exergy becomes equal to the difference of Gibb's potential (free energy) of the system and the same system is at thermodynamic equilibrium. Notice also that exergy depends on intensive state variables of the environment.

5.2. What is exergy? Different interpretations of the exergy concept

Exergy is not conserved unless entropy-free energy is transferred which implies that the transfer is reversible. All processes in reality are, however, irreversible, which means that exergy is lost (and entropy is produced). Loss of exergy and production of entropy are two different descriptions of the same reality, namely, that all processes are irreversible, and we unfortunately always have some loss of energy forms which can do work, to energy forms which cannot do work (heat at the temperature of the environment) (see also Jørgensen, 2001b). So the formulation of the Second Law, using exergy is: *All real processes are irreversible which implies that exergy is inevitably lost. Exergy is not conserved*; while energy, of course, is conserved by all processes according to the First Law. It is therefore wrong (as already mentioned briefly) to speak of the energy efficiency of an energy transfer because it will always be 100%; rather, the exergy efficiency is of interest because it expresses the ratio of useful energy to total energy, which is always less than 100% for real processes. All transfers of energy imply that exergy is lost because energy is transformed into heat at the temperature of the environment.

It is, therefore, of interest for all environmental systems to set up an exergy balance in addition to an energy balance. Our concern is loss of exergy, because here "first class energy" which can do work is lost and replaced by "second class energy" (heat at the temperature of the environment) which cannot do work. So, as presented in Chapter 3, the particular properties of heat and of temperature are a measure of the movement of molecules, and give limitations in our possibilities to utilise energy to do work. Due to these limitations, we have to distinguish between exergy, which can do work, and energy, which cannot do work. The latter may be called *anergy* (see, for instance, Cerbe and Hoffmann, 1996). Therefore, the energy can be represented as a sum of two items:

$$\text{Energy} = \text{exergy} + \text{anergy}. \tag{2.1}$$

In accordance with the Second Law, *anergy* is always positive for any process.

It seems more useful to apply exergy than entropy to describe the irreversibility of real processes as it has the same unit as energy and is an energy form, while the definition of entropy is more difficult to relate to concepts associated with our usual description of reality. In addition, entropy is not clearly defined for systems "far from thermodynamic equilibrium", particularly for living systems (see, for instance, Tiezzi, 2003). Moreover, it should be mentioned that the self-organising abilities of systems are strongly dependent on temperature, as discussed in Jørgensen et al. (1999). Exergy takes the temperature into consideration as the definition shows, while entropy does not. The negative entropy is as discussed in Chapters 2 and 3, i.e. it does not express the ability of the system to do work (we may call it "the creativity" of the system as creativity requires work), but exergy becomes a good measure of "the creativity", which is increasingly proportional with the temperature. Furthermore, exergy facilitates the differentiation between low-entropy energy and high-entropy energy, as exergy is entropy-free energy. These expressions were not properly defined in Chapter 3.

If the two systems have different temperatures, the entropy lost by one system is not equal to the entropy gained by the other system, while the exergy lost by the first system is equal to the exergy transferred and equal to the exergy gained by the other system, provided that the transformation is not accompanied by any loss of exergy. In this case it is obviously more convenient to apply exergy than entropy.

The exergy of the system measures the contrast—i.e. the difference in free energy if there is no difference in pressure and temperature, as may be assumed for an ecosystem or an environmental system and its environment—with the surrounding environment. If the system is in equilibrium with the surrounding environment the exergy is, of course, zero.

Since the only way to move systems away from equilibrium is to perform work on them, and since the available work in a system is a measure of the ability to do it, we have to distinguish between the system and its environment or thermodynamic equilibrium, alias, for instance, an inorganic soup. Therefore it is reasonable to use the available work, i.e. the exergy, as a measure of the distance from thermodynamic equilibrium.

As we know that ecosystems (due to the through-flow of energy) have the tendency to move away from thermodynamic equilibrium losing entropy or gaining exergy and information, we can put forward the following proposition of relevance for ecosystems: *Ecosystems attempt to develop towards a higher level of exergy.*

It is interesting in this context to draw a parallel with the discussion of the development of entropy for the entire Universe. The classic thermodynamic interpretations of the Second Law of Thermodynamics predict that the Universe will develop towards "the heat death", where the entire Universe will have the same temperature, no changes will take place and a final overall thermodynamic equilibrium will be the result. This prediction is based upon the steady increase of the entropy according to the Second Law of Thermodynamics: the thermodynamic equilibrium is the attractor. It can, however, be shown (see Frautschi, 1988; Jørgensen, 2002b) that we are moving away from the thermodynamic equilibrium at a high rate due to the expansion of the Universe.

Due to the incoming energy of solar radiation an ecosystem is able to move away from thermodynamic equilibrium—i.e. the system evolves and obtains more information and organisation.

The ecosystem must produce entropy for maintenance, but the low-entropy energy flowing through the system may be able to more than cover this production of disorder, resulting in an increased order or information of the ecosystem.

One of the main concepts in the thermodynamics of an open system is the decomposition of the total production of entropy into two items: the rate of entropy exchange between the system and its environment and the internal entropy production by the system: $\mathrm{d}S/\mathrm{d}t = \mathrm{d_e}S/\mathrm{d}t + \mathrm{d_i}S/\mathrm{d}t$. An analogous relation can be written for the time derivative of exergy

$$\frac{\mathrm{d(Ex)}}{\mathrm{d}t} = \frac{\mathrm{d_e(Ex)}}{\mathrm{d}t} + \frac{\mathrm{d_i(Ex)}}{\mathrm{d}t}, \tag{2.2}$$

where, in fact, the real process of the system evolution can be represented as the composition of two processes: the first is a forced movement *from* thermodynamic equilibrium when the exergy increases, $d_e(\text{Ex})/dt > 0$, and the second when the system spontaneously moves *to* thermodynamic equilibrium. In the course of the latter the exergy decreases, $d_i(\text{Ex})/dt < 0$, i.e. it is lost. The loss of exergy is a result of such spontaneous irreversible processes within the system as diffusion, heat conduction, turbulence and chemical reactions (as presented in Chapter 3). Eq. (2.2) shows among other things that systems can only maintain a non-equilibrium steady state by compensating the loss of exergy with a positive exergy inflow. Such an inflow induces order into the system. In ecosystems the ultimate exergy inflow comes from solar radiation, and the order induced is, for example, biochemical molecular order. If $d_i(\text{Ex})/dt > |d_e(\text{Ex})/dt|$ (the exergy loss within the system), the system has surplus exergy input, which may be utilised to construct further order in the system, or as Prigogine (1980) calls it: *dissipative structure*. The system will thereby move further away from the thermodynamic equilibrium. Evolution shows that this situation has been valid for the ecosphere on a long-term basis. In spring and summer, ecosystems are in the typical situation that $d_e(\text{Ex})/dt$ exceeds $|d_i(\text{Ex})/dt|$. If $d_i(\text{Ex})/dt < |d_e(\text{Ex})/dt|$, the system cannot maintain the order already achieved, but will move closer to the thermodynamic equilibrium, i.e. it will lose order. This may be the situation for ecosystems during autumn and winter or due to environmental disturbances.

5.3. Thermodynamic machines

The exergy of a system is more probably a measure of the *ability* to perform some work than of the work itself. In order to realise the work, i.e. to bring about a reversible transfer of energy, and thus obtain the maximum work, it is necessary to introduce some auxiliary body (*working body*), which performs a certain reversible, cyclic process. For instance, this process may take place in such a way that the body and the system, between which a direct exchange of energy takes place, are at the same temperature. That is to say, the working body at temperature T is brought into contact with the system at the same temperature T and isothermally gains from it a certain energy dU. Then it is adiabatically cooled to temperature T_0 and gives up this energy at this temperature to the environment at temperature T_0 and finally returns adiabatically to its initial state. During this process the working body does work on external objects (environment). The cyclic process we have described is called the *Carnot Cycle*. Turning now to the calculation of the resulting maximum work we note that the working body need not be considered since it returns to its initial state.

In the Carnot Cycle only the transfer of energy from a warmer system to a colder environment is considered. Therefore, in Eq. (1.4) the terms connected with compression and expansion and the chemical processes can be discarded: $|dA_{\max}| = (T - T_0)dS$. Since $dS = dU/T$, $|dA_{\max}| = [(T - T_0)/T]dU$. The ratio of the work done to the amount of energy used is called the *efficiency coefficient* (or simply *efficiency*) η. The maximum efficiency is equal to

$$\eta_{\max} = \frac{|dA_{\max}|}{dU} = \frac{d(\text{Ex})}{dU} = \frac{T - T_0}{T}. \qquad (3.1)$$

A more convenient value is the so-called *coefficient of utility* u_t, defined as the ratio of the work done to the maximum work which could be obtained in the given conditions. It is obvious that $u_t = \eta/\eta_{max}$. Note that efficiency may be equal to 100% only if $T_0 = 0$, i.e. at absolute zero of temperature. Even the temperature in Space is about 2.7 K!

It would be very useful to set up an exergy balance for any system. It is hardly possible to give general equations to be applied for all systems. However, it may be possible to make some general considerations on the development of exergy balances for a system. Usually this will not involve that the exergy of information is included in the analysis (about the exergy of information, see Sections 5.5 and 5.6), but that the input of first class energy in the form of, for instance, fossil fuel, electricity or other energy forms is calculated and compared with the exergy of the outputs and (or) with the loss of exergy in the form of non-useful heat from the system. Useful heat is always associated with a higher temperature than the environment or the reference state—compare this condition with the Carnot Cycle, where the temperature difference between the warm and cold reservoirs determines the efficiency. It should always be considered whether non-useful heat could be made useful, but this would of course require that the heat can utilise the temperature difference.

A machine based on the Carnot Cycle is naturally named the *thermal* machine. Since solar radiation can be considered as a form of heat, formally we can define a *solar* machine as some kind of thermal one. Following this logic, the working body of the solar machine is "hot" photons with $T = T_s = 5770$ K. These photons have to "heat" the machine up to the same temperature. Under such a condition the transfer of energy from the Sun to the solar machine is possible. In the process of adiabatic cooling down to the temperature of Space ($T_0 = 2.7$ K) the solar machine performs work. It is obvious that its efficiency is almost 100%. An ideal solar machine made from a super refractory alloy, located on board some star cruiser, could attain this efficiency. The second possibility is that the machine consists of a photon gas or plasma, such as may exist on the planet *Solaris* created by the Polish writer Stanislav Lem.

Unfortunately (or perhaps fortunately), photons have been able to (i.e. they really have done it) heat our planet only up to ~290 K, and the effective temperature of photons penetrating through the atmosphere is approximately equal to this value. The climatic machine, which is undoubtedly a solar machine, operates with these photons as a source of energy, the corresponding entropy of which is equal to q_{in}/T_E ($q_{in} \approx 240$ W/m^2 is the energy flux of incoming solar radiation, $T_E \approx 290$ K is Earth's mean temperature) but not q_{in}/T_s. Note that the working body of the atmospheric climatic machine is the air, not photons, and Gibbs has pointed to this!

Let us estimate the efficiency of the climatic machine by taking into account all these considerations. Here we follow conceptually a very interesting work by Petoukhov (1985). The machine includes the troposphere, stratosphere, ocean and the effectively interacting layer of land, i.e. it covers the whole planetary surface Ω. The surface is divided into two (not necessary compact) subsets Ω_+ and Ω_-. In the first domain the system gets energy, and in the second spontaneously loses it. It is natural that the temperature in the first domain, T, is higher than in the second, and also higher than Earth's mean temperature (see also Section 11.6). Note that if the temperature is constant over the whole planet then the climatic machine does not work. Since only Ω_+-domains assimilate solar radiation energy, and in addition the assimilation occurs at the temperature up to which the system is heated by incoming short-wave radiation and at which long-wave radiation is outgoing, dS in the expression for infinitesimal exergy, $|dA_{max}| = d(Ex) = (T - T_0)dS$, is represented as

$$dS = \int_{\Omega_+} \frac{(q_{in} - q_{out})}{T(\omega)} d\omega = \frac{1}{T} \int_{\Omega_+} (q_{in} - q_{out}) d\omega. \tag{3.2}$$

Here T is the mean temperature of the "hot" domain, $T \approx 300$ K, and q_{in} and q_{out} are incoming and outgoing (thermal) radiation fluxes at the system's upper boundary. Since a temperature of the environment T_0 is the temperature of tropopause, where a jump in the temperature gradient takes place, then it is natural to set $T_0 \approx 220$ K. Therefore, the efficiency of the climatic machine *in relation to the absorbed energy* $Q_{abs} = \int_{\Omega_+} (q_{in} - q_{out}) d\omega$ will be equal to the Carnot efficiency $\eta_{max} = (T - T_0)/T = (300 - 220)/300 \approx 26.7\%$. However, the real efficiency, *in relation to the total solar energy* $Q_s = \int_{\Omega} q_{in}\, d\omega$, will be significantly lower; $\eta = u_t \eta_{max}$ where the utility coefficient $u_t = Q_{abs}/Q_s$. Using the mean estimation of these values (Lorenz, 1967): $\int_{\Omega_+} (q_{in} - q_{out}) d\omega / \int_{\Omega} d\omega \approx 15\ \mathrm{W/m^2}$ and $\int_{\Omega} q_{in}\, d\omega / \int_{\Omega} d\omega \approx 240\ \mathrm{W/m^2}$ we get $u_t = 6.25\%$ and $\eta \approx 1.67\%$. It is interesting that this value of efficiency of the Earth's climatic machine is very close not only to Petoukhov's estimation, $\eta \approx 1.25$–1.7%, but also to the maximal empirical estimations cited by Lorenz.

The other type of solar machine is a green leaf, which in the process of photosynthesis creates new biomass. In this case the working body comprises the chemical molecules (chlorophyll, ATP, etc.) transferring the energy of photons into leaves. Since the reaction of photosynthesis is exogenous, the heat warms the system—the "leaf"—which then can be considered as a thermal machine with maximum efficiency $\eta_{max} = (T_{leaf} - T_{air})/T_{leaf}$. The mean daily difference between the temperatures of leaves and the surrounding air under summer conditions in the temperate latitudes is of the order of several degrees (about 3–7°C; Budyko, 1977). To determine it we set that the difference is equal to 5°C, and the temperature of the surrounding air $T_0 = 20$°C, then $\eta_{max} = 5/298 \approx 1.68\%$ (see also Chapter 10).

Let us consider the so-called *chemical* machine, assuming that the work is performed by the system only by means of a change in the number of particles in the field of chemical potentials. Then from Eq. (1.4) we get

$$|dA_{max}| = d(Ex) = \sum_{i=1}^{n} (\mu_i - \mu_i^0) dN_i. \tag{3.3}$$

A significant difference between this machine and the thermal one is that it cannot be characterised by an efficiency coefficient like $\eta_{therm} = (T - T_0)/T$, since the chemical potentials are not at zero. Nevertheless, we keep in mind that the change of internal energy in the case of chemical reactions is $dU = \sum_{i=1}^{n} \mu_i\, dN_i$, then the maximum efficiency of the chemical machine can be defined as

$$\eta_{max} = \frac{\sum_{i=1}^{n} (\mu_i - \mu_i^0) dN_i}{\sum_{i=1}^{n} \mu_i\, dN_i}. \tag{3.4}$$

As an example, we try to estimate the efficiency of such a global chemical machine as the GCC. We assume that the GCC is described only by one variable: the amount of carbon in the living biomass of global vegetation, C. This is close to reality since the global vegetation is the most important part of the GCC. In the initial stage, the environment does work on the system (the GCC) by increasing the equilibrium amount of carbon C_0 up to C, so that $C_0 \rightarrow C = C_0 + \text{NPP}$ where NPP is the annual net production of global vegetation. In the final stage, as a result of spontaneous destructive processes, an equivalent amount of carbon is returned into the environment. Note that the latter takes place only for stationary processes.

We already know that the chemical potential of ith substance with concentration (activity) $[c_i]$ is $\mu_i = \mu_i(0) + RT \ln[c_i]$ where $\mu_i(0)$ is a constant depending on the origin of the substance and its temperature. The concentration may be expressed in any units that are able to describe the total amount of substance in the system, but it is important that the system's volume does not change. We also assume that the GCC is an isothermal process. Then substituting the expression for chemical potential into Eq. (3.4) we have

$$\eta_{\max} = \frac{RT_0 \ln(C/C_0)\mathrm{d}C}{[\mu(0) + RT_0 \ln C]\mathrm{d}C} = \frac{\ln(1 + \text{NPP}/C_0)}{a + \ln(C_0 + \text{NPP})} < \frac{\ln(1 + \text{NPP}/C_0)}{\ln(C_0 + \text{NPP})}, \tag{3.5}$$

where $a = \mu(0)/RT_0 > 0$. The standard estimations for $C_0 \approx 610$ GtC and NPP $\approx$ 60 GtC (den Elzen et al., 1995; 1 Gt $= 10^9$ t), and from Eq. (3.5) we get $\eta_{\max} \approx \ln 1.1/\ln 660 \approx 1.44\%$.

We would like to call your attention to an interesting coincidence: the efficiencies of *natural* machines, such as climatic and photosynthetic machines, and the GCC are all very low and close to each other: approximately 1.5%. This may just be a coincidence, but perhaps it shows some common property of natural systems on our planet?

Now we go from the global to the regional scale. Szargut (1998) calculates the exergy efficiency of technological processes by adding the exergy loss due to the emission of waste to the environment. Application of this approach is indeed recommended, as a highly exergy-effective process generally should not be used if it produces waste products which can harm the environment. As the harm can be expressed as a loss in exergy, the overall exergy efficiency also considering this loss of exergy should, of course, be used in the selection of technologies.

When contaminants, e.g. heavy metals, are widely dispersed, exergy is lost. When lead was used all over the world in petrol to obtain a higher octane number, in the order of 2×10^8 kg of lead was widely dispersed every year. It was even found in the ice pack of Greenland! This "dispersed" state can be considered as a thermodynamic equilibrium. Since the exergy loss is equal to the exergy accumulated within lead ore, $\Delta(\text{Ex}) = RT_0 \times \ln(c_{\text{ore}}/c_{\text{dis}})\Delta C$, where c_{ore} is the lead concentration in ore (0.05 kg/kg ore), c_{dis} a typical concentration of lead in the environment after dispersion (1 μg/kg soil), and ΔC is the total amount of dispersed lead (in moles). If we presume $T_0 = 300$ K then the annual exergy lost can be found as:

$$\Delta\text{Ex} = 8.3 \times 300 \ln(0.05/10^{-9})(2 \times 10^{11}/207) \approx 42.5 \text{ GJ}. \tag{3.6}$$

The exergy efficiencies of many of our production processes are surprisingly low. There has been an increasing interest in reducing the exergy consumption with increasing exergy (energy) cost and decreasing access to low-cost energy. The exergy consumption of a good modern refrigerator is, for instance, only 30–40% of a 20-year-old refrigerator with the same capacity. By more comprehensive use of exergy balances, it will be possible to indicate how to reduce the overall exergy consumption in developed countries very significantly. If, on average, we could increase exergy efficiencies from 40 to 60%, it would imply a reduction of our energy consumption and the associated emissions of pollutants by 33.3%.

5.4. Exergy far from thermodynamic equilibrium

Up to this point we have implicitly assumed that a new non-equilibrium state is located not far from thermodynamic equilibrium. Moreover, the transition to the new state was quasi-stationary, so that the system was at a dynamic equilibrium with the environment. In this case, the infinitesimal change of exergy, d(Ex), was very close to the difference of exergies between the current and initial states, $|\delta A_{max}| = \delta(\text{Ex})$. However, generally speaking it is possible for a system to be moved far from thermodynamic equilibrium where this closeness is not present anymore. How can exergy be calculated in this case? Here we shall follow the method used in the work (Svirezhev, 2001b).

Let the transition be *chemical*, i.e. it occurs in the field of chemical potentials when work is done against them by means of the change in the number of particles. We assume that the infinitesimal change of exergy described by Eq. (3.3) takes place in an infinitesimal time dt, so that the environment is not able to change. As is evident from equality $\mathrm{d}N_i^0 = -\mathrm{d}N_i = -(\mathrm{d}N_i/\mathrm{d}t)\mathrm{d}t$, if derivative $(\mathrm{d}N_i/\mathrm{d}t)$ is finite then for small dt the value of $\mathrm{d}N_i^0$ can be also small. This condition is sufficient for relinquishing the previous assumption about the "great" environment, in particular, considering that several N_i^0 are small. Thus, we can fully give up the above-mentioned local time condition; to do this we must replace differentials in Eq. (3.3) by the corresponding time derivatives:

$$\frac{\mathrm{d(Ex)}}{\mathrm{d}t} = \sum_{i=1}^{n} (\mu_i - \mu_i^0)\frac{\mathrm{d}N_i}{\mathrm{d}t}. \tag{4.1}$$

Keeping in mind the definition of chemical potentials $\mu_i = \mu_i(0) + RT \ln N_i,\ i = 1, \ldots, n$ where N_i can be considered as molar concentrations of corresponding chemical substances, and R is the gas constant, we re-write Eq. (4.1) as $(T = T_0)$

$$\frac{\mathrm{d(Ex)}}{\mathrm{d}t} = RT_0 \sum_{i=1}^{n} \ln\frac{N_i}{N_i^0}\frac{\mathrm{d}N_i}{\mathrm{d}t}. \tag{4.2}$$

There is one more argument on behalf of this replacement. Since the order of change for the chemical potential μ_i^0 with respect to N_i^0 is logarithmic, should the number N_i^0 change, the value of the chemical potential changes significantly more slowly.

By integrating both sides of Eq. (4.2) with respect to time and taking into account that $\mathrm{Ex}(t_0) = 0$ we get

$$\mathrm{Ex}(t) = RT_0 \int_{t_0}^{t} \sum_{i=1}^{n} \ln\frac{N_i(t)}{N_i^0}\frac{\mathrm{d}N_i}{\mathrm{d}t}\mathrm{d}t = RT_0 \sum_{i=1}^{n} \int_{t_0}^{t} \ln\frac{N_i(t)}{N_i^0}\frac{\mathrm{d}N_i}{\mathrm{d}t}\mathrm{d}t$$

$$= RT_0 \sum_{i=1}^{n} \int_{N_i^0}^{N_i} (\ln N_i - \ln N_i^0)\mathrm{d}N_i = RT_0 \sum_{i=1}^{n} \left[N_i \ln\frac{N_i}{N_i^0} - (N_i - N_i^0) \right]. \quad (4.3)$$

Since the gas constant is measured in J/mol K, the number of particles has to be measured in moles. Then exergy is measured in energy units (joules). If we wish—as before—to use the number of particles we must write Boltzmann's constant k instead of R. As a rule, later on we shall omit the factors RT_0, and kT_0, remembering them only when it is necessary.

One can see that $\mathrm{Ex}(t) > 0$ for any $N_i > 0$, except $N_i = N_i^0$, $i = 1, \ldots, n$, when $\mathrm{Ex} \equiv 0$. Also, the value of exergy is determined only for the current and initial states (for this reason the latter will be termed a *reference state*) and does not depend on characteristics of this transition.

If $|\mathrm{d}N_i| \ll N_i^0$, i.e. we do not go far from thermodynamic equilibrium, then Eq. (4.2) is written as

$$\mathrm{d(Ex)} = \sum_{i=1}^{n} \ln\frac{N_i^0 + \mathrm{d}N_i}{N_i^0}\mathrm{d}N_i \approx \sum_{i=1}^{n} \frac{(\mathrm{d}N_i)^2}{N_i^0} \geq 0. \quad (4.4)$$

This means that in moving from thermodynamic equilibrium the system *accumulates* exergy. It is natural that in the course of inverse spontaneous transition to the equilibrium the system *loses* exergy. However, inequality (4.4) must not necessarily be fulfilled far from equilibrium. Later on we shall show that some additional conditions are necessary for this.

It is very important that this definition of exergy does not at all require that the system must be in a dynamic equilibrium with the environment during the whole path, as is required in Section 5.1. Equilibrium at the beginning of movement is sufficient, and formally it is not necessary that the equilibrium is a thermodynamic one. Moreover, it is not necessary that the starting point is a dynamic equilibrium: it is sufficient if the system has the equilibrium somewhere else. We shall discuss this (and many other things) below.

Let us consider the following model of a "chemical" system. We assume that the dynamics (kinematics) of the model are described by the system of ordinary differential equations for concentrations (specific numbers) $\mathbf{c} = \{c_i \geq 0\}$, $i = 1, \ldots, n$

$$\frac{\mathrm{d}c_i}{\mathrm{d}t} = f_i(c_1, \ldots, c_n; \alpha_1, \ldots, \alpha_m), \quad (4.5)$$

where the vector of parameters $\boldsymbol{\alpha}$ describes either the state of the environment or the characteristics of exchange between the system and its environment. However, very often we do not know the concrete form of the right sides in Eq. (4.5); we only know that system (4.5) has a single stable equilibrium $\mathbf{c}^* = \{c_i^* \geq 0\}$, where either its stability (see Chapter 6) or its position depends on parameters. Therefore, let us consider two cases.

Case 1. Let the parameters be $\boldsymbol{\alpha}^0$, and the system is in the *stable* equilibrium $\mathbf{c}^*(\boldsymbol{\alpha}^0) = \mathbf{c}^0$. The change $\boldsymbol{\alpha}^0 \to \boldsymbol{\alpha}^1$ leads to the loss of its stability, but its position is not changed. We identify the stable state $\mathbf{c}^0$ with thermodynamic equilibrium of the system with its environment, when $\mathbf{c}^0 = \mathbf{c}^{\text{env}}$ (the latter is the vector of concentrations in the environment). Then the change $\boldsymbol{\alpha}^0 \to \boldsymbol{\alpha}^1$ may be interpreted as the following: Ecodemon opens a lid at time t_0, and some additional flows of energy come into the system by performing some work on it. The system starts to go away from the state $\mathbf{c}^0$. From the viewpoint of dynamical theory it implies that any small fluctuation leads the system out of the equilibrium, and $\mathbf{c}^0 \to \mathbf{c}(t)$ for $t > t_0$. The transition is accompanied with a dissipation of energy, which is calculated as (see Chapter 3)

$$\text{Diss}(\mathbf{c}^0 \to \mathbf{c}) = RT_0 \int_{t_0}^{t} \sum_{i=1}^{n} \ln\frac{c_i^0}{c_i}\frac{\mathrm{d}c_i}{\mathrm{d}t}\mathrm{d}t = RT_0 \sum_{i=1}^{n} \int_{c_i^0}^{c_i} \ln\frac{c_i^0}{c_i}\mathrm{d}c_i$$

$$= -RT_0 \sum_{i=1}^{n} \left[c_i \ln\frac{c_i}{c_i^0} - (c_i - c_i^0) \right] < 0. \tag{4.6}$$

Here the expression $RT_0 \ln[c_i^0/c_i(t)]$ gives the value of affinity for the transition $c_i^0 \to c_i(t)$ where $c_i^0 = \text{constant}$. This is the affinity of diffusion processes transporting a matter from the environment into the system.

The negativeness of $\text{Diss}(\mathbf{c}^0 \to \mathbf{c})$, i.e. the total value of dissipated energy, implies that the system transition from some initial state to a current one is not spontaneous, but *forced*. We know that all spontaneous processes of return to a stable equilibrium after relatively small internal fluctuations are always accompanied by an increase in entropy. In this case entropy of our open system decreases, since the system consumes free energy from the environment.

In accordance with one of the definitions of exergy, it is equal to the absolute value of the entropy decrease multiplied by the temperature of environment, i.e. $\text{Ex}(\mathbf{c}, \mathbf{c}^0) = -\text{Diss}(\mathbf{c}^0 \to \mathbf{c})$, where the equilibrium $\mathbf{c}^0$ is considered as a reference state. For calculation of exergy we used formula Eq. (4.3), in which N_i and N_i^0 were replaced by c_i and c_i^0, respectively. We also replaced the upper index ($*$) by (0).

We can also say that the system moving far from the equilibrium $\mathbf{c}^0$ *accumulates* the exergy, so that at moment t its exergy, i.e. its potential ability to perform some useful work, is equal to $\text{Ex}(\mathbf{c}(t), \mathbf{c}^0) = -\text{Diss}(\mathbf{c}^0 \to \mathbf{c}(t))$.

Let us assume that at some moment $t = t_1$ our Ecodemon closes the lid, by the same token stopping the energy flow, i.e. it switches $\boldsymbol{\alpha}^1$ again to $\boldsymbol{\alpha}^0$, so that the equilibrium $\mathbf{c}^0$ becomes stable again. The system starts to move to the state $\mathbf{c}^0$, but the movement is spontaneous, not forced, since the inflow of free energy is stopped. Entropy of the system increases. In the course of the transition $\mathbf{c}(t_1) = \mathbf{c}^1 \to \mathbf{c}^0$ the system can do some work on the environment, decreasing the entropy of the latter. However, if to take into account that at time t_1 the system has accumulated storage of low-entropy "good" energy in the form of exergy, which will be dissipated when $\mathbf{c}^1 \to \mathbf{c}^0$, then the transition can be considered as a forced one, and the dissipation of energy is calculated analogously (as was shown above). Since the affinity for $\mathbf{c} \to \mathbf{c}^0$ is $RT_0 \ln[c_i/c_i^0]$,

$$\mathrm{Diss}(\mathbf{c}^1 \to \mathbf{c}^0) = RT_0 \int_{t_1}^{\infty} \sum_{i=1}^{n} \ln\frac{c_i}{c_i^0}\frac{\mathrm{d}c_i}{\mathrm{d}t}\mathrm{d}t = RT_0 \sum_{i=1}^{n} \int_{c_i^1}^{c_i^0} \ln\frac{c_i}{c_i^0}\mathrm{d}c_i$$

$$= -RT_0 \sum_{i=1}^{n} \left[c_i^1 \ln\frac{c_i^1}{c_i^0} - (c_i^1 - c_i^0) \right]. \qquad (4.7)$$

By comparing Eqs. (4.6) and (4.7) we see that $\mathrm{Diss}(\mathbf{c}^1 \to \mathbf{c}^0) = \mathrm{Diss}(\mathbf{c}^0 \to \mathbf{c}^1) = -\mathrm{Ex}(\mathbf{c}^1, \mathbf{c}^0)$. All these equalities imply that

1. Work done on the system by its external environment in the process of forced transition $\mathbf{c}^0 \to \mathbf{c}^1$ is equal to $|A_{01}| = -\mathrm{Diss}(\mathbf{c}^0 \to \mathbf{c}^1)$.
2. In the course of the transition the system accumulates exergy, which is equal to $\mathrm{Ex}(\mathbf{c}^1, \mathbf{c}^0) = |A_{01}| = -\mathrm{Diss}(\mathbf{c}^0 \to \mathbf{c}^1)$.
3. When the system is closed, it returns spontaneously to the stable state $\mathbf{c}^0$; the return is accompanied by the dissipation of exergy in the process of performing work by the system. The transition must be finished in the state $\mathbf{c}^0$: namely, at this moment the work $|A_{10}| = -\mathrm{Diss}(\mathbf{c}^1 \to \mathbf{c}^0) = \mathrm{Ex}(\mathbf{c}^1, \mathbf{c}^0)$ and the whole storage of exergy will become exhausted.

However, there are some remarks about the last point. Strictly speaking, the equilibrium $\mathbf{c}^0$ is attained at infinite time; at finite time the system can come into some vicinity of the equilibrium so that the cycle cannot be closed. As a result, the accumulated exergy does not dissipate fully, and its small part remains within the system before a new cycle starts. In fact, entropy of environment increases by the corresponding quantity. In order to "save" the Energy Conservation Law we can say that the conserved exergy is used by Ecodemon for opening and closing of the lid. By the same token Ecodemon adjusts the evolution of our system. The problem is how to estimate this value if we do not know the concrete form of dynamic equation (4.5). To get inside the problem, let us consider another model.

Case 2. Unlike Case 1, we assume that the system has a single *stable* equilibrium, continuously depending on the parameters $\boldsymbol{\alpha}$. Let the vector $\boldsymbol{\alpha} = \boldsymbol{\alpha}^0$ before the moment t_0 and the system is in the stable equilibrium $\mathbf{c}^*(\boldsymbol{\alpha}^0) = \mathbf{c}^0$. At the moment t_0 the parameters are shifted from $\boldsymbol{\alpha}^0$ to $\boldsymbol{\alpha}^1$. The operation was realised by Ecodemon very quickly, so that the state $\mathbf{c}^0$ does not manage to change. As a result the state $\mathbf{c}^0$ ceases to be in equilibrium and the system begins to move to the new stable equilibrium $\mathbf{c}^*(\boldsymbol{\alpha}^1) = \mathbf{c}^1$. In this case the affinity for the transition $\mathbf{c}(t) \to \mathbf{c}^1$ ($\mathbf{c}(t)$ is a current state) is equal to $RT_0 \ln[c_i/c_i^1]$, and the total dissipation of energy for the transition $\mathbf{c}^0 \to \mathbf{c}^1$ is

$$\mathrm{Diss}_1(\mathbf{c}^0 \to \mathbf{c}^1) = RT_0 \int_{t_0}^{\infty} \sum_{i=1}^{n} \ln\frac{c_i}{c_i^1}\frac{\mathrm{d}c_i}{\mathrm{d}t}\mathrm{d}t = -RT_0 \sum_{i=1}^{n} \left[c_i^0 \ln\frac{c_i^0}{c_i^1} - (c_i^0 - c_i^1) \right]. \qquad (4.8)$$

Generally, as we mentioned above, the transition requires infinite time, but if a finite time t_1 is sufficiently large to get into a small vicinity of $\mathbf{c}^1$, then with a sufficient accuracy we can assume that the equilibrium $\mathbf{c}^1$ is attained in the finite time. In other words, in the course of t_1 the transition process $\mathbf{c}^0 \to \mathbf{c}^1$ must be established.

Let at the moment t_1 the vector $\boldsymbol{\alpha}$ be again shifted from $\boldsymbol{\alpha}^1$ to $\boldsymbol{\alpha}^0$. In this case the equilibrium $\mathbf{c}^1$ ceases to be equilibrium, and the system begins to move to the old stable equilibrium $\mathbf{c}^0$. The total energy dissipated in the course of transition $\mathbf{c}^1 \rightarrow \mathbf{c}^0$ is described by Eq. (4.7), and its absolute value is formally equal to $\mathrm{Ex}(\mathbf{c}^1, \mathbf{c}^0)$.

The latter may be interpreted in the following manner: the exergy is equal to the minimal work, which must be performed in order to *kill* the system or to *destroy* it. Note (this is very important) that the work cannot be done on the system directly; it must be done on its environment. In other words, we cannot destroy the system by a direct impact, but we have to change the environment in a hostile (for the system) way. For that a mechanical work may be used and the system will be destroyed mechanically. Or, this could be a result of a change of the chemical status of the environment (for instance, pollution).

Let us consider the following example. Assume that some poisonous substance impacts on a living system. In order to "poison" the system we have to increase the poison concentration from the basic concentration in a "normal" environment, c^0, to the mortal concentration, c^{m}. For that we have to perform work A_{ch} against the gradient of chemical potentials, $RT_0 \ln(c^{\mathrm{m}}/c^0)$: $A_{\mathrm{ch}} = RT_0[c^{\mathrm{m}} \ln(c^{\mathrm{m}}/c^0) - (c^{\mathrm{m}} - c^0)]$. Then the exergy of the living system (per unique volume) will be equal to A_{ch}. Since the basic concentration for the normal condition, c^0 is usually very low, the term $\ln(c^{\mathrm{m}}/c^0)$ would be sufficiently large and, as a result, the exergy can be also sufficiently large.

By comparing the expressions (4.7) and (4.8) we can see that they differ from each other, i.e. the total energy dissipated in the course of the first half of the cycle when the system moves far from its initial state differs from the same value calculated for the second half when the system returns to its initial state. This is a principal distinction between Case 1, when both values are equal, and Case 2. The reason for the distinction is that in Case 1 the system is simply moving far from some equilibrium, which is saved, while in Case 2 a new stable equilibrium appears far from the initial state, which furthermore is ceasing to be equilibrium. The phenomenon of the appearance of a new (dynamic) equilibrium far from thermodynamic one is a typical character of non-linearity, and we have the right to expect some new effects in such a type of cycles.

So, from the "exergetic" point of view, the system moving away from its initial state to the new equilibrium accumulates the exergy $\mathrm{Ex}(\mathbf{c}^0, \mathbf{c}^1)$, and dissipates the exergy $\mathrm{Ex}(\mathbf{c}^1, \mathbf{c}^0)$ returning to the initial state. In Case 2 the cycle is closed incompletely: a residual of exergy

$$\delta\mathrm{Ex} = \mathrm{Ex}(\mathbf{c}^0, \mathbf{c}^1) - \mathrm{Ex}(\mathbf{c}^1, \mathbf{c}^0) = RT_0 \sum_{i=1}^{n} \left[(c_i^1 + c_i^1)\ln\frac{c_i^1}{c_i^0} - 2(c_i^1 - c_i^0) \right], \qquad (4.9)$$

remains in the system. The residual may be both positive and negative. In the first case this exergy may be used as a "push" for the next cycle, Ecodemon's interference is not needed and the system begins its slow evolution (of course, by means of increasing the entropy of environment). In other words, if $\delta\mathrm{Ex} > 0$, then the system *possesses an ability to evolution*. In the opposite case, if $\delta\mathrm{Ex} < 0$, then the system does not have this ability.

What are the orders of magnitude of the exergy and its residual? In order to answer the question visually we are restricted to (1) the case of one variable, $c(t)$, and (2) the case of small deviation, $c^1 - c^0 = \Delta c$, $|\Delta c| \ll c^0$. Expanding the expressions for $\mathrm{Ex}(\mathbf{c}^0, \mathbf{c}^1)$ and

δEx into a series in the power of Δc and restricting only the first terms of expansion we get:

$$\begin{aligned} \mathrm{Ex}(\mathbf{c}^0, \mathbf{c}^1) &\approx \frac{1}{2c^0}(\Delta c)^2 \approx \frac{1}{2c^1}(\Delta c)^2, \\ \delta\mathrm{Ex} &\approx \frac{1}{6(c^0)^2}(\Delta c)^3 \approx \frac{1}{6(c^1)^2}(\Delta c)^3. \end{aligned} \tag{4.10}$$

It is clear that when we calculate the exergy, we are not leaving the Onsager–Prigogine world with its weak quadratic non-linearity, while the exergy residual is a phenomenon of another, more non-linear world. In practice the phenomenon is unnoticeable when the system is in a vicinity of thermodynamic equilibrium, and the long-term evolution is needed in order to reveal this phenomenon's significant role.

All these results have biological interpretation. Let the system be a living organism and the exergy cycle be its life cycle, which is passed from "dust" to "dust", from detritus to detritus. Then the exergy residual can be interpreted as an amount of information, which must be saved to start the next life cycle, i.e. to maintain evolution. This is nothing more or less than the genetic information contained in the genome. "Soma is mortal, genome is immortal" (Weismann).

Finally, we would like to call your attention to the following: there is a fundamental difference between the types of *Gedanken experiment* in classic thermodynamics and here. If in classic thermodynamics in order to change the state of the system we perform the work on the system, then here in order to obtain the same result we must perform the work on the system's environment.

5.5. Exergy and information

Introducing the new variables $N = \sum_{i=1}^{n} N_i$ and $p_i = N_i/N$ where N is the total number of particles (matter) in the system, we can rewrite expression (4.3) for exergy as

$$\mathrm{Ex} = N\sum_{i=1}^{n} p_i \ln\frac{p_i}{p_i^0} + \left[N \ln\frac{N}{N_0} - (N - N_0)\right]. \tag{5.1}$$

The vector of intensive variables $\mathbf{p} = \{p_1, \ldots, p_n\}$ describes the system composition; N is an extensive variable. The value $K = \sum_{i=1}^{n} p_i \ln(p_i/p_i^0)$ within the constant factor $1/\ln 2$ is the so-called Kullback measure of the increment of information (see Chapter 4).

Let us bear in mind the exact meaning of Kullback's measure. From the information point of view any distribution $\mathbf{p}$ contains a certain quantity of information determined by Shannon's formula. The quantity of information increases as a result of transition from one distribution to another ($\mathbf{p}^0$ to $\mathbf{p}$). It is, namely, this increment of information (per one particle or one unit of matter), which is determined by Kullback's measure. Then the product NK can be interpreted as a measure of the total information, which has been accumulated in the process of transition from some reference state to the current one.

We can present the expression for exergy in the form

$$\mathrm{Ex} = \mathrm{Ex}_{\mathrm{inf}} + \mathrm{Ex}_{\mathrm{mat}} \tag{5.2}$$

where $\mathrm{Ex}_{\mathrm{inf}} = NK(\mathbf{p}, \mathbf{p}^0) \geq 0$ and $\mathrm{Ex}_{\mathrm{mat}} = N \ln(N/N_0) - (N - N_0) \geq 0$, i.e. as the sum of two items: the first is a result of structural changes in the system and the second is caused by a change in the total mass of the system.

We have shown above that exergy increases with time, but this is proved only in the vicinity of thermodynamic equilibrium. It was assumed that this statement is true along the whole system trajectory in the course of the whole system evolution from thermodynamic equilibrium, which can be identified as some "pre-biological" situation, to the current state (Jørgensen, 1992c). Then, if we follow one of the main biological paradigms, namely that "the ontogenesis is always repeating the phylogenesis", we can generalise the previous statement in the form of the so-called "exergy maximum principle": *any ecosystem in the process of evolution towards its climax state tends to increase its own exergy.*

Thus, we postulate that $\mathrm{dEx}/\mathrm{d}t \geq 0$. By differentiating Eq. (5.2), $\mathrm{d(Ex)}/\mathrm{d}t = \mathrm{d(Ex)}_{\mathrm{inf}}/\mathrm{d}t + \mathrm{d(Ex)}_{\mathrm{mat}}/\mathrm{d}t = (\mathrm{d}N/\mathrm{d}t)(K + \ln(N/N_0)) + (\mathrm{d}K/\mathrm{d}t)N$, denoting $\ln(N/N_0) = \xi$ and taking into account that $N > 0$ we get the evolutionary criterion in the form

$$\frac{\mathrm{d}K}{\mathrm{d}t} + (K + \xi)\frac{\mathrm{d}\xi}{\mathrm{d}t} \geq 0. \tag{5.3}$$

If the positiveness of $\mathrm{d}\xi/\mathrm{d}t$ means an increase in the total biomass in the course of evolution then the positiveness of $\mathrm{d}K/\mathrm{d}t$ can be interpreted as an increase in the specific information content (per unit of biomass). It is obvious that if both the total biomass and its specific information content increases then exergy also increases. However, if the total biomass is constant ($\mathrm{d}\xi/\mathrm{d}t = 0$) then the system can evolve only if the information content of its biomass is growing. This could also be interpreted as the growth of diversity. On the other hand, the information content can decrease ($\mathrm{d}K/\mathrm{d}t < 0$), but if the total biomass grows sufficiently quickly ($\mathrm{d}\xi/\mathrm{d}t \gg 1$) then exergy is growing, and the system is evolving. The evolution is carried out if the biomass is decreasing but the information content of the biomass is growing (sufficiently quickly). At last, there is a paradoxical situation when exergy is increasing while the total biomass and its information content are decreasing. If $N < N_0$ then $\xi < 0$, and $\xi(\mathrm{d}\xi/\mathrm{d}t) > 0$. From Eq. (5.3) we have

$$\xi\frac{\mathrm{d}\xi}{\mathrm{d}t} \geq \left|\frac{\mathrm{d}K}{\mathrm{d}t}\right| + K\left|\frac{\mathrm{d}\xi}{\mathrm{d}t}\right|. \tag{5.4}$$

It is obvious that this inequality can be realised if $|\xi| \gg K$ and $|\mathrm{d}K/\mathrm{d}t| \ll 1$, i.e. the information content is sufficiently low ($K \ll 1$) and the process of its further decrease is very slow. Note that the inequality $|\xi| \gg 1$ takes place only if the condition $N \ll N_0$ is fulfilled. In this case we can say that the system is "paying" for its evolution with its own biomass. In the vicinity of thermodynamic equilibrium, at the initial stage of evolution $K \cong \xi \sim 0$. Then from Eq. (5.3) we have $\mathrm{d}K/\mathrm{d}t > 0$, i.e. at this stage, in order to evolve, the system has to increase its own biomass information content.

It is interesting that the exergy maximum principle possesses certain selective properties. In order to show this we consider a partial case when the total biomass is constant. Then the maximum of exergy coincides with the maximum of K, which is attained at faces of the simplex $\sum_{i=1}^{n} p_i = 1; p_i \geq 0, i = 1, \ldots, n$. It is possible to show that $\max_{p_i} K = \max_i [\ln(1/p_i^0)]$. This result could be interpreted in the following way.

The system with constant biomass, which is increasing its own exergy in the course of its evolution, tends to eliminate all the elements (substances) except one, which has a minimal initial concentration. In a "pre-biological" (reference) state, some "living substance" has a minimal concentration. In other words, the system, which increases its own exergy, selects among its components that which had been present in a minimal quantity at the beginning of evolution. But among these elements there are some which are necessary for the maintenance of life and the system must retain them. How can this contradiction be resolved? We can do this by introducing some constraints. For instance, one requirement would be to maintain a certain level of system diversity. Formally this means (in this partial case) that we look for the maximum of K under the additional constraint $H = -\sum_{i=1}^{n} p_i \ln p_i = \text{constant}$. This implies that exergy has to increase while maintaining a certain non-arbitrary level of diversity. We think there is a very deep analogy with Fisher's fundamental theorem of natural selection (see, for instance, Svirezhev and Passekov, 1982).

Until now the connection between exergy and information rests on the fact that one of the important terms in the expression for exergy has the form of one of the main information measures, namely Kullback's measure. However, the connection is rather deeper. We try to show this using a very popular method in statistics, an *urn scheme*.

Let there be n urns corresponding to n different sorts of particles, and each urn contains N_i^0 $(i = 1, \ldots, n)$ particles. We assume that all operations with ith urn do not depend on operations with other urns and, vice versa, operations with others do not influence operations with ith urn. In accordance with Shannon (or Boltzmann), the information (or entropy) contained in this urn is proportional to the logarithm of number by which N_i^0 particles could be ordered. This number is $(N_i^0)!$. Using Stirling's formula: $\ln\lfloor(N_i^0)!\rfloor \approx N_i^0 \ln N_i^0$ for $N_i^0 \gg 1$, we get that the information per one particle of ith urn is equal to $\sigma_i^0 = \lambda \ln N_i^0$ where $\lambda = 1/\ln 2$, so that the information is expressed in bits. It is necessary to explain what is meant by the word "ordering". For this we call on our Ecodemon to help us. In order to determine the number of particles it has to enumerate them. It can do this by associating them with some ordered structure, for instance a row of N_i^0 miniboxes, each of which can contain not more than one particle. When all particles are distributed among all miniboxes, Ecodemon can say that this urn contains N_i^0 particles. To distribute these particles among the miniboxes it can use $(N_i^0)!$ different ways.

There is another model which gives the same results. It is known that the entropy of the system of N_i^0 *identical* particles, as a measure of uncertainty, is equal to $S_i^{(n)} = N_i^0 S_i^{(1)} - k \ln N_i^0!$ where $S_i^{(1)}$ is the entropy of a single particle and k is a dimensionless coefficient (Landau and Lifshitz, 1995). If the system consists of different particles then its entropy is simply equal to the sum of the entropies of each particle. But this entropy could not be a characteristic of substance since the system consisting of different particles cannot represent some type of matter. Note that entropy here is not additive, the proper entropy of a particle decreases by the value $k \ln N_i^0$, that is, equivalent to the rise of information because of the joining of identical particles into some subsystem (urn).

Then we shall change the number of particles in the ith urn by adding or removing the small quantity δN_i^0 ($\delta N_i^0 > 0$ if particles are added and $\delta N_i^0 < 0$ if removed). In the chemical interpretation these particles are molecules of a certain chemical substance, and their change is a result of chemical reactions that occur in the environment.

Thus, this experiment can be considered as a simple statistical model describing change in the intensity and spectral composition of solar radiation as a result of reflection, absorption and spectral transformation of solar radiation by some active surface (see below for details).

Let a new number of particles in ith urn be N_i. Then the information contained in one particle in this new situation will be equal to $\sigma_i = \lambda \ln N_i$, and the change of information (per one particle) as a result of transition from the initial situation to the new one is equal to

$$\sigma_i - \sigma_i^0 = \lambda \ln \frac{N_i}{N_i^0}. \tag{5.5}$$

In order to get the total increment of information, δI_i, for the whole system, we must multiply this value, $\sigma_i - \sigma_i^0$, by the number of new particles, i.e. by δN_i, and must sum their partial terms over all urns

$$\delta I = \sum_{i=1}^{n} \delta I_i = \sum_{i=1}^{n} (\sigma_i - \sigma_i^0)\delta N_i = \lambda \sum_{i=1}^{n} \ln \frac{N_i}{N_i^0} \delta N_i. \tag{5.6}$$

By integrating Eq. (5.6) from initial state N_i^0 to current state N_i we obtain the expression for the total increase of information (in bits):

$$I - I^0 = \lambda \sum_{i=1}^{n} [N_i \ln(N_i/N_i^0) - (N_i - N_i^0)]. \tag{5.7}$$

Comparing Eq. (5.7) with the expression for exergy (formula (4.3)) we can see that

$$\text{Ex} = \frac{kT_0}{\lambda}(I - I^0) \tag{5.8}$$

where the exergy is measured in energy units.

Note that by using the definition of Kullback's measure we can obtain the same expression for exergy in a simpler way. Indeed, the process of the increment of information can be considered as a sum of two independent processes. The first process is a transition from the distribution $\{p_i^0 = N_i^0 / \sum_{i=1}^{n} N_i^0\}$ to the new distribution $\{p_i = N_i / \sum_i N_i\}$, $i = 1, \ldots, n$. This process is accompanied by the following increment of information:

$$I(1) - I_0(1) = \lambda N \sum_{i=1}^{n} p_i \ln \frac{p_i}{p_i^0} = \lambda N K \tag{5.9}$$

where K is Kullback's measure and N the current total number of particles. Note, in this connection, we implicitly assume that the total number of particles is not changed.

The second process is a *small* (namely, in this case these processes can be considered as independent) change in the total number of particles. Using the above arguments for the calculation of the information increment as a result of a change in the number of particles in the ith urn, we can write the following expression for the second process:

$$I(2) - I_0(2) = \lambda \int_{N_0}^{N} \ln \frac{N}{N_0} \mathrm{d}N = \lambda \left[N \ln \frac{N}{N_0} - (N - N_0) \right]. \tag{5.10}$$

The information is additive and summing both these expressions we get

$$\begin{aligned} I - I_0 &= [I(1) - I_0(1)] + [I(2) - I_0(2)] \\ &= \lambda[NK + N \ln(N/N_0) - (N - N_0)]. \end{aligned} \tag{5.11}$$

It is easy to see that this expression and the expression (5.7) do not fundamentally differ from each other.

5.6. Exergy of solar radiation

Solar and heat radiation fluxes (incoming and outgoing), which are incident and reflected from a surface, are described by the continuous spectral distributions $E^{\text{in}}(\nu)$ and $E^{\text{out}}(\nu)$, where ν is a spectral frequency or, if the whole of the spectrum is divided into n spectral intervals $[\nu_i, \nu_{i+1}]$, $i = 1, \ldots, n$, their discrete analogies E_i^{in} and E_i^{out}:

$$E_i^{\text{in}} = \int_{\nu_i}^{\nu_{i+1}} E_0(\nu)\mathrm{d}\nu, \qquad E_i^{\text{out}} = \int_{\nu_i}^{\nu_{i+1}} E_1(\nu)\mathrm{d}\nu \tag{6.1a}$$

Since the energy of one photon is equal to $h\nu$, where h is the Planck constant, $E^{\text{in}}(\nu)/h\nu = n_0(\nu)$ is the number of photons with the frequency ν ("green" photons, "red" photons, etc.) incoming to a single surface during a time unit. Correspondingly, $E^{\text{out}}(\nu)/h\nu = n(\nu)$ is the number of outgoing photons with the frequency ν reflected and irradiated by the surface. Then

$$E_i^{\text{in}} = h\int_{\nu_i}^{\nu_{i+1}} n_0(\nu)\nu\,\mathrm{d}\nu, \qquad E_i^{\text{out}} = h\int_{\nu_i}^{\nu_{i+1}} n(\nu)\nu\,\mathrm{d}\nu. \tag{6.1b}$$

Thus, photons are distinguished from each other by their frequencies. On the other hand, we can consider each photon with the frequency ν as the sum of $\nu\tau$ identical particles where τ is some characteristic time of the considered process. The energy of each particle is equal to h/τ. Note that it is not necessary to seek some physical sense in our virtual particles; this is simply a methodological way to apply the results described below.

In accordance with our assumption the radiation fluxes E_0^i and E_1^i can be presented in the form of some flows of *particles*, $q_i^0 = \int_{\nu_i}^{\nu_{i+1}} n_0(\nu)\nu\tau\,\mathrm{d}\nu$ and $q_i = \int_{\nu_i}^{\nu_{i+1}} n(\nu)\nu\tau\,\mathrm{d}\nu$, so that $E_i^{\text{in}} = (h/\tau)q_i^0$ and $E_i^{\text{out}} = (h/\tau)q_i$.

Keeping in mind that solar radiation, not including energy, is also bringing information, we apply the result of Section 5.5 to this case. According to the information concept, the outgoing radiation contains the whole information about the active surface, which interacts with incoming radiation. We also assume that the transformation of distribution $(E_1^0, \ldots, E_n^0)$ into $(E_1^0, \ldots, E_n^0)$ takes place very fast in comparison with the characteristic time of radiation fluxes. By setting instant values N_i^0 and N_i in Eq. (5.7) to be equal to their flows q_i^0 and q_i and moving to the energy units we obtain the expression for the power of exergy of the solar radiation (Svirezhev and Steinborn, 2001)

$$P_{\text{Ex}} = \sum_{i=1}^{n} [E_i^1 \ln(E_i^1/E_i^0) - (E_i^1 - E_i^0)]. \tag{6.2}$$

Note that later on we shall often use the expressions: energy, exergy (Ex), radiation, information, etc. instead of power, power of exergy (P_{Ex}), radiation, information, etc.

flow and fluxes. Introducing exergy in such a way we can apply the general thermodynamic concepts to the process of interaction between solar radiation and some reflecting, transforming and absorbing surface (e.g. vegetation).

The sum $R = \sum_{i=1}^{n} E_i^0 - \sum_{i=1}^{n} E_i^1 = E^{\text{in}} - E^{\text{out}}$, which is the difference between the total incoming and outgoing radiation, is called the *radiation balance*. In other words, R is the amount of incoming solar energy which is absorbed by a surface (soil, water, vegetation, etc.). Introducing the frequencies $p_i^{\text{in}} = E_i^0/E^{\text{in}}$ and $p_i^{\text{out}} = E_i^1/E^{\text{out}}$, and keeping in mind Kullback's measure, the expression for exergy can be rewritten in the form

$$\text{Ex} = (E^{\text{in}} - R)\left[K + \ln\left(\frac{E^{\text{in}} - R}{E^{\text{in}}}\right)\right] + R \tag{6.3}$$

where $K = \sum_{i=1}^{n} p_i^{\text{out}} \ln(p_i^{\text{out}}/p_i^{\text{in}})$ is Kullback's measure. One can see that the value of exergy can be considered as a function of two *independent* variables, K and R, and one external parameter, E^{in}. The function $\text{Ex} = \text{Ex}(R, K)$ for fixed E^{in} monotonously increases with increasing K, and has the minimum $\text{Ex}^{\text{cr}} = \text{Ex}^{\text{cr}}(R^{\text{cr}}, K) = E^{\text{in}}(1 - \text{e}^{-K})$ with respect to R at $R^{\text{cr}} = E^{\text{in}}(1 - \text{e}^{-K})$. It is clear that $\text{Ex}(0, K) = E^{\text{in}}K$, $\text{Ex}(E^{\text{in}}, K) = E^{\text{in}}$ and $\text{Ex}(0, 0) = 0$.

Undoubtedly, all main conclusions would be obtained from the analysis of this expression but we can do it in a more elegant way if we use its dimensionless form. In addition, the transition to dimensionless variables implies no dependence on the external parameter E^{in}.

We introduce two new definitions. The ratio $\eta_R = R/E^{\text{in}}$ is named the *radiation efficiency*. It is obvious that this value describes a fraction of the total energy absorbed by the surface. (Note: if the radiation balance is negative, η_R must also be negative.) Analogously, the ratio $\eta_{\text{Ex}} = \text{Ex}/E^{\text{in}}$ is called the *exergy efficiency*. By keeping in mind that exergy is a measure of the "useful" work which a system is able to perform, we can say that η_{Ex} is an efficiency coefficient of some "radiative" machine, which is our active surface. The working process of this machine is an interaction of incoming radiation with the active surface. Dividing both sides of Eq. (6.3) by E^{in} and using the notations introduced above we get

$$\eta_{\text{Ex}} = (1 - \eta_R)K + (1 - \eta_R)\ln(1 - \eta_R) + \eta_R. \tag{6.4}$$

The value η_{Ex} is also a function of two *independent* variables, η_R and K, but it does not depend on a parameter. The function η_{Ex} monotonously increases with increasing K and it has the minimum $\eta_{\text{Ex}}^{\text{cr}} = \eta_{\text{Ex}}(\eta_R^{\text{cr}}, K) = 1 - \text{e}^{-K}$ with respect to η_R at $\eta_R^{\text{cr}} = 1 - \text{e}^{-K}$ (see Fig. 5.3). It is obvious that $\eta_{\text{Ex}}(0, K) = K$, $\eta_{\text{Ex}}(1, K) = 1$ and $\eta_{\text{Ex}}(0, 0) = 0$.

In order to analyse some properties of the function $\eta_{\text{Ex}}(\eta_R, K)$ we consider its two-dimensional sections for two values of K (Fig. 5.4). Let $K = K_1$ then the minimum $(\eta_R^{\text{cr}})_1$ divides the curve η_{Ex} into two branches (*ab* and *bc*): the first, where $\eta_R > (\eta_R^{\text{cr}})_1$ and $\eta_{\text{Ex}} < \eta_R$, and the second, where $\eta_R < (\eta_R^{\text{cr}})_1$ and $\eta_{\text{Ex}} > \eta_R$.

The shape of this curve and the inequalities which define its different branches allow us to formulate the following hypothesis:

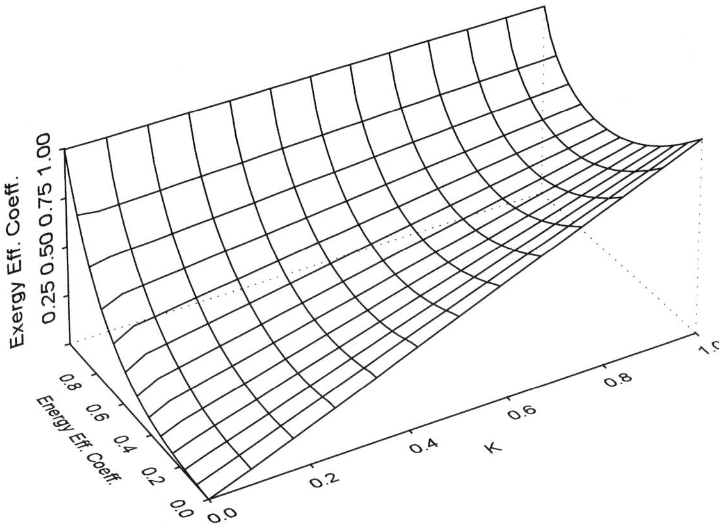

Fig. 5.3. The exergy efficiency η_{Ex} as a function of two variables: η_R and K.

1. On the right-hand branch $(a-b)$ the active surface operates as a classic thermodynamic machine, performing mainly mechanical or chemical work.
2. On the left-hand branch $(b-c)$ the active surface operates as an information machine, producing mainly information.

If the radiation (energy) balance was a measure of the mechanical work of the thermodynamic machine, then the value of Kullback's measure, K, could be a measure of

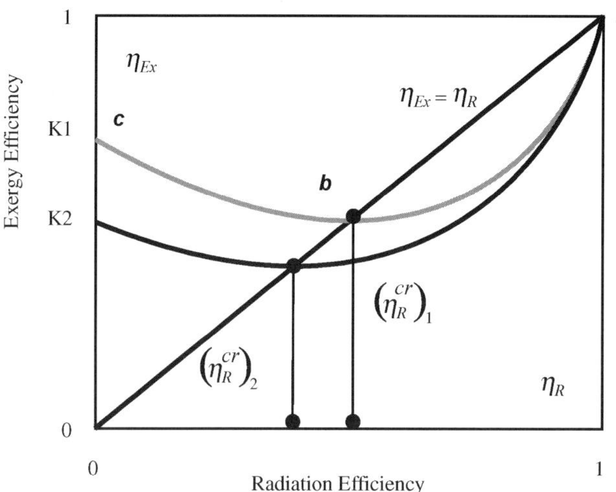

Fig. 5.4. Dependency of η_{Ex} on radiation efficiency coefficient, η_R, for two different values of $K(K_1 < K_2)$.

"information work". It is easy to see that $K \to 0$ then $(\eta_R^{cr})_1 \to 0$, and the curve η_{Ex} has only the "classic" branch. In this case the system works only like a classic thermodynamic machine. Conversely, if $K \to \infty$ then $(\eta_R^{cr})_1 \to 1$, the curve η_{Ex} has only the "non-classic" information branch. The system works purely as an information machine producing an infinite amount of information. These cases are marginal; substantively the active surface is a composition of these two ideal machines and it produces both mechanical (or chemical) work and information.

It is interesting that on the left-hand branch the exergy efficiency increases with a decrease in the radiation efficiency, i.e. with a decrease in energy absorbed by the active surface. At its limit, where the absorbed energy tends to zero, the exergy efficiency tends to a non-zero value, which is equal to K. Moreover, on the left-hand branch the exergy is always more than the radiation balance. If we keep in mind one of the possible interpretations of exergy as the maximal useful work which can be performed by the system, and consider the radiation balance as a portion of external energy consumed by the system, then, formally, we have a situation when the low energy conservation is not fulfilled. At first sight, this is a paradoxical result, but it is an artefact of such a representation of exergy and our assumption about the independency of η_R and K. Indeed, if we write the expression for partial exergy as $\mathrm{Ex}_i = E_i^{out} \ln(E_i^{out}/E_i^{in}) + (E_i^{in} - E_i^{out})$ then it is just as easy to see $\mathrm{Ex}_i \to 0$ for $E_i^{out} \to E_i^{out}$; therefore, the variables η_R and K are not really independent, but the closer η_R is to zero, the more they are dependent. In other words, the model is not completely adequate in the vicinity of $\eta_R = 0$. Nevertheless, the model is working even on the left-hand side. This is explained by the fact that the information can be transferred even where there is a very weak signal. The point is that the information measures are functions of frequencies, but the frequency $p_i = N_i/N$ may tend to a finite value even if N (and N_i, of course) tends to zero. This, by the way, explains the fact that even a weak signal can transfer a relatively large quantity of information.

Conversely, as we can see in Fig. 5.4 on the right-hand branch, when $\eta_R \to 1$, the value of η_{Ex} does not depend on K (two branches, corresponding to different values of K, come together when $\eta_R \to 1$). This is natural since the main performance of the classic machine is not to produce information but only to do mechanical or chemical work.

5.7. How to calculate the exergy of living organic matter?

If we compare the living and dead organic matter then we see that their chemical compositions almost do not differ from one another; moreover, their thermodynamic characteristics such as enthalpies are identical. The same statement is true for the biomass of all living organisms (see Chapter 2). However, if the detrital organic matter is a mixture of carbohydrates, fats and proteins then the living organic matter is an *ordered structure* created from these materials in accordance with a plan given by genetic information. Naturally, both the biomass of living organisms and the mass of detritus could be measured in the same units, since 1 g of biomass and 1 g of detritus have the same chemical composition, but their roles in the ecosystem differ significantly from each other.

Therefore a question arises: what is the difference between the same quantities of detritus and living matter? We postulate that *they differ in the exergy contents*.

The problem is how to calculate the exergy contained in 1 g of living organic matter? We assume that the process of creation of living matter can be represented as the chemical formation of organic molecules from non-organic substances of the environment and simultaneously the ordering of these molecules in such structures as organisms in accordance with their genomes. As a result the specific exergy of living matter (per 1 g) will be

$$\mathrm{ex} = \mathrm{ex}^{\mathrm{chem}} + \mathrm{ex}^{\mathrm{biol}} = T_0(\mathrm{inf}^{\mathrm{chem}} + \mathrm{inf}^{\mathrm{biol}}) \tag{7.1}$$

where $\mathrm{ex}^{\mathrm{chem}}$ is the specific exergy obtained by living matter due to chemical processes, and $\mathrm{ex}^{\mathrm{biol}}$ is the specific exergy as a result of ordering, i.e. the construction of an organism in accordance with genetic plan. In other words, we have to find the correct composition of the enzymes determining the biochemical processes in the organisms. Since exergy has an information interpretation, ex can be represented as a sum of two information quantities multiplied by the temperature of the environment, T_0: the first is the content of "chemical" information and the second is the content of genetic information in 1 g of biomass.

Since there is no difference between the chemical compositions of dead and living organic matters, the first item in Eq. (7.1) can be associated with the exergy of detritus. But how to calculate this? At first sight the problem seems very complex, but as Anton Chekhov said: "The gun hanged on the wall in the first act has to fire in the last act". We have "hung our gun" in Section 2.7 of Chapter 2, where we calculated the standard Gibbs' potential for detritus, ΔG_0. Now it is time to "fire it". If keeping in mind that the exergy accumulated by the system is equal to the increase of its Gibbs potential then

$$\mathrm{ex}^{\mathrm{chem}} = \frac{T_0}{300}\Delta G_0 = \frac{T_0}{300} 18.7\ \mathrm{kJ/g}. \tag{7.2}$$

The appearance of the factor $(T_0/300)$ is explained by the fact that ΔG_0 was calculated at the temperature of 300 K. Since the exergy can be represented in the form

$$\mathrm{ex}^{\mathrm{chem}} = \frac{T_0}{300}\Delta G_0 = T_0\,\mathrm{inf}^{\mathrm{chem}} = T_0\,62.3\ \mathrm{J/Kg}. \tag{7.3}$$

the quantity of information contained in 1 g of partially ordered organic matter of detritus and expressed in the units of thermodynamic entropy is equal to

$$\mathrm{inf}^{\mathrm{chem}} = 62.3\ \mathrm{J/K\ g}. \tag{7.4}$$

It is known that the same information, expressed in bits, $\mathrm{inf}_{\mathrm{bit}}^{\mathrm{chem}}$, will be equal to $\mathrm{inf}_{\mathrm{bit}}^{\mathrm{chem}} = \log_2 w_{\mathrm{d}} = 1.44 \ln w_{\mathrm{d}}$ where w_{d} is the number of possible chemical states of 1 g of detritus. The problem is how to establish the similarity between $\mathrm{inf}^{\mathrm{chem}}$ and $\mathrm{inf}_{\mathrm{bit}}^{\mathrm{chem}}$? Here we are compelled to use the thermodynamic model of ideal solution, i.e. to consider detritus as an ideal solution of organic matter in water. It seems fully unrealistic, but keeping in mind that even less realistic models in physics and chemistry nevertheless give good results, we think that this model should be sufficient. Then

$$\mathrm{inf}^{\mathrm{chem}} = 62.3\ \mathrm{J/K\ g} = \frac{R}{M_{\mathrm{d}}}\ln w_{\mathrm{d}} = \frac{R}{1.44M_{\mathrm{d}}}\mathrm{inf}_{\mathrm{bit}}^{\mathrm{chem}} \tag{7.5}$$

where $R = 8.4$ J/K mol is the gas constant and $M_{\mathrm{d}} \approx 10^5$ is the molecular weight of detritus. Using this relation we can estimate the weight of 1 bit in the units of thermodynamic entropy: 1 bit $= R/1.44M_{\mathrm{d}} \approx 5.8 \times 10^{-5}$ J/K g.

The genetic information is calculated very easily (see Section 4.9, Chapter 4). Living organisms use 20 different amino acids and each gene determines on average the sequence of about 700 amino acids. Therefore, the information $\mathrm{inf}_{\mathrm{bit}}^{\mathrm{biol}}$ can be found from the number of permutations among which the characteristic amino acid sequence for the considered organism has been selected ($\mathrm{inf}_{\mathrm{bit}}^{\mathrm{biol}} = I_{\mathrm{nuc}}$ of Section 4.9):

$$\mathrm{inf}_{i,\mathrm{bit}}^{\mathrm{biol}} = 1.44\ln(20^{700g_i}) = (1.44 \times 700\ln 20)g_i \approx 3000g_i\ \mathrm{bits} \tag{7.6}$$

where g_i is the number of non-nonsense genes in the genome of the organism of some ith species or some taxon (see Table 5.1). Using the value of 1 bit expressed in entropy units we get

$$\mathrm{inf}_i^{\mathrm{biol}} \approx 5.8 \times 10^{-5} \times 3000g_i = 0.174g_i\ \mathrm{J/K\ g}. \tag{7.7}$$

Table 5.1
Numbers of non-repetitive genes in genome (g_i) and weighting factors (β_i) for living organisms (Jørgensen et al., 1995a,b, 1998, 2000)

Organisms	g_i	β_i
Detritus	0	1
Minimal cell	470	2.3
Bacteria	600	2.7
Algae	850	3.4
Yeast	2000	5.8
Fungi	3000	9.4
Annelids	9000	26.2
Moulds	9500	27.6
Plants, trees	10,000–30,000	29–87
Worms	10,500	30.4
Insects	10,000–15,000	29–43
Jellyfish	10,000	29
Zooplankton	10,000–15,000	29–43
Fish	100,000–120,000	281–337
Birds	120,000	337
Amphibians	120,000	337
Reptiles	130,000	365
Mammals	140,000	393
Homo sapiens	250,000	701

Sources: Cavalier-Smith (1985), Li and Grauer (1991) and Lewin (1994).

Finally, the specific exergy of 1 g of biomass of *i*th species is equal to

$$\text{ex}_i = T_0(\text{inf}^{\text{chem}} + \text{inf}_i^{\text{biol}}) = T_0(62.3 + 0.174g_i)\ \text{J/g}. \tag{7.8}$$

The total exergy of ecosystem can be found by summing up the contributions, originating from all *n* components including detritus ($i = 1,\ g_1 = 0$)

$$\text{Ex} = \sum_{i=1}^{n} \text{ex}_i c_i [\text{J}] \tag{7.9}$$

where c_i are the biomasses of corresponding components expressed in grams. By introducing the weighting factor $\beta_i = \text{ex}_i/\text{ex}_1,\ i = 1, \ldots, n$ we are able to cover the exergy for various organisms in the detritus equivalent unit $D = 62.3$ J/K g:

$$\text{Ex} = RT_0 \sum_{i=1}^{n} \beta_i c_i. \tag{7.10}$$

The weighting factor defined as exergy content in relation to detritus (see Table 5.1) might be considered as a quality factor, reflecting how developed are the various groups and to what extent they contribute to the exergy due to their information content, reflected in the computation. This is completely in accordance with Boltzmann (1905), who gave the following relationship for the work, *A*, that is embodied in the thermodynamic information; we have $A = RT \ln w$ where *w* is the number of possible states, among which the information has been selected, i.e. *w* for species is the inverse value of probability spontaneously to obtain the valid amino acid sequence.

The total exergy of an ecosystem *cannot* be calculated exactly, as we cannot measure the concentrations of all the components or determine all possible contributions to exergy in an ecosystem. If we calculate the exergy of a fox, for instance, the above calculations will only give us the contributions coming from the biomass and the information embodied in the genes, but what is the contribution from the blood pressure, the hormonal activity and so on? These properties are at least partially covered by the genes, but is that an entire picture? We can calculate the contributions from the dominant components, for instance, by the use of a model or measurements that cover the most essential components for a focal problem.

Exergy calculated using the above method has some clear shortcomings:

- We apply the model of ideal dilution to such a non-ideal substance as detritus.
- We do not know the non-nonsense gene for all the organisms.
- We calculate the exergy embodied in the proteins (enzymes) only in principle, while there are other components important for life processes. These components are contributing less to the exergy than the enzymes, and the information embodied in the enzymes controls the formation of other components, for instance hormones. It cannot, however, be excluded that these components will contribute to the total exergy of the system.
- We do not include the exergy of the ecological network. If we calculate the exergy of models, the network will always be relatively simple and the contribution coming from the information content of the network is negligible.

- We will always use a simplification of the ecosystem, for instance by a model or a diagram or similar. This implies that we only calculate the exergy contributions of the components included in the simplified image of the ecosystem. The real ecosystem will inevitably contain more components which are not included in our calculations.

It is therefore proposed to consider the exergy found by these calculations as a *relative minimum exergy index* to indicate that there are other contributions to the total exergy of an ecosystem, although they may be of minor importance. In most cases, however, a relative index is sufficient to understand the reactions of ecosystems because the absolute exergy content is irrelevant for the reactions. It is in most cases the change in exergy which is of importance to understand the ecological reactions.

The weighting factors presented in Table 5.1 have been applied successfully in several structurally dynamic models and furthermore in many illustrations of the maximum exergy principle that will be presented in Chapter 12. The relatively good results in the application of the weighting factors, in spite of the uncertainty of their assessment, only seem to be explicable by the robustness of the application of the factors in modelling and other quantifications. The differences between the factors of the microorganisms, the vertebrates and invertebrates, are so clear that it does not matter if the uncertainty of the factors is very high—the results are not affected.

On the other hand, it would be important progress not only from a theoretical point of view to get better weighting factors but it also would enable us to model the competition between species that are closely related.

5.8. Other methods for the exergy calculation

Fonseca et al. (2000) have proposed another method to estimate the weighting factors based on the overall DNA content. It could be argued that all DNA contribute to the information content and that all DNA would have some function, for instance as spare parts for damaged DNA. We slightly modify this method taking into account the pair correlations between amino acids (see Section 4.9, Chapter 4). According to modified Fonseca et al.'s proposal the genetic information contained in the biomass of ith species (group of species, taxon), $\mathrm{inf}_{i,\mathrm{bit}}^{\mathrm{biol}}$, is defined as ($\mathrm{inf}_{\mathrm{bit}}^{\mathrm{biol}} = I_{\mathrm{nuc}}^{(3)}$):

$$\mathrm{inf}_{i,\mathrm{bit}}^{\mathrm{biol}} = 1.44 \frac{L_{\mathrm{nuc}}}{6} \ln 20 \text{ bits} \tag{8.1}$$

where L_{nuc} (n in Section 4.9) is the nucleotide length of single-stranded DNA. If C is the amount of DNA contained in the organism (in picograms per cell) then $L_{\mathrm{nuc}} = (1/2) \times 0.98 \times 10^9 C$ (nucleotides). The weighting factors β_i^{DNA} are calculated by the method described in Section 5.7. In Fonseca et al. (2000) a table can be found with the β-values estimated according to their method. Table 5.2 is our modification of their table.

The advantage of this method is that the C-value is known for the higher number of organisms than the number of non-nonsense genes, which of course is not needed if the estimation method is wrong. If we compare two columns in Table 5.2 then we can see that for some organisms $\beta_i > \beta_i^{\mathrm{DNA}}$ (Bacteria, Fungi, Molluscs, Crustaceans and Fish),

Table 5.2
Lowest DNA contents (C_i) and weighting factors (β_i^{DNA}) for living organisms

Organisms	C_i	β_i^{DNA}	β_i
Detritus	0	1	1
Bacteria	0.0017	1.6	2.7
Algae	0.04	14	3.4
Fungi	0.005	2.7	9.4
Annelids	0.07	24	26.2
Arthropods	–		–
Insects	0.1	34	29–43
Crustaceans	0.35	115	144[a]
Molluscs	0.43	142	287[a]
Gastropods	0.68	224	–
Bivales	1.16	381	–
Echinoderms	0.54	178	144[a]
Chordates	0.20	67	–
Fish	0.39	129	281–337
Amphibians	1.2	394	337
Reptiles	1.5	493	365
Birds	1.7	558	337
Mammals	3.0	985	393
Homo sapiens	2.0	657	701

Sources: Cavalier-Smith (1985), Fonseca et al. (2000).
The values of β_i from Table 5.1 are shown for comparison.
[a] Figures presented by Marques et al. (1997).

for others (Algae, Echinodermms, Amphibians, Reptiles, Birds and Mammals) $\beta_i < \beta_i^{\mathrm{DNA}}$. However, the β^{DNA}-distribution was shifted towards greater values: on average, the DNA-method overestimates the weighting factor. This is easily explained, since the number of non-repetitive genes is, as a rule, less than the total genome size. For instance, the overall C-value for Insect is equal to 0.143 (*D. melanogaster*), but the non-repetitive DNA is 0.102 (71.4%). This latter value of C was used when the weighting factor was calculated for Table 5.2. As a result, $\beta_i \approx \beta_i^{\mathrm{DNA}}$. Another example: the overall C-value for Mammals is about 3.37, but the non-repetitive DNA in *H. sapiens* genome is 2.04 (60.6%). When we use the latter value as the DNA content in Table 5.2, we get very close estimations of weighting factor for *Homo sapiens* ($\beta_i = 701$ and $\beta_i^{\mathrm{DNA}} = 657$). When we do not take into account the reduction of non-nonsense genes then the weighting factors significantly differ from each other ($\beta_i = 393$ and $\beta_i^{\mathrm{DNA}} = 985$ for Mammals).

There is no doubt that the right estimation of β-values should be based on the number of proteins that are controlling the processes in the cells of various organisms. The knowledge of all human genomes is available today, and it has been found that the number of non-nonsense genes is not 250,000 as indicated in Table 5.1 but rather 40,000. On the other hand, it is also clear that the number of amino acids controlled by one gene is more than 700 for *H. sapiens*. It may, for some genes, be as high as 38,000 (Hastie, 2001). The weighting factor in Table 5.1 may therefore still be approximately correct. The importance of the proteins can be seen from the intensive analytical work on finding the composition of the human proteins, or as they are called now, the genetically determined proteins—proteomes. This word is used to underline that the genetically produced proteins are of a particular importance. The great interest in the proteomes is due to their control of the life processes. Many proteins may become the medicine of the future. The application of enzymes in industrial productions is just in its infancy, as there is an enormous potential to control many more industrial processes by enzymes.

On the one hand the key to finding better β-values is the proteomes. On the other hand, our knowledge about the proteomes in various organisms is very limited—more limited than for the number of non-nonsense genes. It may be possible, however, to put together our knowledge about non-nonsense genes, the overall DNA content, the limited knowledge about the proteomes and the evolution tree and see some pattern emerging, which could be used to give better but still very approximate β-values at this stage. For *H. sapiens* it is presumed that 200,000 proteomes are produced by the cells and that they contain about 15,000 amino acids in average (Nielsen, 2003). In this case the β-value would be 12,275, or considerably higher than the value in Table 5.1. However, other values in the table might be changed in a similar manner.

5.9. Why living systems have such a high level of exergy?

A frog weighing 20 g at the temperature of environment 25°C will have the exergy content of $20 \times 62.3 \times 298 \times 337 = 0.125$ GJ, while a dead frog will have only the exergy content of 371 kJ, although they have the same chemical composition, at least until a few seconds after the frog has died. The difference is rooted in the information, or rather in the difference of the useful information. The dead frog possesses the information for a few seconds after its death (the amino acid composition has not yet been decomposed), but the difference between the live and dead frog is the ability to utilise the enormous information stored in its genes and proteom.

The amount of information stored in a frog is surprisingly high. The number of amino acids placed in the right sequence is 8.4×10^7 and for each of the 8.4×10^7 amino acids the number of possibilities is 20. This amount of information is able to ensure reproduction and is transferred from generation to generation which ensures that the evolution can continue because an already favourable combination of properties is conserved through the genes. Because of the very high number of amino acids, 8.4×10^7, it is not surprising that there will always be a minor difference between frogs in the amino acid sequence. It may be a result of mutations or of a minor mistake in the copying process. This variation is important because it gives possibilities to "test" which amino acid sequence gives the best result with respect to survival and growth. The best, representing the most favourable combination of properties, will offer the highest probability of survival and ensure maximum growth so that corresponding genes will therefore prevail. Survival and growth mean more exergy resulting in a larger distance from thermodynamic equilibrium. Exergy could therefore be considered a thermodynamic function, which could be used to quantify Darwin's theory. It is interesting in this context that exergy also represents the amount of energy needed to tear down the system (Svirezhev, 1998b). It means that the more exergy the system possesses the more difficult it becomes to "kill" the system and therefore the probability of survival is higher.

5.10. Summary of the important ecological issues

Introduction of the concept of exergy facilitates the interpretation of the Second Law of Thermodynamic in an ecological context. All ecological processes are irreversible, and exergy (work capacity) is therefore inevitably lost as heat to the environment and, as there is no temperature gradient to utilise, energy that can do work is converted into energy that cannot do work.

The exergy transfer processes in the biosphere can be considered as a Carnot Cycle, driven by the solar radiation. The exergy efficiency of the Carnot Exergy Cycle is, under the given constraints, about 1.5%, irrespective of whatever we consider—the climate cycle (absorption of solar radiation by the atmosphere), the photosynthesis or the GCC. It explains why the photosynthesis has an efficiency as low as about 1.5%. From an evolutionary point of view it may be possible to explain this low efficiency by the factors determining the selection. As energy has been continuously supplied by the sun during the entire evolution, the main competition has been about the elements that are needed to build up living matter because the elements have always been available in a limited amount. The constraints on the exergy efficiency of the solar radiation determine that the exergy efficiency is 1.5%.

Pollution implies a loss of exergy, for instance dispersion or increased concentration of a toxic substance. When the dispersion is global the exergy loss is very significant.

It can be shown that the exergy of an ecosystem corresponds to the amount of energy that is needed to kill the system. It means that the more exergy a system has the more difficult it will be to kill it. Higher exergy levels therefore mean higher overall buffer capacity, which is equal to the sum of resistances against changes. The exergy is, in other words, a useful variable to apply for assessment of ecosystem health.

It can be shown that exergy has two contributions: one determined by the amount of biomass and one covering the information (in the form of Kullback's measure of information). In other words, exergy expresses both the biomass and the information. Higher organisms carry more information and therefore have more exergy than lower organisms with the same biomass. An approximate calculation of living organisms' exergy is possible by Eq. (7.10). The weighing factor β expresses the information content. Note that a second after an organism has died the biochemical composition is still the same as for the living organism, but now the information embodied in the biochemical composition including the amino acid sequence of the enzymes for the life processes cannot be utilised and they are therefore worthless—the weighing factor β becomes 1.0 in Eq. (7.10).

An ecosystem that will develop towards a higher exergy level can only increase the biomass to a certain point (corresponding to the amount of the most limiting elements in its environment). Therefore, a mature system is characterised by attributes that are based on information (higher biodiversity, more complicated network and so on; see Table 2.5), while a system at the early state of ecosystem development has relatively fast growth of the biomass. The level of information can, in principle, increase infinitely, which can explain the continuous evolution.

The efficiency of the exergy gained from the solar radiation is dependent on Kullback's measure of information, K, for the system and the efficiency with which the ecosystem is able to capture the solar radiation. An ecosystem in an early state (see Table 2.5) will only have a low K-value. It implies, with reference to Fig. 5.4, that the active surface (the ecosystem) works as a classical thermodynamic machine, performing mechanical and chemical work (a build up of biomass). The radiation efficiency cannot exceed about 80% due to physical constraints but, by development of a high K value at the time, when a high radiation efficiency has been obtained (a high concentration of biomass that as a parabola captures as much solar radiation as possible), it is possible for the ecosystem to work as an information machine. The point b (see Fig. 5.4) is, by the development of K, moving toward a and b may correspond for a mature ecosystem to a radiation efficiency of more than 80%, which implies that the ecosystem is working as an information machine (information increases, but not the biomass). This is beneficial for further development of the ecosystem because it is hardly possible to do more chemical work for the ecosystem when the elements needed for the construction of biomass are already used up or are there already in the form of living matter. As seen, the shift of the ecosystem between a classic thermodynamic machine doing chemical work and an information machine takes place at the proper state in the ecosystem development.

This chapter has demonstrated clearly that the application of exergy facilitates the thermodynamic explanation of ecosystem reactions; but how can we calculate exergy for an ecosystem? It is hardly possible due to the enormous complexity of ecosystems. It may, however, be possible and useful to calculate the exergy of an ecosystem model. Assessment of ecosystem health by the use of exergy and development of structurally dynamic models, that can account for adaptation and shifts in species composition, require that we can calculate, at least relatively, the exergy for an ecosystem model. If we can calculate the exergy of the components of the model, including the organisms, we can just sum up the contributions to the exergy of the ecosystem model. In principle, we should also add the exergy of the information contained in the ecological network; it is minor compared with the enormous amount of information embodied in the organisms when we are using (simple) models but it may be significant for real ecosystems. It is not difficult to calculate the exergy of living organisms provided that we know the information content. In spite of big progress in genetics, there is still a long way to go (see the alternative methods in Sections 5.8 and 5.9) before we have all the information needed to make more accurate exergy calculations. It is for instance clear that not only the genes determining the amino acid sequence transfer important information for the life processes, but also there are genes that function as the directory and other genes that are important spare parts, although there are also redundant genes. Fortunately, the application of exergy for ecosystem health assessment and for the development of structural dynamic modelling is rather robust and it seems that relative approximate values of exergy are sufficient. To emphasise, we get, by the calculation presented in Section 5.7, only relatively approximate values and it would be preferable to indicate the exergy result as an exergy index. Moreover, several exergy balance calculations can be performed and lead to interesting results, as shown throughout this volume.

Chapter 6
Stability in mathematics, thermodynamics and ecology

Das Sein das Sein des Seinden sei.
Alexander von Humboldt "Flora Freibergiensis"

6.1. Introduction. Stability concepts in ecology and mathematics

At the end of the XVIIIth century, young Alexander von Humboldt in his "Flora Freibergiensis" proclaimed the methodological principle applicable to natural systems: "Das Sein das Sein des Seinden sei" (let the being be the being of the being). From our viewpoint, this is another formulation of the general stability principle: "Only stable systems can exist".

It is intuitively clear that both an ecosystem and a biological community that exist sufficiently long in a more or less invariant state (such a property is often called *persistence*) should possess an intrinsic ability to resist perturbations coming from the environment. This ability is usually termed "stability". It is some general, emergent property (the so-called "scalar invariant") of a system. Apparently, we can observe only *stable* ecosystems, since all *unstable* ecosystems had disappeared in the process of evolution. The environment destroyed them, since they could not *be adapted* to it. (You can see that we introduced the new term: "adaptation". Really, these terms are very close, and we can say that only a stable system is able to be adapted to the environment, i.e. to survive sufficiently long under given environmental conditions.)

In spite of being intuitively clear, "an ability to persist in the course of a sufficiently long time in spite of perturbations" can scarcely be defined in a unique and unambiguous way. The reason is that both the "persistence" and the "perturbations" or "fluctuations" in thermodynamics (as well as the "sufficiently long time") are elements of the idea that need further clarification, to say nothing of the scale of a system under consideration. What is understood by "an ability to persist" and what kind of "perturbations" are relevant? How long can this "sufficiently long time" continue? Different answers to these basic questions and a variety of stability concepts have been proposed and discussed in the literature on mathematical ecology (May, 1973; Svirezhev, 1976, 1987, 2000; Svirezhev and Logofet, 1978, 1995; Jeffries, 1988; Logofet, 1993) and in purely ecological literature (see, for instance, Lewontin, 1969; Usher and Williamson, 1974), and yet few of them have attracted a proper mathematicians' attention (Svirezhev, 1983). The main concepts of

Towards a Thermodynamic Theory for Ecological Systems, pp. 127–152

stability and their application to the problem of complexity are very well shown in the book of Nicolis and Prigogine (1989).

Though the notion of stability seems obvious, it is quite a problem to provide it with a precise and unambiguous definition. In fact, stability can be defined in quite a lot of ways, both in verbal and formal terms, either in ecology or mathematics. While none of the "ecological" meanings of stability can now be recognised as the most fundamental one, mathematics is "luckier", giving rise to the notion of Lyapunov stability. It appears to be inherent in, or substantial for, any further notion of stability—at least within the theory of dynamical systems. And even then, this heavily overloaded term found no established ("stable") definition. For instance, the theory of stability, which can be considered as a branch of applied mathematics and mechanics, uses about 30 different definitions of stability. So, the definition of stability is some "fuzzy" definition. Paraphrasing von Neumann's sentence, we can say that "... nobody knows what stability means in reality, that is why in the debate you will always have an advantage".

6.2. Stability concept in thermodynamics and thermodynamic measures of stability

Among the different definitions, we can select two large classes differing with respect to the requirements coming under the heading of "stability". The first group of requirements concerns preservation of the number of species in a community. A community is stable if the number of member-species remains constant over a sufficiently long time. This definition is the closest to various mathematical definitions of stability, such as those of Lagrange and Poincaré–Lyapunov.

The second group refers rather to populations than to community, which is considered to be stable when numbers of component populations do not undergo sharp fluctuations. This definition is closer to the thermodynamic (or rather, statistical physics) notion of system stability. In thermodynamics (statistical physics) a system is believed to be stable when large fluctuations, which can leave the system far from equilibrium or even destroy it, are unlikely to happen (see, for instance, Landau and Lifshitz, 1995). Evidently, general thermodynamic concepts (for instance, the stability principle associated in the case of closed systems with the Second Law and, in the case of open systems, with Prigogine's theorem) should be applicable to biological (and, in particular, ecological) systems.

There is a very deep connection between thermodynamics as a physical theory and the mathematical theory of stability. One of the most important concepts in the theory of stability is the Lyapunov functions concept. Positive functions, defined in a phase space of a dynamical system, possess the following property: either monotonous increase or monotonous decrease along trajectories. They can be considered as some special class of goal functions. On the other hand, the main thermodynamic laws (the Second Law, Prigogine's theorem) state the similar properties of monotonicity for special functions called potentials, entropy, etc. These functions are Lyapunov functions and thermodynamic laws can be considered as applications of the direct Lyapunov method to special dynamical systems.

It was proved in thermodynamics that in any thermally isolated system for any initial distribution, a single equilibrium distribution is established. Moreover, in accordance with the Second Law the entropy monotonously increases and reaches, at this equilibrium, the maximum S^{max}. Let us introduce the positive function $L = S^{max} - S$ and name it Lyapunov's function. It is obvious that it is equal to zero at the equilibrium and its derivative $\mathrm{d}L/\mathrm{d}t = -\mathrm{d}S/\mathrm{d}t < 0$. If we use the thesaurus of the stability theory, we can say that thermodynamic equilibrium is *globally asymptotically stable*, and the difference between maximal and current values of entropy is Lyapunov's function of the system, which describes such a general property as Lyapunov stability (see below).

If now we keep in mind one of the definitions of exergy, $\mathrm{Ex} = T_0(S^{max} - S)$, and compare it with the definition of L, we can say that exergy is also Lyapunov's function. A negativeness of the exergy derivative in this case means that exergy is dissipated along a trajectory towards thermodynamic equilibrium. It is clear that the opposite inequality for the exergy derivative is possible only if $\mathrm{d}S/\mathrm{d}t < 0$, i.e. the system is open.

In a number of simulation experiments on aquatic ecosystem models, Mejer and Jørgensen (1979) and Jørgensen (1982) observed an increase in exergy in response to certain changes in external conditions, i.e. certain perturbations of the system. The exergy correlated positively with values of "ecological buffer capacities", defined (Mejer and Jørgensen, 1979) as the inverse sensitivity of a state variable to changes in the driving force on a model. Basically, the buffer capacity concept, representing the capability of the ecosystem to absorb perturbations, is a measure of stability with respect to perturbations of a certain kind. Therefore, the idea arises that there may be an internal link between the "increasing exergy" concept and the "Lyapunov stability" concept, the most fundamental one in mathematical stability theory. Heuristically, Lyapunov's function is also natural (as well as consistent) for quantifying how far the current state of a system is from some reference state (see, for example, La Salle and Lefschetz, 1961; Rouche et al., 1977). This idea seems consistent with the concept of exergy, which measures the free energy that a system possesses relative to its environment and shows how far the current state of the system is from the state of thermodynamic equilibrium with the environment.

Lyapunov functions exist for a number of theoretical models of population and community dynamics (e.g. Goel et al., 1971; Svirezhev and Logofet, 1978). These functions look similar to the exergy expressions presented in Chapter 5, which has motivated us to investigate the link between exergy and Lyapunov stability functions. It should be emphasised that Lyapunov functions, depending on their particular forms and properties, may represent either the *stability* or *instability* of a system state.

From the "stability" point of view, it means that thermodynamic equilibrium becomes *unstable*. The illustrations of asymptotic stability and instability are shown in Fig. 6.1a–c. Note that here the equilibrium domain, to which all the trajectories tend (or depart from) at $t \to \infty$, consists of a single point. In the general case, the domain, also called the limiting set, can possess a more complex structure. These sets can be either attracting (Fig. 6.1a) or repulsing (Fig. 6.1b). The attracting set is often simply called the *attractor*. Some intermediate case is possible when trajectories along some directions are attracted, but are repulsed along others (Fig. 6.1c).

In the general case the following situation is possible: if perturbations (fluctuations) are relatively small and do not exceed some threshold, then the system remains in the vicinity of a certain equilibrium; if perturbations exceed it, then the system recedes from it. In this case the equilibrium is *locally stable* but *globally unstable*. In other words, if the system is

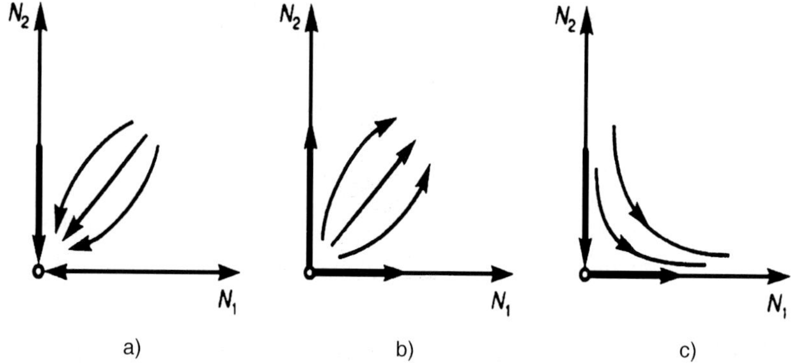

Fig. 6.1. Typical phase patterns in the vicinity of the zero equilibrium state: (a) asymptotic stability (any trajectory tends to zero); (b) instability as covered by Chetaev's instability theorem (any trajectory evolves from zero); and (c) instability in the Lotka–Volterra case (where there is a trajectory that tends to zero).

stable for any initial perturbations, then we talk about global stability, and the attractor is named a *global* attractor. For instance, thermodynamic equilibrium in a thermally isolated system is a global attractor; but as soon as the system is opened the attractor ceases to be global. Moreover, since in an open non-equilibrium system the entropy may increase or decrease, and the Second Law does not already determine the sign of entropy change (besides other function of states), the existence of some general, universal Lyapunov function becomes very problematic (for details see Section 6.4). Immediately the problem of stability for states far from thermodynamic equilibrium arises; a loss of stability (under certain conditions) causes transition phenomena, which can lead to order and structure arising within the system, making it more complex. A typical example of similar phenomenon is the exergy growth, when thermodynamic equilibrium loses stability, and the exergy begins to increase.

A complexity of the system can be seen as the appearance of several locally stable and unstable states. Consider one simple example of a population with two equilibriums: trivial $N_0^* = 0$ and non-trivial $N_K^* = K$. The population dynamics is described by the logistic equation:

$$\frac{dN}{dt} = f(N) = \alpha N\left(1 - \frac{N}{K}\right), \tag{2.1}$$

where N is the population size (or biomass) and α is the intrinsic rate. In Fig. 6.2 the phase portraits of the population are shown for two values of α: $\alpha^+ > 0$ and $\alpha^- < 0$ (see also Section 5.4).

In the first case ($\alpha = \alpha^+ > 0$) the trivial equilibrium $N_0^* = 0$ is unstable, and non-trivial equilibrium $N_K^* = K$ is stable. If $\alpha = \alpha^- < 0$ then the latter becomes unstable, and N_0^* is stable. In this case the trivial equilibrium, which can be associated with the thermodynamic one, is locally stable, and the population always decreases. Let us now consider the open system so the environment can work on the system (population), creating conditions for reproduction, i.e. the shift of intrinsic rate from α^- to α^+. This work is equivalent to some amount of exergy stored by the system. Then the majority of

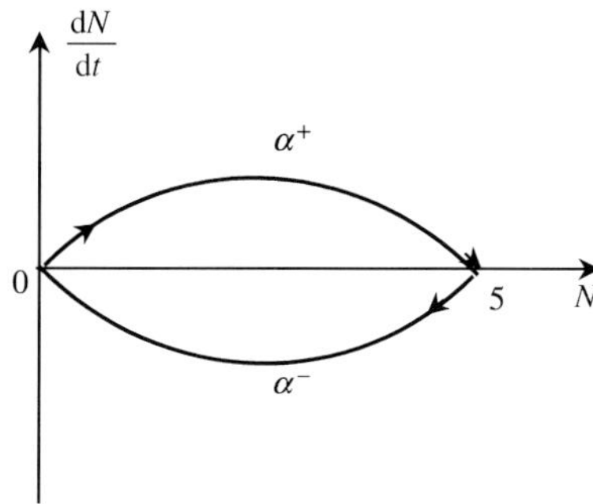

Fig. 6.2. Phase portrait of Eq. (2.1) for two values of α: $\alpha^+ = 1$ and $\alpha^- = -1$; $K = 5$.

the stored exergy is dissipated in order to "organise" a new equilibrium and to maintain it. It is natural that the non-trivial equilibrium can only exist in an open system so the processes of exergy accumulation and dissipation take place simultaneously, balancing each other at the dynamic equilibrium.

In reality, the population always undergoes permanent random perturbations. Since real processes are irreversible, their effect is accumulated, and these small perturbations can "shatter" and destroy the population so that after some time the system will again reach the thermodynamic equilibrium; without perturbations it could exist indefinitely. Constructed in some special way, the Lyapunov function can help us to answer the question: "How long the population within random environment could exist?" (Svirezhev and Logofet, 1978).

Since the Lyapunov function $L(N)$ is a quasi-potential, then $-(\partial L/\partial N) = f(N)$ and after integrating we get

$$L(N) = \alpha\left(\frac{K^2}{6} - \frac{N^2}{2} + \frac{N^3}{3K}\right), \tag{2.2}$$

where the integration constant is determined from the condition $L(K) = 0$. The function is shown in Fig. 6.3.

As a model of random perturbations, we select the "white noise" with the small amplitude $a >> 1$. Let the population be near a stable equilibrium $\overset{\bullet}{N_K} = K$, and Ω_K : $0 < N < \infty$ is the attraction domain of this point. It is natural that, as a measure of stability of this equilibrium against random perturbation, we can take the mathematical expectation of the mean time needed for the trajectory initiated in Ω_K to leave this domain. This value is $\hat{\tau} \sim \exp(2L(0)/a^2) = \exp(\alpha K^2/3a^2)$. It is interesting that this time depends (certainly, with the exception of an intensity of random perturbations) only on the value of the Lyapunov function in the zero equilibrium. We can say that both equilibriums are not independent; and the characteristics of thermodynamic equilibrium significantly influence the stability of dynamics equilibrium even at a long distance.

We continue to consider the problem of Lyapunov functions from the viewpoint of classic thermodynamics. If the entropy is a sufficiently smooth function of the phase variables, then, in the vicinity of equilibrium, the entropy of some open system is

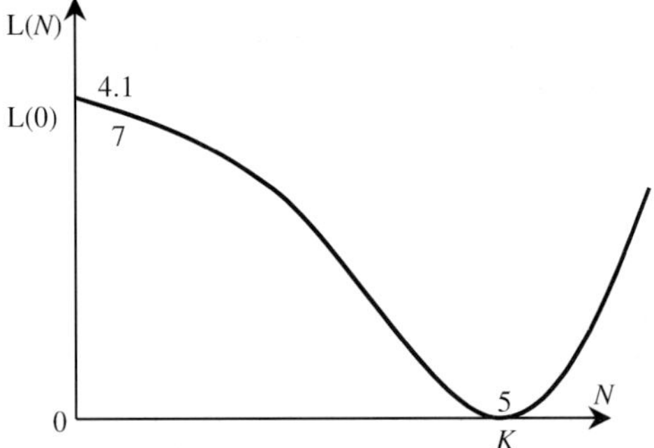

Fig. 6.3. The Lyapunov function (quasi-potential) for logistic equation (2.1) ($\alpha = 1$, $K = 5$).

presented in the form:

$$S = S^{\mathrm{eq}} + (\delta S) + \frac{1}{2}(\delta^2 S). \tag{2.3}$$

Differentiating Eq. (2.3) with respect to time, we get ($\Delta S = S - S^{\mathrm{eq}}$)

$$\frac{\mathrm{d}\Delta S}{\mathrm{d}t} = \frac{\mathrm{d}S}{\mathrm{d}t} = \frac{\mathrm{d}(\delta S)}{\mathrm{d}t} + \frac{1}{2}\frac{\mathrm{d}(\delta^2 S)}{\mathrm{d}t}. \tag{2.4}$$

As usual we represent the difference $\mathrm{d}S$ as the sum of two items, $\mathrm{d}S = \mathrm{d_e}S + \mathrm{d_i}S$, where $\mathrm{d_i}S$ is the entropy change caused by internal processes and $\mathrm{d_e}S$ is the entropy change caused by independent changes in the interaction between the system and its environment. If we assume that the difference $\mathrm{d_i}S$ has a higher order of smallness than $\mathrm{d_e}S$, then

$$\frac{\mathrm{d_e}S}{\mathrm{d}t} = \frac{\mathrm{d}(\delta S)}{\mathrm{d}t} \quad \text{and} \quad \frac{\mathrm{d_i}S}{\mathrm{d}t} = \frac{1}{2}\frac{\mathrm{d}(\delta^2 S)}{\mathrm{d}t}. \tag{2.5}$$

In accordance with the Second Law the entropy within any system must increase (not decrease). Then

$$\frac{\mathrm{d_i}S}{\mathrm{d}t} = \frac{1}{2}\frac{\mathrm{d}(\delta^2 S)}{\mathrm{d}t} \geq 0. \tag{2.6}$$

In thermodynamics the second variation $\delta^2 S$ is presented as a quadratic form of $\delta N_i = N_i - N_i^*$ and if it is negative definite then the corresponding equilibrium is stable (in a thermodynamic sense).

Setting $\delta^2 S = 2L$ we can consider $\delta^2 S$ as the Lyapunov function. In accordance with Lyapunov stability theorem, if $\delta^2 S < 0$ (this inequality is a condition for thermodynamic stability) and $\mathrm{d}(\delta^2 S)/\mathrm{d}t \geq 0$ (this is the Second Law), then the equilibrium is stable (in Lyapunov's sense).

If the considered system is chemical and its variables c_i, $i = 1, \ldots, n$ are molar chemical concentrations, then

$$\delta^2 S = -\frac{1}{T}\sum_{i=1}^{n}\sum_{j=1}^{n}\frac{\partial \mu_i}{\partial c_j}\delta c_i\, \delta c_j, \tag{2.7}$$

where $\mu_i = \mu_i(c_1, \ldots, c_n)$ is the chemical potential of ith component, which in the general case can depend on all other concentrations. For so-called Onsager's systems (see Chapter 3):

$$\gamma_{ij} = \frac{1}{T}\frac{\partial \mu_i}{\partial c_j} = \frac{1}{T}\frac{\partial \mu_j}{\partial c_i} = \gamma_{ji},$$

i.e. these expressions have to be symmetrical.

There is a deep analogy between Onsager's chemical system and the community of competing populations. So, the competition coefficients are analogous to the partial derivatives of chemical potentials. We think that these and other analogies will be helpful if we intend to construct some form of phenomenological thermodynamics of biological communities.

Since L (and also correspondingly, $\delta^2 S$) increases when going further from the stable equilibrium, this may be regarded as a peculiar form of the *Le Chatelier principle*: any displacement from a stable equilibrium increases the competition expenses within the community. It becomes clearer if we re-write the expression for the Lyapunov function as

$$L = -\frac{1}{2}\sum_{i=1}^{n}\sum_{j=1}^{n}\gamma_{ij}(c_i - c_i^*)(c_j - c_j^*) = -\frac{1}{2}\sum_{i=1}^{n}\sum_{j=1}^{n}\gamma_{ij}\,\delta c_i\,\delta c_j, \tag{2.8}$$

where c_i are the equilibrium values.

6.3. Model approach to definitions of stability: formal definitions and interpretations

In contrast with an intuitive understanding of stability typical of the "stability versus diversity" speculations (see Chapter 4), the model approach can provide for quite formal, mathematically rigorous definitions.

Let us assume that we have a "good" enough (from the viewpoint of adequacy and descriptive completeness) mathematical model of a biological community or ecosystem, then stability properties of a real system can be deduced from investigating its model by the mathematical technique of stability theory.

But, as was mentioned above, there are a lot of formal definitions of stability. The task is to decide just what kind of model behaviour should correspond to a stable functioning of the real system and to select those of the mathematical stability definitions which are adequate both to a meaningful, say, "ecological", perception of stability and to the mathematics of the model.

Stability investigations are thus dependent on a particular mathematical model, assuming they are adequate enough for the studied system. As far as the assumption being true, the model approach has an obvious advantage in its prognostic ability, as well as in its capability of relating stability to other systems properties such as the structure and particular mechanisms of functioning. Expressed in formal terms, stability conditions of the model promote a formulation of hypotheses concerning the functioning of the real system.

Besides everything else, the adequacy assumption itself may always be questioned, and sometimes answered at least, in qualitative terms, from the outcome of stability analysis too.

Let the system dynamics and its evolution be described by the system of ordinary differential equations:

$$\frac{dx_i}{dt} = F_i(x_1, \ldots, x_n; \lambda_1, \ldots, \lambda_m), \quad i = 1, \ldots, n \tag{3.1a}$$

or in a vector form:

$$\frac{d\mathbf{x}}{dt} = \mathbf{F}(\mathbf{x}, \boldsymbol{\lambda}), \tag{3.1b}$$

where $\mathbf{x} = \{x_1, \ldots, x_n\}$ and $\mathbf{F} = \{F_1, \ldots, F_n\}$. The $\boldsymbol{\lambda} = \{\lambda_1, \ldots, \lambda_m\}$ is a vector of parameters, which describes an influence of the environment on the system. It is natural that only the environment can change them. The system has an equilibrium $\mathbf{x}^*(\boldsymbol{\lambda})$, so that $\mathbf{F}(\mathbf{x}^*) = 0$. Let S_R be a spherical domain $\|\mathbf{x} - \mathbf{x}^*\| < R$ and H_R be a sphere, which is bounding this domain.

In accordance with Lyapunov's theory (see, for instance, Rouche et al., 1977) the equilibrium $\mathbf{x}^*(\boldsymbol{\lambda})$ of system (3.1) is

(a) *stable*, if for any given ε $(0 < \varepsilon < A)$ such $\eta(\varepsilon)$ $(0 < \eta(\varepsilon) < \varepsilon)$ is found, then the solution $\mathbf{x}(t, \boldsymbol{\lambda})$ of system (3.1) with initial state $\mathbf{x}_0 = \mathbf{x}(0) \in S_\eta$ will never leave the domain S_ε;

(b) *asymptotically stable*, if it is stable in the sense (a), and moreover, if for the certain ε $\mathbf{x}(t, \boldsymbol{\lambda}) \to \mathbf{x}^*(\boldsymbol{\lambda})$ by $t \to \infty$. Later on, if we talk about *stability*, then, as a rule, we take into account the *asymptotic* stability;

(c) *unstable*, if for any arbitrary given ε $(0 < \varepsilon < A)$ and arbitrary $\eta(\varepsilon)$ $(0 < \eta(\varepsilon) < \varepsilon)$ the solution $\mathbf{x}(t, \boldsymbol{\lambda})$ is always found, which starts from some point $\mathbf{x}_0 \in S_\eta$ and attains H_ε at some finite moment of time $t > 0$.

The theorems which are formulated below are based on the following concept: the scalar function $L(\mathbf{x})$ is called *positive (negative) definite in the domain* S_A if, in this domain, $L(\mathbf{x}^*) = 0$ and $L(\mathbf{x}) > 0$ (<0) at any other points of S_A. The total derivative of $L(\mathbf{x})$ with respect to time:

$$\dot{L} = \frac{d}{dt} L[\mathbf{x}(t, \boldsymbol{\lambda})], \tag{3.2}$$

along the trajectory $\mathbf{x}(t, \boldsymbol{\lambda})$ of system (3.1) in the domain S_A is taken into consideration (the so-called Lee derivative), so that

$$\dot{L} = \left(\frac{\partial L}{\partial \mathbf{x}}, \mathbf{F}\right) = \sum_{i=1}^{n} \frac{\partial L}{\partial x_i} F_i. \tag{3.3}$$

Such a type of functions is called the *Lyapunov function* for a system (3.1).

1. *Stability theorem*. If for some S_A such a positive definite function $L(\mathbf{x})$ is found, the derivative $\dot{L}$ along the trajectories of Eq. (3.1) is a non-positive definite function, then the equilibrium $\mathbf{x}^*(\boldsymbol{\lambda})$ of system (3.1) is stable.

2. *Asymptotic stability theorem.* If the derivative $\dot{L}$ is a negative definite function in S_A, then the equilibrium $\mathbf{x}^*(\boldsymbol{\lambda})$ is asymptotically stable.
3. *Instability theorem.* If some positive definite in S_A function $L(\mathbf{x})$ exists, the derivative $\dot{L}$ along the trajectories of Eq. (3.1) is also a positive definite (except, maybe, the point $\mathbf{x}^*(\boldsymbol{\lambda})$), then the equilibrium $\mathbf{x}^*(\boldsymbol{\lambda})$ is unstable. Moreover, we can hit any surface H_ε ($\varepsilon < A$) moving along a trajectory, which starts at any arbitrary point of S_ε (except the equilibrium point).

   All these theorems have been formulated and proved by A. Lyapunov. We will also use the following additions.
4. *Chetaev theorem about instability* (Chetaev, 1955). Note that the previous instability theorem has one principal defect: the derivative $\dot{L}$ has to be positive definite in the entire domain S_A. Meanwhile, in order to detect the instability (if it takes place), it is sufficient to detect even one unstable trajectory in an arbitrary small vicinity of equilibrium. For this we have to know the behaviour of trajectories in some part of S_A, not in the entire domain. If in an arbitrary small vicinity of equilibrium a domain exists in which $L(\mathbf{x}) > 0$ ($L(\mathbf{x}) = 0$ on its boundary) and $\dot{L}(\mathbf{x}, \boldsymbol{\lambda}) > 0$ at all its points, then the equilibrium is unstable.
5. *Barbashin–Krasovsky supplement about asymptotic stability* (Barbashin, 1967). In the stability theorem: (a) $S_A = S_\infty$ and (b) $L \rightarrow \infty$ together with $\|\mathbf{x}\|$, then all trajectories converge to the equilibrium $\mathbf{x}^*(\boldsymbol{\lambda})$.

The last theorem is a simple conclusion of a theorem about the instability of equilibrium at infinity.

Note that the stability conditions depend on parameters $\boldsymbol{\lambda}$; therefore, the whole domain of parameters definition could be divided into different sub-domains, in each one of which the system is either stable or unstable.

6.4. Thermodynamics and dynamical systems

Returning to Section 6.3 we again consider a dynamical system (3.1): $\mathrm{d}x_i/\mathrm{d}t = F(x_1, \dots, x_n; \lambda_1, \dots, \lambda_m)$, $i = 1, \dots, n$. It will be shown in Chapters 7 and 8 that the dynamic equations in ecology can be interpreted from the thermodynamic point of view. If we want to develop a phenomenological thermodynamics for the system describing these equations, then the Lee derivative of the corresponding Lyapunov function can be considered as the entropy production, which is also called a *function of dissipation*, Diss. In phenomenological thermodynamics $\mathrm{Diss} = \sum_{i=1}^{n} X_i J_i$ (Chapter 2) where X_i are generalised thermodynamic forces and J_i are generalised thermodynamic fluxes. Formally, we can assume $X_i = \partial L/\partial x_i$ and $J_i = \mathrm{d}x_i/\mathrm{d}t$, i.e. generalised fluxes are equal to components of the vector (phase) field $\mathbf{x}$. Then $\mathrm{Diss} = \sum_{i=1}^{n} (\partial L/\partial x_i)(\mathrm{d}x_i/\mathrm{d}t) = \dot{L}$, i.e. the Lee derivative of the Lyapunov function can be considered as a dissipative function for the systems, described by Eq. (3.1). Thus, there is a certain connection between the dynamical theory and the irreversible thermodynamics. One can say that the dynamical systems with global non-negative production of entropy have sufficiently simple topological construction of phase space. In other words, the systems, which have Lyapunov functions, have also a sufficiently simple topology of phase space. However, in the general case, there is no constructive algorithm which allows obtaining Lyapunov functions for any dynamical systems. Keeping in mind our thermodynamic analogies, we can assume that a complex behaviour of dynamical systems is determined by a "patchy" structure of phase space with respect to the non-negative production of entropy. The "patchiness" means that, in some domains of phase space, the production of entropy is positive and in others it is negative (i.e. the production of negentropy takes place). Finally, there are domains in which the entropy production is equal to zero. In other words, in this case, the general Lyapunov function does not exist.

Continuing our analogy, we can say that if $\dot{L} = \mathrm{d}L/\mathrm{d}t$ is the entropy production, then the value L may be considered as the entropy (to within some arbitrary item). The analogy will be complete if the expression $\mathrm{d}L = \sum_{i=1}^{n} (\partial L/\partial x_i)\mathrm{d}x_i$ is a full differential that does not always occur. However, if $\partial^2 L/\partial x_i\, \partial x_j = \partial^2 L/\partial x_j\, \partial x_i$ then $\mathrm{d}L$ will be a full differential, and L (as the entropy) is a *function of state*. It is easy to see that all the so-called *separable* Lyapunov functions, which are represented as $L(x_1, \ldots, x_n) = \sum_{i=1}^{n} L_i(x_i)$, have full differentials, since

$$\partial^2 L/\partial x_i\, \partial x_j = \partial(\partial L_i(x_i)/\partial x_i)/\partial x_j = 0, \qquad \partial^2 L/\partial x_j\, \partial x_i = \partial(\partial L_j(x_j)/\partial x_j)/\partial x_i = 0.$$

If we remember the definition of exergy, $\mathrm{Ex} = T(S_i^{\mathrm{eq}} - S)$, where S is the entropy and S_i^{eq} is its value at some ith (thermodynamic) equilibrium. Then under isothermal condition $T = \mathrm{const}$, we can interpret the Lyapunov function as exergy, $L = \mathrm{Ex}$. The exergy is positive, and if its derivative is also positive, then it means that the thermodynamic equilibrium will be repulsing. We can say that by receding from it the system is accumulating the exergy. In the opposite case, when the exergy derivative is negative, the system approaching the thermodynamic equilibrium spends its storage of exergy. Since $\mathrm{d}(\mathrm{Ex})/\mathrm{d}t \sim -\mathrm{d}S/\mathrm{d}t$ then the curious linguistic pair arises:

the system produces the entropy = the system spends the exergy,

the system produces the negentropy = the system accumulates the exergy.

Although at first glance this is not more than a purely philological game, there is, however, a deep sense in this game. In fact, when we deal with closed systems, which are under the action of the Second Law, then the entropy is always increasing, i.e. it is always produced. But as soon as we have encountered an open system, we have seen that the entropy might not only increase but also even decrease, and its derivative could be negative. This contradiction could be resolved in two different ways. The first is to assume that the entropy may be negative (negentropy), then the decrease of entropy corresponds to the increase of negentropy. In this case the system produces the negentropy. The second way is to define the antinomy at the level of derivatives, for instance, "some values are accumulated" versus "some values are spent or dissipated". As follows from the above considerations, this value may be the exergy.

Note that in the classic dynamical theory all the limiting sets (attracting or repulsing) are manifolds, i.e. they have a simple topological structure (point, line, curve, cycle, torus). Our thermodynamic speculations allow us to formulate the following statement: *dynamical system (3.1) has not any attracting set, which is not a manifold, if and only if the Lyapunov function* $L(\mathbf{x}) \not\equiv 0$ $[\mathrm{d}L(\mathbf{x})/\mathrm{d}t \geq 0]$ *exists for system (3.1).*

It became clear (Lorenz, 1963) that even for the system of three ordinary differential equations with quadratic non-linearity in the corresponding phase space, there is an attracting invariant set, which is bounded and connected. It differs from both a point and a limiting cycle, and its value is equal to zero. Such a type of set is called a "strange attractor" (Ruelle and Tackens, 1971). Locally, a strange attractor is the direct product of a two-dimensional manifold and some Cantor set. The fate of the representing point is explicitly unpredictable within the attractor, i.e. its trajectories will be irregular and chaotic. The system behaviour will be very sensitive in relation to initial values. Another simple example of ecological system with chaotic behaviour will be considered in Chapter 7.

It is possible that the dynamical systems, which demonstrate such a type of "strange" behaviour, do not have the corresponding Lyapunov functions. In other words, if the system in certain domains of phase space produces entropy, and negentropy in others, then the behaviour of the system may become chaotic and irregular. Certainly, instead of entropy we can use exergy. Then the last sentence may be reformulated as: if in some domains of phase space the system accumulates exergy, in others the system dissipates it, then its behaviour may be chaotic and irregular.

It is known that the Lyapunov function can be constructed for any linear system. On the other hand, all the limiting sets of linear systems are manifolds.

And finally, let the considered dynamical system be a system of gradient type, i.e. $\mathrm{d}\mathbf{x}/\mathrm{d}t = \mathrm{grad}\, \Phi(\mathbf{x})$. The Lyapunov function exists for these systems, and is: $L = (\mathrm{grad}\, \Phi, \mathrm{grad}\, \Phi) \geq 0$. Therefore, gradient systems do not have strange attractors. We shall deal with such a type of equations in Chapter 8, when we consider the dynamics of competing populations in a certain way transformed space. Since the Svirezhev–Shahshahani transformation (Burger, 2000; see also Section 6.5) used for this does not change types of singularities, there are no any strange attractors in the real space if they were absent in the transformed one. From this, it follows that the dynamics of the community of competing populations are always regular; the community cannot principally have irregular and chaotic trajectories.

6.5. On stability of zero equilibrium and its thermodynamic interpretation

Let us consider the following dynamical system:

$$\frac{\mathrm{d}N_i}{\mathrm{d}t} = F_i(N_1, \ldots, N_n), \quad i = 1, \ldots, n, \tag{5.1}$$

where vector $\mathbf{N} = \{N_1, \ldots, N_n\}$ is non-negative, i.e. belongs to the positive orthant $\mathbf{P}^n$ of Euclidian space $\mathbf{E}^n$. We also assume that all F_i are analytical with respect to their variables, i.e. they are expanded to Taylor's series in the vicinity of any point of phase space. This is a typical form of equations describing the dynamics (kinetic) of a system; each state, in turn, is described either by the number of individuals (populations, communities) or by the number of molecules (chemical concentrations). It is natural that all these values are non-negative. In particular, in mathematical ecology the values N_i are interpreted as either biomasses or densities of corresponding species, age cohorts and other ecological groups constituting a biological community. If we assume that the system occupies some fixed spatial volume, then the values N_i can be considered as either biomass densities per volume unit or a volume concentration of some substances. Later on, we choose the first interpretation, namely, the value N_i is the biomass of ith species in biological community.

The trivial equilibrium $\mathbf{N}^* = 0$, which corresponds to the full absence of biological "particles" in the system, is naturally interpreted as thermodynamic equilibrium. Since $\mathbf{N}^* = 0$ is an equilibrium then $F_i(0, \dots, 0) = 0$, $i = 1, \dots, n$. It means that a Taylor-series expansions of F_i in a vicinity of $\mathbf{N}^* = 0$ has to begin with linear terms, so that

$$F_i \approx \sum_{j=1}^{n} a_{ij} N_j. \tag{5.2}$$

A natural question arises: under what kind of conditions will the trivial equilibrium be unstable? In other words, what kind of conditions has to be fulfilled in order for the system to be able to leave the thermodynamic equilibrium and begin to recede far from it, i.e. for the life to arise within the system?

It seems that the problem's solution is very simple when using the Lyapunov instability theorem. However, it is not the case, since the equilibrium lies on a boundary of admissible domain (positive orthant), and any negative variations which lead out of it are senseless. In other words, keeping in mind the definition of Section 6.3, we cannot surround the equilibrium point by spheres which are fully situated in the admissible domain. But the thing which is impossible in the usual phase space of population numbers becomes possible in the space of new variables $z_i = \pm\sqrt{2N_i}$, $i = 1, \dots, n$. This is the so-called Svirezhev–Shahshahani transformation, which transforms the positive orthant into the entire phase space. In this connection, a boundary singularity is transformed into an internal one, and the origin of coordinates (the boundary point of positive orthant) becomes an internal point.

Therefore, there are no obstacles to use the Lyapunov instability theorem for the transformed system (5.1), which is

$$\frac{\mathrm{d}z_i}{\mathrm{d}t} = \frac{F_i(z_1^2, \dots, z_n^2)}{z_i}, \quad i = 1, \dots, n, \tag{5.3}$$

where $\mathbf{z} \in \mathbf{E}^n$, i.e. it could be used for the entire phase space. Since the origin of coordinates $\mathbf{z} = 0$ is an equilibrium point then

$$\lim_{\|\mathbf{z}\| \to 0} \frac{F_i(z_1^2, \dots, z_n^2)}{z_i} = 0, \quad i = 1, \dots, n. \tag{5.4}$$

This condition has to be fulfilled for *any* path along which the representative point moves to the origin of the coordinates. This is possible if the functions $F_i(N_1, \dots, N_n)$ are represented in the form $F_i = N_i f_i(N_1, \dots, N_n)$, where f_i are again analytical with respect to its variables. Then $a_{ii} \neq 0$, $a_{ij} = 0$ for all $i \neq j$ in expansion (5.2). In this case, Eq. (5.2) is written as

$$\frac{dz_i}{dt} = \frac{1}{2} z_i f_i(z_1^2, \ldots, z_n^2), \quad i = 1, \ldots, n. \tag{5.5}$$

It is obvious that the function $L = \sum_{i=1}^{n} z_i^2/2$ is equal to zero at the point $\mathbf{z}^* = 0$ and positive definite in the entire phase space. If the derivative dL/dt is also positive definite in some domain S_A, then the thermodynamic equilibrium is unstable.

Returning to our real variables, N_i, we get $L = \sum_{i=1}^{n} N_i$, and the Lyapunov function is nothing more or less than the total number (biomass) of communities, N. All Lyapunov's spheres $H_c : \|\mathbf{z}\| = c$ in the transformed space correspond to the simplexes $\Sigma_c : N = \sum_{i=1}^{n} N_i = c$ in the real space of numbers. The surfaces of constant biomass are the surfaces of constant "energy" in the transformed "mechanistic" space.

The derivative of the Lyapunov function is the rate of biomass, $dL/dt = dN/dt$. Then we can say that, in accordance with Chetaev instability theorem, the increase of total biomass, even if along a single trajectory of the system, is a sufficient condition for the instability of thermodynamic equilibrium to arrive. If the total biomass increases in its entire vicinity S_A, in accordance with Lyapunov instability theorem, then any trajectory of the system which starts at any arbitrary point of S_ε ($\varepsilon < A$) (except the thermodynamic equilibrium point) attains the sphere H_ε. In the real phase space of species biomasses this geometric result can be interpreted in the following manner.

It is clear that sooner or later the total biomass of community is stabilised at some equilibrium level N^*. The level depends on the flow of external energy and resource, which can be assimilated and dissipated by the system. The simplex $\sum_{i=1}^{n} N_i^* = N^*$ bounds the domain S_A, in which the derivative dN/dt is positive. Then any trajectory started within the simplex will, in any case, reach the simplex $\sum_{i=1}^{n} N_i = N^* - \varepsilon$, which will differ arbitrarily little from the equilibrium simplex.

Finally, we can say that although the results of the chapter seem trivial at first sight, the fact that the total biomass is the Lyapunov function for thermodynamic equilibrium allows us to consider this value as a macroscopic variable of the system.

6.6. Stability of non-trivial equilibrium and one class of Lyapunov functions

Let system (5.1) have a single non-trivial equilibrium $\mathbf{N}^*$, which is situated within orthant $\mathbf{P}^n$. We shall consider the following class of functions, which may be candidates for Lyapunov functions (Svirezhev, 1998b):

$$L = \sum_{i=1}^{n} N_i^* \varphi(N_i/N_i^*) \text{ or, if } \xi_i = N_i/N_i^*, \; L = \sum_{i=1}^{n} N_i^* \varphi(\xi_i), \tag{6.1}$$

where the function $\varphi(\xi)$ possesses the following properties:

$$\varphi(1) = \varphi'(1) = 0; \quad \varphi'' > 0 \text{ for any } \xi \geq 0. \tag{6.2}$$

Since $\mathbf{N} \in \mathbf{P}^n$ then the vector $\vec{\xi} = \{\xi_1, \ldots, \xi_n\}$ also belongs to the positive orthant. It is obvious that the point $\xi_i = 1$, $i = 1, \ldots, n$ corresponds to the equilibrium $\mathbf{N}^*$. In other words, the function $\varphi(\xi)$ has to be convex for positive ξ. It is obvious that $L(\mathbf{N}^*) = 0$; therefore, the first variation of L in the vicinity of $\mathbf{N}^*$ is equal to

$$\delta L = \sum_{i=1}^{n} \frac{\partial L}{\partial N_i} \delta N_i = \sum_{i=1}^{n} \frac{\partial \varphi}{\partial \xi_i} \delta N_i, \tag{6.3}$$

then $\delta L(\mathbf{N}^*) = 0$ for any arbitrary variations $\delta \mathbf{N}^*$. By calculating the second variation we get

$$\delta^2 L = \frac{1}{2} \sum_{i=1}^{n} \sum_{j=1}^{n} \frac{\partial^2 L}{\partial N_i \, \partial N_j} \delta N_i \, \delta N_j = \frac{1}{2} \sum_{i=1}^{n} \varphi''(\xi_i) \frac{(\delta N_i)^2}{N_i^*} > 0 \tag{6.4}$$

for any non-zero variations $\delta \mathbf{N}$. Thus, L is a convex function of $\mathbf{N}$, which has an isolated minimum at the point $\mathbf{N}^*$ and monotonous increases with increase in the norm $\|\mathbf{N}\|$ in the orthant $\mathbf{P}^n$. In accordance with Lyapunov stability theorem the equilibrium $\mathbf{N}^*$ is asymptotically stable in some domain $S_A \subseteq \mathbf{P}^N$ if the derivative $\mathrm{d}L/\mathrm{d}t$ taken along the trajectories of Eq. (5.1) in S_A will be negative:

$$\frac{\mathrm{d}L}{\mathrm{d}t} = \sum_{i=1}^{n} \varphi'(\xi_i) F_i(N_1, \ldots, N_n) \leq 0, \tag{6.5}$$

where the equality is fulfilled only at $\mathbf{N}^*$. On the other hand, the equilibrium $\mathbf{N}^*$ is unstable in S_A, if the derivative $\mathrm{d}L/\mathrm{d}t$ is positive, except $\mathbf{N}^*$ where it is equal to zero (Lyapunov instability theorem). Generally speaking, in accordance with Chetaev instability theorem, it is sufficient for instability if $\mathrm{d}L/\mathrm{d}t$ would be positive even along a single trajectory of Eq. (5.1) in a small vicinity of equilibrium.

Note that, in Eq. (5.1), if $F_i(N_1, \ldots, N_n) = N_i f_i(N_1, \ldots, N_n)$, i.e. the origin of coordinates is also equilibrium, then only the interior of orthant can be considered as the domain S_A. It becomes evident by expression (6.5) for $\mathrm{d}L/\mathrm{d}t$: in this case the derivative $\mathrm{d}L/\mathrm{d}t$, in addition to the non-trivial equilibrium, is equal to zero also at the point $\mathbf{N} = 0$.

Let us consider one partial form of the function φ: $\varphi(\xi_i) = N_i^*(1 - \xi)^2$. It is easily seen that $\varphi(\xi_i)$ possesses all the necessary properties in order to be the Lyapunov function. In fact, $\varphi(1) = 0$, $\varphi'(1) = 0$, $\varphi''(\xi_i) = 2N_i^* > 0$. Then the Lyapunov function is written as (see Eq. (6.1))

$$L = \left[\sqrt{\sum_{i=1}^{n} (N_i - N_i^*)^2} \right]^2, \tag{6.6}$$

i.e. the value of L may be considered as a square of Euclidian distance between a current state of the system and its equilibrium. If this distance decreases with time (i.e. the derivative $\mathrm{d}L/\mathrm{d}t < 0$ along any system's trajectory), then in accordance with the Lyapunov stability theorem we have hoped that the system goes to the stable equilibrium. Note that I speak of "hope", since we have to test all the trajectories (or a statistically "sufficient" large number of them) in order to be able to speak correctly about stability of equilibrium. If, on the contrary, this distance increases along some trajectory, then (in accordance with Chetaev instability theorem) the equilibrium is unstable and this trajectory recedes from it.

All these results can be easily interpreted from the thermodynamic point of view. Indeed, in thermodynamics the value of L is proportional to the mean square of fluctuations around equilibrium or the *power of fluctuations*. Therefore, we can state that if the power of fluctuations decreases with time, then the system goes to stable equilibrium. Note that if the movement in the direction of this equilibrium can be called an *evolution of the system*, then the foregoing statement can be reformulated in the following way: the power of fluctuations decreases during the process of the system's evolution. Since the state $\mathbf{N}^*$ is a *goal of evolution* then, if we understand the evolution in this sense, the Lyapunov function L will be a *goal function*.

In the opposite case, when the equilibrium is unstable, the power of fluctuations increases with time. We can say that the growth of fluctuations in the system, when it leaves thermodynamic equilibrium, is a sufficient condition for its destruction, since upon their action, sooner or later the system will come to the thermodynamic equilibrium again.

6.7. Lyapunov function and exergy

Let the function φ be $\varphi(\xi) = \xi - \ln \xi - 1$, then $\varphi(1) = 0$; $(\varphi' = 1 - \xi^{-1})_{\xi=1} = 0$; $\varphi'' = 1/\xi^2 > 0$, and the corresponding Lyapunov function will be

$$L = \sum_{i=1}^{n} [(N_i - N_i^*) - N_i^* \ln(N_i/N_i^*)]. \tag{7.1}$$

When we introduced such a kind of Lyapunov function then, generally speaking, we assumed that the equilibrium $\mathbf{N}^*$ was non-trivial. It is necessary that the function L and its Lee derivative do not have any singularities at the equilibrium point. However, it is easy to see that both the first and the second derivatives of φ tend to infinity as $\xi \to 0$. On the other hand, apparently, there are many cases when such a sort of Lyapunov function can be used even for $\mathbf{N}^* = 0$. Namely, there is a similar case to what we have here when

$$\lim_{\mathbf{N}^* \to 0} \left\{ L = \sum_{i=1}^{n} [(N_i - N_i^*) - N_i^* \ln(N_i/N_i^*)] \right\} = \sum_{i=1}^{n} N_i = N.$$

Comparing this result and the results of Section 6.4, we can see that in the limiting case, the Lyapunov function belonging to this type coincides with the same one for trivial equilibrium.

If we re-write Eq. (7.1) as $L = \sum_{i=1}^{n} [N_i^* \ln(N_i^*/N_i) - (N_i^* - N_i)]$ and compare this expression and the expression for exergy, suggested in Chapter 5, we can see that these expressions are equivalent (within units, in which the species numbers are measured), but only when assuming that the reference state is equivalent to the current state of the system (5.1) for which the Lyapunov function (7.1) is defined (see definition of exergy). Using frequencies $p_i = N_i/N$ and $p_i^* = N_i^*/N^*$ we re-write the expression for L in the form:

$$L = N^* K(\mathbf{p} \to \mathbf{p}^*) + N^* \ln(N^*/N) - (N^* - N), \tag{7.2}$$

where $K(\mathbf{p} \to \mathbf{p}^*) = K_{pp^*} = \sum_{i=1}^n p_i^* \ln(p_i^*/p_i) \geq 0$ is Kullback's measure, which is equal to the increment of information when the distribution $\mathbf{p}$ is transformed to $\mathbf{p}^*$.

Let us assume that the general dynamics of the system could be presented as a sum of the fast dynamics of the total biomass N and the slow dynamics of community composition $\mathbf{p}$. Then the total biomass is quickly established at the equilibrium level, so that $N = N^*$ and $L = N^* K_{pp^*} = NK_{pp^*}$. The last result may be interpreted in the following way: K_{pp^*} is a specific value of information store per one unit of biomass in the community, which is spent when the current structure of the community evolves to the stable equilibrium $\mathbf{p}^*$. The other interpretation is: K_{pp^*} is an information distance from a current state to equilibrium.

Let the trajectory be started at some initial point $\mathbf{N}_0$: $\{N_1^0, \ldots, N_n^0\}$ (or $(\mathbf{p}_0, N_0)$: $\mathbf{p}_0$: $\{p_1^0, \ldots, p_n^0\}$; $N_0 = \sum N_i^0$). Then the Lyapunov function $L(\mathbf{N}_0, \mathbf{N}^*)$ can be considered as a distance from the initial point $\mathbf{N}_0$ to the non-trivial equilibrium $\mathbf{N}^*$. If the equilibrium is stable then the trajectory has to move to it, and the distance has to be reduced. However, another interpretation is also possible here. If the initial point is situated somewhere near the thermodynamic equilibrium at the simplex $\sum_{i=1}^n N_i^0 = N_0$, the point $\mathbf{N}^*$ is far from it so that $N_0/N^* = \varepsilon << 1$ and, keeping in mind that the exergy is almost equal to zero in a vicinity of thermodynamic equilibrium, then the value $L(\mathbf{N}_0, \mathbf{N}^*)$ can be considered as *the exergy which was accumulated by the system in the course of evolution from thermodynamic to dynamic equilibrium.* The evolution is a forced movement, which occurs on account of the exchange between the system and its environment. *Spontaneous processes, which are accompanied by the exergy dissipation, shift the dynamic equilibrium to the thermodynamic one.* Note that when we talk about the evolution we imply the following process.

By representing the expression for L as

$$L(\mathbf{N}_0, \mathbf{N}^*) = \mathrm{Ex} = N^*\left[K_{p_0p^*} + \ln(1/\varepsilon) - 1 + \varepsilon\right] = \mathrm{ex}\, N^*, \tag{7.3}$$

we see that the value $\mathrm{ex} = K_{p_0p^*} + \ln(1/\varepsilon) - 1 + \varepsilon \approx K_{p_0p^*} + \ln(1/\varepsilon)$ can be considered as a *specific* exergy per individual or one unit of biomass, which consists of two items: the first is determined by the increment of information ($K_{p_0p^*}$) caused by the evolution of composition (implicitly assuming that the total biomass is not changed), and the second is determined by the increment of information, caused by the growth of the total biomass by $(1/\varepsilon)$ times under the permanent composition.

6.8. One more Lyapunov function

Let us consider the following function $\varphi(\xi) = \xi \ln \xi - \xi + 1$. It is obvious that $\varphi(1) = 0$; $(\varphi')_{\xi=1} = (\ln \xi)_{\xi=1} = 0$; $\varphi'' = 1/\xi > 0$, and the corresponding Lyapunov function is

$$L(\mathbf{N}^*, \mathbf{N}) = \sum_{i=1}^{n}\left[N_i \ln(N_i/N_i^*) - (N_i - N_i^*)\right]. \tag{8.1}$$

If we again compare Eq. (8.1) and the expression for exergy, we can see that the latter coincides with the formally defined Lyapunov function. However, there is a difference between this exergy and the exergy defined in Section 6.7: if in the previous case, a reference state, which corresponds to thermodynamic equilibrium, was a current (initial) state and not at a dynamic equilibrium, then a reference state is a dynamic equilibrium, close to a thermodynamic one. If in the first case, a *goal of evolution* is the dynamic equilibrium, which is far from the thermodynamic one, then in the second case, a goal of evolution is not defined. The system is simply going away from thermodynamic equilibrium, which is an unstable dynamic one in this case. The value

$$L(\mathbf{N}^*, \mathbf{N}) = NK_{p^*p} + N\ln(N/N^*) - (N - N^*) = N\left[K_{p^*p} + \ln(1/\varepsilon) - 1 + \varepsilon\right], \quad (8.2)$$

where $K_{p^*p} = \sum_{i=1}^{n} p_i \ln(p_i/p_i^*)$ and $\varepsilon = N^*/N$, can also be interpreted as a measure of how the system has receded from thermodynamic equilibrium. The Kullback measure K_{p^*p} is equal to the increment of information in the course of evolution of the structure when it is transformed from some prime pattern to the current one. If we assume $\varepsilon \ll 1$ then the value $\mathrm{ex} = K_{p^*p} + \ln(1/\varepsilon)$ also may be considered as a *specific* current exergy.

Logically, we may say that the origin of life can be considered as the loss of stability for thermodynamic equilibrium and the movement of the system away from it along one trajectory. In this case (in accordance with Chetaev instability theorem), if the exergy increases along this trajectory, i.e. the inequality $\mathrm{d}L/\mathrm{d}t = \mathrm{d(Ex)}/\mathrm{d}t > 0$ takes place, then the thermodynamic equilibrium is unstable. It is easy to see that this is the other formulation for Jørgensen's maximal principle. (Note that here we implicitly assume that the reference state, i.e. the thermodynamic equilibrium, is also one of the possible equilibriums of the considered dynamical system.)

In spite of an outer similarity between these two definitions of exergy, there is a deep difference between them. The point is that in the first case, we deal with a *teleological* system, which has a goal function; the system "knows" about the final target of evolution (non-trivial equilibrium) and tends to it. We assume implicitly that the system could tear away from a sphere of attraction of the thermodynamic equilibrium, and then it can move to the known target. In the second case, the target is to tear from the sphere, the furthest is not important. Later on, we shall consider the problem in detail.

Finally, it is necessary to note that the "second" exergy has a logarithmic singularity when one or several $N_i \to 0$ when $\mathrm{Ex} \to \infty$. The problem of how to regularise the singularity will also be considered later on. Note that one method of regularisation was already suggested in Section 5.5 of Chapter 5.

6.9. What kind of Lyapunov function we could construct if one or several equilibrium coordinates tend to zero

In Section 6.8 the exergy was defined as

$$\mathrm{Ex} = L = \sum_{i=1}^{n} \{N_i \ln(N_i/N_i^*) - (N_i - N_i^*)\}, \quad (9.1)$$

where the equilibrium $\mathbf{N}^*$ is considered as a reference state and interpreted as a thermodynamic equilibrium corresponding to some "pre-biological" state. The chain of these arguments is logically faultless, except for one fact: when no life exists, at the thermodynamic equilibrium, all the concentrations concerning the living matter must be equal to zero. Then the corresponding items in expression (9.1) for exergy tend to infinity and by the same token are "blocking up" an influence of other components. In order to by-pass the difficulty, Jørgensen (1992c) suggested the so-called concept of "inorganic soup", when these concentrations are very close to zero but, nevertheless, differ from zero. Another by-pass method using a "genetic" paradigm was described in Section 6.5. Certainly, in this case the corresponding items are finite; however, their influence remains prevalent. Perhaps there is a deep biological sense in this, but we would like to suggest here another by-pass based on a formal application of Lyapunov's theory.

Let the considered system include not only biotic components, N_k, $k \in \omega$, but also abiotic ones, N_s, $s \in \bar{\omega}$, for instance some chemical elements or substances. We assume that ω and $\bar{\omega}$ are some subset of the set of indices so that $\omega \cup \bar{\omega} = [1, 2, \ldots, n]$. It is obvious that at an *exact* thermodynamic equilibrium $N_k^* = 0$, $k \in \omega$ and $N_s^* \neq 0$, $s \in \bar{\omega}$. Instead of the exergy defined by Eq. (9.1) we shall consider the function:

$$\hat{\mathrm{E}}x = \sum_{k \in \omega} N_k + \sum_{s \in \bar{\omega}} \left[N_s \ln \frac{N_s}{N_s^*} - (N_s - N_s^*) \right]. \tag{9.2}$$

Since only the positive variations, δN_k^*, $k \in \omega$, are admissible then the second differential $\delta^2 \hat{\mathrm{E}}x$ will be positive definite, $\hat{\mathrm{E}}x(\mathbf{N}^*)$, and the function $\hat{\mathrm{E}}x(\mathbf{N})$ is the Lyapunov function. If the exergy, defined in such a way, increases along even one trajectory, then the reference state is unstable.

We illustrate this by a simple example. Suppose there is a system consisting of a living biomass, N, and an inorganic resource, C. Then the exergy will be $\hat{\mathrm{E}}x = C \ln(C/C^*)(CC^*)N$.

Being given the dynamic equations for N and C in the form describing a simplest biological cycle, $\mathrm{d}C/\mathrm{d}t = -\alpha CN + mN$, $\mathrm{d}N/\mathrm{d}t = \alpha CN - mN$, so that $N + C = C^*$, we get: $\mathrm{d}\hat{\mathrm{E}}x/\mathrm{d}t = \ln(C/C^*)(\mathrm{d}C/\mathrm{d}t) + \mathrm{d}N/\mathrm{d}t = ((\mathrm{d}N/\mathrm{d}t)(1 - \ln(C/C^*)) > 0$.

Since $N = C^* - C < C^*$ for any $N > 0$ then the second multiplier is always positive. In order to increase the exergy, it is necessary that

$$\mathrm{d}N/\mathrm{d}t = N(\alpha C - m) = N\left[(\alpha C^* - m) - \alpha N \right] > 0.$$

In the vicinity of $N = 0$ this is fulfilled if $C^* > m/\alpha$. This means the following: in order for a biological cycle to start "turning", an initial value of "turned" matter will be more than the ratio (m/α). With a small amount of matter when exergy does not increase, the cycle would not turn!

Of course, the example is trivial, and the same result could be achieved in a different way, without the use of the exergy concept. But in our opinion, the general idea is more important than a concrete result in this case.

6.10. One more ecological example

Note that we have forgotten about another "trivial" idea, i.e. how to by-pass the singularity of exergy in the "trivial" equilibrium of ecological models: we can simply "shift" the coordinates' system of phase space (Jørgensen et al., 1995a,b). As a result, the trivial equilibrium becomes non-trivial. We shall illustrate the method with the help of a simple example.

Consider the simple point model of a pond used by Mejer and Jørgensen (1978) for illustration of the exergy concept. The model has the following equations (this is a typical *chemostat* model):

$$\frac{\mathrm{d}P_\mathrm{s}}{\mathrm{d}t} = (P_\mathrm{in} - P_\mathrm{s})\frac{Q}{V} - (\mu - m)P_\mathrm{a}, \tag{10.1}$$

$$\frac{\mathrm{d}P_\mathrm{a}}{\mathrm{d}t} = \left(\mu - m - \frac{Q}{V}\right)P_\mathrm{a},$$

where the state variables P_s and P_a are concentrations of soluble and algal-bound phosphorus. P_in is the soluble phosphorus concentration in the inflow, Q is the rate of outflow (or dilution rate), V is the volume; $\mu = \mu_\mathrm{max}P_\mathrm{s}/(K + P_\mathrm{s})$ is the P-uptake rate and m is the rate of remineralisation. An obvious constraint for such a model is the positiveness of state variables. By summing these equations, we obtain the equation for the total phosphorus $P_\mathrm{tot} = P_\mathrm{s} + P_\mathrm{a}$:

$$\frac{\mathrm{d}P_\mathrm{tot}}{\mathrm{d}t} = (P_\mathrm{in} - P_\mathrm{tot})\frac{Q}{V}. \tag{10.2}$$

We assume that the total amount of phosphorus is conserved in the system. It is possible if $V << Q$, and the equilibrium with respect to P_tot is established very quickly. Then the system of equations (10.1) is reduced to one equation:

$$\frac{\mathrm{d}\xi}{\mathrm{d}t} = \left[\mu(\xi) - m - \frac{Q}{V}\right](P_\mathrm{a}^\mathrm{eq} + \xi), \tag{10.3}$$

where the reaction coordinate ξ is defined by any one of the conditions:

$$P_\mathrm{a} = P_\mathrm{a}^\mathrm{eq} + \xi, \quad P_\mathrm{s} = P_\mathrm{s}^\mathrm{eq} - \xi,$$

while P_a^eq and P_s^eq designate concentrations at the thermodynamic equilibrium. If the inorganic soup represents the state of thermodynamic equilibrium for living nature, we have $P_\mathrm{a}^\mathrm{eq} \approx 10^{-50}$ gP/m^3 from estimates by Morowitz (1968). Thus, even at the beginning of organic evolution, which corresponds to the thermodynamic equilibrium state of $\xi = 0$ in Eq. (10.3), the concentration of organic matter is not zero but a negligibly small quantity. Therefore, if we are precise, $P_\mathrm{a} \geq P_\mathrm{a}^\mathrm{eq}$. This minor adjustment of the constraints must not affect the dynamic behaviour of a model in other regions of the phase space. Therefore, we have to transfer the origin of our coordinate system to the *thermodynamic equilibrium point* while retaining the same phase portrait as before, within the shifted positive semi-axes.

In the general case of n living components with the thermodynamic equilibrium state $\mathbf{N}^\mathrm{eq}$, the positive orthant, $\mathbf{P}^n$, of the n-dimensional space must be transformed to $\tilde{\mathbf{P}}^n =$

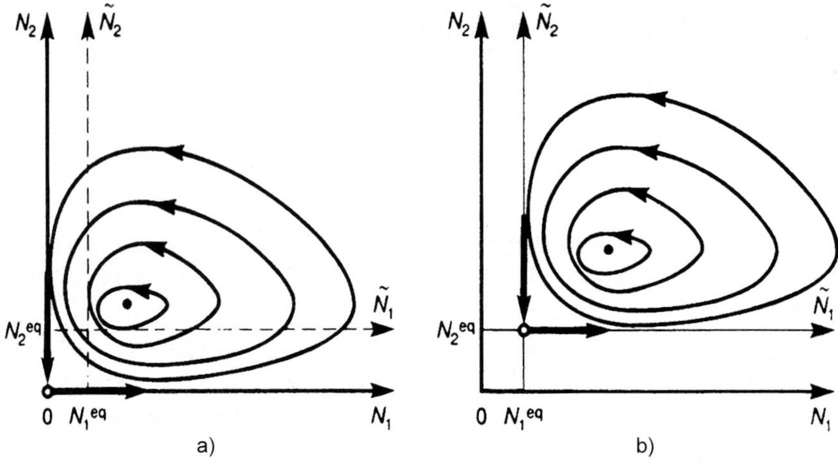

Fig. 6.4. Phase portrait of a prey–predator type system consisting of equations for two populations: (a) formal change of variables $\tilde{N}_i = N_i - N_i^{\mathrm{eq}}$ retains the invariance of the positive orthant $\mathbf{P}^n$: $N_i > 0, i = 1,\dots,n$, (b) "old" equations for "new" variables result in the invariance of the "shifted" orthant $\mathbf{P}^n$: $N_i > N_i^{\mathrm{eq}}$.

$\mathbf{P}^n - \mathbf{N}^{\mathrm{eq}}$. Practically, this means replacing the state variables of the model by their shifted values.

Note that the above procedure does not correspond merely to the formal change of variables, $\tilde{N}_i = N_i - N_i^{\mathrm{eq}}$, $i = 1,\dots,n$, since the latter keeps the phase portrait unchanged (see Fig. 6.4a), whereas the purpose is to have it shifted and invariant for the transformed equations of the model (Fig. 6.4b). Since the values of N_i^{eq} are negligibly small, the dynamics of the model far from the equilibrium are not affected by this transformation.

Applying this general idea to the phosphorus model, described by Eq. (10.3), we replace the state variable P_{a} by $P_{\mathrm{a}} - P_{\mathrm{a}}^{\mathrm{eq}}$ and obtain

$$\frac{\mathrm{d}\xi}{\mathrm{d}t} = \left[\tilde{\mu}(\xi) - m - \frac{Q}{V}\right]\xi. \tag{10.4}$$

Furthermore, $\tilde{\mu}(\xi) \approx \mu(\xi)$ as $P_{\mathrm{s}} >> P_{\mathrm{s}}^{\mathrm{eq}}$. It has to be determined now whether the exergy function is the Lyapunov function by using the equilibrium $\xi = 0$, corresponding to the thermodynamic equilibrium, as the reference state. We have

$$\mathrm{Ex} = RT_0\left[(P_{\mathrm{a}}^{\mathrm{eq}} + \xi)\ln\frac{P_{\mathrm{a}}^{\mathrm{eq}} + \xi}{P_{\mathrm{a}}^{\mathrm{eq}}} + (P_{\mathrm{s}}^{\mathrm{eq}} - \xi)\ln\frac{P_{\mathrm{s}}^{\mathrm{eq}} - \xi}{P_{\mathrm{s}}^{\mathrm{eq}}}\right] \tag{10.5}$$

when $\mathrm{Ex}(0) = 0$ and $\mathrm{Ex}(\xi) > 0$ when $\xi > 0$. The derivative

$$\partial\mathrm{Ex}/\partial\xi = RT_0 \ln[(P_{\mathrm{a}}^{\mathrm{eq}} + \xi)P_{\mathrm{s}}^{\mathrm{eq}}/(P_{\mathrm{s}}^{\mathrm{eq}} - \xi)P_{\mathrm{a}}^{\mathrm{eq}}] \tag{10.6}$$

is always positive (since ξ is always non-negative), except at the point $\xi = 0$ where it vanishes. Note that $\partial^2\mathrm{Ex}/\partial\xi^2 = RT_0[(P_{\mathrm{a}}^{\mathrm{eq}} + \xi)^{-1} + (P_{\mathrm{a}}^{\mathrm{eq}} + \xi)^{-1}] > 0$, whenever $\xi \leq P_{\mathrm{s}}^{\mathrm{eq}}$, so that function $\mathrm{Ex}(\xi)$ has its local minimum at point $\xi = 0$: $\mathrm{Ex}(0) = 0$ (which is also a global minimum in accordance with thermodynamic theory).

It follows from Eqs. (10.5) and (10.6) that the derivative of $\mathrm{Ex}(\xi)$ by virtue of Eq. (10.4), i.e. along the trajectories, is defined as

$$\frac{\mathrm{d(Ex)}}{\mathrm{d}t} = \frac{\partial \mathrm{Ex}}{\partial \xi}\frac{\mathrm{d}\xi}{\mathrm{d}t} = RT_0 \ln \frac{(P_{\mathrm{a}}^{\mathrm{eq}} + \xi)P_{\mathrm{s}}^{\mathrm{eq}}}{P_{\mathrm{a}}^{\mathrm{eq}}(P_{\mathrm{s}}^{\mathrm{eq}} - \xi)}\left[\mu(\xi) - m - \frac{Q}{V}\right]\xi > 0 \tag{10.7}$$

everywhere in some finite domain (excluding the point $\xi = 0$, in which the model equation gives a positive increase in the concentration of algal phosphorus, i.e. whenever $\mu(\xi) > m + Q/V$.

Thus, as seen, exergy $\mathrm{Ex}(\xi)$ possesses all the basic properties of the Lyapunov function; the derivative has the same sign as the function itself. According to Chetaev instability theorem, $\mathrm{Ex}(\xi)$ verifies *local instability* at the equilibrium state $\xi(t) \equiv 0$. This means that any initial deviation, however small, from the thermodynamic equilibrium state $t = 0$ will cause an increase in Ex. Such behaviour of model trajectories is consistent with the concept that life began as small fluctuations in the vicinity of thermodynamic equilibrium and a gradual movement away from it (Schrödinger, 1944).

Notice that we have only used (1) the invariance of the phase orthant for the model equations and (2) the mass conservation principle of the system to test the "Lyapunov" properties of exergy. We have not used dynamic equations (10.1) beyond property $\mu(\xi) > m + Q/V$.

Hence, we may expect these properties to occur in a sufficiently wide class of ecological models possessing the above-mentioned characteristics. Such models could be referred to as *exergical at zero*, with the idea that model trajectories go away from thermodynamic equilibrium and exergy increases along these trajectories.

6.11. Problems of thermodynamic interpretation for ecological models

We are here dealing with a principal distinction between ecological models and those of theoretical mechanisms or chemical kinetics, where the minimum of a potential function (or the minimum of a thermodynamic potential) corresponds to a stable steady state. The dynamics of such systems are determined by the use of potential functions, while for ecological models (especially models of mathematical ecology), such a simple form of dynamic equations becomes practically unacceptable so that the equations have to be chosen by other considerations. These represent, in essence, a phenomenological description of the ecosystem. Nevertheless, in Chapter 7 we try to give a thermodynamic justification for the basic equations of mathematical ecology. We think that, after this procedure, we would be easily able to interpret the obtained results.

From the standpoint of thermodynamics, any ecological system should be exergical at zero but, on the other hand, the exergy function of the classical Lotka–Volterra "prey–predator" model:

$$\frac{\mathrm{d}N_1}{\mathrm{d}t} = \alpha N_1 - \gamma N_1 N_2, \quad \frac{\mathrm{d}N_2}{\mathrm{d}t} = \gamma N_1 N_2 - m N_2 \tag{11.1}$$

is not the Lyapunov function. To write the expression for exergy, we should first expand system (11.1) with non-living environmental variable N_0 (N_1 and N_2 are the biomasses of prey and predator) such that the mass conservation law holds: $N_0 + N_1 + N_2 = N = \text{const}$. Then, bearing in mind the modification to which any model should be subject in order to be considered an exergical system at zero, we have the following expression for exergy according to definition:

$$\mathrm{Ex} = RT_0\left[(N_1 + N_1^{\mathrm{eq}})\ln\frac{N_1 + N_1^{\mathrm{eq}}}{N_1^{\mathrm{eq}}} + (N_2 + N_2^{\mathrm{eq}})\ln\frac{N_2 + N_2^{\mathrm{eq}}}{N_2^{\mathrm{eq}}} + N_0\ln\frac{N_0}{N_0^{\mathrm{eq}}}\right]. \tag{11.2}$$

Differentiating Eq. (11.2) along trajectories of Eq. (11.1) results in

$$\frac{\mathrm{d(Ex)}}{\mathrm{d}t} = RT_0\left\{\ln\left(1 + \frac{N_1}{N_1^{\mathrm{eq}}}\right)\frac{\mathrm{d}N_1}{\mathrm{d}t} + \ln\left(1 + \frac{N_2}{N_2^{\mathrm{eq}}}\right)\frac{\mathrm{d}N_2}{\mathrm{d}t} - \left(\frac{\mathrm{d}N_1}{\mathrm{d}t} + \frac{\mathrm{d}N_1}{\mathrm{d}t}\right)\ln\frac{N - N_1 - N_2}{N - N_1^{\mathrm{eq}} - N_2^{\mathrm{eq}}}\right\}. \tag{11.3}$$

It is possible to prove (Jørgensen et al., 1995a,b) that this derivative is neither positive nor negative and, furthermore, in any small vicinity of $(N_1^{\mathrm{eq}}, N_2^{\mathrm{eq}})$ such points $(N_1, N_2) > (N_1^{\mathrm{eq}}, N_2^{\mathrm{eq}})$ exist, so that $\mathrm{d(Ex)}(N_1, N_2)/\mathrm{d}t < 0$.

These observations can be explained from general Lyapunov stability concepts, since Chetaev instability theorem actually covers only the cases that, in a certain sense (that of time reversion, changing t to $-t$), are opposite to those of asymptotic stability, i.e. to phase patterns of the generalized topological node type (Fig. 6.1a and b). However, the prey–predator case corresponds to neither of these two extreme types of phase patterns (Fig. 6.1c).

Generally speaking, the "prey–predator" model is a curious example of a dissipative system having (as a conservative system) an integral, $(N_1)^m(N_2)^\alpha \exp[-\gamma(N_1 + N_2)] = \text{const}$. Maybe it is more correct to speak of a through-flow (not a dissipative) system, a proper existence of which is maintained by a permanent flow of energy through it. Such systems are well known in mechanics (gyroscopes). Let us calculate the function of dissipation for this system $\mathrm{Diss} = A_1(\mathrm{d}N_1/\mathrm{d}t) + A_2(\mathrm{d}N_2/\mathrm{d}t)$, where $A_1 = RT_0\ln(N_1/N_1^*)$ and $A_2 = RT_0\ln(N_2/N_2^*)$ are the affinities of prey and predator with respect to their equilibriums $N_1^* = m/\gamma$ and $N_2^* = \alpha/\gamma$, and the derivatives are taken from Eq. (11.1). Then

$$\mathrm{Diss} = RT_0\left\{\alpha N_1\ln\frac{N_1}{N_1^*} - mN_2\ln\frac{N_2}{N_2^*} - \gamma N_1N_2\left(\ln\frac{N_1}{N_1^*} - \ln\frac{N_2}{N_2^*}\right)\right\}. \tag{11.4}$$

It is easily shown that in the domain, lying above the line $N_2 = (\alpha/m)N_1$, the function of dissipation is negative, and it is positive in the domain lying below (Fig. 6.5); it is identically equal to zero along this line.

Since the function of dissipation is the entropy production, we can interpret the "prey–predator" system as some entropy machine, which produces entropy along the part of trajectory abc and releases it along the part cda, disturbing formally the Second Law.

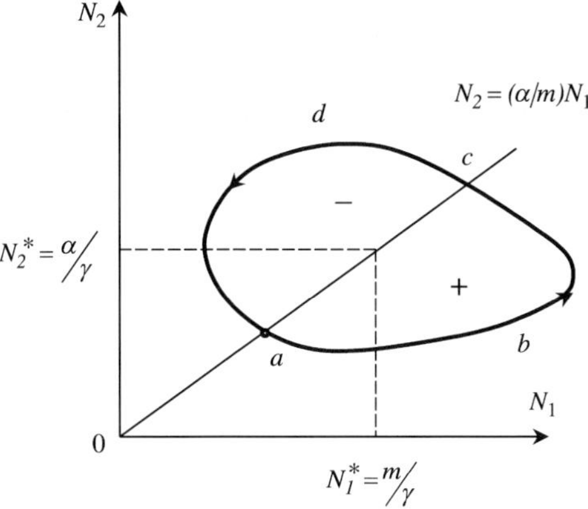

Fig. 6.5. Phase portrait of the "prey–predator" system: (+)-domain, in which the system produces entropy; (−)-domain, in which the system releases it.

However, if we integrate the function of dissipation along the entire close trajectory abcda, then as a result we get a zero, i.e. if locally the entropy can both increase and decrease along the close trajectory, then globally, on average, the entropy does not change.

If we keep in mind that the derivative of exergy is equal to the function of dissipation taken with the opposite sign, then all these results can be re-formulated in "exergical" terms. For instance, we can say that along the part cda the exergy is accumulated, and it is dissipated along abc. As we have already said this derivative is either positive or negative along the entire cycle. On the whole, when the cycle is completed, then the value of exergy is returned to its initial value.

We do not use the missing "exergical" property in a model as an argument against the model in general. Many theoretical models do not claim to be adequate in all regions of the phase space and, in particular, not in a critical region such as in the vicinity of zero equilibrium. Most models were developed to provide a reasonable description in a region where a biological community or ecosystem was believed to function normally, for example in the vicinity of a *non-trivial or feasible steady state*, $N^* > 0$ (with plausible values of state variables).

We could see that when we tried to define a correspondence between exergy and Lyapunov function, the principal difficulty is how to define a reference state in the case of exergy and an equilibrium state in the Lyapunov case. Despite all the similarity between expressions for exergy and Lyapunov function, there are principal differences in their meanings. The expression for exergy contains values of state variables at the thermodynamic equilibrium point $\mathbf{N}^{eq}$, whereas the Lyapunov function is referred to the steady state $\mathbf{N}^*$. As accepted theoretically, $\mathbf{N}^*$ is associated with a state to which the ecosystem normally evolves or at which it has its normal functions. In principle, such a state is irreversible to the thermodynamic equilibrium one.

However, to represent the system evolution as a sequence of stages, where at every one of them the system evolves from one unstable to another stable equilibrium, we can consider the unstable equilibrium as some intermediate reference state. The point when the system, at the previous stage, reaches the stable equilibrium can be considered as a new thermodynamic equilibrium for a closed at a given moment system. When the system is open, this equilibrium becomes unstable as a result of interaction between the system and its environment, and the next stage of evolution to a new stable equilibrium is starting, while the systems exergy begins to increase as well.

The conclusions of this section concern the behaviour of a system with the same values for its internal and external parameters (i.e. coefficients in model equations). However, as stated in the introduction, the "exergy principle" considers a slow evolution of ordering or structure of the system in the sense that the parameters may also change in addition to the state variables. Could the exergy principle indicate the direction in which such reorganisation will evolve? Could it predict the change in system parameter in response to perturbations? These questions are discussed in Chapters 7 and 8.

6.12. Complexity versus stability

Among ecologists it is almost taken as an axiom that ecosystems which are more complex in structure are more stable. There is a wide variety of arguments in favour of this thesis, supported by observations of real ecosystems.

Thus, for instance, laboratory systems of only two species, the predator and prey, most usually prove to be unstable; population explosions of pests are more typical for agroecosystems than for natural ecosystems, and their effect is more disastrous when the crops are monocultural. In contrast, the communities of rain forests, rich in specific composition and interspecific connections, demonstrate very stable functioning: there are no population explosions and population oscillations are much less pronounced in forests of the sub-arctic zone with less species diversity and relatively greater populations. It is believed that ecosystems of complex structures are more stable under perturbations of environmental factors and the random oscillations in populations of some species, whereas more simple structures, such as the sub-arctic fauna communities, e.g. when subject to sharp population oscillations, are unable to dampen out the perturbations.

Thus, the greater stability of natural ecosystems versus agroecosystems could be attributed to the longer co-evolution period of species composing natural ecosystems. The greater stability of trophic communities compared to the communities of the sub-arctic zone can be explained by the destabilising effects of sharp oscillations in climatic conditions.

But practically all of these facts and observations could also have another interpretation, not appealing to ecosystem complexity. We may remark that all these speculations are not only based on strong and correct definitions of complexity and its measure, but also rest on the intuitive idea that an increase in such ecosystem characteristics—the number of species and trophic levels—makes the number of interspecific connections and their strength comply with higher structural complexity. And though we do not have a formal definition for ecosystem complexity, which would be similar in universality, say, to the definition of stability via stability of equilibrium in an appropriate model, still the mentioned ecosystem characteristics are explicitly present in systems of model equations, thus enabling one to judge the influence of these characteristics on the *model* stability. The problem is thereby transmitted to the field of dynamical models of ecosystems, where the corresponding analysis is usually referred to as *complexity versus stability*.

The stability analysis of models of the various ecosystems has shown (see, for instance, Svirezhev and Logofet, 1978) that higher complexity might:

- reduce the probability of stability,
- increase this probability and
- in no way affect stability.

Using the terminology of the well-known physicist, we shall not "obscure this already well-confused subject further". The only obvious conclusion is that within a framework of mathematical models, there is no use in looking for a *unique* relation between complexity and stability that in the particular case is determined by peculiarities of structures under consideration and the specific character of mathematical formulations.

6.13. Summary of the ecological important issues

The concept of ecosystem stability has several meanings expressed as resilience, resistance, persistence, Lyapunov stability, thermodynamic stability and buffer capacity. The concept is under all circumstances multi-dimensional because the question is: stable in what context? It is also important to quantify the stability. Resilience, meaning ability to return to normal, can hardly be quantified, while resistance and buffer capacity can be quantified. They express what can be considered the inverse sensitivity, defined as the change of a state variable relative to the change caused by the forcing functions (impacts). The higher buffer capacity and the more resistance, the more change in the forcing functions (impact) to change the system. The change becomes multi-dimensional because the change of the system can be described by several (many) state variables. It can be shown that, by the use of model results, exergy and the sum of several buffer capacities are well correlated.

It has been shown in this chapter that exergy has all the properties of the Lyapunov function and the derivate has the same sign as the function itself. Exergy and biomass are both zero at thermodynamic equilibrium, but the system will move towards any point with higher exergy according to Lyapunov stability provided that the new point is stable. Exergy can be considered accumulated by the system in the course of the evolution from thermodynamic equilibrium to a dynamic equilibrium. According to Chetaev instability theorem, exergy has local instability at the thermodynamic equilibrium. This implies that any initial deviation, even a small one, from the thermodynamic equilibrium will cause increase of exergy. The system will move further away from thermodynamic equilibrium. It is also consistent with Schrödinger (1944) that life began as small fluctuations in the vicinity of thermodynamic equilibrium and a gradual movement away from it took place afterwards.

As previously discussed there is no simple relationship between the complexity of the system and the stability. The *spectrum*, not necessarily the size, of buffer capacities is, however, wider the higher the biodiversity is because the presence of more species must give increased probability that one of the species is at least able to cope with the focal problem.

Chapter 7
Models of ecosystems: thermodynamic basis and methods. I. Trophic chains

So, naturalists observe, a flea
Has smaller fleas that on him prey;
And these have smaller still to bite 'em;
And so proceed ad infinitum.
Jonathan Swift. Poetry, a Rhapsody.

"Great fleas have little fleas upon their backs to bite 'em,
And little fleas have lesser fleas, and so ad infinitum.
And the great fleas themselves, in turn, have greater fleas to go on;
While these again have greater still, and greater still, and so on."
De Morgan: A Budget of Paradoxes.

"Chaos had arisen in the Universe, before all things."
Hesiod

7.1. Introduction

From a thermodynamic point of view, any ecosystem is a typical open system, which exists only due to the permanent exchange of energy and matter between the ecosystem and its environment. As a rule it is at *dynamic* equilibrium, and this equilibrium is far from *thermodynamic* equilibrium. The latter can be interpreted as the state of death when there is not any living matter in the ecosystem. In fact, if we interrupt all energy and matter flows both into the system and out of it then the ecosystem will perish.

If also keeping in mind the classic "ecological" tradition going back to Lindeman (1942), in which an ecosystem is described by flows of energy and matter both within the ecosystem and between the ecosystem and its environment (which type of system is a field of interest for phenomenological thermodynamics), then the first idea which comes to mind is to directly apply classic thermodynamic methods and models for description of the ecosystem. But here we collide immediately with the problem of reductionism.

In fact, thermodynamics is a physical science, while in ecology there are certainly some conformities which cannot be reduced to physical laws. Strictly speaking, unless we consider ecological systems as physical–chemical ones there are no prohibitions in

Towards a Thermodynamic Theory for Ecological Systems, pp. 153–188

principle to applying thermodynamic concepts in ecology. However, as soon as we take into account any purely ecological conformities then we come up against a very serious problem: there is not a direct homomorphism between models (in a broad sense) in thermodynamics and models in ecology. At the same time, since we do not know even one example when physical laws were "cancelled" in ecology then all ecological laws have to be compatible with physical principles. Therefore, physical methods can be used in studying some properties of ecosystems, and the conclusions by analogy with physical systems (if used with a proper degree of caution) may be helpful for the ecologist.

7.2. General thermodynamic model of ecosystem

An ecosystem takes up all matter (nutrients) necessary for it from its environment; it gets free energy from solar energy and organic food, and returns matter producing entropy (in general in the form of heat). In the process, organisms (both individuals and combined into populations and biological communities) permanently dissipate the energy in order to maintain their structures (which might be called *dissipative*), to develop and to evolve. A measure of this dissipated energy is entropy, which has to be exported out of the ecosystem, since a system which accumulates entropy cannot survive in principle. Hence, the access to free energy is a central problem of existence for any living organism. *The struggle for life is first of all a struggle for free energy*!

Let the considered ecosystem consist of chemical components with numbers of moles $c = \{c_1, \dots, c_m\}$ and living organisms belonging to n different species with numbers $N = \{N_1, \dots, N_n\}$. The ecosystem occupies some volume V.

The ecosystem considered as an entire system has to satisfy the conditions of energy and entropy balance, i.e. for this the First and Second Laws have to be fulfilled. In addition, the temporal derivatives for internal energy U and entropy S are connected with each other by the Gibbs fundamental equation (see Chapter 2):

$$\frac{\mathrm{d}U}{\mathrm{d}t} = T\frac{\mathrm{d}S}{\mathrm{d}t} - p\frac{\mathrm{d}V}{\mathrm{d}t} + \sum_{i=1}^{n} \mu_i^{\mathrm{b}} \frac{\mathrm{d}N_i}{\mathrm{d}t} + \sum_{k=1}^{m} \mu_k \frac{\mathrm{d}c_k}{\mathrm{d}t}. \tag{2.1}$$

Here, T is the temperature, p is the pressure, and V is the ecosystem volume; μ_k is the chemical potential of kth component and μ_i^{b} is the analogue of chemical potential for ith species. In the last case, living individuals are considered as some virtual particles.

The free energy available to the ecosystem is equal to $F = U - TS$; hence, if $T = \mathrm{const}$ then

$$\frac{\mathrm{d}F}{\mathrm{d}t} = \frac{\mathrm{d}U}{\mathrm{d}t} - T\frac{\mathrm{d}S}{\mathrm{d}t}. \tag{2.2}$$

In order to write the First and Second Laws for the ecosystem, we represent the change of thermodynamic potentials, as usual, as the sum of two items: the item which corresponds to internal processes (i), and the item which corresponds to processes of exchange between the system and its environment (e):

$$\frac{\mathrm{d}U}{\mathrm{d}t} = \frac{\mathrm{d_e}U}{\mathrm{d}t} + \frac{\mathrm{d_i}U}{\mathrm{d}t}, \quad \frac{\mathrm{d}S}{\mathrm{d}t} = \frac{\mathrm{d_e}S}{\mathrm{d}t} + \frac{\mathrm{d_i}S}{\mathrm{d}t}, \quad \frac{\mathrm{d}F}{\mathrm{d}t} = \frac{\mathrm{d_e}F}{\mathrm{d}t} + \frac{\mathrm{d_i}F}{\mathrm{d}t}. \tag{2.3}$$

Then the First and Second Laws are written as

$$\frac{\mathrm{d_i}U}{\mathrm{d}t} = 0; \quad \frac{\mathrm{d_i}S}{\mathrm{d}t} > 0, \tag{2.4}$$

and from Eq. (2.2) for $T = \text{const}$ we have

$$\frac{\mathrm{d_i}F}{\mathrm{d}t} = -T\frac{\mathrm{d_i}S}{\mathrm{d}t} < 0. \tag{2.5}$$

It is clear that for a "normal" ecosystem the entropy, on average, should not increase:

$$\frac{\mathrm{d}S}{\mathrm{d}t} = \frac{\mathrm{d_e}S}{\mathrm{d}t} + \frac{\mathrm{d_i}S}{\mathrm{d}t} \leq 0.$$

Therefore, $(-\mathrm{d_e}S/\mathrm{d}t) \geq (\mathrm{d_i}S/\mathrm{d}t) > 0$, i.e. the export of entropy should not be less than the entropy production within the ecosystem. If the ecosystem is functioning at a constant temperature and either constant volume or pressure, then this relation is represented as

$$\frac{\mathrm{d_e}F}{\mathrm{d}t} \geq T\frac{\mathrm{d_i}S}{\mathrm{d}t} > 0 \text{ or } \frac{\mathrm{d_e}G}{\mathrm{d}t} \geq T\frac{\mathrm{d_i}S}{\mathrm{d}t} > 0, \tag{2.6}$$

i.e. if isothermal processes take place in the ecosystem they require the input of either free energy or enthalpy.

Let us consider one more concrete model, the so-called *model of "dilute solution"* (Feistel and Ebeling, 1981; Mauersberger, 1981). In particular, we apply the Planck theory of a dilute solution when living organisms are considered as molecules of different chemical substances submerged into an environment, which is considered as a solvent. About the latter it is assumed that it is at thermodynamic equilibrium and characterised by the numbers of substrate moles $c_1, \ldots, c_m$, the given chemical potentials of substrate $\mu_1, \ldots, \mu_m$, the given temperature T and the pressure p (Fig. 7.1).

The chemical components are considered as the solvent with a surface which coincides with the surface of the ecosystem, the volume of which is equal to V_s. The pressure within the ecosystem is equal to $p + p_s$. Continuing our analogy we can say that the "ecological" pressure p_s corresponds to the osmotic pressure of a solution, and living organisms are "dilute solution" in the sense of Planck's theory.

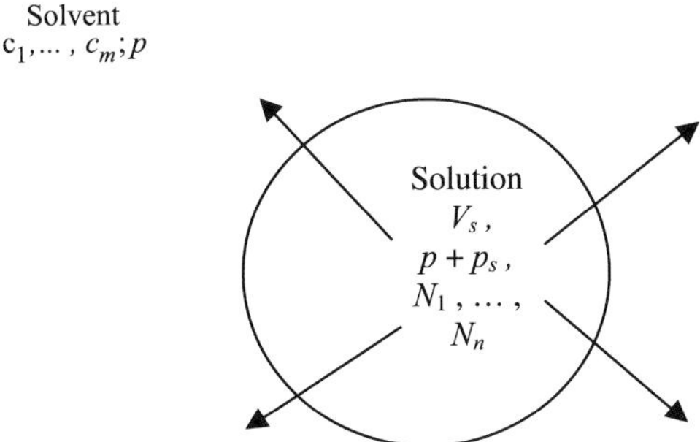

Fig. 7.1. The "dilute solution" as a model of ecosystem.

We also assume that the numbers of individuals of ith species ($i = 1, \ldots, n$) within the volume V_s (areal) are subjected to the following phenomenological equations:

$$\frac{dN_i}{dt} = f_i(N_1, \ldots, N_n), \quad i = 1, \ldots, n. \tag{2.7}$$

It is clear that an arbitrary extensive thermodynamic potential Φ of our system has to depend on the number of moles of substrate $c_1, \ldots, c_m$, the number of individuals of each species $N_1, \ldots, N_n$, the pressure p and the temperature T, so that

$$\Phi = \Phi(c_1, \ldots, c_m; N_1, \ldots, N_n; p, T). \tag{2.8}$$

The main assumption for the theory of the dilute solution is that the potential Φ can be represented as the sum of two items:

$$\Phi = \Phi_0 + \Phi_s, \quad \text{where } \Phi_0 = \Phi(c_1, \ldots, c_m; 0, \ldots, 0; p, T).$$

In this case, the fundamental Gibbs equation for the living subsystem is written as

$$\frac{dU_s}{dt} = T\frac{dS_s}{dt} - p_s\frac{dV_s}{dt} + \sum_{i=1}^{n} \mu_i^s \frac{dN_i}{dt}. \tag{2.9}$$

Here the index "s" indicates belonging to the living subsystem and μ_i^s is the chemical potential of ith species.

Since $U_s = 0$ for $N_1 = \cdots = N_n = 0$ then the linearised form of U_s is written as

$$U_s = \sum_{i=1}^{n} N_i u_i(c_1, \ldots, c_m; p, T) \tag{2.10}$$

where the energy of individual u_i is represented as the sum of kinetic (e_i^k) and potential (e_i^p) energy, and also the energy of interaction between the individual and substrate $(\sigma_i T)$: $u_i = e_i^k + e_i^p + \sigma_i T$. Assume that the individuals are in thermal equilibrium with the environment, and their heat capacity is σ_i. For entropy, the Planck approximation is written as

$$S_s = \sum_{i=1}^{n} N_i\left[s_i(c_1, \ldots, c_m; p, T) - k \ln\frac{N_i}{V_s}\right] \tag{2.11}$$

where k is Boltzmann's constant and s_i are the specific entropies. If $\pi_i = N_i/N$ and N is the total number of individuals then the entropy per individual is

$$\frac{S_s}{N} = \sum_{i=1}^{m} \pi_i s_i - k\sum_{i=1}^{m} \pi_i \ln \pi_i - k \ln\frac{N}{V_s}. \tag{2.12}$$

If all specific entropies are approximately equal, $s_i \approx s$, then

$$\frac{S_s}{N} = s - k\sum_{i=1}^{m} \pi_i \ln \pi_i - k \ln\frac{N}{V_s}. \tag{2.13}$$

It is easy to see that the value of S_s/N reaches a maximum for uniform distribution $\pi_i = 1/n : \max(S_s/N) = s - k \ln(N/V_s) + k \ln n$; and this maximum grows with an increase of the number of species, but decreases with growth of the total number. If only one species is dominant in the ecosystem, for instance $\pi_1 \approx 1, \pi_2 = \cdots = \pi_n \approx 0$, then the entropy will be minimal: $\min(S_s/N) \approx s - k\ln(N/V_s)$. However, if the specific entropies strictly differ from each other, then the distribution where the entropy attains maximum will also significantly differ from the uniform one.

For free energy we obtain:

$$F_s = U_s - TS_s = \sum_{i=1}^{n} N_i \left[e_i^k + e_i^p + \sigma_i T + kT \ln \frac{N_i}{V_s} - Ts_i \right]. \quad (2.14)$$

Knowing free energy we can calculate an "ecological" pressure

$$p_s = -\frac{\partial F_s}{\partial V_s} = \frac{kT}{V_s} \sum_{i=1}^{n} N_i = \frac{kTN}{V_s}. \quad (2.15)$$

Since in ecosystems the number of "particles" is much less than 10^{23}, then this pressure is vanishingly small.

For the chemical potential of ith species we have

$$\mu_i^s = \frac{\partial F_s}{\partial N_i} = \mu_i^s(0) + kT \ln \frac{N_i}{V_s} \quad (2.16)$$

where $\mu_i^s(0) = e_i^k + e_i^p + \sigma_i T + kT - Ts_i$. Using this expression for chemical potential, we can represent free energy and Gibbs potential in a standard form:

$$F_s = \sum_{i=1}^{n} (\mu_i^s - kT)N_i; \quad G_s = \sum_{i=1}^{n} \mu_i^s N_i. \quad (2.17)$$

It is obvious that if the input of free energy is interrupted then, according to the Second Law, the system tends to thermodynamic equilibrium, in which both F and G have a minimum. The minimum has to be reached under additional constraints, which are given by kinetic equations for "particles". These equations are standard equations of mathematical ecology

$$\frac{dN_i}{dt} = F_i(N_1, \ldots, N_n); \quad i = 1, \ldots, n. \quad (2.18)$$

For instance, these could be the Lotka–Volterra equations.

Since we associate thermodynamic equilibrium with the death of living matter, then the tendency to this is described by the limit transition $N_i \to 0$. Therefore, in the vicinity of thermodynamic equilibrium we can neglect all non-linear terms in Eq. (2.18) and vary the values of N_i independently. As at thermodynamic equilibrium

$$G_s = \sum_{i=1}^{n} \mu_i^s N_i = \min \text{ then}$$

$$\delta G_s = \sum_{i=1}^{n} \delta\mu_i^s N_i + \sum_{i=1}^{n} \mu_i^s \delta N_i = 0 \text{ at the zero point.}$$

As soon as we try to do the limit transition $N_i \to 0$ we obtain:

$$\lim_{N_i \to 0}\left(\sum_{i=1}^{n} \delta\mu_i^{\mathrm{s}} N_i\right) = kT \sum_{i=1}^{n} \delta N_i;$$

$$\lim_{N_i \to 0}\left(\sum_{i=1}^{n} \mu_i^{\mathrm{s}} \delta N_i\right) = \lim_{N_i \to 0}\left[\sum_{i=1}^{n} (\mu_i^{\mathrm{s}}(0)\delta N_i + kT \ln \frac{N_i}{V_{\mathrm{s}}} \delta N_i\right].$$

We see that the first limit vanishes only if $\sum_{i=1}^{n} \delta N_i = 0$, i.e. the total number of "particles" of the ecosystem is constant. The second limit is equal to (negative) infinity: we again meet the logarithmic singularity as in Chapter 6. In chemical thermodynamics all these paradoxes are resolved rather simply (see, for instance, Landau and Lifshitz, 1995), but how can we do this in our model? For this we have to keep in mind that the population sizes N_i have to be varied in accordance with linearised "ecological" equation (2.18): $\delta N_i = \alpha_i N_i \delta t$. The latter means that in the vicinity of thermodynamic equilibrium all the species grow exponentially, but each with its own exponent α_i. If this condition is fulfilled then both free energy and Gibbs potential (free enthalpy) have a minimum at zero. Note that the constraint is very strong. In particular, from this follows that the function F_i in Eq. (2.18) has to be represented as $F_i = N_i f_i(N_1, \ldots, N_n)$ where $f_i(0, \ldots, 0) = \alpha_i > 0$ (compare with Section 6.5).

There is also another method to solve the problem. Assume that:

(a) The matter conservation law is valid: $\sum_{i=1}^{n} \delta N_i = 0$.
(b) All organisms have a similar chemical composition: $\mu_1^{\mathrm{s}}(0) \approx \cdots \approx \mu_n^{\mathrm{s}}(0) \approx \mu^{\mathrm{s}}(0)$.
(c) In the vicinity of zero the distribution of "particles" is close to uniform:

$$N_1 \approx \cdots \approx N_n \approx N/n \text{ so that } kT \sum_{i=1}^{n} \ln \frac{N_i}{V_{\mathrm{s}}} \delta N_i \approx kT \ln \frac{N}{nV_{\mathrm{s}}} \sum_{i=1}^{n} \delta N_i = 0.$$

This means that we have to assume that in the vicinity of thermodynamic equilibrium the living "particles" already exist, at least in small quantities. In other words, we come again to the concept of "inorganic soup".

In this state let $N_i = N_i^{\mathrm{eq}}$ be extremely small and $(\partial F_{\mathrm{s}}/\partial N_i)_{T,V_{\mathrm{s}}} = (\partial G_{\mathrm{s}}/\partial N_i)_{T,p_{\mathrm{s}}} = \mu_i^{\mathrm{eq}}$ be equilibrium values of chemical potential, which are determined by the chemical composition of organisms.

Then

$$\mu_i^{\mathrm{s}} = \mu_i^{\mathrm{eq}} + kT \ln \frac{N_i}{N_i^{\mathrm{eq}}}, \tag{2.19}$$

$$G_{\mathrm{s}} = \sum_{i=1}^{n} \mu_i^{\mathrm{eq}} N_i + kT \sum_{i=1}^{n} N_i \ln \frac{N_i}{N_i^{\mathrm{eq}}}. \tag{2.20}$$

It is interesting that if $\mu_i^{\mathrm{eq}} = 0$; then we immediately get

$$G_{\mathrm{s}} = kT \sum_{i=1}^{n} N_i \ln \frac{N_i}{N_i^{\mathrm{eq}}} = \mathrm{Ex}. \tag{2.21}$$

As a rule, ecosystems are always far from thermodynamic equilibrium, so that $N_i >> N_i^{eq}$, whence $\mu_i^s >> kT$; $F_s >> NkT$; $G_s >> NkT$. These inequalities are not fulfilled for gases and solutions studied by classic thermodynamics; therefore, they express a fundamental difference between physical–chemical and ecological multiparticles systems.

Note that if we know the solution of "ecological" equation (2.18) then in the framework of the "dilute solution" model we can calculate all the thermodynamic functions using formulas (2.14)–(2.17).

7.3. Ecosystem's organisation: trophic chains

The ecosystem structure, presenting the transfer of energy from one species to another trapped in food, linked together by prey–predator (or resource–consumer) type relations, is known in ecology as the *trophic chain*. In every successive transfer a significant part of energy (70–80%) is lost, being spent on respiration and heat. This restricts the number of "links" in the chain usually to four or five, though in all ecosystems there either already exist (if only in negligible numbers), or are introduced from outside, the specimens which could form the next trophic level (*panspermia* hypothesis). When the amount of energy coming into the ecosystem abruptly increases, the possibility of forming a new level using the generative material available is realised, and a new trophic level appears and establishes itself in the chain.

Trophic chains are not isolated from each other, but interweave to form a trophic web. An example of such a trophic web is displayed in Fig. 7.2.

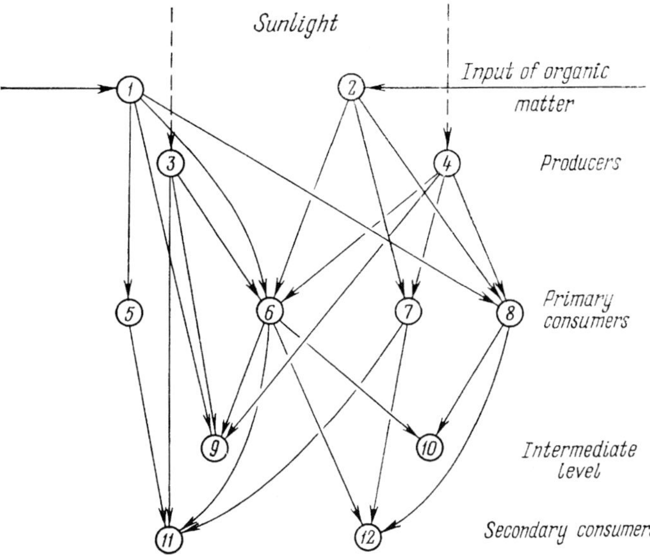

Fig. 7.2. Part of a trophic web in a stream ecosystem in South Wales. (From Jones, J.R.E., 1949. J. Anim. Ecol. 18(2), 142–159.)

In order to judge about the direction of the energy transfer in ecosystems which is realised along trophic chains and the stability of these processes (and hence about their possible evolution), it is desirable to have some energetic criteria for these inferences. It is natural that the first idea is to apply the methods of the thermodynamics of irreversible processes.

Processes in ecological systems occur, as usual, under the conditions of constant temperature, volume and pressure. In this case, the function of free energy F and Gibbs potential G play the most important role. Since the physical volume of living organisms does not practically change in the process of energy exchange (slowly increasing as a result of energy accumulation and growth) then ($p\mathrm{d}V \approx V\mathrm{d}p \approx 0$) $\mathrm{d}U \approx \mathrm{d}H$ and $\mathrm{d}G \approx \mathrm{d}F$.

Consider a simplified scheme of energy exchange between different trophic levels in the chain with length n (Patten, 1968). We assume that each level is an open thermodynamic system, so that

$$\mathrm{d}F_k = \mathrm{d_e}F_k + \mathrm{d_i}F_k = \mathrm{d}U_k - T\mathrm{d}S_k = \mathrm{d}U_k - T\mathrm{d_e}S_k - T\mathrm{d_i}S_k; \tag{3.1}$$

$$\mathrm{d_e}F_k = \mathrm{d}U_k - T\mathrm{d_e}S_k; \quad T\mathrm{d_e}S_k = \mathrm{d}U_k - \mathrm{d_e}F_k$$

Because each trophic level exists, it has to export entropy, $\mathrm{d_e}S_k < 0$; $|\mathrm{d_e}S_k| \geq \mathrm{d_i}S_k > 0$, and we have

$$\mathrm{d_e}F_k \geq \mathrm{d}U_k + T\mathrm{d_i}S_k. \tag{3.2}$$

Since in our case (isothermal system at constant pressure) $T\mathrm{d_e}S_k = \mathrm{d}H_k - \mathrm{d_e}G_k$ then

$$\mathrm{d_e}G_k \geq \mathrm{d}H_k + T\mathrm{d_i}S_k. \tag{3.3}$$

Thus, in order to provide the export of entropy out of the kth trophic level, the inflow of free energy (enthalpy) from the environment has to overlap both the change of enthalpy of the trophic level and its expenditure for the internal entropy production, when the energy is dissipated in different decay processes, such as metabolism and dying-off processes. The equality in Eq. (3.3) holds in the case of dynamic equilibrium.

At the first autotrophic level, the inflow of free enthalpy, $\mathrm{d_e}G_1$, corresponds to accumulated solar energy. Note that $\mathrm{d_e}G_1$ does not necessarily have to be solar energy. It may be the chemical energy of detritus for a so-called "detritus trophic chain". If we assume that the Liebig "Limiting Factors Principle" is valid, then even for the first autotrophic level the value of $\mathrm{d_e}G_1$ may also be the chemical energy of some limiting nutrients. At the other levels, the free enthalpy input is determined by the part of enthalpy transferred from the previous level.

Let us represent the change of total enthalpy, $\mathrm{d}H_k$, at the kth level, as $\mathrm{d}H_k = \mathrm{d}H'_k + \lambda_k\mathrm{d}H''_k + (1-\lambda_k)\mathrm{d}H''_k$, where $\mathrm{d}H'_k$ is the enthalpy of the increment of biomass at this level and $\mathrm{d}H''_k$ is the enthalpy transferred to the next $(k+1)$th level. The $(1-\lambda_k)$th part of this value is lost in the process of transfer, so that only λ_kth part reaches the next level: $\mathrm{d_e}G_{k+1} = \lambda_k\mathrm{d}H''_k$. Assume that the lost parts of energy, $T\mathrm{d_i}S_{k,k+1} = (1-\lambda_k)\mathrm{d}H''_k$ are dissipated; therefore, they have to be included into the total entropy production when all these levels are joined into a single ecosystem.

Thus, conditions (3.3) for the entire trophic chain are written as

$$\begin{aligned}
&d_eG_1 \geq dH_1' + \lambda_1 dH_1'' + (1-\lambda_1)dH_1'' + Td_iS_1,\\
&d_eG_1 = \lambda_1 dH_1'' \geq dH_2' + \lambda_2 dH_2'' + (1-\lambda_2)dH_2'' + Td_iS_2,\\
&\dots\dots\dots\dots\dots\dots\dots\dots\dots\dots\dots\dots\dots\dots\dots\\
&d_eG_k = \lambda_{k-1} dH_{k-1}'' \geq dH_k' + \lambda_k dH_k'' + (1-\lambda_k)dH_k'' + Td_iS_k,\\
&\dots\dots\dots\dots\dots\dots\dots\dots\dots\dots\dots\dots\dots\dots\dots\\
&d_eG_n = \lambda_{n-1} dH_{n-1}'' \geq dH_n' + \lambda_n dH_n'' + (1-\lambda_n)dH_n'' + Td_iS_n.
\end{aligned} \tag{3.4}$$

Since the chain ends with the nth link then $dH_n'' = 0$. By summing these inequalities, we obtain

$$\begin{aligned}
d_eG_1 &\geq \sum_{k=1}^{n} dH_k' + \sum_{k=1}^{n}(1-\lambda_k)dH_k'' + T\sum_{k=1}^{n} d_iS_k\\
&= \sum_{k=1}^{n} dH_k' + T\sum_{k=1}^{n}(d_iS_k + d_iS_{k,k-1}).
\end{aligned} \tag{3.5}$$

From this expression we can see that the part of the free enthalpy of solar energy accumulated by autotrophic organisms is spent on increasing both its biomass and the biomass of other levels forming part of the chain (it is equal to $\sum_{k=1}^{n} dH_k'$), and the other part, $T\sum_{k=1}^{n}(d_iS_k + d_iS_{k,k+1})$, is irreversibly lost in the form of metabolic heat and dead organic matter.

Certainly, we can complicate system (3.4) by taking into consideration the energy inflows from adjacent chains and outflows to them, i.e. by considering so-called *branching trophic chains*. However, we shall not follow this path, since this is not fundamental to our analysis.

By representing Eqs. (3.4) and (3.5) in the infinitesimal form we get

$$\frac{d_eG_1}{dt} \geq \sum_{k=1}^{n} \frac{dH_k'}{dt} + T\sum_{k=1}^{n} \frac{d_i(S_k + S_{k,k+1})}{dt}. \tag{3.6}$$

Using the standard representation of the entropy production: $(dS/dt) = (d_eS/dt) + (d_iS/dt)$ and taking into account that $(d_eS/dt) = -(1/T)(dH_1/dt)$ and $(d_iS/dt) = (d_i(S_k + S_{k,k+1})/dt)$ $(d_eS/dt) = -(1/T)(d_eG_1/dt)$ we obtain:

$$\frac{dS}{dt} = \frac{d_eS}{dt} + \frac{d_iS}{dt} \leq -\sum_{k=1}^{n} \frac{dH_k'}{dt}. \tag{3.7}$$

If we measure the total biomass of the trophic chain, $N = \sum_{k=1}^{n} N_k$, in enthalpy units then $\sum_{k=1}^{n}(dH_k'/dt) = (dN/dt)$, and Eq. (3.7) is represented as

$$(dS/dt) + (dN/dt) \leq 0. \tag{3.8}$$

To increase the total biomass, the entropy has to decrease; moreover, the rate of entropy decrease has to be higher than the rate of biomass growth: $|dS/dt| \geq (dN/dt)$. In the steady state (dynamic equilibrium), when the processes of energy inflow and utilisation balance each other, $(dS/dt) = 0$, then $(dN/dt) = 0$, i.e. the total enthalpy of the system contained in the biomass of the ecosystem does not change, although the enthalpy (biomass) of each level separately may change. The decrease of entropy in the course of the ecosystem evolution implies the accumulation of free enthalpy (energy); conversely, the increase of entropy can be interpreted as the free enthalpy dissipation.

One can see that these considerations give us a simplified scheme of energy balance in ecosystems. At the same time, in order to conclude that the transfer of energy from kth to $(k+1)$th level is accompanied by some losses, and the increase of free energy at the acceptor level has not to exceed its decrease at the donor level, it is sufficient to use the matter and energy conservation laws; it is not necessary to apply the methods of non-equilibrium thermodynamics. This would be justified if, prior to estimating the rate of energy dissipation in the system, we could predict the direction of its transfer. For this we would need to know in turn the values of velocity and moving forces of the trophic transfer of energy. Of course, if the increment of biomass at each trophic level is known, there is not a problem to estimate these velocities, but that is concerned with moving forces, which is a problem.

It seems that if we know the energy contents of a biomass unit at the level of donor, H^{d}, and acceptor, H^{a}, then the moving force $X = H^{\mathrm{d}} - H^{\mathrm{a}}$. There are well-known methods of thermochemistry which allow the enthalpy of 1 g of dry biomass to be determined by the thermal effect in the reaction of a burnt sample. Then, determining, for instance, the values of enthalpy differences $X_1 = \Delta H_1 = H^{\mathrm{d}} - H^{\mathrm{a1}}$ and $X_1 = \Delta H_1 = H^{\mathrm{d}} - H^{\mathrm{a1}}$, which correspond to the energy trophical transfers from donor d to two different acceptors a1 and a2, we obtain the estimation for moving forces. If the velocities of energy transfers $J_1(\mathrm{d} \rightarrow \mathrm{a1})$ and $J_2(\mathrm{d} \rightarrow \mathrm{a2})$ are also known, the functions of dissipation can be found: $\beta_1 = X_1 J_1$ and $\beta_2 = X_2 J_2$. They describe the process of energy transfer from the trophic donor to each of the two acceptors.

Nevertheless, if we want to forecast the direction of energy in irreversible trophic transfers, this method could be applied only in an ideal case where the chemical compositions of 1 g of the biomasses of the donor and acceptors are identical. This would be equivalent to the statement that the character and direction of the trophic transfer are only determined by the difference in the energy contents of the biomass unit between different components of the ecosystem. A lot of ecological data bear witness against such a suggestion and, on the contrary, the change of energy contents in biomasses is said to be a result of the combined action of many factors, just as the growth of organisms is the final result in the course of the whole complex of cell metabolic processes (see, for instance Tables 2.2–2.4 in Chapter 2).

This means that equal values of energy contained in the biomasses of interacting species are not actually equivalent in a biological sense, so that the values of the energy content of biomass units cannot be used to define the moving forces in trophic transfer. It is appropriate to draw an analogy with the thermodynamic questions of the growth and development of organisms discussed above (Chapter 3). Obviously, the final state of a system in both cases depends on actions within a whole complex of processes for which we now have no adequate description in terms of chemical kinetics. Even if such a description is performed, then the forecast of a system's behaviour over time based on it would depend, in the first place, on the character of kinetic factors of regulations. Calculation of the functions of dissipation using methods of non-equilibrium thermodynamics for systems far from equilibrium could only be important for estimation of their energetic effectiveness and plays no "forecasting role".

However, apart from the energy content there is one more value that characterises biomass: exergy (see Chapter 5). How could we apply the exergy concept to determine the direction of energy transfer between the kth and $(k+1)$th levels? For instance, if we now introduce the specific exergies ex^k and ex^{k+1} of the corresponding biomasses then we can define the active energy transfer in the form of biomass flow from kth to $(k+1)$th level as $q_{k,k+1} \sim \ln(\mathrm{ex}^{k+1}/\mathrm{ex}^k)N_k N_{k+1}$. It is obvious that in this case the flow of matter is always directed to such a level where the biomass has a higher specific exergy.

7.4. Dynamic equations of the trophic chain

Let the biomass of the species belonging to the kth trophic level be equal to N_k. A standard estimation of the energy contents of a biomass unit in ecology is the measurement of biomass in terms of carbon units (the quantity in grams of carbon contained in the biomass) and then its multiplication by the enthalpy of 1 g of carbon (~ 10 kcal $= 42$ kJ). From this point of view the concepts of biomass and energy are equivalent. The existence of the trophic chain is maintained by an inflow of solar energy q_1, accumulated at the first autotrophic level as new biomass. Note (as we already mentioned above) that the flux q_1 does not necessarily have to be a flux of solar energy; it may be either the flux of chemical energy of detritus or the chemical energy flux of some limiting nutrient (resource). It is very important that the flux is regulated, $q_1(N_1, R)$, $\partial q_1/\partial N_1 \geq 0$, i.e. a feedback exists in the system. The parameter R is either the external flux of solar radiation or the current value of the resource. One part of the energy is spent within the trophic level, lost in different decay metabolism and dying-off processes (mortality), $D_1(N_1)$. Another part, in turn, is spent in maintaining reproduction (accumulated in seeds, eggs, support of growth, reaching maturity, etc.) and the growth of population, $\mathrm{d}N_1/\mathrm{d}t$. The last, residual part $q_{12}(N_1, N_2)$ is partially transferred to the next level. Then the law of energy (matter) conservation for the first level is presented as

$$q_1(N_1) = \frac{\mathrm{d}N_1}{\mathrm{d}t} + q_{12}(N_1, N_2) + D_1(N_1). \tag{4.1}$$

If the transfer is realised without any losses, and the flux of energy accumulated at the second level, $q_2 = q_{12}$, then the energy conservation law for the second level will be

$$q_{12}(N_1, N_2) = \frac{\mathrm{d}N_2}{\mathrm{d}t} + q_{23}(N_2, N_3) + D_2(N_2). \tag{4.2}$$

Finally, for the chain of n length we have

$$q_1(N_1) = \frac{\mathrm{d}N_1}{\mathrm{d}t} + q_{12}(N_1, N_2) + D_1(N_1),$$

$$q_{12}(N_1, N_2) = \frac{dN_2}{dt} + q_{23}(N_2, N_3) + D_2(N_2),$$

$$\ldots\ldots\ldots\ldots\ldots\ldots\ldots\ldots\ldots\ldots\ldots\ldots\ldots\ldots\ldots\ldots\ldots\ldots\ldots\ldots$$

$$q_{k-1,k}(N_{k-1}, N_k) = \frac{dN_k}{dt} + q_{k,k+1}(N_k, N_{k+1}) + D_k(N_k), \; k = 3, \ldots, n-1, \tag{4.3}$$

$$\ldots\ldots\ldots\ldots\ldots\ldots\ldots\ldots\ldots\ldots\ldots\ldots\ldots\ldots\ldots\ldots\ldots\ldots\ldots\ldots$$

$$q_{n-1,n}(N_{n-1}, N_n) = \frac{dN_n}{dt} + D_n(N_n).$$

Finally, we get the balance (with respect to matter or energy) dynamic equations of the trophic chain as:

$$\frac{dN_1}{dt} = q_1(N_1) - q_{12}(N_1, N_2) - D_1(N_1),$$

$$\frac{dN_2}{dt} = q_{12}(N_1, N_2) - q_{23}(N_2, N_3) - D_2(N_2),$$

$$\ldots\ldots\ldots\ldots\ldots\ldots\ldots\ldots\ldots\ldots\ldots\ldots\ldots\ldots\ldots\ldots\ldots\ldots\ldots\ldots$$

$$\frac{dN_k}{dt} = q_{k-1,k}(N_{k-1}, N_k) - q_{k,k+1}(N_k, N_{k+1}) - D_k(N_k), \; k = 3, \ldots, n-1, \tag{4.4}$$

$$\ldots\ldots\ldots\ldots\ldots\ldots\ldots\ldots\ldots\ldots\ldots\ldots\ldots\ldots\ldots\ldots\ldots\ldots\ldots\ldots$$

$$\frac{dN_n}{dt} = q_{n-1,n}(N_{n-1}, N_n) - D_n(N_n).$$

As one can see from these equations, the direction of energy movement given here is not concluded from thermodynamics. This means that, as we assumed beforehand, the input of chemical resource from the environment and the transfer of matter between levels are a result of active transport. In this case, there are conjugating reactions within the system, which have to provide energy for the transfer processes.

By summing Eq. (4.4) we get $\left(\sum_{k=1}^{n} N_k = N\right)$:

$$\frac{dN}{dt} = q_1(N_1) - \sum_{k=1}^{n} D_k(N_k). \tag{4.5}$$

The sum $\sum_{k=1}^{n} D_k(N_k)$ includes both the proper metabolisms of organisms of the trophic chain and the outflow of dead organic matter. If the energy spent by the metabolism is immediately transformed into heat, then the energy contained in the dead organic matter is either transformed into heat within the system (as a result of its decay) or exported out of the system. In the first case, the sum $\sum_{k=1}^{n} D_k(N_k)$ can be interpreted as the rate of entropy production within the system, $T(d_iS/dt)$. The intensity of the "entropy pump" sucking the entropy out of the system is equal to the rate of accumulation of solar energy, $T(d_eS/dt) = -q_1(N_1)$. Before a consideration of the second case, we present each item of the sum as $D_k(N_k) = D_k^m(N_k) + D_k^d(N_k)$. The term $D_k^m(N_k)$ describes the rate of production of metabolic heat, while the term $D_k^d(N_k)$ describes the mortality at kth level. Then the rate of entropy production within the system $T(\mathrm{d_i}S/\mathrm{d}t) = \sum_{k=1}^{n} D_k^{\mathrm{m}}(N_k)$, but the intensity of the "entropy pump" will be

$$T(\mathrm{d_e}S/\mathrm{d}t) = -\left[q_1(N_1) - \sum_{k=1}^{n} D_k^{\mathrm{d}}(N_k)\right].$$

We say that the chain as a whole is in a dynamic equilibrium when these entropy fluxes are balanced. Then (we shall denote this state by the upper index (^{eq})):

$$\begin{aligned} q_1[N_1^{\mathrm{eq}}(t)] &= \sum_{k=1}^{n} D_k[N_k^{\mathrm{eq}}(t)], \\ \sum_{i=1}^{n} [N_k^{\mathrm{eq}}(t)] &= \mathrm{const} = N^{\mathrm{eq}}. \end{aligned} \tag{4.6}$$

The latter relation means that when the trophic chain is in dynamic equilibrium with its environment, then its total biomass has to be constant (although the biomasses of each trophic level may fluctuate).

We can see that the conditions of stationary state do not depend on the choice of hypothesis about the fate of dead organic matter. It is possible to show that other results would not fundamentally differ from each other. Therefore, in order to avoid overloading our account with superfluous details, we shall consider only the case when the dead organic matter remains within the system, when

$$T\frac{\mathrm{d_i}S}{\mathrm{d}t} = \sum_{k=1}^{n} D_k(N_k). \tag{4.7}$$

Until now we have implied a dynamic equilibrium of the entire chain when the latter was considered as a single thermodynamic system. However, stronger conditions of equilibrium could be formulated if every level is considered as some separate system. In this case, the equilibrium conditions are: $(\mathrm{d}N_1/\mathrm{d}t) = \cdots = (\mathrm{d}N_n/\mathrm{d}t) = 0$, from which it follows that ($N_i^{\mathrm{eq}} = N_i^* = \mathrm{const}$):

$$\begin{aligned} &q_1(N_1^*) = q_{12}(N_1^*, N_2^*) + D_1(N_1^*), \\ &q_{12}(N_1^*, N_2^*) = q_{23}(N_2^*, N_3^*) + D_2(N_2^*), \\ &\cdots\cdots\cdots\cdots\cdots\cdots\cdots\cdots\cdots \\ &q_{k-1,k}(N_{k-1}^*, N_k^*) = q_{k,k+1}(N_k^*, N_{k+1}^*) + D_k(N_k^*), \\ &\cdots\cdots\cdots\cdots\cdots\cdots\cdots\cdots\cdots \\ &q_{n-1,n}(N_{n-1}^*, N_n^*) = D_n(N_k^*). \end{aligned} \tag{4.8}$$

These equations give us the co-ordinates of equilibrium.

7.5. Prigogine-like theorems and the length of trophic chain

If we keep in mind now the Prigogine theorem (see Chapter 3) then in the application to our case its main statement says that at a stable dynamic equilibrium the value $\sum_{k=1}^{n} D_k(N_k)$ is minimal, and also the minimum is positive. It is obvious that $D_k(N_k)$ have to be monotonously increasing functions and $D_k(0) = 0$. Then the minimum is attained with certainty at the origin of co-ordinates $\{N_k^{\min} = 0, k = 1,\dots,n\}$ if there are no constraints except their non-negativeness. However, the minimum is zero, which contradicts Prigogine's theorem. Therefore, the minimum has to be found under conditions of stationary state. Therefore, at the dynamic equilibrium $N^{eq} = \{N_k^{eq}, k = 1,\dots,n\}$ where $N \in P^n$, $P^n : \{N_k \geq 0, k = 1,\dots,n\}$ is a positive orthant,

$$\sum_{k=1}^{n} D_k(N_k^{eq}) = \min_{N \in P^n} \sum_{k=1}^{n} D_k(N_k) \tag{5.1}$$

under constraints (4.4).

In this formulation the problem is too general; in order to obtain some understandable results we simplify it. For this we are restricted to the case $n = 3$ and $q_1 = q = \text{const}$. The latter implies that the inflow of energy or a resource does not depend on the state of the autotrophic link. The functions $D_k(N_k)$ are represented as $D_k(N_k) = (r_i + m_i)N_i = \mathrm{d}_i N_i$, where r_i and m_i are the coefficients of respiration (metabolism) and mortality. The inflows between trophic levels are represented in the bilinear form: $q_{12} = \alpha_1 N_1 N_2$; $q_{23} = \alpha_2 N_2 N_3$. Then Eqs. (4.4) and (4.5) are written as

$$\begin{aligned} \frac{\mathrm{d}N_1}{\mathrm{d}t} &= q - \alpha_1 N_1 N_2 - d_1 N_1, \\ \frac{\mathrm{d}N_2}{\mathrm{d}t} &= N_2(\alpha_1 N_2 - \alpha_2 N_3 - d_2), \\ \frac{\mathrm{d}N_3}{\mathrm{d}t} &= N_3(\alpha_2 N_2 - d_3). \end{aligned} \tag{5.2}$$

$$\frac{\mathrm{d}N}{\mathrm{d}t} = q - \sum_{k=1}^{3} d_k N_k = q - \hat{d}N \tag{5.3}$$

where $\hat{d} = \sum_{k=1}^{n} d_k p_k$ is the mean value of coefficients d_k, $k = 1, 2, 3$.

Elementary calculations show that the system has three equilibrium points:

(a) $N_1^* = q/d_1$, $N_2^* = N_3^* = 0$;
(b) $N_1^* = d_2/\alpha_1$, $N_2^* = (q/d_2) - (d_1/\alpha_1)$, $N_3^* = 0$ and
(c) $N_1^* = q\alpha_2/(\alpha_1 d_3 + \alpha_2 d_1)$, $N_2^* = d_3/\alpha_2$, $N_3^* = q\alpha_2/(\alpha_1 d_3 + \alpha_2 d_1) - (d_2/\alpha_2)$.

These equilibriums correspond to the chains with lengths 1, 2, and 3, respectively. The first chain is stable if $q < d_1 d_2/\alpha_1$, the second chain is stable if $d_1 d_2/\alpha_1 < q < d_1 d_2/\alpha_1 + d_2 d_3/\alpha_2$, and the third is stable if $q > d_1 d_2/\alpha_1 + d_2 d_3/\alpha_2$. By the same token the axis q is divided into non-overlapping intervals, in each of them a chain of fixed length can exist (see Fig. 7.4). The general result for any length was described in Svirezhev and Logofet (1978).

If we assume that the main statement of Prigogine's theorem about the minimisation of the internal entropy production in the stable equilibrium takes place in our case, then we have to prove that the value $\sigma = d_1 N_1 + d_2 N_2 + d_3 N_3$ is minimal in each stable equilibrium of Eq. (5.2).

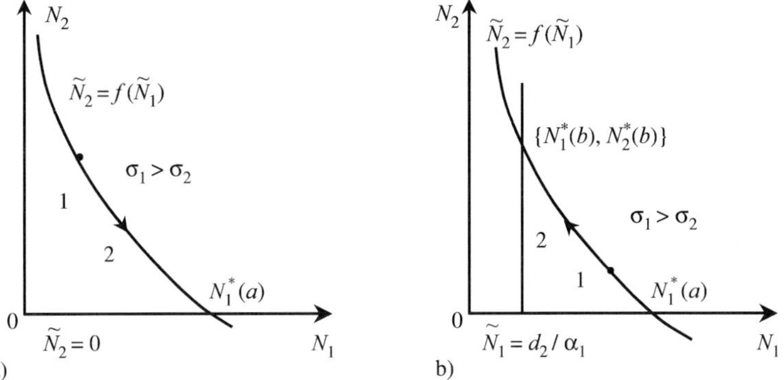

Fig. 7.3. To the proof of the σ minimality. $\tilde{N}_2 = f(\tilde{N}_1) = (1/\alpha_1)(q/\tilde{N}_1 - d_1)$, $N_1^*(\mathrm{a}) = q/d_1$, $N_1^*(\mathrm{b}) = d_2/\alpha_1$, $N_2^*(\mathrm{b}) = (q/d_2) - (d_1/\alpha_1)$; (a) $q < q_1^*$, (b) $q > q_1^*$.

Let $q < q_1^* = d_1 d_2/\alpha_1$, then only the state (a) that lies at the point of intersection of isoclines $\tilde{N}_2 = (1/\alpha_1)(q/\tilde{N}_1 - d_1)$ and $\tilde{N}_2 = 0$ (Fig. 7.3a) is stable.

If we could prove that moving along the first isocline where $\mathrm{d}N_1/\mathrm{d}t = 0$ to the equilibrium $N_1^*(\mathrm{a}) = q/d_1$, we decrease the value of σ, then, by the same token, we would prove that σ is minimal at this point. The proof is elementary. Indeed, along the first isocline

$$\sigma = d_1 \tilde{N}_1 + \frac{d_2}{\alpha_1} q \frac{1}{\tilde{N}_1} - \frac{d_1 d_2}{\alpha_1},$$

and, if $\tilde{N}_1 < N_1^*(\mathrm{a})$, $q < d_1 d_2/\alpha_1$, then $\partial\sigma/\partial\tilde{N}_1 < 0$.

Now let $d_1 d_2/\alpha_1 + d_2 d_3/\alpha_2 = q_2^* > q > q_1^* = d_1 d_2/\alpha_1$, then only the state (b), which lies at the point of intersection of isoclines $\tilde{N}_2 = (1/\alpha_1)(q/\tilde{N}_1 - d_1)$, $\tilde{N}_1 = d_2/\alpha_1$ and $\tilde{N}_3 = 0$ (Fig. 7.3b) is stable. In this case, although we shall move along the first isocline in the opposite direction from point $N_1^*(\mathrm{a}) = q/d_1$ (which now becomes unstable) towards the stable equilibrium (b), the value of σ will again decrease, reaching its minimum at equilibrium. Analogously we can prove the same statement for the equilibrium (c).

We have thus proved some analogue of Prigogine's theorem about the minimisation of the internal entropy production in the stable equilibrium (this statement can be proved for any n). However, in spite of the fact that the formulations of Prigogine's original theorem and our Prigogine-like theorem are very similar, they differ significantly from one another. If the original theorem was proved only for the system with a single dynamic equilibrium, then our Prigogine-like theorem deals with a system possessing several multiple equilibriums, which replace each other as a result of pre-determined bifurcations. Let us consider this process in more detail.

Assume that the value of inflow q slowly increases, so that one of these equilibriums is established rather quickly in relation to the "slow" time. If we now calculate the dynamics of the internal entropy production in relation to "slow" time, i.e. in the equilibrium, then

$$T(\mathrm{d_i}S/\mathrm{d}t) = \sigma^* = \sum_{k=1}^{3} d_k N_k^* = q. \tag{5.4}$$

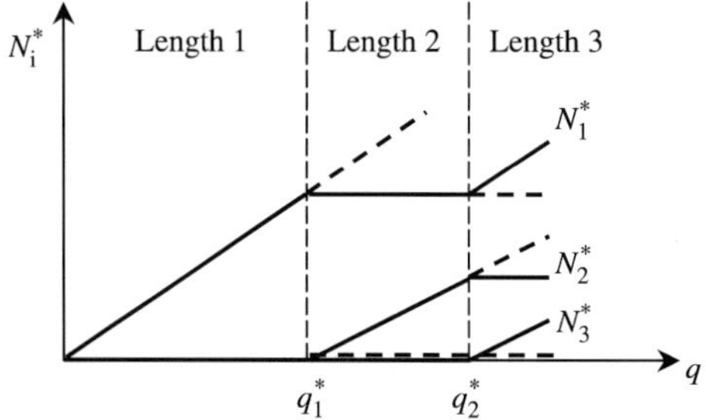

Fig. 7.4. The bifurcation diagram of equilibrium solutions: — stable branch, – – unstable branch.

It is interesting that this value is constant for a given chain of any length (it follows from Eq. (5.3) if $q = \text{const}$); it is determined only by characters of the energy exchange between the system and the environment and does not depend on the internal system parameters.

In Fig. 7.4 a so-called "bifurcation diagram" is shown.

We can see that when the value of inflow q passes over the first critical number $q_1^* = d_1 d_2/\alpha_1$ then the equilibrium (a) loses its stability, but the new stable equilibrium (b) arises. The solution is branched; instead of one stable branch, the two branches appear. If in the vicinity of a critical number the system remains on the old branch then $\sigma = d_1 N_1^*(\text{a}) + d_2 N_2 > q$. In order to return to the minimal value of $\sigma^* = q$, the system has to be reconstructed and the chain has to increase its length from one to two links. Moreover, the further growth of q does not increase the value of the first level: it remains constant and all the increments of q are spent in maintaining the second level, the biomass of which is growing with the growth of q.

Calculating the total biomass of the chain:

(a) $N^* = q/d_1;\ q \leq d_1 d_2/\alpha_1;$
(b) $N^* = N_1^* + N_2^* = (q/d_2) + (d_2 - d_1)/\alpha_1;$
(c) $N^* = N_1^* + N_2^* + N_3^* = [q(\alpha_1 + \alpha_2)/(\alpha_1 d_3 + \alpha_2 d_1)] + (d_3 - d_2)/\alpha_2,$

we see that the total biomass increases with the growth of q (see Fig. 7.5).

Let us return to Eq. (5.3), from which follows that $N^* = q/\hat{d}$, and if $N^* \rightarrow \max$ then $\hat{d} \rightarrow \min$. It is obvious that the d_k/T can be considered as a specific (per one unit of biomass) internal production of entropy at the kth level. Then $\hat{d}/T = (1/T)\sum_{k=1}^{n} d_k p_k$ is the mean entropy production, averaged over the whole chain. Therefore, we can formulate the following statement: *if the mean specific value of the internal entropy production in a given chain tends to its possible minimum then the total chain biomass tends to its possible maximum*. The proof is based on the statement proven earlier that the value of σ decreases along the trajectories towards a stable equilibrium. Indeed, since $\sigma = \hat{d}N$ then $(\mathrm{d}\sigma/\mathrm{d}t) = (\mathrm{d}\hat{d}/\mathrm{d}t)N + \hat{d}(\mathrm{d}N/\mathrm{d}t) < 0$; since $(\mathrm{d}N/\mathrm{d}t) > 0$ then $(\mathrm{d}\hat{d}/\mathrm{d}t) < 0$.

A similar statement can also be made for the "slow" evolution when the inflow q is slowly growing, but only for "ordered" systems, in which $d_1 > d_2 > d_3$. Only in this case $\partial\hat{d}/\partial q < 0$ (in the first interval, when $0 < q \leq d_1 d_2/\alpha_1$, $\hat{d} = d_1 = \text{const}$), i.e. the value of $\hat{d}$ decreases with the growth of q. If the "order" was violated (as occurs, for instance, in the

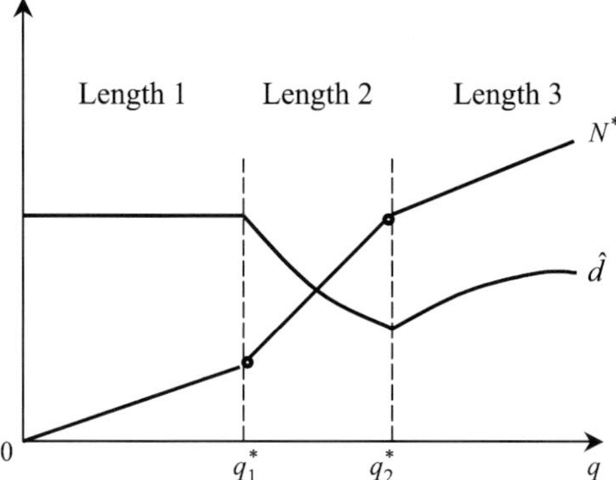

Fig. 7.5. "Slow" evolution of the total biomass (N) and the mean value of the specific entropy production ($\hat{d}$) as a function of inflow q: $\alpha_0 = \alpha_1 = 1$; $d_1 = 3$, $d_2 = 1$, $d_3 = 2$.

example of Fig. 7.5, where $d_1 > d_2$; $d_2 < d_3$) then in the corresponding interval the value of $\hat{d}$ increases.

As we shall show below, it is this kind of "disordering" which can cause the system to behave in a complex, "non-thermodynamic", even chaotic way.

7.6. The closed chains with conservation of matter. Thermodynamic cost of biogeochemical cycle

All ecosystems live only at the expense of a permanent flow of energy through the system. The energy is permanently losing value; it cannot be accumulated in its original (for instance, solar) form within the system. The situation with matter is different. Ecosystems *do not consume* matter, as they do with low-entropy incoming energy, but they *exchange* matter with the environment by means of biogeochemical cycles. Therefore, we can consider the joint system "ecosystem + environment" to be closed with respect to matter. In other words, if we measure biomass in units of matter (carbon, nitrogen, phosphorus, etc.) then we can assume that the total amount of matter contained in the biomass and the environment is constant.

Of course, we immediately encounter the "problem of scale" here. For instance, in relation to carbon, the whole atmospheric carbon is considered as an environmental one; therefore only a change in the state of the global ecosystem could be significant for atmospheric carbon. The dynamics of local ecosystems have no influence. This is explained by the high mobility of CO_2: all local emissions become mixed up in the atmosphere in the course of 3–4 months. Because of this carbon is practically never a limiting element. The mobility of such nutrients as nitrogen and phosphorus is relatively low; their environmental concentrations are often determined by local conditions, and they are very often limiting factors. All these arguments allow us to assume the existence of corresponding local laws of matter conservation for these elements. Their content in the environment and in living biomass is thus naturally strictly negatively correlated.

The balanced equations for the system should be written as

$$\frac{dN_0}{dt} = \sum_{k=1}^{n} D_k(N_k) - q_{01}(N_0, N_1),$$
$$\frac{dN_1}{dt} = q_{01}(N_0, N_1) - q_{12}(N_1, N_2) - D_1(N_1),$$
$$\cdots\cdots\cdots\cdots\cdots\cdots\cdots\cdots\cdots\cdots \tag{6.1}$$
$$\frac{dN_k}{dt} = q_{k-1,k}(N_{k-1}, N_k) - q_{k,k+1}(N_k, N_{k+1}) - D_k(N_k),\ k = 2, \ldots, n-1,$$
$$\cdots\cdots\cdots\cdots\cdots\cdots\cdots\cdots\cdots\cdots$$
$$\frac{dN_n}{dt} = q_{n-1,n}(N_{n-1}, N_n) - D_n(N_n).$$

where N_0 is the concentration of the selected element (for instance, nitrogen) in the environment, N_k is the same in the biomass of kth level, and $D_k(N_k)$ represents only dying-off biomass. It follows from these equations that

$$N_0 + \sum_{k=1}^{n} N_k = A = \text{const} \text{ or } N_0 = A - \sum_{k=1}^{n} N_k. \tag{6.2}$$

The general theory of such systems was developed by Svirezhev and Logofet (1978). As we have already done above, we restrict the analysis to three levels that is sufficient for understanding and interpretation.

Assuming that $N_0 = A - N_1 - N_2 - N_3$, $q_{k-1,k} = \alpha_{k-1} N_{k-1} N_k$ and $D_k = d_k N_k$; $k = 1, 2, 3$ system (6.1) is written as

$$\frac{dN_1}{dt} = N_1[(\alpha_0 A - d_1) - \alpha_0 N_1 - (\alpha_0 + \alpha_1) N_2 - \alpha_0 N_3],$$
$$\frac{dN_2}{dt} = N_2(\alpha_1 N_1 - \alpha_2 N_3 - d_2), \tag{6.3}$$
$$\frac{dN_3}{dt} = N_3(\alpha_2 N_2 - d_3).$$

The parameter A (the total amount of matter in the system) is naturally considered as a bifurcation parameter.

The separatrices portrait of Eq. (6.3) is shown in Fig. 7.6.

The trivial equilibrium $N_0^* = \{0, 0, 0\}$, which corresponds to thermodynamic equilibrium when there is no living matter in the system, is stable if $A < A_0^* = d_1/\alpha_0$; if the inverse inequality holds then it is unstable. The unstable manifold is the separatrix S_1 outgoing from N_0^*; its equation is $(dN_1/dt) = N_1(\alpha_0 A - d_1 - \alpha_0 N_1)$. It is incoming in the next equilibrium $N_1^* = \{N_1^* = A - (d_1/\alpha_0), 0, 0\}$ (see Fig. 7.6). It is clear that $N_1^* > 0$ for $A > A_0^*$. For $A > A_1^* = A_0^* + (d_2/\alpha_1) = (d_1/\alpha_0) + (d_2/\alpha_1)$ the equilibrium N_1^* is unstable; its unstable manifold is the separatrix S_2 situated in the plain (N_1, N_2), as shown in Fig. 7.6.

Consider the other equilibrium $N_2^* = \{N_1^*, N_2^*, 0\}$ where $N_1^* = d_2/\alpha_1$, $N_2^* = (\alpha_0/(\alpha_0 + \alpha_1))(A - A_1^*)$. For $A > A_1^*$ the equilibrium N_2^* is situated within the positive

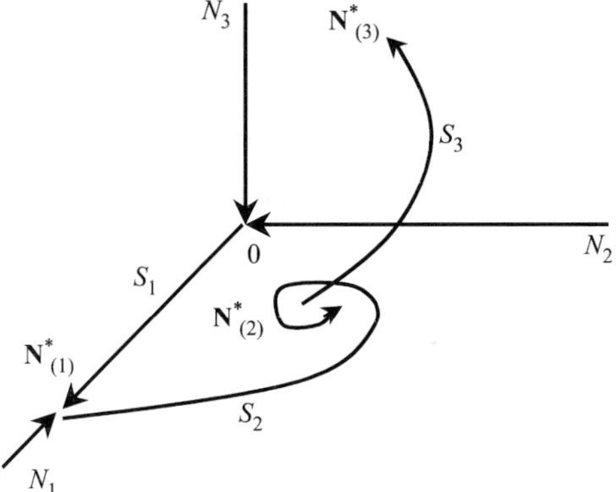

Fig. 7.6. Separatrices S_1, S_2 and S_3 of system (6.3).

quadrant of the plain (N_1, N_2). It loses its stability when $A > A_2^* = A_1^* + d_3(\alpha_1/\alpha_0\alpha_2) = (d_1/\alpha_0) + (d_2/\alpha_1) + (d_3/\alpha_2)((\alpha_0 + \alpha_1)/\alpha_0)$; its unstable manifold is the separatrix S_3 outgoing from N_2^* inwards the positive 3D orthant (see Fig. 7.6). Since system (6.3) does not have cycles (oscillations) then the separatrices S_2 and S_3 come into the equilibriums N_2^* and N_3^*, where (Svirezhev, 1987)

$$N_1^* = \frac{\alpha_2}{\alpha_1 + \alpha_2}(A - A_2^*) + \frac{\mathrm{d}_2}{\alpha_1}, \quad N_2^* = \frac{\mathrm{d}_3}{\alpha_2}, \quad N_3^* = \frac{\alpha_1}{\alpha_1 + \alpha_2}(A - A_2^*) \tag{6.4}$$

We see that the dynamics of system (6.3) is very simple: the critical bifurcation numbers A_0^*, A_1^* and A_2^* divide the axis A into the domains in each of which either there is no chain at all, or a chain of fixed length (one, two, three) can exist (Fig. 7.7). This is a typical bifurcation picture, which is very similar to that shown in Fig. 7.4: one solution loses its stability, a new stable solution arises, etc. when the bifurcation parameter passes over the critical value of A: A_0^*, A_1^* and A_2^*.

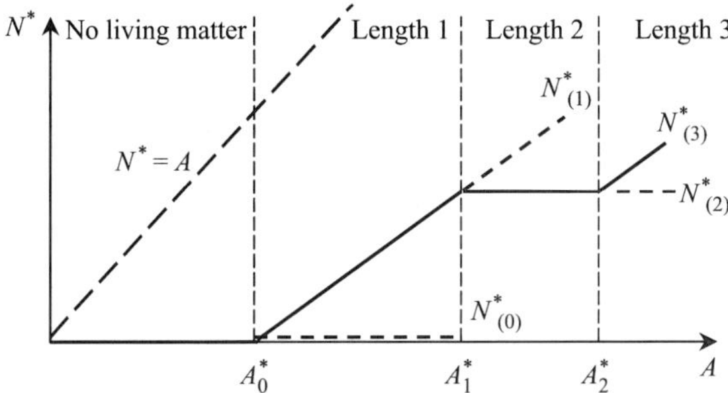

Fig. 7.7. The bifurcation diagram of equilibrium solutions for the total biomass: — stable branch, – – unstable branch.

If now we calculate the total chain biomass:

$$N^*_{(0)} = 0, \quad N^*_{(1)} = A - A^*_0, \quad N^*_{(2)} = \frac{\alpha_0}{\alpha_0 + \alpha_1}(A - A^*_1) + \frac{d_2}{\alpha_1},$$
$$N^*_{(3)} = (A - A^*_2) + \frac{d_2}{\alpha_1} + \frac{d_3}{\alpha_2}, \qquad (6.5)$$

we can see that it increases with the growth of A and number of links, but the biomass increases slower than the total amount of matter A (Fig. 7.7).

From a thermodynamic point of view the thermodynamic equilibrium corresponds to the case when there is no living biomass matter ($N^*_0 = A$). The appearance of living biomass implies that some thermodynamic machine such as the biogeochemical cycle began to work, and the system started its evolution from thermodynamic equilibrium; in the system, exergy began to accumulate. In the ideal case, the process has to continue until all the matter is transferred from a non-living to a living form. This would be possible if the residence time of biomass is equal to infinity, but this is not the case: this time is in principle finite, and such a kind of infinity is provided by the biogeochemical cycles of corresponding elements.

It is obvious that if the total amount of matter in the environment is equal to zero then it is senseless to talk about a turnover of matter, even if there is an inflow of energy. However, from the point of view of linear thermodynamics, it is sufficient for a very small quantity of matter to appear ($A \sim \varepsilon \ll 1$) in order to make the cycle work. But our system is non-linear in principle, and the cycle cannot work before the amount of matter in the environment exceeds the critical value of $A^*_0 = d_1/\alpha_0$. It is natural that this value is determined by characteristics of the system realising the matter turnover. Just after this moment the system begins to move far from thermodynamic equilibrium with a velocity equal to the growth rate of A. Note that although the slopes of A and N^* are the same, they are shifted from each other by the value of $A^*_0 = d_1/\alpha_0 = 1/\alpha_0\tau_1$. The τ_1 is implied as some mean residence time of living organisms forming the first (autotrophic) level. By expressing this in terms of energy units the value of A^*_0 can be considered as the *cost* of biological turnover for a simplest chain consisting of a single level, whereas the τ_1 is the *time* of turnover. Note that the dynamics of such a system is very poor and simple and because of that we are not yet far from thermodynamic equilibrium.

There is one curious interpretation of this result. Let A be the total amount of nitrogen in assimilated form which is turned over in the global biogeochemical cycle of this element. However (it follows from the result), for living matter to arise and the nitrogen cycle to start up it is necessary to have some initial finite quantity of ammonium and nitrates; note that the atmospheric nitrogen is biologically inert. Apparently, in the "pre-biosphere" these compounds could be formed as a result of thunderstorms from atmospheric nitrogen. It is evident that their quantity was very low, and in order for the turnover to begin to function the threshold $A^*_0 = 1/\alpha_0\tau_1$ has also to be low. For this a living matter has to possess a very high rate of biomass production (the α_0 is large) and a very low rate of dying-off of biomass (the τ_1 is also large). A good candidate for this role could be photosynthesising micro-organisms (for instance, green algae). And only after the non-biological processes (thunderstorms) have made a sufficient quantity of nitrogen compounds, which are necessary for living matter in the simplest ecosystems to arise and exist, could the global cycle of nitrogen and other biogenic elements by the same token start.

The next increase of A results in a growth of the length of the trophic chain that can be interpreted as a complication of the ecosystem structure. The set of bifurcation values $A_1^*, A_2^*, A_3^*, \ldots$ can be considered in this case as the costs of sequential complication of structure.

There is one more serious problem here: the criterion of the choice of branch when the solution branches after passing over a bifurcation point. Above we used some thermodynamic criteria for this, but we have to say frankly that this is not obtained very well here. The passage over the first critical point $A = A_0^*$ is accompanied by the branching of stable solution $N_{(0)}^* = 0$, $N_0^* = A$ into two branches: stable $N_{(1)}^* = A - A_0^*$ and unstable $N_0^* = A_0^*$ (see Fig. 7.7). It seems at first sight that the branch corresponding to the greater biomass can be selected as a true stable path of evolution. However, when we consider the passage over the second critical point $A = A_1^*$, we see that greater total biomass goes to the unstable branch $N_{(1)}^* = A - A_0^*$; the stable branch

$$N_{(2)}^* = \frac{\alpha_0}{\alpha_0 + \alpha_1}(A - A_1^*) + \frac{d_2}{\alpha_1}$$

corresponds to a lesser value than the total biomass ($N_{(1)}^* > N_{(2)}^*$ for $A > A_1^*$). So this idea can be discounted.

We already dealt with the problem of weighing biomasses at different levels: it was assumed that these biomasses have different exergy. The exergy is ranked in relation to the energy content (enthalpy) of detritus. In the framework of our model the value of N_0 could be considered as a mass of detritus. Then the total exergy of the system "chain biomass + detritus" expressed in detritus units can be written as

$$\mathrm{Ex}(A) = N_0^* + \mathrm{ex}_1 N_1^* + \mathrm{ex}_2 N_2^* + \mathrm{ex}_3 N_3^*. \tag{6.6}$$

In accordance with the principle of exergy growth in the process of system evolution we postulate that *the system always selects the branch (and correspondingly the structure of length of the chain) on which the exergy is greater than another.* For instance, in the right vicinity of the first critical point $\mathrm{Ex}_{(0)} = A < \mathrm{Ex}_{(1)} = A_0^* + \mathrm{ex}_1(A - A_0^*)$ or $(A - A_0^*) < \mathrm{ex}_1(A - A_0^*)$. If $A > A_0^*$ then $\mathrm{ex}_1 > 1$. In other words, if the first trophic level is existing and stable then the specific exergy of its biomass must be greater than one.

Let us now consider the case of the second critical point A_1^*. In its right vicinity where $A > A_1^*$

$$\mathrm{Ex}_{(1)} = A_0^* + \mathrm{ex}_1(A - A_0^*) < \mathrm{Ex}_{(2)}$$

$$= A - \frac{\alpha_0}{\alpha_0 + \alpha_1}(A - A_1^*) - \frac{d_2}{\alpha_1} + \mathrm{ex}_1 \frac{d_2}{\alpha_1} + \frac{\mathrm{ex}_2 \alpha_0}{\alpha_0 + \alpha_1}(A - A_1^*),$$

or

$$(\mathrm{ex}_1 - 1)(A - A_1^*) < (\mathrm{ex}_2 - 1)\frac{\alpha_0}{\alpha_0 + \alpha_1}(A - A_1^*);$$

whence it immediately follows that $\mathrm{ex}_2 > 1 + (1 + (\alpha_1/\alpha_0))(\mathrm{ex}_1 - 1)$.

For the third critical point we have

$$\mathrm{Ex}_{(2)} < \mathrm{Ex}_{(3)} = A_2^* - A_1^* + \mathrm{ex}_1\left[\frac{\alpha_2}{\alpha_1 + \alpha_2}(A - A_2^*) + \frac{d_2}{\alpha_1}\right] + \mathrm{ex}_2\frac{d_3}{\alpha_2} + \frac{ex_3\alpha_1}{\alpha_1 + \alpha_2}(A - A_2^*),$$

or

$$ex_3 > 1 + \left(1 + \frac{\alpha_2}{\alpha_1}\right)\left[\frac{\alpha_2}{\alpha_1 + \alpha_2}(\mathrm{ex}_1 - 1) + \frac{\alpha_0}{\alpha_0 + \alpha_1}(\mathrm{ex}_2 - 1)\right]$$

for $A > A_2^*$.

A typical example of a simple trophic chain is the chain

Phytoplankton → zooplankton → fish.

Such a type of ecosystem (a warm ecosystem in Silver Springs, FL) was analysed by Svirezhev and Logofet (1978). Using H. Odum's experimental data they estimated the ratios $(\alpha_1/\alpha_0) \approx 10$, $(\alpha_2/\alpha_1) \approx 8.5$, $(\alpha_2/\alpha_0) \approx 80$. Substituting these ratios into the corresponding inequalities we get:

$$\mathrm{ex}_2 > 1 + 10(\mathrm{ex}_1 - 1), \qquad \mathrm{ex}_3 > 1 + 9.5[0.9(\mathrm{ex}_1 - 1) + 0.09(\mathrm{ex}_2 - 1)].$$

Looking in Table 5.1 we find: $\mathrm{ex}_1 = 3.4$ for phytoplankton, $\mathrm{ex}_2 = 29\text{–}43$ for zooplankton, and $\mathrm{ex}_3 = 281\text{–}337$ for fish. By comparing these values with our inequalities we can see that they are satisfied:

$$\mathrm{ex}_1 = 3.4 > 1, \qquad \mathrm{ex}_2 = 29 \div 43 > 1 + 11 \times 2.4 = 26.4,$$

$$\mathrm{ex}_3 = 281 \div 337 > 1 + 9.5[0.9 \times 2.4 + 0.09(28 \div 42)] \approx 45 \div 57.$$

The bifurcation diagram for the total exergy is shown in Fig. 7.8.

7.7. Complex behaviour: cycles and chaos

At first sight it seems that the dynamics of the trophic chain remains relatively simple even if its length is 4, 5, ..., i.e. the value of A is rather large, and the system is sufficiently far from thermodynamic equilibrium. But this is not true: even in a chain of length 4, when $A > A_3^{\mathrm{cr}} > A_3^*$, cycles and oscillations may arise, i.e. a complex dynamic behaviour arises (Svirezhev, 1987). The cycle encloses the equilibrium $N_4^* = \{N_1^*, N_2^*, N_3^*, N_4^*\}$, which loses its stability for $A > A_3^{\mathrm{cr}}$. In this case, we have the right to expect that the exergy, which monotonously increases at a certain rate up to this point, has to begin to increase at a lesser rate (an unstable branch corresponds to lesser exergy).

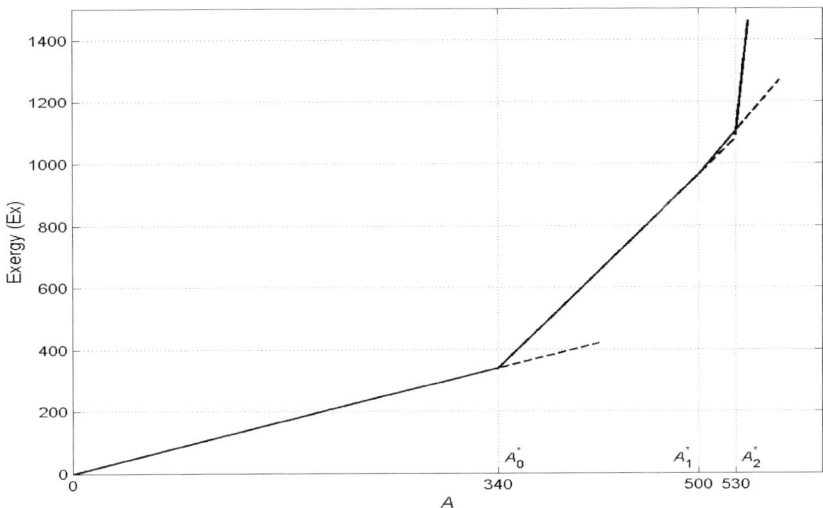

Fig. 7.8. The bifurcation diagram when the equilibrium solutions are expressed in exergy units. Real data about the Silver Springs ecosystem are used. — stable branch, – – unstable branch.

Up to this point we have studied systems with weak non-linearities, when the flows from one level to another were proportional to the product of their biomasses. However, in reality we deal with more complex functions, for instance, $q_{k,k+1} = V_k(N_k)N_{k+1}$ where the so-called "trophic functions" $V_k(N_k)$ seem to look as shown in Fig. 7.9.

It is possible to prove that chains with such types of trophic functions cycles arise even in the simplest chain with a length of two (and naturally also in longer chains). Apparently, the cyclic dynamics is typical for all closed trophic chains if there is a so-called "enrichment effect", i.e. an abrupt increase in the total amount of nutrients. This point of view explains such dynamic phenomena as oscillations of the biomasses of phytoplankton, zooplankton and fish in closed lakes and ponds as a result of the so-called "anthropogenic eutrophication", when the fertilisers washed out from fields and household waste waters

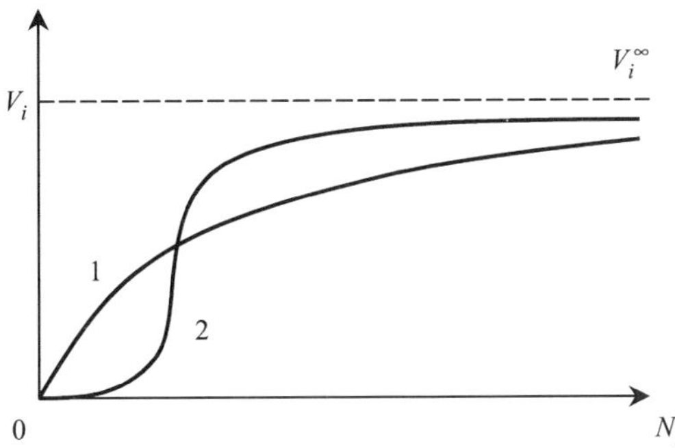

Fig. 7.9. Types of trophic functions: 1—hyperbolic, 2—s-shaped.

increase abruptly the amount of nutrients (nitrogen and phosphorus) in these water-body systems. Since the specific mean exergy (the total exergy divided by the total amount of matter in the system) in this moment has to decrease, then the fracture on the curve of monotonously growing exergy can be a good indicator to warn about this ecological hazard (for details, see Section 7.8).

The next complex dynamic effect may be "dynamic chaos" (Svirezhev, 1983, 1987). Note that in mathematical ecology this is the rule rather than an exotic exception; the problem is how to make up an adequate model to describe it. For instance, a closed, three-level trophic chain with non-linear trophic functions can be considered, perhaps, as the simplest object that demonstrates dynamic chaos. A model of this system can be represented as:

$$\begin{aligned}\frac{dN_1}{dt} &= N_1\left[-d_1 + \alpha_0(A - N_1 - N_2 - N_3) - \frac{V_1(N_1)}{N_1}N_2\right],\\ \frac{dN_2}{dt} &= N_2\left[-d_2 + V_1(N_1) - \frac{V_2(N_2)}{N_2}N_3\right],\\ \frac{dN_3}{dt} &= N_3[-d_3 + V_2(N_2)],\quad V_k(N_k) = \alpha_k N_k/(K_k + N_k),\quad k = 1,2.\end{aligned} \tag{7.1}$$

This system, where

$$d_1 = 0.1;\ d_2 = d_3 = 0.2;\ 0.30 \le \alpha_0 = \alpha \le 0.37;\ \alpha_1 = \alpha_1 = 1;\ K_1 = K_2 = 5,$$

was studied numerically; the total amount of matter in the system, A, and the per capita rate of resource consumption by the species of the first trophic level, α, were selected as bifurcation parameters. In the plane (A, α) the two curves $A_{+1}(\alpha)$ and $A_\infty(\alpha)$ were constructed, which separate the domains with different dynamic behaviour (Fig. 7.10). The values of these curves at the point $\alpha = 0.34$ are equal to $A_{+1}(0.34) = 36.251$ and $A_\infty(0.34) = 36.24$.

If we now move along the line ab (at constant $\alpha = 0.34$) in the direction of increase of the parameter A then the behaviour of the system varies by means of the doubling of cycles from regular to stochastic regimes. The transition does not occur directly: there exists a

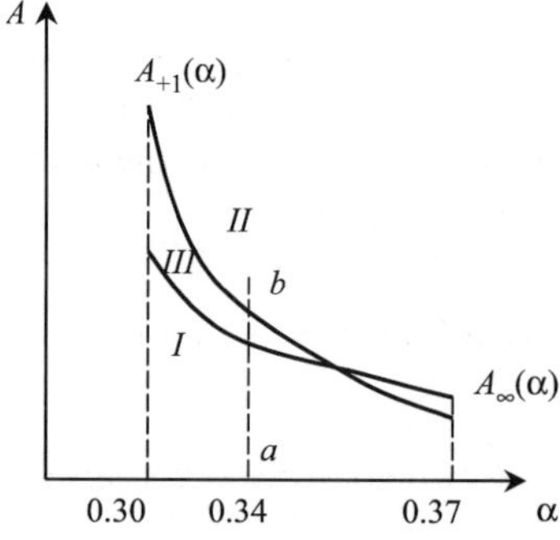

Fig. 7.10. The domains of different dynamic behaviours: I—regular, II—stochastic, III—"pre-stochasticity" domain.

"pre-stochasticity" domain III, in which there are both regular and stochastic trajectories, and parallel with the "strange attractor" which results in the doubling of cycles, there is a stable limit cycle. This regime, also named as "pre-turbulence" or "metastable chaos", is characterised by the existence of both chaotic and regular trajectories, which are attracted to the stable cycle. The latter are very similar to chaotic trajectories; but in spite of this they are regular. It is interesting that the pre-stochasticity regime is typical for Lorenz's attractor, but in the latter the process of stochastisation differs from Feigenbaum's mechanism of the doubling of cycles. Our "ecological strange attractor" occupies an intermediate position between Lorenz's and Feigenbaum's attractors. It is curious that the phase volume of our system contracts; in other words, our system (like Lorenz's) is dissipative in spite of the conservation of matter taking place within it.

By moving further along *ab* we get into the domain of stochasticity, where chaos arises. This is domain II, or more correctly, its lower boundary, as its upper one and its geometry were not studied in detail. The chaotic trajectories are shown in Fig. 7.11.

Let us set the question: would the property of dynamic stochasticity be conserved with the growth of the chain length? It has been proven that the existence of a strange attractor for chains of arbitrary length (longer than three) follows from the existence of the strange attractor for a closed trophic chain of a length of three (Svirezhev, 1987).

In contrast, there are only regular trajectories (cycles) in open chains, in which a dead biomass is neither decomposed nor returned to the resource compartment. Naturally there is a constant inflow of an external resource, which is equal to zero for the closed chain. Now let the chain be partly closed, i.e. a certain part of matter contained in the dead biomass returns to the resource level. A model of such a system can be represented, for instance, as

$$\frac{\mathrm{d}N_0}{\mathrm{d}t} = q - V(N_0)N_1 + \lambda \sum_{k=1}^{3} d_k N_k, \quad 0 < \lambda < 1,$$

$$\frac{\mathrm{d}N_k}{\mathrm{d}t} = -d_k N_k + V_{k-1}N_k - V_k N_{k+1}, \quad V_k = \frac{\alpha_k N_k}{K_k + N_k}, \qquad k = 1, 2, 3. \tag{7.2}$$

It is interesting to find such a threshold value λ^* of the "closure" parameter λ ($\lambda = 0$ for open system and $\lambda = 1$ for fully closed one) so that if $\lambda < \lambda^*$ then the dynamic chaos arises in system (7.2).

Naturally, other formulations are also possible. We believe that models of mathematical ecology constitute a favourable field to look for different strange attractors and dynamic stochastic behaviour corresponding to them, but it is necessary to remember the biological adequacy of the model being used.

7.8. What kind of exergy dynamics takes place when the enrichment and thermal pollution impact on the ecosystem?

Let us carry out the following two numerical experiments with system (7.1):

1. We shall effect a slow, quasi-stationary increase of the total amount of matter in the system (A) starting from the value of $A < d_1/\alpha_0 = 0.1/0.34 \approx 0.29$, when there

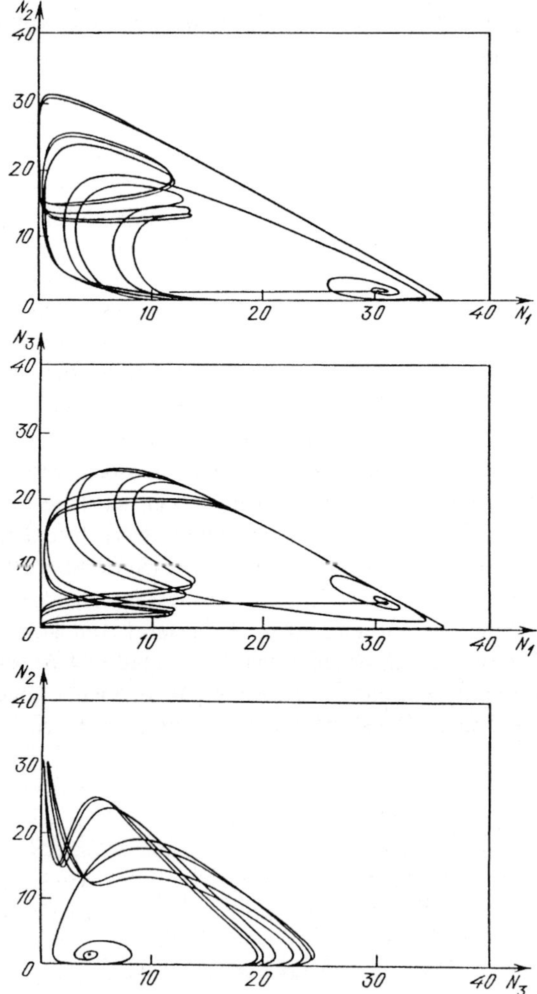

Fig. 7.11. An approach of the non-periodic trajectory of Eq. (7.1) to the strange attractor: $A = 36.5$; $\alpha = 0.34$.

is no biomass at all, until $A = 36.5$ when dynamic chaos arises; $\alpha_0 = \text{const} = 0.34$. In the same way we simulate the enrichment phenomenon in closed water-body systems.

2. At the constant value of $A = 36.5$ we shall effect a slow, quasi-stationary increase in the rate of nutrient uptake by phytoplankton (α_0) starting from the value of $\alpha_0 < d_1/A = 0.1/36.5 \approx 0.27 \times 10^{-2}$, when there is no biomass, until $\alpha_0 = 0.34$, when dynamic chaos arises. In the same way we simulate the effect of thermal pollution, since the value of α_0 increases with the growth of temperature. The α_0 can be interpreted as $\alpha_0 = \mu_{\max}^{\text{ph}}/K_{\text{ph}}$ where $\mu_{\max}^{\text{ph}}$ is the maximal growth rate of phytoplankton and K_{ph} is the Michaelis–Menten constant in the expression for the growth rate $\mu^{\text{ph}} = \mu_{\max}^{\text{ph}} N_0/(K_{\text{ph}} + N_0)$.

If we interpret the chain as consisting of phytoplankton, zooplankton and fish then the total exergy is represented as

$$\mathrm{Ex}(t) = [A - N_1(t) + N_2(t) + N_3(t)] + 3.4N_1(t) + 38N_2(t) + 330N_3(t). \tag{8.1}$$

Since the exergy defined by Eq. (8.1) depends on the value of A (this can mask dynamic effects connected with exergy) then it may be more correct to use such a value as the *specific mean exergy*, $\hat{\mathrm{e}}\mathrm{x} = Ex/A$. It is obvious that this value is dimensionless and does not explicitly depend on A. Another macroscopic characteristic is the total biomass of the chain, $N(t) = N_1(t) + N_2(t) + N_3(t)$, but instead of that we shall use such a dimensionless value as $\eta = N/A$, which could be termed as a "utilisation coefficient". Indeed, the greater is the coefficient η, the greater part of the matter is concentrated in the biomass. At the limit when the η is close to one, practically the whole matter is in the biomass, and the nutrient which is formed as a result of dead organic matter is almost instantly consumed by phytoplankton. In this case we can say that the system is almost ideally adapted to the environment.

If the value of η determines the *degree* of adaptation then the value of êx determines the *quality* of adaptation. In other words, if the η is an index of the efficiency of the trophic chain then the êx is an index of its organisation or an index of biomass quality. Indeed, if the specific exergy determines the quality of one biomass unit of the corresponding species, then the greater its mean value is, the higher is the percentage of "quality" biomass in the total chain biomass.

The specific mean exergy (in dimensionless units) is calculated with respect to time until a stationary regime for each given value of A is established (Fig. 7.12). Until $A <$ 9.455 (the first picture) the dynamics remains very simple and regular and the equilibrium is very quickly established so that there is not a problem to estimate the limit value of êx. One can see that it increases with the growth of A (see Fig. 7.14a). At $A = 9.447$ a first cycle arises: the system begins oscillating. The specific mean exergy also oscillates (the second picture), but the oscillations are very simple. Starting with this value of A the êx begins decreasing; moreover, the decrease is monotonous in relation to the increase of A (Fig. 7.14a). Finally at $A = 36.24$ the dynamic chaos arises, but the behaviour of êx (also chaotic) is simpler than the behaviour of the state variables N_1, N_2, and N_3 (compare the third section of Figs. 7.11 and 7.12). The specific mean exergy continues decreasing. Note when we speak about the dependence of êx on A we imply that we operate with the temporal mean of the value, obtained as a result of averaging over one of the limit sets: point, cycle or strange attractor. For this we average the êx over a rather long piece of established trajectory. All the same methods are used when we deal with the utilisation coefficient η and the dependences of êx and η on α_0.

Analogous pictures are constructed for the utilisation coefficient η (Fig. 7.13).

One can see that these pictures are qualitatively similar to Fig. 7.12 (the same regular behaviour for $A = 5$, oscillations at $A = 9.447$ and non-periodic, irregular impulses at $A = 36.5$, which is characteristic for chaos), but all dynamic effects are apparently weaker than in the previous case. As the value of A increases the temporal mean of η very quickly reaches saturation at the level very close to one (Fig. 7.15a). The saturation occurs when the system dynamics becomes sufficiently complex (oscillations and chaos). However, in

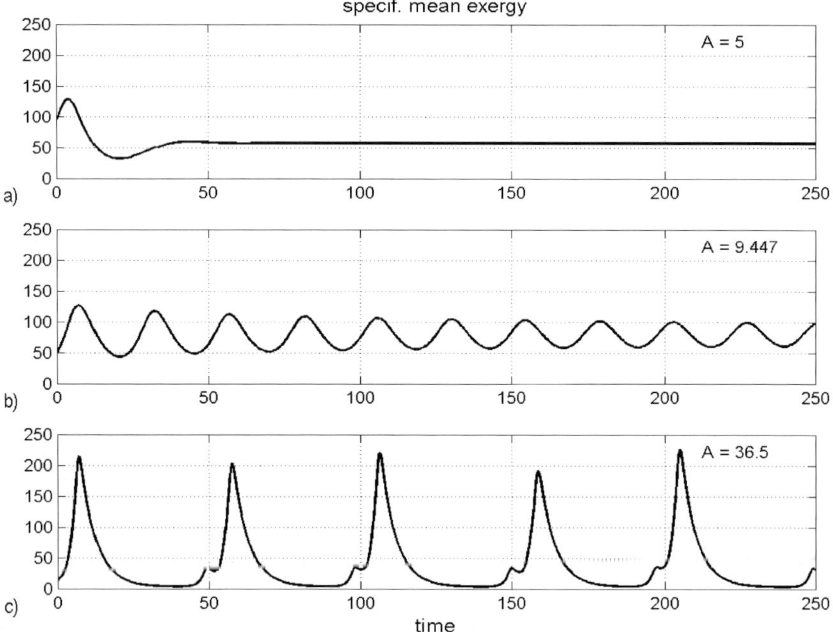

Fig. 7.12. Dynamics of the specific mean exergy ($\hat{\text{ex}} = Ex/A$) for three different dynamic regimes: (a) regular regime without oscillations ($A = 5$), (b) doubling oscillations ($A = 9.447$), (c) chaos ($A = 36.5$). The exergy and the total amount of matter are measured in dimensionless units.

contrast to the previous case the occurrence of complex dynamics is difficult to detect if the graph $\eta(A)$ is used.

Let us once more look at Fig. 7.15a. We can see that when there is chaos then the utilisation coefficient is very close to one (about 0.98–0.99). This implies that the resource is almost completely used; it suggests that the complex dynamics of ecosystems (doubling oscillations and, especially, chaos) is one form to which they adapt, allowing them to use (almost completely) a trophic resource. However, this adaptation is paid for by a fall in biomass quality: the fall of exergy is accompanied by a reduction of fish biomass and increase of the biomass of plankton (especially phytoplankton). A compromise between these tendencies is reached at the boundary between the quiet regular dynamics ($A < 9.447$) and the situation when the dynamics is suddenly complicated ($A > 9.447$). At the "boundary" point $A = 9.447$, on the one hand, the degree of resource utilisation is already rather high; on the other hand, the biomass quality (the specific mean exergy) is *still* high (at this point the $\hat{\text{ex}} = 75.3$ is maximal). As we have already mentioned, the sharply delineated fracture on the graph in Fig. 7.14a can be used as a detector to indicate that we have to expect the occurrence of new regimes in the ecosystem.

It seems that this state is optimal: there is an almost complete utilisation of the resource here, and everything is quiet. However, at any time we have to expect the appearance of such annoying (from the anthropocentric point of view) events as big oscillations and

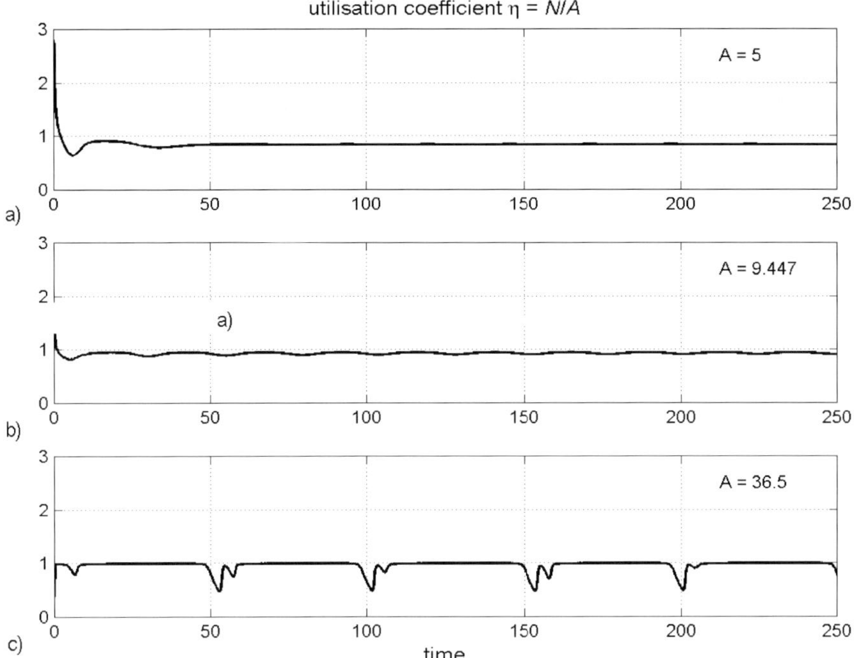

Fig. 7.13. Dynamics of the utilisation coefficient ($\eta = N/A$) for three different dynamic regimes: (a) regular regime without oscillations ($A = 5$), (b) doubling oscillations ($A = 9.447$), (c) chaos ($A = 36.5$). The total biomass and the total amount of matter are measured in dimensionless units.

chaotic behaviour of species biomasses. The chaos is especially annoying because of the basic unpredictability of drastic changes in the ecosystem state. Let us compare the two pictures showing the dynamics of êx(t) and $\eta(t)$ (Figs. 7.12c and 7.13c). The peaks of êx and falls of η, although occurring simultaneously, are irregular. Although at these times the fish biomass increases sharply (the êx is sharply increasing) the increase is of a short duration. At the same time the degree of utilisation falls sharply by almost half (see Fig. 7.13c). In the second experiment we vary only the coefficient $\alpha_0 = \alpha$ at the fixed total amount of matter, $A = 36.5$. The qualitative dynamics of the system is similar to that obtained in the previous experiment. As the value of α increases the system passes through the regular regime without oscillations (the corresponding α-interval is very short; even for $\alpha = 0.01$ long-periodic oscillations arise). In contrast to the previous case the pre-stochasticity regime is relatively long: it corresponds approximately to the interval $\alpha \in [0.1, 0.3]$. At last, a state of chaos arises beginning with $\alpha \approx 0.33$. However, the results which were obtained for the dependencies of êx and η and shown in Figs. 7.14b and 7.15b differ from the previous ones, especially for the specific mean exergy. If the utilisation coefficient reaches its saturation at approximately $\alpha = 0.035$ when the oscillations already dominate in the dynamics of the ecosystem (that, in general, coincides with the previous case), then the specific mean exergy attains its maximum just when the pre-stochastical regime arises. Just after this, the value of êx begins to decrease very slowly. The change of dynamic regimes is very weakly

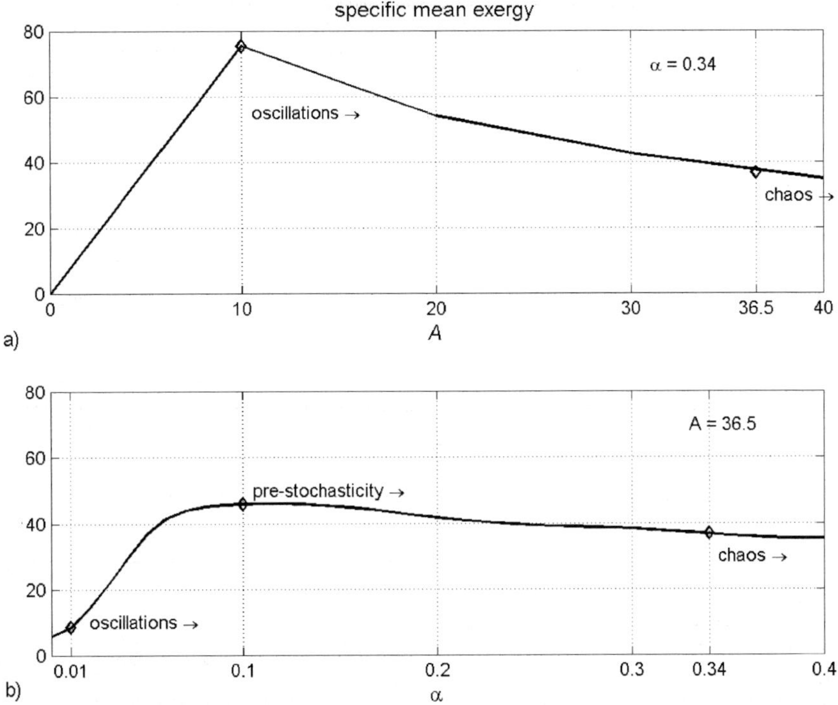

Fig. 7.14. The specific mean exergy as a function of the total amount of matter (a), and the rate of nutrient uptake by phytoplankton (b).

delineated. Finally, we can say that the dynamic reaction of the considered ecosystem is weaker in relation to the thermal pollution than to the enrichment.

Note, finally, that very similar results were obtained by Jørgensen (1995a) when he studied numerically two models of eutrophications with two and three trophic levels, respectively. These models also have complex dynamic behaviours (cycles and pre-stochasticity regime), and the behaviour of exergy, which is defined analogously, is very similar to that considered above. The only difference is that the maximum rate of zooplankton growth, μ_{max}, is considered as a bifurcation parameter.

7.9. Embodied energy (emergy)

In Section 7.8 we demonstrated that the weighing of biomasses in relation to exergy has allowed us to obtain interesting results. Generally speaking, the idea lies on the surface: the problem is how to select reasonable and ecologically meaningful weighing coefficients. One concept of this type was introduced by Odum (1983a,b) and attempts to account for the energy required in the formation of organisms in different trophic levels. The idea is to correct energy flows for their quality. Energies of different types are converted into equivalents of the same type by multiplying by the energy transformation ratio. For example, fish, zooplankton and phytoplankton can be compared by multiplying

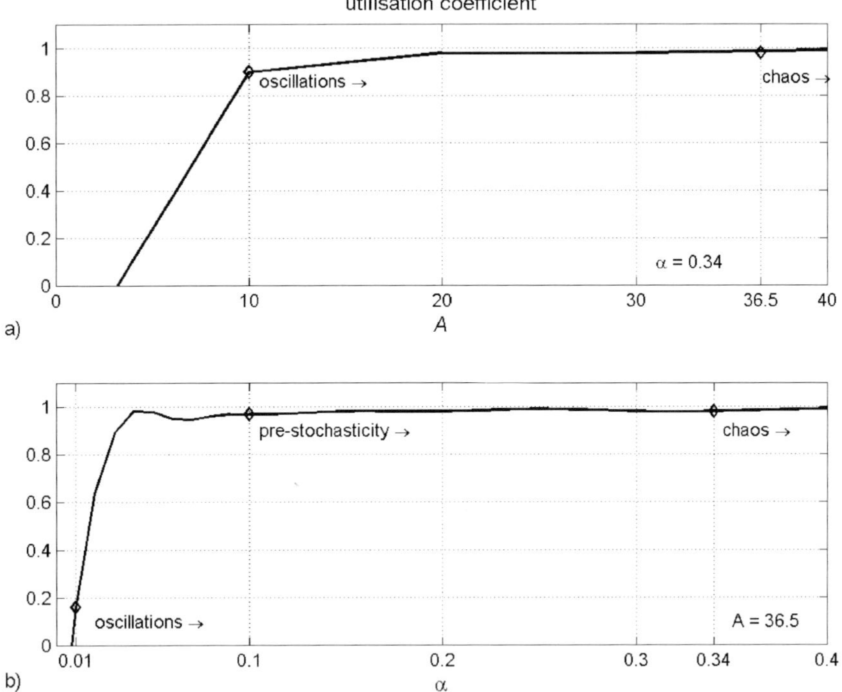

Fig. 7.15. The utilisation coefficient as a function of the total amount of matter (a), and the rate of nutrient uptake by phytoplankton (b).

their actual energy content by their solar energy transformation ratios. The more transformation steps there are between two kinds of energy, the greater is the quality and the greater is the solar energy required to produce a unit of energy (Joule) of that type. When one calculates the energy of one type that generates a flow of another, this is sometimes referred to as the *em*bodied en*ergy* (*emergy*) of that type.

Fig. 7.16 presents the concept of embodied energy in a hierarchical chain of energy transformation and Table 7.1 gives embodied energy equivalents for various types of energy.

H.T. Odum reasons that surviving systems develop designs that receive as much energy amplifier action as possible. The energy amplifier ratio is defined in Fig. 7.17 as the ratio of output *B* to control flow *C*. H.T. Odum suggests that in surviving systems the amplifier effects are proportional to embodied energy, but full empirical testing of this theory still needs to be carried out in the future.

One of the properties of high-quality energies is their flexibility. Whereas low-quality products tend to be special, requiring special uses, the higher-quality part of a web is of a form that can be fed back as an amplifier to many different units throughout the web. For example, the biochemistry at the bottom of the food chain in algae and microbes is diverse and specialised, whereas the biochemistry of top animal consumer units tends to be similar and general, with services, recycles and chemical compositions usable throughout.

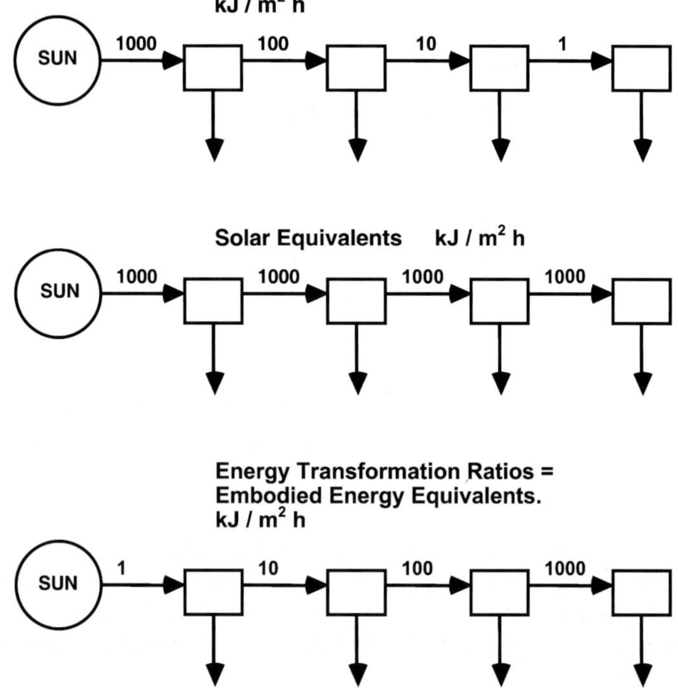

Fig. 7.16. Energy flow, solar equivalents and energy transformation ratio (embodied energy equivalents) in a trophic chain (from Jørgensen, 2002b).

Hannon (1973, 1979, 1982) and Herendeen (1981) applied energy intensity coefficients as the ratios of assigned embodied energy to actual energy to compare systems with different efficiencies.

The difference between embodied energy flows and power, defined as in phenomenological thermodynamics

$$\text{Power} = \sum_{k=1}^{n} (\Delta G_k / N_k)(\mathrm{d}N_k / \mathrm{d}t) \tag{9.1}$$

Table 7.1
Embodied energy equivalents for various types of energy

Type of energy	Embodied energy equivalents
Solar energy	1
Winds	315
Gross photosynthesis	920
Coal	6800
Tide	11,560
Electricity	27,200

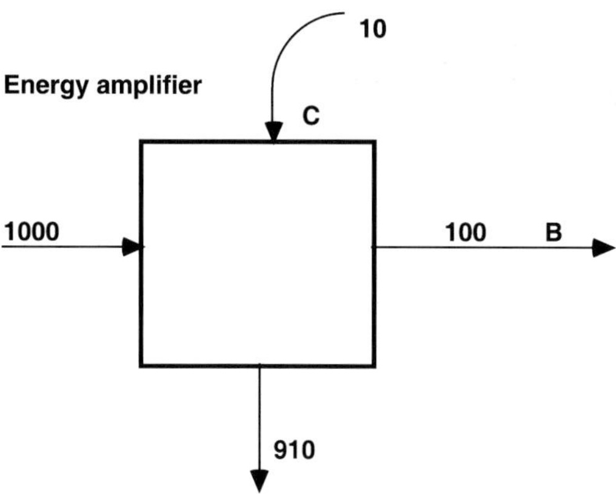

Fig. 7.17. The energy amplifier ratio, R is defined as the ratio of output B to control flow C. It means that $R = 10$ in this case (from Jørgensen, 2002b).

where the generalised flow or the ecoflow, indicated as $(\mathrm{d}N_k/\mathrm{d}t)$, is multiplied with the specific useful work, i.e. $(\Delta G_k/N_k)$, simply seems to be a conversion to solar energy equivalents of the free energy. The increase in biomass in Eq. (9.1) is a conversion to the free energy flow, and the definition of embodied energy is a further conversion to solar energy equivalents.

Embodied energy is, as seen from these definitions, determined by the biogeochemical energy *flow* into an ecosystem component, measured in solar energy equivalents. The stored emergy, Em, per unit of area or volume to be distinguished from the emergy flows can be found from:

$$\mathrm{Em} = \sum_{k=1}^{n} \mathrm{em}_k N_k \tag{9.2}$$

where the specific emergy, em_k, is the quality factor, which is the conversion to solar equivalents, as illustrated in Table 7.1 and Fig. 7.16, and N_k is the concentration expressed per unit of area or volume.

The emergy concept could be formally explained from the point of view of equilibrium thermodynamics. In fact, in a statistical equilibrium system without any constraints (except perhaps the conservation of the total number of particles) the equilibrium distribution is uniform. If we consider the trophic chain as such a system then the state $N_1^* = \cdots = N_n^*$ must be in equilibrium, but this contradicts trophic pyramids observed in reality. This contradiction can be resolved by weighing the biomasses in such a way that the equilibrium condition holds for the weighed biomasses. The specific emergy could be such a weighing factor, so that

$$\mathrm{em}_1 N_1^* = \mathrm{em}_2 N_2^* = \cdots = \mathrm{em}_n N_n^*. \tag{9.3}$$

The calculations by Eq. (9.2) reduce the difference between stored emergy and stored exergy, which as we showed can also be found with a good approximation as the sum of concentrations multiplied by a quality factor. The quality factor for exergy accounts for the information embodied in the various components of the system (detailed information is given in Chapter 5), while the quality factor for emergy accounts for how much solar energy it has cost to form the various components. Emergy thereby calculates how much solar energy (which is our ultimate energy resource) it has cost to obtain one unit of biomass of various organisms, while exergy accounts for how much "first class" energy (energy which can do work) the organisms possess as a result of the complex interactions in an ecosystem. Both concepts attempt to account for the quality of the energy. Emergy does this by looking into the energy flows in the ecological network to express the energy costs in solar equivalents. Exergy does it by considering the amount of information (which also contains first class energy able to do work, as was shown in Chapter 5) that the components have embodied.

The differences between the two concepts may be summarised as follows:

(1) Emergy has no clear reference state, which is not needed as it is a measure of energy flows, while exergy is defined relative to the environment (see also Chapter 5).
(2) The quality factor of exergy is based on the content of information, while the quality factor for emergy is based on the cost in solar equivalents.
(3) Exergy is better anchored in thermodynamics and has a wider theoretical basis.
(4) The quality factor em_k may be different from ecosystem to ecosystem, and in principle it is necessary in each case to assess the quality factor based on an energy flow analysis, which is sometimes cumbersome to make.

The quality factors listed in Table 7.1 or in Brown and Mc Clanahan (1992) may be used generally as good approximations. The quality factors used for computation of exergy (see Chapter 5) require knowledge of the non-nonsense genes (information content) of various organisms, which is sometimes surprisingly difficult to assess. A number of exergy quality factors have been found. From a theoretical point of view they can be used generally. Further comparisons of the two concepts will be presented in Chapter 12.

In his book "Environmental Accounting—Emergy and Environmental Decision Making" (Odum, 1996), H.T. Odum has used calculations of emergy to estimate the sustainability of the economy of various countries. As emergy is based on the cost in solar equivalents, which is the only long-term available energy, it seems to be a sound first estimation of sustainability.

7.10. Summary of the ecological important issues

We are now able to describe how an ecosystem is working from a thermodynamic point of view. The maintenance of the ecological structure and the organisms forming the structure of the ecosystem requires a supply of exergy, which is lost as heat to the environment. If more exergy is supplied than needed for maintenance of the status quo, the surplus exergy is stored as more biomass, the more effective and complex network, or more information. It is stored to meet an unexpected period of shortage and to increase the sum of buffer capacities. The structure was named by Prigogine the dissipative structure, because it is permanently dissipating energy (converting exergy to anexergy—"destroying" exergy). It can be shown to be a consequence of the thermodynamic laws and a few other realistic assumptions (for instance, all organisms have a similar chemical composition, which biochemistry has shown us is the case) that an ecosystem will have a considerable number of various organisms with a very high chemical potential (many complex biochemical compounds with a high content of information, for instance hormones) and therefore also a high content of free energy. The latter is consistent with Boltzman's statement: the struggle for life is first of all a struggle for free energy.

It is possible to show that a Prigogine-like theorem is valid for a trophic chain in a mature ecosystem. An early-stage ecosystem (see Table 2.5) will quickly try to build as much biomass as possible, because the biomass will, as a parabolic antenna, capture as much electromagnetic photons (solar radiation) as possible, i.e. capture as much exergy as possible. The mature system on the other hand already captures the possible amount of solar radiation (about 75–80%) and has therefore in its further development to attempt to save exergy. The entropy production will tend to its possible minimum and the biomass of the chain will tend to its possible maximum. Prigogine's thermodynamics seems valid for mature ecosystems, i.e. ecosystems that have reached a dynamic "almost" steady state. It may also be expressed by exergy terms: the trophic chain tends to minimise the exergy spent for maintenance in order to have more surplus exergy to store in the ecosystem. It can even be shown that in a trophic chain the specific exergy (exergy/biomass) tends to increase through a food chain. Zooplankton has higher specific exergy (higher β-value) than phytoplankton and fish again has higher specific exergy than zooplankton. Nature has a certain degree of randomness that implies that this rule is general but exceptions can of course be found.

When an ecosystem starts at an early state, the appearance of living biomass implies that some thermodynamic machine such as the biochemical cycles begins to work. The system starts to evolve away from the thermodynamic equilibrium: more and more exergy is stored/accumulated in the system and the process continues until all the matter is transferred from non-living to living matter. As for biomass the residence time is not infinite; biochemical cycling of the corresponding elements has to take place. As the system moves away from thermodynamic equilibrium, it can be shown that the trophic chain meets bifurcation points and that the system selects the branch that gives the highest exergy.

Which among many possibilities does a trophic chain select in its effort to move as far away from thermodynamic equilibrium as possible? To answer this question, two useful concepts are introduced: degree of adaptation = the ratio of living biomass to total organic matter and quality of adaptation = specific mean exergy = the total exergy of the system/total organic matter of the system. The maximum specific exergy = the quality of adaptation is reached at a certain level of organic matter where the utilisation of the organic matter (the degree of adaptation) is close to one. Further, an addition of organic matter implies that the quality of adaptation = the specific mean exergy is declining and the utilisation coefficient is only increased slightly (see Figs. 7.14 and 7.15). This is consistent with Kaufman's statement: biological systems operate at the edge of chaos to be able to utilise the available resources at the optimum. An overexploitation of the resources will involve oscillations that will become more and more violent and finally the system will become chaotic. An attempt to increase the specific exergy beyond what the system can bear implies an overexploitation of the lower levels in the food chain and a chaotic system is the result. An ecosystem finds, in other words, a suitable balance of the various levels in the trophic chain.

Emergy measures the cost of energy expressed as solar equivalents, while exergy gives the result of the solar radiation in a form of energy that can do work. It is interesting (see, for instance, Jørgensen, 2002b) that natural systems usually have a low emergy/exergy ratio, while man-made systems have a high ratio emergy/exergy. Natural systems are, in other words, more effective in getting exergy out of solar radiation. This is also consistent with the Prigogine-like theorem. It is recommendable to apply both emergy and exergy as ecological indicators. Emergy gives the cost, exergy the result and the ratio gives the efficiency.

Chapter 8

Models of ecosystems: thermodynamic basis and methods. II. Competition and trophic level

Struggle for Life is a permanent reality in the Nature and the Society.
T. Malthus

8.1. Introduction

While in Chapter 7 vertically structured ecosystems, with interactions between trophic levels, were considered, this chapter is devoted to the interactions between species on a trophic level, which, in turn, is considered as a horizontally structured system. From all the five theoretically possible types of interaction between species on one level (symbiosis, commensalisms, amensalism, neutralism and competition, see also Chapter 9) we shall only consider competition, since in nature it is the most widespread type. The latter is a non-linear process. For instance, if we keep in mind the well-known *competitive exclusion principle* by G.F. Gause, which forbids co-existence of two species with similar ecological requirements in one habitat, we immediately get a non-linear operator that takes us outside the boundaries of linear thermodynamics. In fact, the third Caratheodory Axiom of Thermodynamics states that intensive thermodynamic variables of a joined system are equal to mean values of partial variables of constituent subsystems. It is easy to see that Gause's principle contradicts this axiom. Nevertheless, some thermodynamics statements and theorems, and in particular Prigogine-like theorems, can be valid and proved for this case.

8.2. Thermodynamics of a competing community

So, a trophic level is considered as a horizontally structured biological community of competing populations ("competing community"). Their number (number of species) varies in a wide interval, from one to a hundred, but they are all connected by a relation of competition. From the thermodynamic point of view the competition is a typical dissipative process, producing entropy. Thus, the competing community is a typical dissipative structure, the stability of which is maintained by permanent export of entropy from the system into its environment, so that the competing community is also an open system.

Towards a Thermodynamic Theory for Ecological Systems, pp. 189–219

Let us consider (as in Chapter 7) an open system, within which an equilibrium in relation to temperature and pressure, $T, p =$ constant. Thus, such irreversible processes as heat transport, viscous flows, gravitational sedimentation, etc. are not included in the system. But equilibrium in relation to the distribution of matter or particles, which are capable of a chemical (and *biological*) interaction, was not established. It is evident that within such a system the entropy cannot increase by means of heat transfer between the compartments with different temperatures. The entropy can increase only by means of chemical reactions, biological interactions (like competition), transfers of mass (particles, individuals) between different phases and compartments of the system and, generally speaking, by means of any process which can be characterised by the change of chemical and biological potentials. As far as the processes of heat exchange between the system and its environment are concerned, we assume that they are in equilibrium.

In this case the function of free energy F and Gibbs' potential G play the most important role. The value $\mathrm{d}G \approx \mathrm{d}F = \mathrm{d}U - T\,\mathrm{d}S$ is a fraction of the internal energy $\mathrm{d}U$, which can be transferred either into work or spent on the change of the number of particles N_i. The term $T\,\mathrm{d}S$, where S is the entropy, corresponds to a fraction of $\mathrm{d}U$, which is transferred into heat. Since the characteristic functions $F = F(T, v, N_1, \ldots, N_n)$ and $G = G(T, p, N_1, \ldots, N_n)$, where the vector $\mathbf{N} = \{N_1, \ldots, N_n\}$ is the vector of numbers (or moles) of chemical molecules or biological individuals, save the potential properties even for the systems with changed composition $N_1, \ldots, N_n$, their exact differentials are written as

$$(\mathrm{d}F)_{v,T} = (\mathrm{d}G)_{p,T} = \sum_{k=1}^{n} \mu_k \,\mathrm{d}N_k \tag{2.1}$$

where $\mu_k = (\partial F/\partial N_k)_{v,T} = (\partial G/\partial N_k)_{p,T}$. By integrating Eq. (2.1) we get

$$G(N_1, \ldots, N_n) - G(N_1^0, \ldots, N_n^0) + \sum_{k=1}^{n} \int_{N_k^0}^{N_k} \mu_k \,\mathrm{d}N_k. \tag{2.2}$$

We consider a competing community consisting of n species populations; the population size of each kth population is equal to N_k. If the specific energy of kth specimen is e_k then the energy contents of kth population will be equal to $e_k N_k$. The population is maintained by either the accumulation of solar energy (if the autotrophic level is considered) or the consumption of biomass of the previous trophic level. The corresponding inflow will be denoted by q_k^{in}. Note that (as we talked about in Chapter 7) even in the case of the autotrophic level this inflow is not necessarily the flux of solar energy: this may be the inflow of some limiting nutrient. It is very important that these inflows depend on the state of the population, i.e. $q_k^{\mathrm{in}}(N_k, R)$ and $\partial q_k^{\mathrm{in}}/\partial N_k \geq 0$, where R is the concentration of some "resource" in the "donor", which can be associated with the environment. The energy balance of kth population can be represented as

$$q_k^{\mathrm{in}}(N_k, R) = M_k(N_k) + D_k(N_k) + e_k(\mathrm{d}N_k/\mathrm{d}t + F_k(N_1, \ldots, N_n) + q_k^{\mathrm{out}}(N_k, \mathrm{St}), \quad k = 1, \ldots, n. \tag{2.3}$$

Here we repeat the arguments which were used in Chapter 7. In the course of one unit of time the system receives $q_k^{\text{in}}(N_k, R)$ units of energy. One part of energy $M_k(N_k)$ is spent within the population for different metabolic decay processes. As a result of dying-off and mortality processes another part of energy, $D_k(N_k)$, is transferred into dead organic matter. If the energy spent by metabolism immediately transforms into heat then the energy contained in the dead organic matter either transforms into heat (as a result of its decomposition) within the system or leaves the system. In turn, the part $e_k(\text{d}N_k/\text{d}t)$ is spent to maintain reproduction and the population growth. It is obvious that both competition between specimens belonging to different populations (interspecific competition) and competition between individuals of the same population (intraspecific competition) are accompanied by a dissipation of some part of the energy, $F_k(N_1, \ldots, N_n)$. The dissipation increases (or, at least, does not decrease) with increase in the size of any population, so that $\partial F_k/\partial N_j \geq 0$; $k, j = 1, \ldots, n$. Finally, a residual part $q_k^{\text{out}}(N_k, \text{St})$ is transferred to another system, an "acceptor" (in particular, to the next trophic level). The last value depends on both the size of kth population and the state of the acceptor, St.

Without loss of generality we can measure the number of species or their biomasses in energy (enthalpy) units (for instance, evaluating biomasses in carbon units we can use the equivalent 1 gC = 41.8 kJ). Formally it means that we can set $e_i = 1$ in Eq. (2.3) and in the further equations.

By summing these equations from 1 to n we get the energy balance for the entire level:

$$\sum_{k=1}^{n} [q_k^{\text{in}}(N_k, R) - q_k^{\text{out}}(N_k, \text{St})] = \sum_{k=1}^{n} k[M_k(N_k) + D_k(N_k)] + \sum_{k=1}^{n} F_k(N_1, \ldots, N_n) + \frac{\text{d}}{\text{d}t} \sum_{k=1}^{n} N_k. \tag{2.4}$$

If the total biomass of trophic level, $N = \sum_{k=1}^{n} N_k$, is associated with internal or free energy of the system then its change under constant temperature has to be proportional to the change of total entropy taken with inverse sign, $-(\text{d}S/\text{d}t)$, so that

$$\frac{\text{d}N}{\text{d}t} = -T\frac{\text{d}S}{\text{d}t}. \tag{2.5}$$

One of the main problems in the thermodynamics of open systems is how to divide the total entropy production, $(\text{d}S/\text{d}t)$, by the two parts, $(\text{d}_\text{e}S/\text{d}t)$ and $(\text{d}_\text{i}S/\text{d}t)$, the first of which is the change of entropy caused by different exchange processes between the system and its environment, and the second is the internal entropy production. Note that a successful solution of the general problem is often dependent on a successful choice of such a type of division.

The sum $\sum_{k=1}^{n} F_k(N_1, \ldots, N_n)$ is the total expenditure (in energy units) for competition; the sum $\sum_{k=1}^{n} [M_k(N_k) + D_k(N_k)]$ includes both the metabolism of all organisms of the trophic level and the total outflow of dead organic matter. At first sight, it seems that both the sum $\sum_{k=1}^{n} F_k$, which is equal to the rate of energy dissipation caused by competition, and the sum $\sum_{k=1}^{n} (M_k + D_k)$, which also describes the energy dissipation processes, have to be included in a general expression for the internal entropy production $\mathrm{d_i}S/\mathrm{d}t$, so that

$$T\frac{\mathrm{d_i}S}{\mathrm{d}t} = \sum_{k=1}^{n}(M_k + D_k) + \sum_{k=1}^{n} F_k. \tag{2.6}$$

Then the intensity of the "entropy pump", sucking the entropy out of the system, is equal to the difference between the total inflow and outflow of energy, taken with the opposite sign, i.e.

$$T(\mathrm{d_e}S/\mathrm{d}t) = -\sum_{k=1}^{n}(q_k^{\mathrm{in}} - q_k^{\mathrm{out}}). \tag{2.7}$$

However, everything is not so simple, and the problem is to determine the entropy of ensemble, community, etc.

It is obvious that the flows q_k^{in}, q_k^{out}, M_k and D_k are determining the states of the single isolated kth populations, and their balance is the current values of net free energy accumulated by each population. In other words, the specific balance $b_k = (q_k^{\mathrm{in}} - q_k^{\mathrm{out}} - M_k - D_k)/N_k$ is the capability of kth specimen to assimilate and to keep the external energy (or resource), which is later spent on both the population growth and competition. Then the value $B = \sum_{k=1}^{n} b_k N_k = \sum_{k=1}^{n}(q_k^{\mathrm{in}} - q_k^{\mathrm{out}} - M_k - D_k)$ can be considered as some integral characteristic of exchange processes. However, the isolated populations, by their nature, do not create such a structure as the competing community. The latter is formed by the competition, which is maintained by the energy dissipation with the rate $F = \sum_{k=1}^{n} F_k$. Taking into account all these arguments we can write:

$$T(\mathrm{d_e}S/\mathrm{d}t) = \sum_{k=1}^{N} B_k, \qquad T(\mathrm{d_i}S/\mathrm{d}t) = F. \tag{2.8}$$

The choice of such a type of division could be explained from the purely physical point of view. In fact, the metabolic processes are very fast (in comparison with the processes of formation and evolution of the community). Since we consider a partially equilibrium system, which is in the state of heat equilibrium with its environment, the metabolic heat has to dissipate very quickly. This process reduces an effective value of consumed free energy and does not influence the competing processes. In this case we can consider the metabolic processes as exchange processes.

The decomposition of dead organic matter is also a dissipative process. If it happens outside the system then there are no problems, and the mortality can be considered as an exchange process. In the opposite case, when the dead organic matter remains within the system (and decomposes there), it seems that the decomposition has to be included in internal entropy production. However, for this, we have to include in the system a variable, which has to describe the state of dead organic matter. In this case, we are not prepared to do it; we will have to consider the mortality as an exchange process.

By keeping in mind one of the basic statements of the thermodynamics of open systems, that at the dynamic equilibrium $(\mathrm{d_e}S/\mathrm{d}t) + (\mathrm{d_i}S/\mathrm{d}t) = 0$, we immediately obtain from Eq. (2.4) that at the dynamic equilibrium (we shall denote this state by the upper index (eq)):

$$\text{(a)} \quad \sum_{k=1}^{n} [q_k^{\text{in}}(N_k^{\text{eq}}, R) - q_k^{\text{out}}(N_k^{\text{eq}}, \text{St})] - \sum_{k=1}^{n} [M_k(N_k^{\text{eq}}) + D_k(N_k^{\text{eq}})] = \sum_{k=1}^{n} F_k(N_1^{\text{eq}}, \dots, N_n^{\text{eq}}); \tag{2.9}$$

$$\text{(b)} \quad \sum_{k=1}^{n} N_k^{\text{eq}} = \text{constant}.$$

The latter relation means that when the trophic level is in a dynamic equilibrium with its environment and other external systems, its total biomass has to be constant (although the biomasses of each population may fluctuate; compare with Chapter 7). We can see that the conditions for a stationary state do not depend on the choice of hypothesis about the division of total entropy production.

All these results are absolutely correct but they are too general. Their concretisation of all functions, used in the description of energy and entropy balance, can give us the results that will be interpreted in a more interesting manner.

It is obvious that all the functions $q_k^{\text{in}}(N_k, R)$, $q_k^{\text{out}}(N_k, \text{St})$, $M_k(N_k)$, $D_k(N_k)$ at the point $N_k = 0$ are also equal to 0. Therefore, they can be represented in the following forms:

$$\begin{aligned} q_k^{\text{in}}(N_k, R) &= N_k f_k^{\text{in}}(N_k, R), \\ q_k^{\text{out}}(N_k, \text{St}) &= N_k f_k^{\text{out}}(N_k, \text{St}), \\ M_k(N_k) &= N_k m_k(N_k), \\ D_k(N_k) &= N_k d_k(N_k). \end{aligned} \tag{2.10}$$

All the values which are denominated by small letters may be interpreted as *specific* ones, i.e. they are values per one individual.

In order to concretise the competition function $F_k(N_1, \dots, N_n)$ we need a model of competition. For this we use the so-called *collision* concept, which is very popular both in the mathematical ecology (Lotka–Volterra equations) and in chemical kinetics. We assume that the energy expenditures for competition are the results of energy dissipation during the act of collision between one pair of individuals. If one individual belongs to the kth species and another belongs to the jth species then the amount of dissipated energy is equal to δ_{kj}. This interaction is symmetric, i.e. $\delta_{kj} = \delta_{jk}$. Collisions of three and more individuals are neglected. We also assume that the system considered (community of the same trophic level) is *well mixed*, i.e. any collision is a result of random pair choice from a large ensemble of $\tilde{N} = \sum_{k=1}^{n} \tilde{N}_k$ individuals. (Here we temporarily return to the primary interpretation of N_k as a number of particles or individuals that will be denominated by " $\sim$ ".) In this case the number of collision between kth and jth individuals is equal to $\tilde{N}_i \tilde{N}_j$. The process of choice takes place once during one temporal unit. Since the pair collision of kth and jth individuals is accompanied by the dissipation of δ_{kj} units of energy then the total dissipation for kth population will be equal to

$$F_k(\tilde{N}_1, \ldots, \tilde{N}_n) = \sum_{j=1}^{n} \delta_{kj} \tilde{N}_k \tilde{N}_j = \tilde{N}_k \sum_{j=1}^{n} \delta_{kj} \tilde{N}_j.$$

By returning to energy units, $N_k = e_k \tilde{N}_k$ we get

$$F_k(N_1, \ldots, N_n) = \sum_{j=1}^{n} \frac{\delta_{kj}}{e_k e_j} N_k N_j = N_k \sum_{j=1}^{n} \gamma_{kj} N_j \tag{2.11}$$

where $\gamma_{kj} = \delta_{kj}/e_k e_j$. Since $\delta_{kj} = \delta_{jk}$, $\gamma_{kj} = \gamma_{jk}$. By substituting all these expressions into Eq. (2.1) we immediately get the *kinetic* equation for the population number:

$$\frac{\mathrm{d}N_k}{\mathrm{d}t} = N_k \left\{ [f_k^{\text{in}}(N_k, R) - f_k^{\text{out}}(N_k, \mathrm{St})] - [m_k(N_k) + d_k(N_k)] - \sum_{j=1}^{n} \gamma_{kj} N_j \right\}, \tag{2.12}$$

$$k = 1, \ldots, n.$$

In order to complete the system we have either to supplement equations for the new variables R and St or to consider them as parameters. In the latter case we have to formulate additional hypotheses about the connection between the system and its environment (for instance, other external systems can be considered as the environment). However, if the connection is very tight, i.e. if not only the state of the system depends on the state of environment but also vice versa, then a consideration of the trophic level as some isolated system becomes very problematic: we have to include both the level and its environment (the resource or previous level and the next level) into the whole system. Nevertheless, we can separate our level if we assume that

1. A resource "pool" is much larger than the level, so that the consumption of resource by the level does not practically change the amount of resource in the pool. Such a situation is typical for the autotrophic level with respect to solar energy. In other words, the system interacts with an infinite resource pool. There is a "bon mot" for this: an interaction "continent–island". It is obvious that in this case $R \approx R^* = \text{constant}$.
2. The state of a given level does not practically influence the state of other systems, then $f_k^{\text{out}} = f_k^{\text{out}}(N_k)$; or, the reversed influence is so low that we can neglect it ($f_k^{\text{out}} = 0$). We shall use the latter assumption in our further deliberations.

By denoting $q_k^{\text{in}}(N_i, R^*) = f_k(N_k)$, $m_k(N_k) + d_k(N_k) = h_k(N_k)$ system (2.12) is re-written as

$$\frac{\mathrm{d}N_k}{\mathrm{d}t} = N_k \left\{ [f_k(N_k) - h_k(N_k)] - \sum_{j=1}^{n} \gamma_{kj} N_j \right\}, \qquad k = 1, \ldots, n. \tag{2.13}$$

The system is very similar (at least, in the part which describes competition between species) to the Lotka–Volterra equations. This is not surprising since we also used the collision concept.

8.3. Community trajectory as a trajectory of steepest ascent

In this section we describe one method (the method of steepest ascent), which allows us to represent the trajectories of kinetic equation (2.13) as some extremals.

Let us assume that the continuous and differentiable function $W(N_1, ..., N_n)$ belonging to the positive orthant $\mathbf{P^n}$ is introduced in such a way that:

$$\frac{dN_k}{dt} = N_k \frac{\partial W}{\partial N_k}, \qquad k = 1, ..., n. \tag{3.1}$$

Calculating the derivative of W with respect to time along the trajectories of Eq. (3.1), $\mathbf{N}(t) = \{N_1(t), ..., N_n(t)\}$, we have

$$\frac{dW}{dt} = \sum_{k=1}^{n} \frac{\partial W}{\partial N_k} \frac{dN_k}{dt} = \sum_{k=1}^{n} N_k \left(\frac{\partial W}{\partial N_k} \right)^2 \geq 0. \tag{3.2}$$

Obviously, $dW/dt = 0$ only at a point where either all $\partial W/\partial N_k = 0$, or all $N_k = 0$, or, in the intermediate case, when several $\partial W/\partial N_k = 0$ and for other indices $N_s = 0$, i.e. at stationary points (equilibriums) of system (3.1). From Eq. (3.2) it follows that the function $W[\mathbf{N}(t)]$ always increases along the trajectories of Eq. (3.1) and reaches its local maximum at the equilibrium $\mathbf{N}^*$ only if $\mathbf{N}^*$ is asymptotically stable. In fact, since trajectories of the dynamical system (3.1) are dense within some domain Ω, to which the point $\mathbf{N}^*$ belongs, for each point $\mathbf{N}$ neighbouring $\mathbf{N}^*$, a trajectory $\mathbf{N}(t) \rightarrow \mathbf{N}^*$ going through $\mathbf{N}$ could be found. The function W is continuous and steadily increasing, and it follows that $W(\mathbf{N}^*) > W(\mathbf{N})$ along the trajectory $\mathbf{N}(t)$. Hence,

$$W(\mathbf{N}^*) = \max_t W(\mathbf{N}(t)) = \max_{\mathbf{N} \in \Omega} W(\mathbf{N}). \tag{3.3}$$

On the other hand, if the function $W(N_1, ..., N_n)$ has an isolated maximum at some stationary point $\mathbf{N}^* = \{N_1^*, ..., N_n^*\} \in \mathbf{P^n}$ then the state $\mathbf{N}^*$ is asymptotically stable, since the function $L(\mathbf{N}) = W(\mathbf{N}^*) - W(\mathbf{N})$ is the Lyapunov function for system (3.1) in this case. This fact is obvious enough, since $L \geq 0$ in a domain which contains $\mathbf{N}$ ($L = 0$ only at $\mathbf{N}^*$) and its time derivative along system trajectories is $dL/dt = -dW/dt \leq 0$ with $dL/dt = 0$ only at the point $\mathbf{N}^*$.

The sufficient condition for the maximum of W is the following: the quadratic form

$$d^2 W(\mathbf{N}^*) = \sum_{k=1}^{n} \sum_{j=1}^{n} \left(\frac{\partial^2 W}{\partial N_k \partial N_j} \right)_{\mathbf{N}^*} dN_k \, dN_j, \tag{3.4}$$

(the second differential of W) is positive definite. In this case W is strictly concave (convex upwards) at a neighbourhood of $\mathbf{N}^*$. If the second differential is positive definite for any $\mathbf{N} \in \mathbf{P^n}$ then the local property of strict concavity becomes global, i.e. it takes place for the whole positive orthant. And as far as the orthant $\mathbf{P^n}$ also represents a convex set, W has a single isolated maximum in this set (on compact subsets or on the boundary). Hence, its local maximum is at the same time global, and system (4.1) has a single stable equilibrium. Moreover, if $|W| \rightarrow \infty$ as $|\mathbf{N}| \rightarrow \infty$, this equilibrium is stable for any initial trajectory displacement within $\mathbf{P^n}$ (globally stable, absolutely stable), thus suggesting that any system trajectory initiated within $\mathbf{P^n}$ approaches the equilibrium as $t \rightarrow \infty$.

If the maximum of W is not attained inside the positive orthant $\mathbf{P^n}$, it will be attained on its boundary, in the appropriate coordinate hyperplanes. This implies that one or several equilibrium coordinates will be zero. In the ecological interpretation it means that in the process of ecosystem evolution one or several species are to be eliminated, since the new equilibrium state with a fewer number of species is stable again.

We see that, generally speaking, many characters of dynamic behaviour of the system are determined by the landscape of the function W. For instance, such general properties as the existence of equilibriums, their coordinate and their stability can be defined in terms of its landscape topography: the existence of peaks and valleys, and their locations. However, we do not know what kind of path in the landscape corresponds to the trajectory of the system movement? In order to answer this question, first we have to introduce some definitions.

A trajectory which goes along a gradient of the surface $W(\mathbf{N})$ is called a *trajectory of steepest ascent*. If this trajectory initiates at the point $\mathbf{N}_0$, then it is the shortest way from this point to the top of a peak. The standard equations of steepest ascent are represented in a vector form as $\mathrm{d}\mathbf{N}/\mathrm{d}t \sim \mathrm{grad}\,\mathbf{N}$, and they differ from Eq. (3.1). By applying the Svirezhev–Shahshahani transformation $z_k = \pm 2\sqrt{N_k}$; $k = 1, \ldots, n$, which transfers the positive orthant $\mathbf{P^n}$ into the full coordinate space $\mathbf{R^n_z}$, we now consider the movement in $\mathbf{R^n_z}$. In the new coordinates $W = W(\mathbf{z}) = W(z_1, \ldots, z_n)$ system (3.1) is written as $\mathrm{d}z_k/\mathrm{d}t = \partial W/\partial z_k$, $k = 1, \ldots, n$, or, in a vector form, $\mathrm{d}\mathbf{z}/\mathrm{d}t = \mathrm{grad}\,W$, i.e. trajectories of Eq. (3.1) will be trajectories of steepest ascent in the special space $\mathbf{R^n_z}$. Since in $\mathbf{R^n_z}$ $\mathrm{d}W/\mathrm{d}t = \sum_{k=1}^{n} (\partial W/\partial z_k)(\mathrm{d}z_k/\mathrm{d}t) = \sum_{k=1}^{n} (\partial W/\partial z_k)^2$, the square velocity of movement along trajectories of steepest ascent may be defined as

$$v^2 = \mathrm{d}W/\mathrm{d}t = \sum_{k=1}^{n} (\partial W/\partial z_k)^2 = \sum_{k=1}^{n} [(\partial W/\partial N_k)(\partial N_k/\partial z_k)]^2 = \sum_{k=1}^{n} [(\partial W/\partial N_k)(\pm N_k)]^2.$$

From this expression it could be seen that, as we approach equilibrium, in which either $\partial W/\partial N_k = 0$ or $N_k = 0$, the velocity is permanently reduced although it can be sufficiently high when far from this equilibrium.

It is interesting that $W(\mathbf{z})$ in $\mathbf{R^n_z}$, where $\partial W(\mathbf{z})/\partial z_k = 0$, corresponds to the stationary points of $W(\mathbf{N})$ with $N_k = 0$ in $\mathbf{P^n}$. But it is not necessary that $\partial W(\mathbf{N})/\partial N_k = 0$; more often these derivatives are negative. In other words, the stationary points of $W(\mathbf{N})$ situated at the boundary of $\mathbf{P^n}$ correspond to the internal stationary points of $W(\mathbf{z})$ in $\mathbf{R^n_z}$. It is obvious that all these points are at the same time the equilibriums of system (3.1) and $\mathrm{d}z_k/\mathrm{d}t = \partial W/\partial z_k$, $k = 1, \ldots, n$. Notice, as well, that all these points are defined by the necessary conditions of the extremum for the functions $W(\mathbf{N})$ and $W(\mathbf{z})$. But the number of equilibriums of this system exceeds the number of equilibriums of the system (3.1)! For instance, if the internal non-trivial equilibriums for both the systems are single then the system (3.1) has the maximal number equal to $(2^n - 1)$ of all possible stationary points with $n, n-1, \ldots, 2, 1$ non-zero coordinates. On the other hand, any point which differs from these only in signs of non-zero components will be stationary for the system $\mathrm{d}z_k/\mathrm{d}t = \partial W/\partial z_k$, $k = 1, \ldots, n$. The number of such points is, evidently, equal to $2^n + n2^{n-1} + \cdots + C_n^k 2^k + \cdots + 2n = 3^n - 1$. Thus, the stationary points of the system $\mathrm{d}z_k/\mathrm{d}t = \partial W/\partial z_k$ are derived from the stationary points of system (3.1) as their symmetric reflections with respect to all sorts of coordinate hyperplanes and the origin, while the values of the function $W(\mathbf{z})$, are identical in symmetric points. We illustrate all the reasoning by the following example (see also Fig. 8.1).

Let $W = a - bN$. In this case the positive orthant coincides with the positive semi-axis $[0, \infty]$, and the maximum W is attained at the point $N^* = 0$. At this point $\partial W/\partial N = -b < 0$. Introducing the new variable $z = \pm 2\sqrt{N}$ we get $W = a - (b/4)z^2$, which is now presented for the entire axis $[-\infty, \infty]$. This function has a maximum at the same zero point, but the derivative $\partial W/\partial z = -(b/2)z$ at this point is equal to zero.

In order to illustrate the symmetry of stationary points we consider the second example (Fig. 8.2). Let $W = N(1 - N)$. It is obvious that W has an internal maximum at the point $N^* = 0.5$, which is equal to 0.25. By going over to the variable $z = \pm 2\sqrt{N}$ we get $W = z^2(4 - z^2)/16$. This function is symmetrical with respect to the origin $z = 0$, and it has two symmetrical maximums at the points $z^*_{1,2} = \pm\sqrt{2}$ (that correspond to $N^* = 0.5$). Both maximums are equal to 0.25. It is easy to see that both $\partial W/\partial N = 1 - 2N$ at the point $N^* = 0.5$ and $\partial W/\partial z = z(2 - z^2)/4$ at the two points $z^*_{1,2} = \pm\sqrt{2}$ are equal to zero, i.e. the extremum necessary conditions are fulfilled for both these cases.

Revenons à nos moutons and try to find the function W, which will be a function of steepest ascent for kinetic equations (2.13). As one of many possibilities let us consider the following function

$$W = \sum_{k=1}^{n} \int_{N_k^0}^{N_k} [f_k(N_k) - h_k(N_k)]\mathrm{d}N_k - \frac{1}{2}\sum_{k=1}^{n}\sum_{j=1}^{n} \gamma_{kj} N_k N_j. \tag{3.5}$$

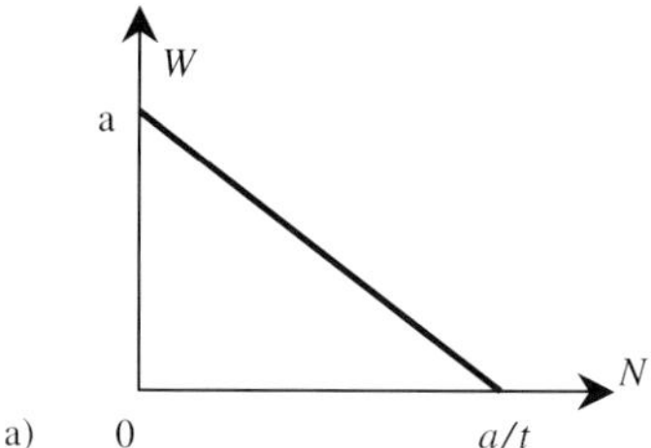

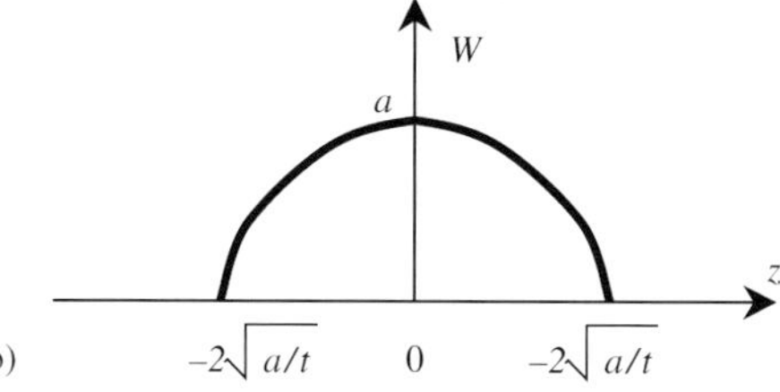

Fig. 8.1. The function $W = a - bN$ in the natural (a) and transformed (b) coordinates.

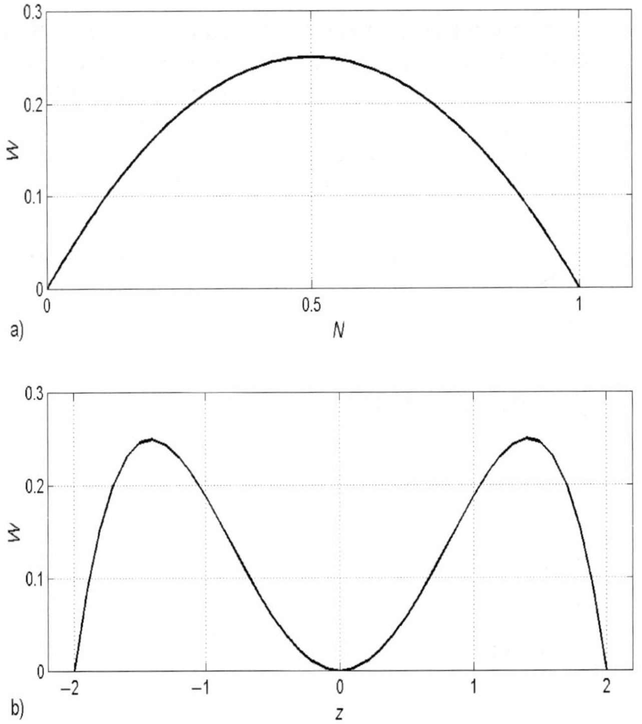

Fig. 8.2. The function $W = N(1 - N)$ in the natural (a) and transformed (b) coordinates.

Calculating the partial derivatives of W with respect to N_k and taking into account that $\gamma_{kj} = \gamma_{jk}$ we get:

$$\frac{\partial W}{\partial N_k} = [f_k(N_k) - h_k(N_k)] - \sum_{k=1}^{n} \gamma_{kj} N_j. \tag{3.6}$$

It is easy to see that the system of kinetic equations (2.13) describing the thermodynamics of trophic level is represented in the form of Eq. (3.1). It means that the function W given by Eq. (3.5) grows along the trajectories of Eq. (2.13) and reaches its maximum at the stable equilibrium of system (2.13).

8.4. Extreme properties of the potential W and other potential functions. Entropy production and Prigogine-like theorem

All these results can be interpreted from the thermodynamic point of view. Since all biomasses are measured in energy (enthalpy) units, the concepts of biomass and energy are equivalent. The existence of a competing community is determined by the resource inflow, $f - h = \sum_{k=1}^{n} N_k[f_k(N_k) - h_k(N_i)]$, which is accumulated by the system, increasing its internal energy. Kinetic equations (2.13) are the consequence of the First Law, while they can be used as a dynamic model of the competing community. This allows us to study such a general property of it, such as stability, by establishing the analogies between the dynamic and thermodynamic criteria. In addition, in order to judge a direction of the system's evolution and its stability (and hence, its possible evolution) it is desirable to have energy and entropy criteria of evolution. It is natural that the first thought would be to apply methods of thermodynamics of irreversible processes.

In our case the potential function W could be used as some analogue of the thermodynamic potential G for such an open system as the competing community or trophic level. In fact, as for $\gamma_{kj} = \gamma_{jk}$,

$$\frac{\partial}{\partial N_k}\left[\frac{1}{2}\sum_{k=1}^{n}\sum_{j=1}^{n}\gamma_{kj}N_kN_j\right] = \sum_{j=1}^{n}\gamma_{kj}N_j, \tag{4.1}$$

then W is represented as

$$W = \sum_{k=1}^{n}\int_{N_k^0}^{N_k}\left[f_k(N_k) - h_k(N_k) - \sum_{j=1}^{n}\gamma_{kj}N_j\right]\mathrm{d}N_k, \tag{4.2}$$

and its full differential will be

$$\mathrm{d}W = \sum_{k=1}^{n}\left[f_k(N_k) - h_k(N_k) - \sum_{j=1}^{n}\gamma_{kj}N_j\right]\mathrm{d}N_k. \tag{4.3}$$

We assume that the functions $f_k(N_k)$ and $h_k(N_k)$ do not have any singularities at the point $\mathbf{N} = \{N_1^0, \ldots, N_n^0\}$. Then $W(\mathbf{N}^0) = 0$, and this initial state can be taken as some reference level for the thermodynamic potential W. It is natural to assume that $W = 0$ at the *thermodynamic* equilibrium. As follows from common sense, life does not exist at the thermodynamic equilibrium, so that all N_k^0 have to be equal to zero, i.e. $\mathbf{N}^0 = 0$. Note that the origin of coordinates $\mathbf{N} = 0$ is the trivial solution to Eq. (3.1), i.e. $\mathbf{N}^* = 0$ is also a *dynamic* equilibrium.

If we look at formula (4.2) we see that $W \to -\infty$ when $\mathbf{N} \to \infty$ and f_k and h_k are bounded. Therefore, there is the open domain $\Omega \subseteq \mathbf{P^n}$, within which the function W is positive, i.e. $W(\mathbf{N}) > 0$ for any $\mathbf{N} \in \Omega$, except the origin of coordinates where $W(0) = 0$. The thermodynamic equilibrium is unstable if $f_k(0) > h_k(0)$ even for single $k \in [0, 1]$. If $f_k(0) < h_k(0)$ for all k, then the equilibrium is stable. The latter has a very simple interpretation: if the metabolism and mortality of individual species are not compensated by external energy flows then all populations are eliminated. In this case $\Omega = \varnothing$, and $W(\mathbf{N}) < 0$ for any positive $\mathbf{N}$. It is natural to assume that the domain of positiveness for W defines a domain in which the concepts of linear thermodynamics are applicable. It is easy to prove that the non-trivial equilibrium $\mathbf{N}^*$ also belongs to the domain Ω. Since the proof is cumbersome, we illustrate it by a simple example for $n = 1$.

Let f_1 and h_1 be constants. Then $W = (f_1 - h_1)N_1 - \frac{1}{2}\gamma_{11}N_1^2$; $f_1 > h_1$. One can see that $W(0) = 0$ and $W(N_1) > 0$ if $N_1 \in (0, N_1^{\mathrm{cr}})$, where $N_1^{\mathrm{cr}} = 2(f_1 - h_1)/\gamma_{11}$. On the other hand, at the non-trivial equilibrium $(\partial W/\partial N_1)^* = (f_1 - h_1) - \gamma_{11}N_1^*$ and $N_1^* = (f_1 - h_1)/\gamma_{11} = N_1^{\mathrm{cr}}/2$, i.e. $N_1^* \in \Omega : (0, N_1^{\mathrm{cr}})$.

Finally, we can say that when such an open system as a community of competing populations tends to its *dynamic* non-trivial equilibrium then the thermodynamic potential, defined by the function W, monotonously increases, attaining its maximum at the equilibrium. At the *thermodynamic* trivial equilibrium the thermodynamic potential is minimal.

Since the change of W is caused by biological processes, it is natural to name the potential function W a *biological potential of community*. We formally assume that there is such a non-negative function $G = G(\mathbf{N})$ $(G(0) = 0)$ that can play the role of Gibbs' thermodynamic potential. By expanding G into the Taylor series in the vicinity of zero and restricting two first terms we obtain:

$$G \approx \sum_{k=1}^{n} \left(\frac{\partial G}{\partial N_k} \right)_0 N_k + \frac{1}{2} \sum_{k=1}^{n} \sum_{j=1}^{n} \left(\frac{\partial^2 G}{\partial N_k \, \partial N_j} \right)_0 N_k N_j. \tag{4.4}$$

Since the function G has to be a full differential,

$$\frac{\partial^2 G}{\partial N_i \, \partial N_j} = \frac{\partial^2 G}{\partial N_j \, \partial N_i},$$

and the differential is represented as

$$\mathrm{d}G = \sum_{k=1}^{n} \left[\left(\frac{\partial G}{\partial N_k} \right)_0 + \sum_{j=1}^{n} \left(\frac{\partial^2 G}{\partial N_k \, \partial N_j} \right)_0 N_j \right] \mathrm{d}N_k. \tag{4.5}$$

Comparing this expression with expression (4.3) we see that they coincide if

$$f_k(N_k) - h_k(N_k) = \left(\frac{\partial G}{\partial N_k} \right)_0 = \mathrm{constant}_i \qquad \text{and} \qquad \left(\frac{\partial^2 G}{\partial N_k \, \partial N_j} \right)_0 = \gamma_{kj}.$$

The first condition implies that if we would like to remain within classic thermodynamics frameworks not far from thermodynamic equilibrium then the flow functions q_k^{in}, q_k^{out}, M_k and D_k must be linear functions of N_k, i.e. the functions f_k and h_k must be constant. In this case Eq. (2.13) is transformed into the classic Lotka–Volterra equations:

$$\frac{\mathrm{d}N_k}{\mathrm{d}t} = N_k \left(\varepsilon_k - \sum_{k=1}^{n} \gamma_{kj} N_j \right). \tag{4.6}$$

where $\varepsilon_k = f_k - h_k = \mathrm{constant}$.

The first condition of dynamic (but not thermodynamic!) equilibrium, $\mathrm{d_i}S/\mathrm{d}t = -\mathrm{d_e}S/\mathrm{d}t$, which has been already formulated in Section 8.2 (see Eq. (2.8)) in a sufficiently general form, is now written as $(N_i^{\mathrm{eq}} = N_i^*)$

$$\sum_{k=1}^{n} N_k^* (f_k - h_k) = \sum_{k=1}^{n} \sum_{j=1}^{n} \gamma_{kj} N_k^* N_j^*. \tag{4.7}$$

The internal production of entropy, $\mathrm{d_i}S/\mathrm{d}t$, is equal to

$$\frac{\mathrm{d_i}S}{\mathrm{d}t} = \frac{1}{T}\sum_{k=1}^{n}\sum_{j=1}^{n}\gamma_{kj}N_kN_j, \tag{4.8}$$

and the entropy export out of the system into its environment, $\mathrm{d_e}S/\mathrm{d}t$, is

$$\frac{\mathrm{d_e}S}{\mathrm{d}t} = -\frac{1}{T}\sum_{k=1}^{n}N_k(f_k - h_k). \tag{4.9}$$

(Note that we consider isothermal processes, i.e. when the temperature is constant.) Since the outflow of entropy, $\mathrm{d_e}S/\mathrm{d}t$, is negative, we can associate this flow with the flow of Schrödinger's *negentropy* from environment into the system.

One of the important characteristics of community is its total number of individuals, or biomass $N = \sum_{k=1}^{n} N_k$. Summing Eq. (2.13) in our case we obtain (see also Eq. (2.5))

$$\frac{\mathrm{d}N}{\mathrm{d}t} = \sum_{k=1}^{n} f_kN_k - \sum_{k=1}^{n}\left(h_k + \sum_{j=1}^{n}\gamma_{kj}N_j\right)N_k = -T\left(\frac{\mathrm{d_e}S}{\mathrm{d}t} + \frac{\mathrm{d_i}S}{\mathrm{d}t}\right) = -T\left(\frac{\mathrm{d}S}{\mathrm{d}t}\right) \tag{4.10}$$

where S is the total entropy of the system. This equation allows us to give a thermodynamic interpretation to different stages through which the system passes in the course of its succession. At the initial state, when N_k are still small, both the free energy inflow and the internal entropy production are also small. Therefore, the absolute values of the rates of biomass and entropy change, $|\mathrm{d}N/\mathrm{d}t|$ and $|\mathrm{d}S/\mathrm{d}t|$, will also be low, but in order for the total biomass to grow it is necessary for the entropy to decrease. It is possible, if the system could possess a powerful energy inflow with low entropy. At this stage the exchange processes, which are described in linear terms, play the main role. And the power of the entropy pump, $|\mathrm{d_e}S/\mathrm{d}t|$, exceeds significantly the internal entropy production. Now the rate of decrease of the total entropy grows, which promotes the increase of $(\mathrm{d}N/\mathrm{d}t)$. But with the growth of N the inhibiting role of quadratic terms, which describe the internal entropy production, becomes more and more tangible. The rate of increase of the total biomass decelerates until it becomes equal to zero. At the same time the internal entropy production becomes equal to the power of the entropy pump.

By integrating both sides of Eq. (4.10) with respect to time and assuming that the integration starts from the thermodynamic equilibrium, when $N = 0$ and $S = S_0$ we get a relation between the total biomass and entropy, or exergy:

$$N(t) = T(S_0 - S(t)) = \mathrm{Ex}(t). \tag{4.11}$$

We can say that if the entropy of the community is minimal then its total biomass is maximal. Correspondingly, the exergy of community will also be maximal. In other words, any well-organised competing community has a maximal biomass and its exergy is maximal.

Note that since $\mathrm{d}N = \sum_{k=1}^{n} \mathrm{d}N_k$, the total biomass N is the exact differential as well as the total entropy S.

Let us keep in mind that the biological potential of community,

$$W = \sum_{k=1}^{n} (f_k - h_k)N_k + \frac{1}{2}\sum_{k=1}^{n}\sum_{j=1}^{n} \gamma_{kj}N_kN_j,$$

monotonously increases and tends to maximum. This result also allows us to do a sensible interpretation. The value $\mathrm{Gr}(\mathbf{N}) = \sum_{k=1}^{n}(f_k - h_k)N_k$ in essence accounts for the rate of biomass gain in the case that competition and any kind of limitation by resources are absent, and the growth is determined only by the physiological fertility and natural mortality of the organisms. Therefore it is natural to define Gr as the *reproductive potential* of the community. The value $\mathrm{Diss} = \sum_{k=1}^{n}\sum_{j=1}^{n}\gamma_{kj}N_kN_j$ may be used as a measure of the *rate of energy dissipation* resulting from inter- and intraspecific competition. Therefore we shall refer to Diss as the *total expenses of competition*. Hence the increase in Diss in the process of evolution may be interpreted as the goal of the community to maximise the difference between its reproductive potential and the total expenses of competition. This goal can be achieved in several ways: either the reproductive potential is maximised at fixed expenses of competition (*r-strategy*) or the competition expenses are minimised for a limited reproductive potential (*K-strategy*). There may also be some intermediate cases.

The main statement of Prigogine's theorem is: at the stable dynamic equilibrium the value of $\mathrm{d_i}S/\mathrm{d}t$, i.e. the internal entropy production, has to be minimal, and this minimum has to be positive. However, as is seen from Eq. (4.8), the minimum of $\mathrm{d_i}S/\mathrm{d}t$ is attained at the origin of coordinates $\mathbf{N}^* = 0$ (certainly, if there are not any constraints except the non-negativeness of N_k). The stationary conditions (4.7) are also satisfied at $\mathbf{N}^* = 0$, i.e. the trivial solution $\mathbf{N}^* = 0$ is equilibrium. We have mentioned above that this equilibrium is *thermodynamic*, although it is also a *dynamic* equilibrium of kinetic equations. This minimum is equal to zero, which contradicts Prigogine's theorem. To resolve the contradiction, at least one equilibrium coordinate has to differ from zero. Then the requirement for positiveness of the minimum becomes understandable: Prigogine's theorem deals with dynamic equilibriums, which differ from thermodynamic ones.

By analogy with the main statement of Prigogine's theorem we postulate:

At the stable non-trivial dynamic equilibrium $\mathbf{N}^* \in \mathbf{P^n}$

$$\left(T\frac{\mathrm{d_i}S}{\mathrm{d}t}\right)^* = \sum_{k=1}^{n}\sum_{j=1}^{n}\gamma_{kj}N_k^*N_j^* = \min_{\mathbf{N}\in\mathbf{P^n}}\left[T\frac{\mathrm{d_i}S}{\mathrm{d}t_i} = \sum_{k=1}^{n}\sum_{j=1}^{n}\gamma_{kj}N_kN_j\right], \tag{4.12}$$

under constraint (4.7): $\sum_{k=1}^{n} N_k^*(f_k - h_k) = \sum_{k=1}^{n}\sum_{j=1}^{n}\gamma_{kj}N_k^*N_j^*$.

We shall assume that the minimum is attained at some internal point of $\mathbf{P^n}$. If it is attained on some orthant faces, so that $\mathbf{N}^* = \{N_s^* > 0, s \in \omega_k; N_l^* = 0, l \in \bar{\omega}_k\}$ where ω_k is the set of k indices because the corresponding equilibrium coordinates are non-zero, $\omega_k \cup \bar{\omega}_k = 1, \ldots, n$, then the minimum will be internal in relation to the sub-orthant $\mathbf{P^k}$: $\{N_s \geq 0, s \in \omega_k\}$. Therefore, the problem can always be reduced to the problem of internal minimum, which will be considered below.

It is not evident that the statement is true for our case. Let us test it, but first we introduce the following notations:

$$\sum_{k=1}^{n} N_k h_k = h, \qquad \sum_{k=1}^{n} N_k f_k = f,$$

$$\sum_{k=1}^{n}\sum_{j=1}^{n} \gamma_{kj} N_k N_j = \Gamma, \qquad \sum_{j=1}^{n} \gamma_{kj} N_j = \Gamma_k.$$

Using the method of Lagrange multipliers we find the minimum of function $Z = T\{(\mathrm{d_i}S/\mathrm{d}t) + \lambda[(\mathrm{d_i}S/\mathrm{d}t) - q + h]\}$ where λ is the Lagrange multiplier. Then the necessary conditions of minimum under constraints will be:

$$\frac{\partial Z}{\partial N_k} = 2\Gamma_k^* + \lambda(2\Gamma_k^* - f_k + h_k) = 0, \qquad k = 1, \ldots, n. \tag{4.13}$$

Multiplying both sides of Eq. (4.13) by N_k^* and summing all these equations we get the formula for the determination of λ: $2\Gamma^* + \lambda\Gamma^* = 0$, whence $\lambda = -2$. Then the minimum necessary condition is:

$$-h_k - 2\Gamma_k^* + 2f_k = 0, \qquad k = 1, \ldots, n. \tag{4.14}$$

Comparing these conditions with the necessary conditions for the maximum W that are written in our case as $f_k - h_k - \Gamma_k^* = 0$, $k = 1, \ldots, n$, we can see that they coincide. The sufficient condition of minimum $\mathrm{d_i}S/\mathrm{d}t = (1/T)\sum_{k=1}^{n}\sum_{j=1}^{n}\gamma_{kj}N_kN_j$ is a positive definiteness of this quadratic form, but this is also the sufficient condition for the maximum W. Since the entropy production within the system is always positive, $\mathrm{d_i}S/\mathrm{d}t > 0$, the corresponding quadratic form has to be positive definite in the entire positive orthant. This is the condition of uniqueness of the minimum. Thus, this Prigogine-like theorem is applicable to our case.

Note that instead of the problem of *minimisation for the internal entropy production* we can consider the problem of *maximisation for the negentropy flow*. Formally the problem is formulated so as to find the maximum of linear form $\mathrm{Gr} = f - h$ under the constraint $f - h = \Gamma$. The latter is the condition of equilibrium. This is the well known *duality principle* in the optimisation theory (Handbook of Mathematical Economy, 1991). Since $\mathrm{d_i}S/\mathrm{d}t$ has to be positive, the constraint can be fulfilled only if $\mathrm{Gr} = -\mathrm{d_e}S/\mathrm{d}t > 0$, i.e. we really deal with the negentropy. Using the method of Lagrange multipliers we find minimum of the function $Z_e = R + \lambda(R - \Gamma)$ where λ is the Lagrange multiplier (we use the notations that were introduced above). Then the necessary conditions of minimum under constraints will be:

$$\frac{\partial Z_e}{\partial N_k} = f_k - h_k + \lambda(2\Gamma_k^* - f_k + h_k) = 0, \qquad k = 1, \ldots, n. \tag{4.15}$$

Multiplying both the sides of Eq. (4.15) by N_i^* and summing all these equations we get: $q - h + \lambda\Gamma^* = 0$, whence $\lambda = (q - h)/\Gamma^* = -1$. Then the necessary conditions of maximum will be: $2(f_k - h_k - 2\Gamma_k^*) = 0$, $k = 1, \ldots, n$, i.e. they coincide with the necessary conditions for the minimum of $\mathrm{d_i}S/\mathrm{d}t$. Hence, we proved that at the equilibrium the flow of negentropy into the system is maximal. It is possible (and it is the principal difference from the situation of classic Prigogine's theorem) because this flow depends on the system's state and, by the same token, it can be controlled by the system.

If we assume that all the populations compete for the single resource R, then the resource compartment can be considered as the system environment, and the resource flow into the system can be identified with the negentropy flow from the environment into the system. We also assume that if the rate of resource uptake is equal to $\sum_{k=1}^{n} \alpha_k RN_k$, then $(\mathrm{d_e}S/\mathrm{d}t) = -\sum_{k=1}^{n} \alpha_k RN_k$. Note that in all our considerations we assume a constancy of the environment that, in turn, implies a constancy of resource. It is possible if an external reservoir containing the resource is very large in comparison with the ecosystem (the latter is a standard assumption for different thermodynamic considerations) and an inverse influence of the system on its environment can be neglected. Then $R \approx R^* =$ constant, and $f_k(R) - h_k(R) = \alpha_k R \approx \alpha_k R^*$ are also constant. In this case all the previous considerations and results are conserved. The coefficients ε_k in the Lotka–Volterra system (4.6) have to be $\varepsilon_k = \alpha_k R^*$. However, it is interesting to consider a situation when we cannot neglect the inverse influence; then we have to supplement the system (4.6) with one more equation for R.

If we assume that the resource is restored by a constant inflow q_R then the equation will be:

$$\frac{\mathrm{d}R}{\mathrm{d}t} = q_r - \sum_{i=1}^{n} \alpha_i RN_i. \tag{4.16}$$

A constancy of resource concentration is maintained if the constraint $\sum_{k=1}^{n} \alpha_k N_k = a =$ constant is fulfilled. As was proved earlier, the form has to be maximum at the dynamic equilibrium (under constraint $\sum_{k=1}^{n} \sum_{j=1}^{n} \gamma_{kj} N_k N_j =$ constant). Since $R^* = q_r/a$, the maximisation of a is equivalent to the minimisation of R^*. These results could be interpreted in the following, somewhat speculative manner.

If such a system as the biological community of competing populations in a process of interaction with its environment tends to a stable equilibrium with non-zero values of population sizes then at the equilibrium:

- the system tends to minimise the internal entropy production system (*Prigogine's theorem*);
- the system tends to arrange its structure and interaction with the environment in such a way that the negentropy flow out of a system into an environment will be maximal (*the principle of maximum for the negentropy flow*);
- the concentration of resource in the environment tends to be minimum (*the principle of maximal utilisation*).

In conclusion we show how the methods of phenomenological thermodynamics can be used for the description of a competing community.

It is obvious that the thermodynamic equilibrium is interpreted as "not a life", i.e. all $N_k^0 = 0$. Therefore, the values $\operatorname{grad} N_i \cong N_i - N_i^0 = N_i$ could be considered as the generalised thermodynamic forces X_k. We assume that generalised thermodynamic flows can be presented in a linear form $J_k = \sum_{j=1}^{n} \gamma_{kj} X_j = \sum_{j=1}^{n} \gamma_{kj} N_j$ where $\gamma_{kj} = \gamma_{jk} \geq 0$.

These are well-known Onsager's reciprocal relations. In this formalism the value of dissipated energy, which is proportional to the entropy production caused by irreversible processes within the system, is equal to

$$\mathrm{Diss} = T\frac{\mathrm{d_i}S}{\mathrm{d}t} = \sum_{k=1}^{n} J_k X_k = \sum_{k=1}^{n}\sum_{j=1}^{n} \gamma_{kj} N_k N_j. \tag{4.17}$$

In accordance with Prigogine's theorem, this value—named the *function of dissipation*—is minimal at the stable dynamic equilibrium. It is obvious that this statement is equivalent to the statement of the above-proved Prigogine-like theorem. In this formalism the reciprocal Onsager coefficients correspond to the competition coefficients in Lotka–Volterra equations, and they are equal to specific energy, which is lost in the single act of competitive collision. The latter always takes place, since a competition is a typical irreversible process leading to the dissipation of energy and the loss of biomass.

8.5. The system of two competing species

In order to illustrate all these theoretical results we consider one partial case, namely two competing species. The corresponding equations are ($f_{1,2} - h_{1,2} = \varepsilon_{1,2}$):

$$\begin{aligned}\frac{\mathrm{d}N_1}{\mathrm{d}t} &= N_1(\varepsilon_1 - \gamma_{11}N_1 - \gamma_{12}N_2),\\ \frac{\mathrm{d}N_2}{\mathrm{d}t} &= N_2(\varepsilon_2 - \gamma_{21}N_1 - \gamma_{22}N_2).\end{aligned} \tag{5.1}$$

Let N_1 and N_2 be microscopic variables, then the reproductive potential $\mathrm{Gr} = \varepsilon_1 N_1 + \varepsilon_2 N_2$ and the function of dissipation $\mathrm{Diss} = \gamma_{11}N_1^2 + \gamma_{12}N_1N_2 + \gamma_{21}N_2N_1 + \gamma_{22}N_2^2$ are also macroscopic ones. However, only in the case $n = 2$ can system (5.1) be written in the variables, Gr and Diss, since their number is equal to the number of phase variables, i.e., two. We maximally simplify the problem setting $\varepsilon_1 = \varepsilon_2 = 1$ and $\gamma_{11} = \gamma_{22} = 1$, $\gamma_{12} = \gamma_{21} = \gamma$, so that the system state will depend only on the single parameter γ, which can be considered as a bifurcation parameter. It is easily seen that the system has three equilibriums: trivial $N_1^* = N_2^* = 0$; semi-trivial $N_1^* = 1, N_2^* = 0$ and $N_1^* = 0, N_2^* = 1$; non-trivial $N_1^* = N_2^* = 1/(1+\gamma)$, where the latter is stable if $\gamma < 1$, and unstable if $\gamma > 1$. The value $\gamma = 1$ is critical. At this time the semi-trivial equilibriums are stable (the trivial equilibrium is always unstable), and the system (depending on the initial state) comes either to one or another equilibrium. In this case the function of dissipation is $\mathrm{Diss} = D = N_1^2 + 2\gamma N_1 N_2 + N_2^2$ and the reproductive potential is equal to the total size of the community, $\mathrm{Gr} = N = N_1 + N_2$. After simple calculations we obtain:

$$\begin{aligned}\frac{\mathrm{d}N}{\mathrm{d}t} &= N - D,\\ \frac{\mathrm{d}D}{\mathrm{d}t} &= 2D - (3+\gamma)ND + (1+\gamma)N^3.\end{aligned} \tag{5.2}$$

We shall study this system analytically. It is easy to see that Eq. (5.2) has the trivial equilibrium $N^* = D^* = 0$, and two non-trivial ones, which are determined from the equation $(1+\gamma)(N^*)^2 - (3+\gamma)N^* + 2 = 0$: $N^* = D^* = 1$ and $N^* = D^* = 2/(1+\gamma)$. The trivial equilibrium is always unstable; the equilibrium $N^* = D^* = 1$ is stable if $\gamma > 1$, and unstable if $\gamma < 1$. The latter case corresponds to the semi-trivial equilibriums $N_1^* = 1$, $N_2^* = 0$ and $N_1^* = 0$, $N_2^* = 1$ in the phase space $\{N_1, N_2\}$. At last, the equilibrium $N^* = D^* = 2/(1+\gamma)$ is stable if $\gamma < 1$, and unstable if $\gamma > 1$. The phase portrait of the system (5.2) is shown in Fig. 8.3 for two values of γ: $\gamma = (1/2) < 1$ and $\gamma = 2 > 1$. Note that from a biological point of view only domains of the phase plane, those corresponding to the non-negative values of the variables N_1 and N_2, make sense. They are non-negative if the inequalities

$$\begin{aligned} &\frac{1+\gamma}{2}N^2 \le D \le N^2, &&\text{for } \gamma < 1, \\ &N^2 \le D \le \frac{1+\gamma}{2}N^2, &&\text{for } \gamma > 1 \end{aligned} \tag{5.3}$$

hold. As is seen in Fig. 8.3 these inequalities cut out a very narrow domain of the positive orthant, where only the thermodynamic evolution of the system is possible. If we look at Fig. 8.3a, then we see that the upper boundary $D = N^2$ (the curve $0a1b$) contains both the unstable node (0,0) and the saddle point (1,1), and the curve $D = N^2$ is the unique (so-called *homoclinical*) trajectory, going out from the origin of the coordinates (it corresponds to thermodynamic equilibrium) and coming into the saddle point.

The latter corresponds to the equilibrium containing only single species (either the first or the second). In this equilibrium the system can be ambiguous in time but always leaves it, moving along the separatrix (1,2). The latter comes into the stable equilibrium $D^*(2) = N^*(2) = 2/(1+\gamma) = 4/3$. It is interesting that the transition along the curve $0a1$ does not depend on γ, i.e. it is true for any system with $\gamma < 1$. The next transition along the separatrix depends naturally on γ. Other possible transitions from thermodynamic equilibrium strictly to dynamic stable equilibrium 2 can be realised along the trajectories, lying lower than $D = N^2$. It is interesting that in this case the system begins its evolution as a community of two populations, while in the previous case it started as a single population. Also note that along these trajectories the dissipation is less and the total size is greater than on $D = N^2$. Clearly, the system passes from thermodynamic to stable dynamic equilibrium $(0 \to 1)$ with minimal current values of D and maximal values of N along the lower boundary $D = ((1+\gamma)/2)N^2$ $(0c2d)$.

So, there are two extreme ways leading from zero point to stable equilibrium: the first is along $0a12$, and the second is along $0c2$. The latter seems more "optimal" from both the thermodynamic and ecological points of view, since current values of dissipation are minimal and the current total size is maximal. What is an "optimality" of the first way? Our hypothesis is: if one excludes the ambiguous residence time in the saddle point, then *the transition time will be minimal for the first way and maximal for the second one* (certainly, we consider the transition between vicinities of these points). The system pays for its reliability by the slowness of its evolution.

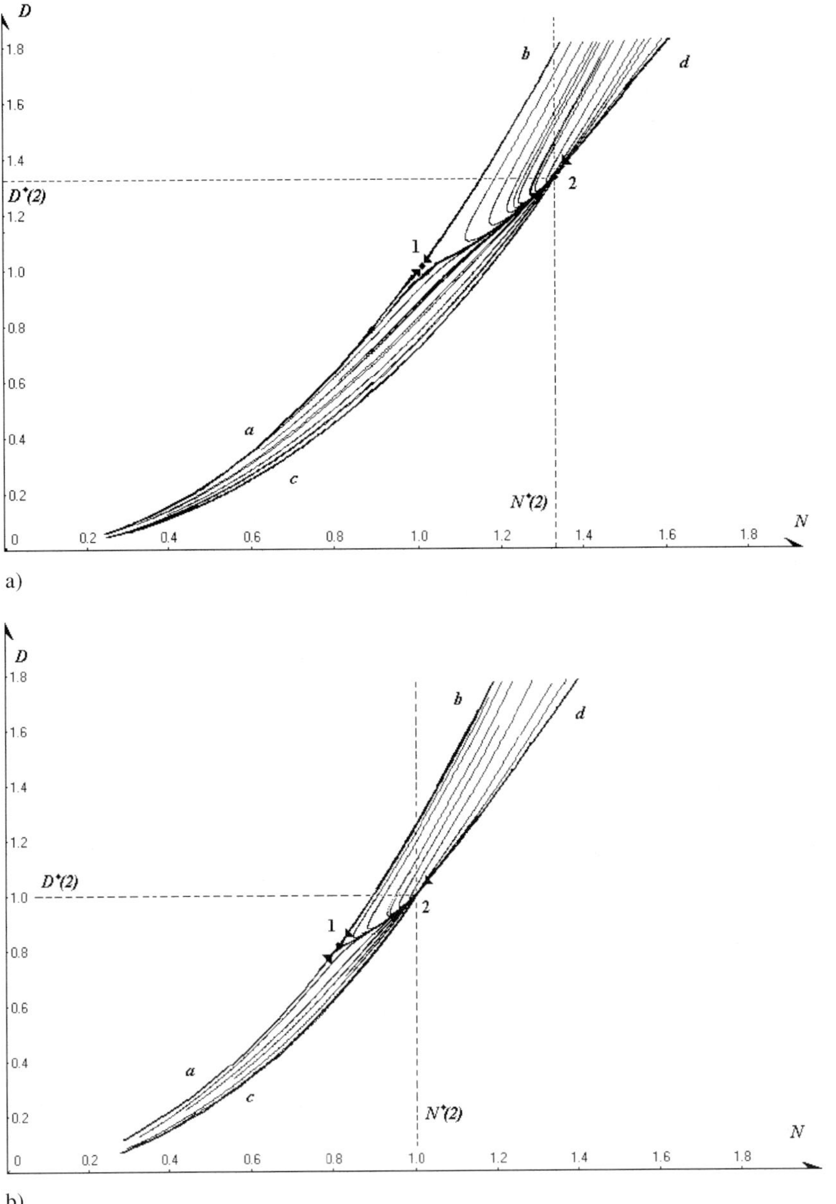

Fig. 8.3. The phase portrait of Eq. (5.2) in the space $\{D, N\}$ for two values of γ: $\gamma = 0.5$ (a) and $\gamma = 1.5$ (b). D is the function of dissipation, and N the total size of community.

Look at the upper part of the graph where $D > D^*(2)$. If we move along the line $N = N^*(2)$ towards the stable point, then the value of D decreases, illustrating by the same token the Prigogine principle (minimum of entropy production). If we now consider the movement along the line $D = D^*(2)$ towards the stable point, then the value

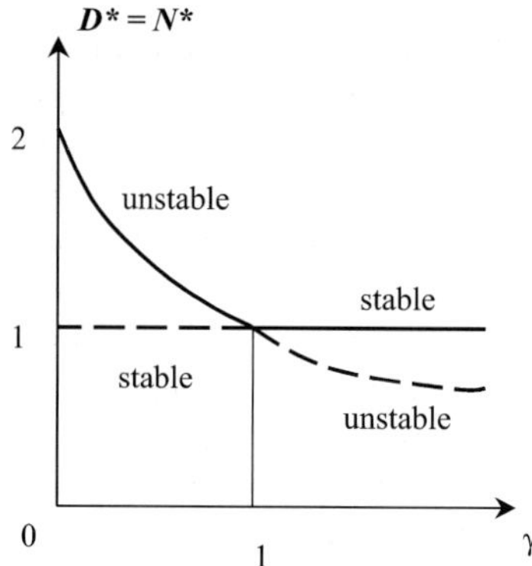

Fig. 8.4. Bifurcation diagram for Eq. (5.2).

of N increases. The latter illustrates our principle (maximum of exergy, which is formulated as a maximum of the total biomass in this case). This principle is dual to Prigogine's one.

Fig. 8.3b can be interpreted analogously, but in this case the upper boundary ($0a1b$) is described by the expression $D = ((1+\gamma)/2)N^2$, and the lower boundary ($0c2d$) by the expression $D = N^2$. Stable point 2 corresponds to the state with a single population, saddle point 1 to the community of two populations. The bifurcation diagram for system (5.2), i.e. the dependence of equilibrium values of D^* and N^* on the parameter γ, is shown in Fig. 8.4. Since $D^* = N^* = 2/(1+\gamma)$ and $D^* = N^* = 1$, we get two curves in the figure.

Within the interval $0 < \gamma < 1$ the values of D and N are greater on the stable branch of solution than on the unstable branch. At the bifurcation point $\gamma = 1$ the type of solution changes, and for $\gamma > 1$ a "degenerated" community (single population) becomes stable but, as before, D and N are greater on the stable branch. At first sight we have obtained a contradiction with Prigogine's principle: the value of dissipation is higher on the stable branch. But this is not the case; this is a consequence of its incorrect extension to the *parametric* domain. Of course, we could save the situation keeping in mind that in our case we minimise the function of dissipation under certain constraints, but…the elegance of the original formulation disappears. At the same time the exergy principle is valid: *On the stable brunch the exergy is higher than on the unstable one.*

8.6. Phenomenological thermodynamics of interacting populations

It is necessary to say at the beginning that there were a lot of different attempts to apply the methods of statistical mechanics directly to the macroscopic description of ecosystems (Kerner, 1957, 1959; Polischuk, 1971; Alexeev, 1975), but they have not brought great success. However, as was shown in the brilliant Khinchin's work (1943), a statistical mechanics could not only describe purely physical systems, but the approach can also be used for the construction of certain *phenomenological* theories of biological communities.

It was shown above that the total biomass of community, N, could be considered as one of its macroscopic variables. It is also obvious that this single variable is not sufficient for the full macroscopic description of the system, since the knowledge about the value of $N(t)$ does not give us any information about the further evolution of a community: it is necessary to know its temporal derivative. Therefore, we take into consideration the vector of *Malthusian parameters* or *functions* $\boldsymbol{\mu} = \{\mu_1, \ldots, \mu_n\}$.

We know that the concept of extensive and intensive variables plays a very important role in thermodynamics. It is obvious that for extensive values a concept of density or a specific value can be introduced, while this cannot be done for intensive values. In biological models, N_i and N are considered as some extensive variables, but the vector of composition (or structure) $\mathbf{p} = \{p_1, \ldots, p_n\}$ is the intensive variable. Then a dynamic model of a community can be presented as

$$\frac{dN_k}{dt} = \mu_k(N_1, \ldots, N_n)N_k, \qquad k = 1, \ldots, n, \tag{6.1}$$

or, with the variables $\mathbf{p}$ and N as

$$\frac{dp_k}{dt} = p_k(\mu_k - \hat{\mu}), \qquad \hat{\mu} = \sum_{k=1}^{n} \mu_k p_k. \tag{6.2}$$

$$\frac{dN}{dt} = \hat{\mu}N. \tag{6.3}$$

It is obvious that the scalar product of vectors $\boldsymbol{\mu}$ and $\mathbf{p}$, $\hat{\mu} = (\boldsymbol{\mu}, \mathbf{p})$, is the *mean Malthusian parameter* (MMP). If, in addition, the Malthusian parameters (functions) depend only on p_i then the system dynamics does not depend on its "total mass", i.e. we could increase (or decrease) the total biomass N by several times but the equation for N does not change. In other words, in this case two systems with the same structures but with different total biomasses will be dynamically equivalent. From this point of view, such systems are very "thermodynamic", since their dynamics depend neither on their size nor on their total biomasses. Moreover, since all thermodynamics identities (potentials) must be homogenous functions of the first order, the μ_k must depend only on p_i. Thus, if we would like to remain within the thermodynamics framework we have to accept the assumption. In this case, if the equations for $\mathbf{p}$ and N are separated from each other, then the equilibriums with respect to $\mathbf{p}$ (we shall name them "structural" ones) and N can be established independently so that, for instance, in a structural equilibrium the total size can either change or remain constant. Later on we shall only consider evolution of the community structure, since knowing its dynamics we can always calculate the dynamics of its total size.

Formally, the dynamics description in terms of frequencies is equivalent to an assumption about the constancy of the total size, $N =$ constant. Then the equations for the population sizes N_k are written in the form (6.2) where p_k are replaced by N_k.

We postulate (without proofs) the following statement:

A community of interacting populations evolves in such a manner that its MMP always increases attaining its maximum at the stable structural equilibrium (Fisher's principle).

Before emerging from ecological problems, we shall perform a short but very fruitful digression into the similar field of evolutionary genetics (Svirezhev and Passekov, 1982).

The main subject is the population, which can be considered as the community consisting of n^2 sub-populations of "genotypes". Every one of them is denoted by the pair $\{k,j\}$; $k,j = 1,\dots,n$; the size of the corresponding sub-population is equal to N_{kj}. In evolutionary genetics the Malthusian parameters μ_{kj} are named as *fitness of genotypes*. Assume that the genotypes $\{i,j\}$ and $\{j,i\}$ are identical so that $\mu_{ij} = \mu_{ji}$ and $N_{ij} = N_{ji}$. The MMP (or the mean fitness of population) will be equal to $\hat{\mu} = \sum_{i,j=1}^{n} \mu_{ij}u_{ij}$ where $u_{ij} = N_{ij}/\sum_{i,j=1}^{n} N_{ij}$ are the genotype frequencies. The analogue of Eq. (6.2) is written in the form

$$\frac{du_{kj}}{dt} = u_{kj}(\mu_{kj} - \hat{\mu}), \qquad \hat{\mu} = \sum_{k,j=1}^{n} \mu_{kj}u_{kj}. \tag{6.4}$$

The principle of the MMP maximum in evolutionary genetics is called the "Fundamental Theorem of Natural Selection"; one of its formulations states that "under the pressure of natural selection the mean fitness of population increases, attaining its maximum at the stable equilibrium". The theorem has been proved by Sir Robert Fisher.

Let μ_{kj} be constant and u_{kj} be independent, then

$$\frac{d\hat{\mu}}{dt} = \sum_{k=1}^{n}\sum_{j=1}^{n} \mu_{kj}\frac{du_{kj}}{dt} = \sum_{k=1}^{n}\sum_{j=1}^{n} u_{kj}(\mu_{kj} - \hat{\mu})^2 \geq 0, \tag{6.5}$$

i.e. the value of $\hat{\mu}$ is always increasing, tending to its maximum in the space of genotype frequencies. Since $\hat{\mu}$ is a linear form of u_{kj} the maximum of $\hat{\mu}$ is attained at one of the cones of simplex $\sum_{i,j=1}^{n} \mu_{ij}u_{ij} = 1$: the population with maximal μ_{kj} survives, the others are eliminated during the process of natural selection. We see that, in this case, the principle of maximum of the MMP, which is even proved here (not only postulated), leads to the trivial result; in order to obtain some more or less meaningful results we must assume that there are some kind of constraints or connections in the system: either within the vector space of $\boldsymbol{\mu}$ or within the vector space of $\mathbf{u}$ (or $\mathbf{p}$) or between the first and the second spaces.

Such a type of connection in evolutionary genetics is described by the Mendel laws connecting the frequencies of genotypes in the parents and off-spring populations so that

$$u_{kj} = \left(\sum_{s=1}^{n} u_{ks}\right)\left(\sum_{s=1}^{n} u_{sj}\right). \tag{6.6}$$

Thus, the genotype frequencies are not independent; the genotype structure of population is changed not only by natural selection, but also as a result of Mendelian splitting, which generates structural constraints within the space of genotypes. Then $\max_{u_{kj}} \hat{\mu}$ can be attained within the simplex, i.e. a polymorphous state of population may exist.

The principle of maximum for the MMP is a natural generalisation of the Fisher Theorem of Natural Selection. Note that this principle is a hypothesis, postulate, credo;

in general, everything is possible. But this is not a theorem, requiring proof, although its variety in evolutionary genetics is a theorem.

If we follow the principle then among all virtual structures we can select a single one which will satisfy this principle; so we announce that this structure really exists. In this case it is not necessary to know the equations describing the local dynamics of the system; and the biological laws, which define the dynamics, are expressed through some geometrical properties of the spaces of $\boldsymbol{\mu}$ and $\mathbf{p}$ and extreme properties of some functions, describing global characteristics of the community.

It is well known that Onsager's reciprocal relations are the basis of any phenomenological thermodynamics. In this case these relations have introduced the linear relation between $\boldsymbol{\mu}$ and $\mathbf{p}$:

$$\mu_k = \sum_{k=1}^{n} \mu_{kj} p_j, \qquad \mu_{kj} = \mu_{jk}; \;\; k,j = 1, \dots, n. \tag{6.7}$$

Substituting Eq. (6.7) into Eq. (6.2), we get

$$\frac{\mathrm{d}p_k}{\mathrm{d}t} = p_k(\mu_k - \hat{\mu}), \tag{6.8}$$

$$\mu_k = \sum_{j=1}^{n} \mu_{kj} p_j, \qquad \hat{\mu} = \sum_{k=1}^{n} \sum_{j=1}^{n} \mu_{kj} p_k p_j.$$

Formally Eqs. (6.8) coincide with the classic Fisher–Haldane–Wright equations of evolutionary genetics. What is the reason for this coincidence? What is the biological sense of Onsager's linear relations (6.7)? Note that Mendel's relations (6.6) are valid if the population heredity is determined by a single gene with n multiple alleles. Then, instead of the n^2 sub-populations of genotypes with frequencies u_{kj}, we can consider the n sub-populations of alleles with frequencies $p_k = \sum_{j=1}^{n} u_{kj}$. In this case the relation (6.6) is represented as $u_{kj} = p_k p_j$. By substituting this relation into Eq. (6.4), after simple transformations we obtain Eq. (6.8), i.e. the Fisher–Haldane–Wright equations. The genetic sense of Onsager's hypothesis is: if the values of $\mu_k = \sum_{j=1}^{n} \mu_{kj} p_j$ are the mean fitness of kth alleles then these equations do not differ from the general equations for the community of n competing populations that are the populations of *alleles*. The genotype is interpreted as a *collision* of two alleles, and its fitness is interpreted as a result of the collision.

By calculating $(\mathrm{d}\hat{\mu}/\mathrm{d}t)$ along trajectories of Eq. (6.8) we get

$$\frac{\mathrm{d}\hat{\mu}}{\mathrm{d}t} = \frac{\mathrm{d}}{\mathrm{d}t}\left(\sum_{k=1}^{n} \sum_{j=1}^{n} \mu_{kj} p_k p_j \right) = 2 \sum_{k=1}^{n} p_k (\mu_k - \hat{\mu})^2 \geq 0, \tag{6.9}$$

i.e. the mean fitness of population tends to maximum in the process of evolution. This is again the well-known Fisher's Fundamental Theorem of Natural Selection. How can the result be interpreted from the thermodynamic viewpoint?

We represent the fitness coefficients in the form $\mu_{kj} = F(1 - s_{kj})$ where F is the mean number of off-spring per one individual, and s_{kj} is the probability for the genotype $\{k,j\}$ to survive until its reproductive age and to produce the vital posterity. Then the number of off-spring, equal to $FN \sum_{k=1}^{n} \sum_{j=1}^{n} s_{kj} p_i p_j = N(F - \hat{\mu})$, must perish during one generation in order for the process of genetic evolution to continue. This is the necessary condition for the continuity of genetic evolution and its cost! The lost of individuals is a typical irreversible process with a biomass dissipation, and the value $\mathrm{Diss} = N(F - \hat{\mu})$ can be considered entirely as the function of dissipation or the entropy produced within the system. In accordance with Prigogine's theorem, $\mathrm{Diss} = (\mathrm{d_i}S/\mathrm{d}t) \rightarrow \min_{\mathbf{p}} (\mathrm{d_i}S/\mathrm{d}t) > 0$, and the gene structure of population, $\mathbf{p}$, tends to the stable equilibrium, $\mathbf{p}^*$. Since the evolutionary dynamics does not depend on N, and F can be always chosen sufficiently large (so that $F > \hat{\mu}^*$), the statement about $\min_{\mathbf{p}} (\mathrm{d_i}S/\mathrm{d}t)$ is equivalent to the statement about $\max_{\mathbf{p}} \hat{\mu}$. This means that in the thermodynamic theory of evolution Prigogine's theorem about the minimum of entropy production is equivalent to Fisher's Fundamental Theorem of Natural Selection in the classic evolutionary genetics.

8.7. Community in the random environment and variations of Malthusian parameters

Assume that the community is embedded into the random environment, whose random fluctuations cause fluctuations of Malthusian parameters, without affecting numbers or frequencies (Svirezhev, 1991). This is a natural assumption, if it is taken into account that in reality the fluctuations of such environmental factors as temperature, humidity, salinity, etc. affect in the first place the intrinsic growth rates of populations, i.e. their Malthusian parameters. Then, implicitly, by means of changing their growth rates, they change the population sizes or biomasses. When looking at models used in ecological modelling, we can see that they all describe the dependence on the environment in the form of dependence of the relative growth rates on the environmental factors.

Let $\mu_k = \mu_k^0 + \xi_k$ where ξ_k is the normally distributed random value with the zero arithmetic mean and the variance τ_{kk}. Note that the hypothesis of normality is one of the most popular hypotheses about the nature of randomness in natural sciences. It works very well if the random fluctuations are the result of the impact of a lot of weakly correlated random factors. Arbitrarily this is valid for ecosystems, therefore we accept the hypothesis.

If all μ_k would be statistically independent, then for the description of the values ξ_k the knowledge about the variances τ_{kk} would be enough, but we assume that they are *statistically dependent*, i.e. they are connected to each other by certain influences of the environment on the community. Then we already have to consider the random vector $\boldsymbol{\xi} = \{\xi_1, \ldots, \xi_n\}$, about which we assume that it is again normally distributed with zero mean but with covariation matrix $\mathbf{T} = \|\tau_{kj}\|$. Since the MMP $\hat{\mu}$ is a linear function of random vector $\boldsymbol{\mu}$, the $\hat{\mu}$ is also normally distributed with the mean $\hat{\mu}_0 = (\boldsymbol{\mu}_0, \mathbf{p})$ and the variance $\hat{\tau} = (\mathbf{p}, \mathbf{Tp}) = \sum_{k=1}^{n} \sum_{j=1}^{n} \tau_{kj} p_k p_j$. In this case the following probabilistic statement is valid:

> the inequality $\hat{\mu} \geq 0$ holds with the probability equal to $\Phi(\hat{\mu}_0/\sqrt{\hat{\tau}})$, where $\Phi(a)$ is the probability integral $\Phi(a) = (1/\sqrt{2\pi}) \int_{-\infty}^{a} \mathrm{e}^{-x^2/2} \mathrm{d}x$.

If we are now given, as is usually done in statistics, some confidence level of probability δ^* (for instance, $\delta^* = 95\%$), to which the certain quantile a^* of normal distribution (so that $\delta^* = \Phi(a^*)$) corresponds, then the MMP $\hat{\mu} \geq \hat{\mu}_0 - a^*\sqrt{\hat{\tau}}$ with the probability δ^*.

If $\hat{\mu} \geq 0$ then the total size of a community will never decrease. If the community structure (the vector $\mathbf{p}$) and $\hat{\mu}_0$ are fixed then the fulfilment of relation $\hat{\mu}_0 = a^*\sqrt{\hat{\tau}}$ means those random fluctuations of the species Malthusian parameters do not bring about a decrease in the total community size. However, this statement is only fulfilled with a certain probability $\delta^* = \Phi(\hat{\mu}_0/\sqrt{\hat{\tau}})$. The probability can be considered as a measure of the community stability. Since the values a^* and δ^* are connected one-to-one with each other, later on we shall also use a^* as a measure of stability.

Let the principle of the maximum MMP be valid, so that it can be realised with the help of the structured change. In our case the maximum of the MMP was attained at such $\mathbf{p} = \mathbf{p}^*$, for which the lower bound of $\hat{\mu}$ has attained its maximum with the fixed probability δ^*, so that $\max_{\mathbf{p}} (\inf \hat{\mu}) = \hat{\mu}_0 - a^*\sqrt{\hat{\tau}}$. We assume that the transition $\mathbf{p} \to \mathbf{p}^*$, which occurs at the fixed $\boldsymbol{\mu}_0$ and $\mathbf{T}$, is the system's adaptation to the given environmental conditions. At this time, the total size may be either changeable or constant. Generally speaking, the latter statement is not completely correct but this is a reasonable approximation if the total number of species is large. In this case the total size varies insignificantly, even if variations of several p_k are very large. We shall name such an adaptive change of the structure as *fast system evolution*. In the course of the fast evolution the system quickly passes from any state to the state with $\max_{\mathbf{p}} (\inf \hat{\mu})$ where $\mathbf{p} = \mathbf{p}^*$, by the same token adapting to the environment. Note that the lower bound may be both positive and negative depending on a^*, i.e. the total size, even at the optimal state, may both increase and decrease under the influence of random perturbations. Certainly, we can always select a^* in such a way that $\hat{\mu}^* \geq 0$ (do not forget about the probabilistic origin of this inequality!). If $\hat{\mu}^* \geq 0$ then we can always state that the system is not degenerated, i.e. the total size of the community does not decrease. Therefore, it is natural to require: $\inf \hat{\mu}^* = 0$. The value of probability, equal to $\delta^* = \Phi(a^*)$, corresponds to the condition; it also determines the stability reserve.

All these statements allow us to formulate the following concept of *stable structural equilibrium* and *stability reserve*.

At the stable structural equilibrium:

$$\text{(a)} \qquad \inf(\hat{\mu}^*) = \max_{\mathbf{p}} (\hat{\mu}_0 - a^*\sqrt{\hat{\tau}}), \tag{7.1}$$

$$\text{(b)} \qquad \inf(\hat{\mu}^*) = 0 \qquad \text{when } a^* = \hat{\mu}_0/\sqrt{\hat{\tau}^*}. \tag{7.2}$$

Here $\hat{\mu}^* = \hat{\mu}_0 - a^*\sqrt{\hat{\tau}^*}$, where the notation ($*$) points to the fact that the value of $\hat{\tau}$ is taken at the point $\mathbf{p}^*$, i.e. at the point where $\inf(\hat{\mu})$ reaches its maximum. Therefore, the community at the stable structural equilibrium is described by the vector $\mathbf{p}^*$ (microscopic variables) and the value of $\delta^* = \Phi(a^*)$ (a macroscopic variable), which could be interpreted as a *probability of existence* of this equilibrium. Note once again that the probability does not depend on the total size of the community.

Consider a problem that is dual to problems (7.1) and (7.2): *find* $\max_{\mathbf{p}} \delta^*$ *under constraint* $\inf(\hat{\mu}^*) = 0$, i.e. in that way we find the most probable structure. The problem is equivalent to the problem of $\max_{\mathbf{p}} a^*$ by virtue of the monotonicity of the probability integral. By virtue of (7.2) $a^* = \hat{\mu}_0/\sqrt{\hat{\tau}}$, and we can postulate the original problem as: how to *find* the $\max_{\mathbf{p}} (\hat{\mu}_0/\sqrt{\hat{\tau}})$. These dual problems have the same solution, whence a very important conclusion follows: *The structural equilibrium with maximal MMP is the most probable.*

In order to find the distribution of species frequencies corresponding to the stable structural equilibrium we have to solve the problem of maximisation defined in (7.1) and (7.2). At the first stage we do not take into account the condition of non-negativeness of frequencies. By applying the Lagrange multipliers method we shall find a maximum of the function

$$F' = \sum_{k=1}^{n} \mu_k^0 p_k - a^* \sqrt{\sum_{k=1}^{n} \sum_{j=1}^{n} \tau_{kj} p_k p_k} + \lambda \left(\sum_{k=1}^{n} p_k - 1 \right), \tag{7.3a}$$

or, in the vector–matrix notations ($\mathbf{e} = \{1, ..., 1\}$ is a unique vector)

$$F' = (\boldsymbol{\mu}_0, \mathbf{p}) - a^* \sqrt{(\mathbf{p}, \mathbf{Tp})} + \lambda[(\mathbf{e}, \mathbf{p}) - 1]. \tag{7.3b}$$

We have already used this method, so the procedure is standard, and therefore without unnecessary details we get the final results at once.

The community structure at the stable structural equilibrium is defined by the vector

$$\mathbf{p}^* = \frac{\mathbf{T}^{-1} \mu_0}{(\mathbf{e}, \mathbf{T}^{-1} \mu_0)}. \tag{7.4}$$

The probability of this equilibrium is equal to

$$\delta^* = \Phi\left(\sqrt{(\mathbf{T}^{-1} \mu_0, \mu_0)} \right). \tag{7.5}$$

Here $\mathbf{T}^{-1} = \|\tau_{kj}^{-1}\|$ is the matrix reciprocal to the covariance matrix $\mathbf{T} = \|\tau_{kj}\|$. Both are positive definite.

Generally speaking, the value $(a^*)^2 = (\mathbf{T}^{-1} \mu_0, \mu_0)$ can be used as a measure of probabilistic stability of the equilibrium. In the space $\{\mu_1^0, ..., \mu_n^0\}$ the condition $(a^*)^2 = (\mathbf{T}^{-1} \mu_0, \mu_0) =$ constant defines an ellipsoid, and if we now assume that the Malthusian parameters change in a quasi-stationary way and their evolution is such that their trajectories always belong to the ellipsoid, then the probability δ^* does not change. We shall name such types of processes as *isostable*.

Generally speaking, as is seen from formula (7.4) several frequencies (let $p_s^* < 0$, $s \in [1, n]$) can be negative. In this case, since the function inf $\hat{\mu}$ and the simplex Σ: $p_k \geq 0$, $\sum p_k = 1$ are convex, the maximum must be situated on the border and is unique. Then the optimal solution $(\mathbf{p}^*)'$ is defined as:

1. For these s we set $(p_s^*)' = 0$.
2. In the matrix $\mathbf{T}$ we cross out the corresponding sth rows and columns by reducing it to the matrix $\mathbf{T}'$.

3. In the vector $\boldsymbol{\mu}_0$ we cross out the corresponding sth elements reducing it to the $(\boldsymbol{\mu}_0)'$.
4. For these new matrix and vector we calculate the reduced new optimal vector $(\mathbf{p}^*)'$.

If several components of the new optimal vector are negative again, then the process is repeated.

Principally the process has to be converged to the situation with positive component. The measure of stability in this case is equal to $[(a^*)^2]' = ((\mathbf{T}^{-1})'(\mu_0)', (\mu_0)')$, where the index $(')$ corresponds to the finish of the process.

Let $(\tau_k^{-1})_\mu = \sum_{j=1}^n \tau_{kj}^{-1} \mu_j^0$ then $p_k^* = (\tau_k^{-1})_\mu / \sum_{k=1}^n (\tau_k^{-1})_\mu$, as follows from Eq. (7.4). The necessary and sufficient conditions, that the maximum lies within the simplex Σ, are

$$(\tau_k^{-1})_\mu = \sum_{j=1}^n \tau_{kj}^{-1} \mu_j^0 > 0; \qquad k = 1, \dots, n. \tag{7.6}$$

Note that the appearance of negative frequencies points to the fact that the optimisation of the structure of community, i.e. its adaptation, requires the elimination of corresponding species out of the community.

Until now we did not assume any constraints for the MMPs of different species, μ_k^0. In all the previous considerations we have also implicitly assumed that $a^* > 0$, i.e. the probability of existence of the stable structural equilibrium $\delta^* > 1/2$. But it could be that $\hat{\mu}_0^* < 0$, then in order to satisfy the equality $\inf(\hat{\mu}_0^*) = 0$, i.e. $\hat{\mu}_0^* - a^* \sqrt{\hat{\tau}} = 0$, we have to assume that $a^* < 0$. Theoretically it is possible, since $a^* = \pm\sqrt{(\mathbf{T}^{-1}\mu_0, \mu_0)}$, we can choose the negative sign $(-)$. Then the probability $\delta^* < 1/2$, and it is *minimal*, so that the corresponding structures are naturally named *low-probability* ones. Any other structure will be more probable, therefore the low-probability structural equilibrium must be unstable and when necessary it must be disintegrated with elimination of some species (since the minimum is reached only within the simplex and the maximum can be reached only at the borders). From Eq. (7.4) it follows that

$$\hat{\mu}_0^* = \frac{(\mu_0, \mathbf{T}^{-1}\mu_0)}{(\mathbf{e}, \mathbf{T}^{-1}\mu_0)}, \tag{7.7}$$

and the value of $\hat{\mu}_0^*$ will be negative only if $(\mathbf{e}, \mathbf{T}^{-1}\mu_0) < 0$, or

$$\sum_{k=1}^n (\tau_k^{-1})_\mu = \sum_{k=1}^n \sum_{j=1}^n \tau_{kj}^{-1} \mu_j^0 < 0. \tag{7.8}$$

Thus, the condition defines the border between structures with high and low probabilities of existence. If we use the geometric interpretation within the space $\{\mu_1^0, \dots, \mu_n^0\}$, then relation (7.8) defines the plane which passes through the coordinate origin and divides the entire set of $\boldsymbol{\mu}_0$ into two subsets corresponding to probable and low-probability structures.

It is interesting that if the function $\inf(\hat{\mu})$ has a minimum then its maximum can be situated only in one of the simplex Σ corners, where the measure a^* will be maximal. In other words, the stable structural equilibrium for low-probability structures is the state with single sth species, for which $a_s^* = \mathbf{max}_k\,(\mu_k^0/\sqrt{\tau_{kk}})$. Hence, in the low-probability structure all species (except one) are eliminated.

If we turn again to the geometric interpretation, then the conditions

$$(\tau_k^{-1})_\mu = \sum_{j=1}^{n} \tau_{kj}^{-1} \mu_j^0 = 0; \qquad k = 1, \ldots, n, \tag{7.9}$$

give the family of n planes passing through an origin of coordinates. They cut such a cone out of the set of $\boldsymbol{\mu}_0$ that if the vector $\boldsymbol{\mu}_0$ lies within the cone then there is no species which would be eliminated out of community.

In order to illustrate visibly all these results we consider the elementary community consisting of a minimal number of species making it still possible to speak about interaction, i.e. about two species.

So, let the two-species community be with $p_1 = p$, $p_2 = 1 - p$, $\mu_1^0 = \mu_1$, $\mu_2^0 = \mu_2$ and the covariation matrix

$$\mathbf{T} = \begin{Vmatrix} \tau_{11} & \tau_{12} \\ \tau_{12} & \tau_{22} \end{Vmatrix},$$

where $\tau_{11}\tau_{22} \geq (\tau_{12})^2$ and $\tau_{11} + \tau_{22} \geq 2\tau_{12}$.

The value p^* given the community structure is (see Eq. (7.4)):

$$p^* = \frac{\tau_{22}\mu_1 - \tau_{12}\mu_2}{\tau_{22}\mu_1 + \tau_{11}\mu_2 - \tau_{12}(\mu_1 + \mu_2)}. \tag{7.10}$$

Using conditions (7.6), which in our case are written as

$$\tau_{22}\mu_1 - \tau_{12}\mu_2 > 0; \qquad \tau_{11}\mu_2 - \tau_{12}\mu_1 > 0, \tag{7.11}$$

we immediately obtain in the plane $\{\mu_1, \mu_2\}$ a domain of co-existence of the two interacting populations for given statistical characteristics. For a more visible interpretation we pass from a covariation to a correlation matrix where

$$\tau_{11} = \sigma_1^2, \qquad \tau_{22} = \sigma_2^2, \qquad \tau_{12}/\sqrt{\tau_{11}\tau_{22}} = \rho,$$

and where σ_1^2 and σ_2^2 are variances of μ_1 and μ_2, and $\rho(-1 \leq \rho \leq +1)$ is a coefficient of correlation between them. Then the conditions of co-existence are represented in the form:

$$\mu_1 > \rho\left(\frac{\sigma_1}{\sigma_2}\right)\mu_2; \qquad \mu_2 > \rho\left(\frac{\sigma_2}{\sigma_1}\right)\mu_1. \tag{7.12}$$

In Fig. 8.5a and b the domains defined by these inequalities are shown for the cases of positive and negative correlation. One can see that the negative correlation between the Malthusian parameters of species significantly increases the possibility of their co-existence.

Line α : $\mu_1 > \rho\left(\frac{\sigma_1}{\sigma_2}\right)\mu_2;$

Line β : $\mu_2 > \rho\left(\frac{\sigma_2}{\sigma_1}\right)\mu_1.$

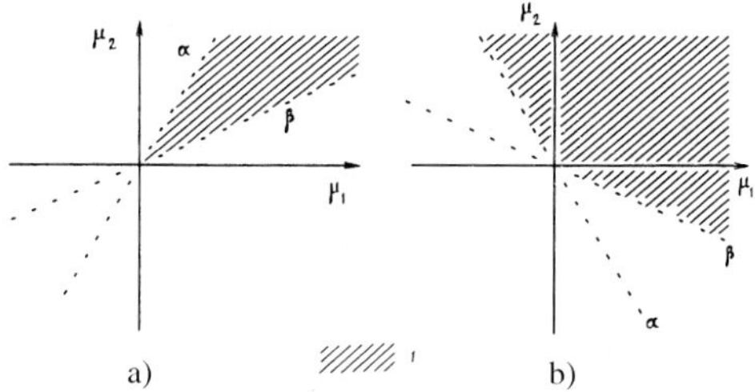

Fig. 8.5. Domains of co-existence (1, domain of ecological stability): (a) $\rho > 0$, (b) $\rho < 0$.

As shown above, the condition $\sum_{k=1}^{n}(\tau_k^{-1})_\mu > 0$, which in our case is written as $\tau_{22}\mu_1 + \tau_{11}\mu_2 > \tau_{12}(\mu_1 + \mu_2)$ or

$$\left(\frac{\sigma_2}{\sigma_1}\right)\mu_1 + \left(\frac{\sigma_1}{\sigma_2}\right)\mu_2 > \rho(\mu_1 + \mu_2), \tag{7.13}$$

is sufficient for existence of the maximum of $\inf(\hat{\mu}_0)$. Therefore, if $\mu_1 < \rho(\sigma_1/\sigma_2)\mu_2$ then the maximum is attained at the point $p^* = 0$, if $\mu_2 < \rho(\sigma_2/\sigma_1)\mu_1$, then at the point $p^* = 1$. The equation $((\sigma_2/\sigma_1) - \rho)\mu_1 + ((\sigma_1/\sigma_2) - \rho)\mu_2 = 0$ in the plane $\{\mu_1, \mu_2\}$ defines the line which divides the entire set of μ_1 and μ_2 into two subsets corresponding to the maximum and minimum of $\inf(\hat{\mu})$ (Fig. 8.6).

Line a: $\mu_1 = \rho(\sigma_1/\sigma_2)\mu_2$;

Line b: $\mu_2 = \rho(\sigma_2/\sigma_1)\mu_1$;

Line c: $\mu_1((\sigma_2/\sigma_1) - \rho) + \mu_2((\sigma_1/\sigma_2) - \rho) = 0$;

Line d: $\mu_1\sigma_2 = \mu_2\sigma_1$.

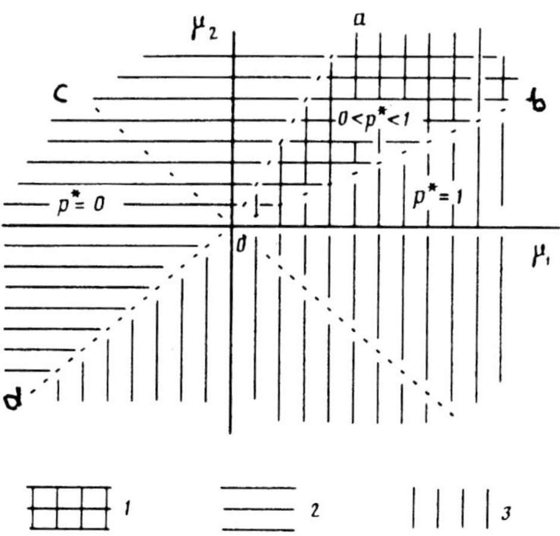

Fig. 8.6. Domains with different structures: (1) $0 < p^* < 1$; (2) $p^* = 0$; (3) $p^* = 1$. The second and third domains are also the domains of the low probability of the co-existence of two species.

The measure of stability for the state with $p^* = 0$ is equal to $a^* = \mu_2/\sigma_2$, with $p^* = 1$ to $a^* = \mu_1/\sigma_1$. The line d: $\mu_1\sigma_2 = \mu_2\sigma_1$ divides the set of μ_1 and μ_2 (except the cone $a0b$) into two subsets, such that if $\mu_1\sigma_2 < \mu_2\sigma_1$ then the state with $p^* = 0$ is stable and vice versa.

A state of the system depends on three parameters, for instance the coefficient of correlation, ρ, the ratio of mean values, $x = \mu_1/\mu_2$, and the ratio of variance, $F^2 = \sigma_1^2/\sigma_2^2$. Note that there are few developed statistical procedures for their estimation by the observation data. Therefore, as it seems to us, the representation of the different domains of these variables will be more interesting, and we show them in the plane $\{\rho, F\}$ for fixed x (Fig. 8.7).

In these variables

$$p^* = \frac{x - \rho F}{x + F^2 - \rho(1 + x)F},$$

and the measure of stability $(a^*)^2$ is different in different domains.

In domain 1:

$$(a^*)^2 = \lambda^2 \frac{x^2 - 2\rho x F + F^2}{1 - \rho^2};$$

In domain 2:

$$(a^*)^2 = \lambda^2 F^2;$$

In domain 3:

$$(a^*)^2 = \lambda^2 x^2.$$

Comparing Figs. 8.6 and 8.7 we see that in the space of variables which are describing less the absolute values of Malthusian parameters as the statistical connection between them, the domain of co-existence is much larger than in the space $\{\mu_1, \mu_2\}$. For instance, if the correlation between both Malthusian parameters is negative then they are always co-existing. The point A is a singular point, which belongs to all three domains. This means that in its small vicinity either both species exist or one of them is eliminated.

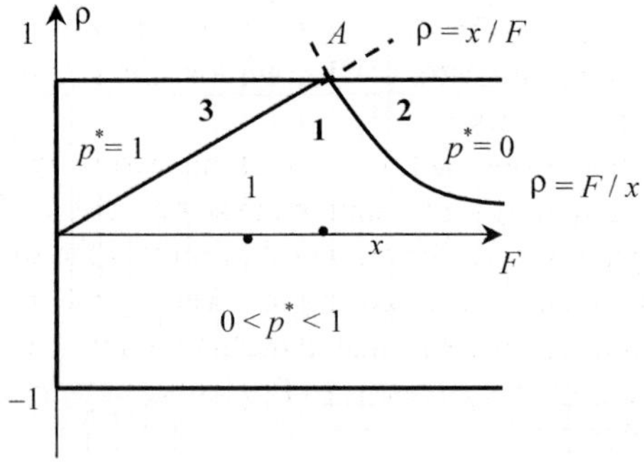

Fig. 8.7. Domains with different structures in the plane $\{\rho, F = \sigma_1/\sigma_2\}$: (1) $0 < p^* < 1$; (2) $p^* = 0$; (3) $p^* = 1$.

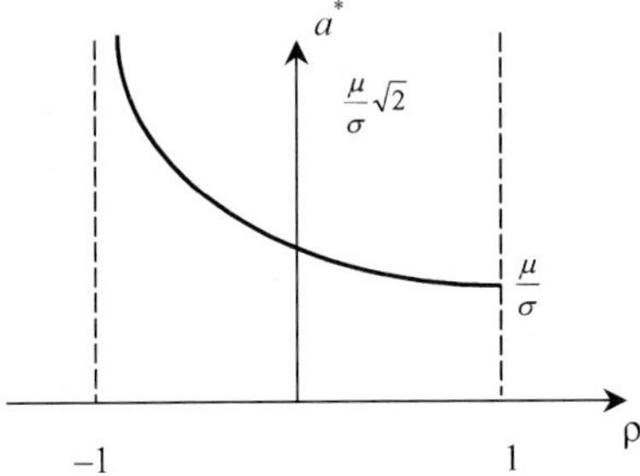

Fig. 8.8. Dependence of the measure of stability, $(a^*)^2$, on the coefficient of correlation, ρ, for two identical species.

Let both the species be identical: $\sigma_1 = \sigma_2 = \sigma$ and $\mu_1 = \mu_2 = \mu$. In this case

$$x = F = 1, \qquad p^* = 1/2 \qquad \text{and} \qquad (a^*)^2 = \left(\frac{\mu}{\sigma}\right)^2 \frac{2}{1+\rho}.$$

A dependence of the measure of stability, $(a^*)^2$, on the coefficient of correlation, ρ, is shown in Fig. 8.8. This figure illustrates one old ecological principle: *physiologically close species, the reactions of them on the environmental perturbations being similar, cannot co-exist*. This is a consequence of monotonous decrease of $(a^*)^2$ with the growth of ρ.

Finally, we would like to state the following. In experimental ecology we usually deal with the long time-series of observations of population sizes or biomasses of species belonging to one trophic level (for instance, various species of phytoplankton). Using these data we can estimate their statistical characteristics as their MMPs and the corresponding covariation matrix. Knowing these values we can estimate the measure of stability of such a type of community in the given environment. It is natural that all these estimations will be correct if the community is quasi-stationary.

8.8. Summary of the ecological important issues

Trophic chains are a simplification. Therefore Chapter 8 examines the thermodynamics of more complex systems: communities with competition between species. A Prigogine-like theorem can be applied again: the exergy of the community will be maximal. Furthermore, the resources in the environment tend to minimum indicating that the resources have a maximal utilisation. The system may play on r-strategists or on K-strategists to obtain these goals.

Moreover, communities of interacting populations evolve in such a manner that their MMPs always increase, attaining maximum at the stable structural equilibrium (this is the so-called Fisher principle after R. Fisher, who has proved the theorem). The Malthusian parameters are named the fitness of genetypes. The population with maximal fitness survives; the other populations are eliminated during the process of natural selection. It is shown that Fisher's theorem is equivalent to Prigogine's theorem in the thermodynamic theory of natural selection.

In conclusion, Prigogine's theorem also seems valid for communities and the general steady development towards a better fitness is consistent with Prigogine's theorem.

Chapter 9
Thermodynamics of ecological networks

Apparently, a network is one of the most widespread forms of organisation of Space in Nature. Note that it can be as real a geometric space as any other; for instance, a space of trophic possibilities.

À propos, what is the dimension of a network? What dimension has a planar cobweb?

From table-talks in Karrebæk.

9.1. Introduction

The network approach is very popular in ecology. Flows of energy, matter and information are described over a network of nodes and links. It began with Lindeman's (1942) classic work. More recently, Patten has devoted much attention to this problem (Patten, 1982, 1991; Patten et al., 1990). Chapter 7 of Jørgensen's (2002b) book "Integration of Ecosystem Theories: A Pattern" contains a very good overview of the concept and an extensive list of references. On the other hand, the so-called "compartmental models" are very popular in biology: see, for example, Straskraba and Gnauk (1985). These models represent typical *linear* dynamic structures (systems of ordinary differential equations) that correspond to specific graphic structures. In this way they can be interpreted as trophic networks. It is obvious that trophic chains (see Chapter 7) are a special case of trophic networks, which, in turn, can be considered as a combination of several trophic chains.

If we look at numerous publications on biogeochemical cycles in ecosystems we can see that, typically, the systems are presented by a flow-chart diagram with storage of certain elements inside compartments and flows between them.

We also know that the non-linear Volterra equations are classic objects of mathematical ecology and they are often used for the description of ecological dynamics. The following question emerges: "Can we construct some non-contradictory combinations from these two approaches?" Later on (in Section 9.7) we shall attempt to answer the question.

The network description is also popular in biological and chemical sciences. In particular, a typical example of a network description is the description of a metabolic cycle in biophysics (Rubin, 1999). A classic work on the network approach in biophysics is "Network thermodynamics: dynamic modelling of biophysical systems" by Oster et al. (1973). Aoki (1988, 1992) has applied the entropy and exergy concepts to ecological networks at steady state.

Towards a Thermodynamic Theory for Ecological Systems, pp. 221–241

In Chapter 4, we already dealt with the hierarchy of descriptions of biological communities (or ecosystems) when the first level of description was simply a list of species, and where if n species were indistinguishable then the information per one species was equal to $I_1 = k \ln n$, where k is either Boltzmann's constant or its analogue. In the next level of description each species possessed such a property as the number of specimens N_i. Then the information per one specimen was equal to $I_2 = -k\sum_{i=1}^{n} p_i \ln p_i;\ p_i = N_i/\sum_{i=1}^{n} N_i$. It is obvious that $\max_{p_i} I_2 = \ln n$ and $I_2 \leq I_1$. This can be expected, since the "specific" information I_1 and I_2 are calculated for different "agents": species in the first case and specimens in the second. Further complication of the description will lead to a more and more detailed description of specimen characteristics, such as their specific biomass, size, volume, energy, exergy, etc. However, beginning with the first level we could be using another approach, bearing in mind that any ecosystem is not only an entity of different individuals but also a combination of interactions both between different species and between them and the environment. The first obvious step, so common in ecology, is to take a qualitative description of the trophic network as a graph with vertices corresponding to species and arcs indicating the pathways of energy and matter in the system. In this case, we say that any two species are connected by a *trophic relationship*. Such a type of graph is called "directed" since the directions of energy and matter flows are given. Since all species are exchanging energy and matter with the environment to "complete" the description we have to define the environment as one or several quasi-species. For instance, the environment can be represented as two quasi-species: the first one combines all flows from the environment into the system (solar radiation, nutrients, etc.) while the second combines all flows out of the system (metabolism, dying-off, etc.).

The representation of ecosystem structure as a trophic graph does not exhaust the entire set of interspecies relations. For instance, in the ecosystem structure we could select a group of species as a *trophic level*. These species are not connected by trophic relations, but they are usually either competing for resources (the *relation of competition*) or co-operating in their utilisation (the *relation of mutualism*, or *symbiosis*).

The classification of all sorts of interspecific relations is based on an idea formulated by Velimir Khlebnikov (1910), a prominent Russian poet. He classified the interaction between species not by their mechanisms in the specimen level (in this case the classification would become boundless), but according to the effect of interaction on the state (in particular, on the size) of both interacting populations. By interpreting Khlebnikov's idea in modern terms we can say that, for instance, if the population growth of one species inhibits the growth of the other, and the growth of the second stimulates the growth of the first, we have the "prey–predator" (or "host–parasite", or "herbivorous–plant") relation. We shall denote the effect of one of the species on another by one of the signs: + (stimulating), − (inhibiting) or 0 (neutral). The full list of pair-wise relations will then consist of the following six types:

++ symbiosis, or mutualism;

−+ prey–predator (and the like);

$- -$ competition;

$+ 0$ commensalism;

$- 0$ amensalism;

$0\ 0$ indifference (there is no relation).

Many times before we said that the system does not affect the environment; therefore, the interaction between them can be considered either as commensalism or as amensalism or as indifference.

If we enumerate all $(n+1)$ species of the ecosystem (including the $(n+1)$th quasi-species "environment"), the overall structure of pair-wise interactions both within the system and between the system and the environment can be described by an $(n+1) \times (n+1)$ sign-matrix of interactions, **S**, where each (i,j)th element $(i \neq j)$ is equal to $+$, $-$ or 0. The sign (or no sign) indicates the qualitative result of the impact of the jth species upon the ith one.

In order to "complete" the description we have to define the diagonal elements of **S**. For this we have to understand how the state of each species (in particular, its size) impacts its own self-growth. Here, there are three possibilities: the state of species can stimulate its growth, it can inhibit growth and at least it can produce a neutral reaction. These three reactions can be denoted by $+$, $-$ and 0, respectively. For the environment we assume the neutral self-interaction. By the same token we have defined the diagonal elements of **S**.

It is obvious that the number of elements of the matrix **S** is equal to $L_S = (n+1)^2$. Since each element can be in one of the three states $(+, -, 0)$ then the total number of different matrices **S** is $n_S = 3^{(n+1)^2} \approx 10^{n^2/2}$. For instance, if the number of species $n = 10$ then $n_S \approx 10^{50}$. This is a huge number! Matrix **S** can be considered as a single word of length L_S written in a three-letter alphabet; then the specific information (per one species) contained in **S** is equal to $I_S = k(n+1)\ln 3 \approx 1.1k(n+1)$. Comparing this expression and the one for $I_1 = k \ln n$ (the information contained in the list of species), we see that $\lim_{n\to\infty} (I_S/I_1) \approx \lim_{n\to\infty} [n/(\ln n)] = \infty$. This implies that if the number of species is growing then the quantity of information which is needed for a qualitative description of the trophic network increases faster than the species diversity.

9.2. Topology of trophic network and qualitative stability

The concept of "qualitative stability" is defined in intuitive rather than formal terms for a system of interacting compartments of any kind. This means that a system (or more correctly, its equilibrium) holds its stability under any of those quantitative variations in the strength of linkages between the compartments, as long as qualitatively the interactions in the system remain unchanged. It is clear that in this case the stability of matrix **S** is to be analysed.

The appeal that this concept has to an ecologist is obvious: while it is always hard to estimate quantitatively the strength of linkages, it is much easier to make a qualitative conclusion about the type of intra- and interspecies relationships for all species. Then from this qualitative knowledge, i.e. from the signs of interactions between each pair of species, it may be possible to speculate about the stability in the whole class of "qualitatively similar" ecosystems. The latter implies that all these ecosystems have the same *S*ign *D*irected *G*raph (SDG) that corresponds to the matrix **S**. As an example, let us consider two **S**-matrices and the corresponding SDGs (Figs. 9.1a,b and 9.2a,b). They describe trophic chains with length 5 that differ from one another only by the following. In the first chain, a self-regulated species belongs to the first trophic level while in the second chain it belongs to the third level.

We can say that the system is qualitatively stable if (Svirezhev and Logofet, 1978; Logofet, 1993)

1. There cannot be any self-stimulated species, and at least one species must be self-limited.
2. There are no relations of competition $(--)$ and mutualism $(++)$.
3. There are no directed loops (or cycles) of length 3 or more in the system.

This is probably the most severe restriction for ecosystems. In particular, it excludes all "omnivory" cases where a predator feeds on two prey species, one of which is also food to another, or, in more general terms, when a predator feeds on more than one trophic level.

4. The system must contain some m "prey–predator" pairs (i.e. 2-cycles) such that the rest $(n - 2m)$ species are self-regulated (i.e. 1-cycle).

Assume that we can extract from the overall graph several sub-graphs that contain only prey–predator relations and isolated vertices (so-called *predation graphs*). If all these predation substructures are qualitatively stable, then the system as a whole will also be qualitatively stable. In order to test this statement it is sufficient to study special topological properties of these substructures. This procedure is called a "colour test".

We colour all self-regulated vortices (for instance, black) keeping others uncoloured (white). We say that "a predation graph passes the colour test if there exist white vertices, any one of which are linked to at least another one white vertex; and if a black vertex is linked to a white one, then it is also linked to at least another one white vertex". If such colouring is impossible then the graph *fails* the test. In this latter case the predation community is qualitatively stable.

Let us consider from this point of view the trophic chains shown in Fig. 9.2a,b. Both of them can be described by predation graphs. In case (a) the colour test is failed since the first

$$\begin{Vmatrix} -1 & 1 & 0 & 0 & 0 \\ -1 & 0 & 1 & 0 & 0 \\ 0 & -1 & 0 & 1 & 0 \\ 0 & 0 & -1 & 0 & 1 \\ 0 & 0 & 0 & -1 & 0 \end{Vmatrix} \qquad \begin{Vmatrix} 0 & 1 & 0 & 0 & 0 \\ -1 & 0 & 1 & 0 & 0 \\ 0 & -1 & -1 & 1 & 0 \\ 0 & 0 & -1 & 0 & 1 \\ 0 & 0 & 0 & -1 & 0 \end{Vmatrix}$$

a) b)

Fig. 9.1. **S**-matrices describing species relations in the system.

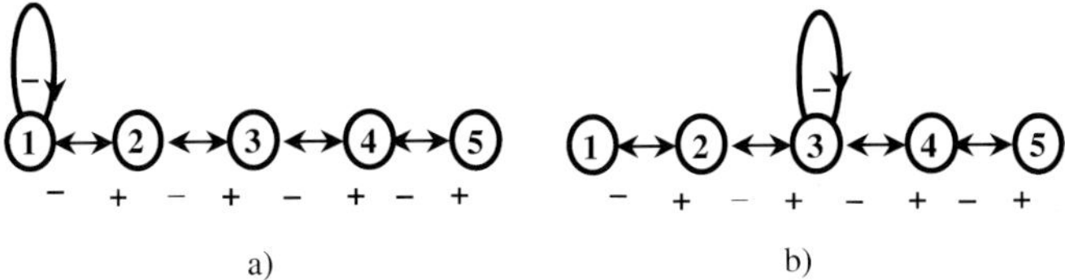

Fig. 9.2. SDG (trophic graphs) corresponding to the sign-matrices in Fig. 9.1.

black vortex is linked to only one white vortex; therefore, this chain is qualitatively stable, i.e. if equilibrium exists then it is always stable for any values of flows between trophic levels. In case (b) the chain is qualitatively unstable since, as is easy to see, it passes the colour test. In other words, any trophic chain self-regulated either with the first or end trophic level (or both) is always stable, no matter what the values of flows between trophic levels.

9.3. Dynamic models of trophic networks and compartmental schemes

A typical example of a graphic representation of a "trophic network" is shown in Fig. 9.3. This is the "flows–storages" diagram (or compartmental scheme) of carbon flows and storages in a bog ecosystem in the temperate zone.

Let x_i be a state variable, i.e. the storage of matter within the ith compartment (we shall consider only matter flows); f_{kj} is the flow from kth to jth compartment; q_i and y_i are the in- and outflow into and out of the ith compartment. Then the general "flows–storages" diagram looks like the one in Fig. 9.4.

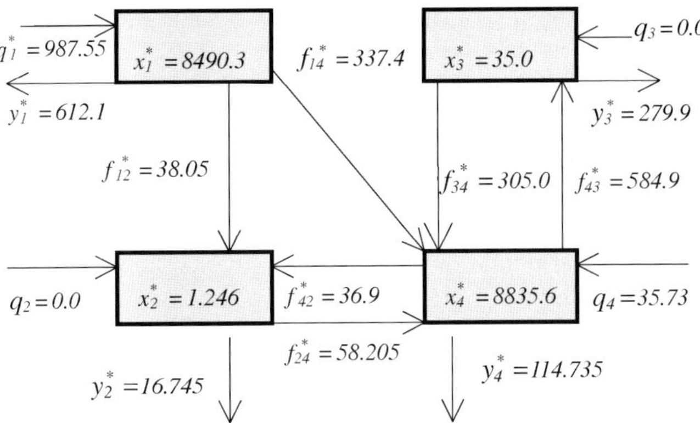

Fig. 9.3. Carbon cycle of a bog in the temperate zone (Zavalishin and Logofet, 2001). Storages in g C/m^2, flows in g C/m^2 year. Storages: x_1—plants, x_2—animals, x_3—bacteria and fungi, x_4—dying-off biomass, excepting peat. Inflows: q_1—assimilation and photosynthesis, q_4—inflow with precipitation and from other ecosystems. Outflows: y_1—plant metabolism and consumption by phytophagous of other ecosystems, y_2—animal metabolism, y_3—metabolism of bacteria and fungi, y_4—outflow with sink, formation of peat and abiotic oxidation.

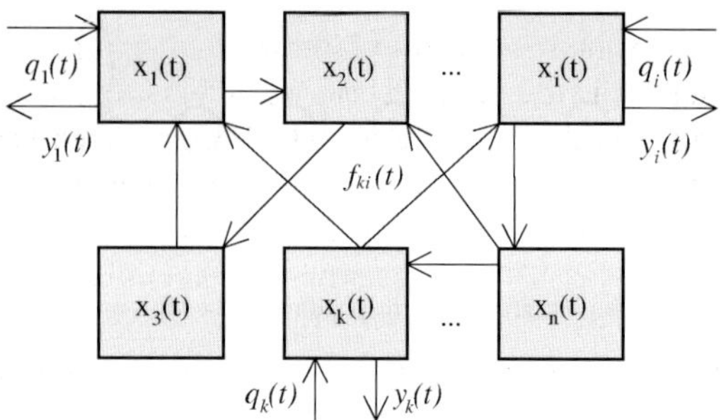

Fig. 9.4. The general "flows–storages" diagram.

In turn, the matter balance is represented in the form of differential equations

$$\frac{dx_i}{dt} = q_i - y_i + \sum_{k \neq i} (f_{ki} - f_{ik}); \quad i, k = 1, \ldots, n, \tag{3.1}$$

or, by introducing the vector notations

$$\frac{d\mathbf{x}}{dt} = \mathbf{q}(\mathbf{x}) - \mathbf{y}(\mathbf{x}) + \mathbf{f}(\mathbf{x}) \tag{3.2}$$

where $\mathbf{x} = \{x_1, \ldots, x_n\}$; $\mathbf{y} = \{y_1, \ldots, y_n\}$; $\mathbf{q} = \{q_1, \ldots, q_n\}$; $\mathbf{f} = \{f_1, \ldots, f_n\}$ and $f_i = \sum_{k \neq i} (f_{ki} - f_{ik})$.

What kind of conformity is there between the system of Eq. (3.1) and the diagrams in Figs. 9.3 and 9.4? What could be said about the general properties of the system, if only a picture with flows and storages is known?

It is necessary to note that as a rule we deal with a static picture, a "snap-shot" of a dynamical system; therefore, we implicitly assume that the system is at steady state. It is obvious in this case

$$\mathbf{q}^* + \mathbf{f}^* = \mathbf{y}^*, \tag{3.3}$$

i.e. flows must be balanced.

Considering the environment as "zero" compartment we can denote $q_i = f_{0i}$, $y_i = f_{i0}$. By also setting $f_{ii} = 0$, Eq. (3.1) is written as

$$\frac{dx_i}{dt} = Q_i^{\text{in}} - Q_i^{\text{out}}, \quad i = 1, \ldots, n \tag{3.4}$$

where $Q_i^{\text{in}} = \sum_{j=0}^{n} f_{ji}$ is the sum of *all* flows coming into the ith compartment and $Q_i^{\text{out}} = \sum_{j=0}^{n} f_{ij}$ is the sum of *all* flows going out of it. Then the equivalent of Eq. (3.3) will be $(Q_i^{\text{in}})^* = (Q_i^{\text{out}})^*$, $i = 1, \ldots, n$. Clearly the sum $Q = \sum_{i=0}^{n} Q_i^{\text{in}} = \sum_{i=0}^{n} Q_i^{\text{out}} = \sum_{i=0}^{n} \sum_{j=0}^{n} f_{ji}$ is the total flow of matter flowing through the system.

If the "dimension" of the network (number n) is not high then we can try to apply the methods developed by the theory of differential equations (if, certainly, we know how to construct these equations by using the flows diagram, but more about this later on). However, what do we do if the dimension is very high? One idea is to apply a thermodynamic approach. A direct application of classic thermodynamic concepts to these systems is difficult, since they differ from ensembles of weakly interacting particles (we have already written about it). Nevertheless, we will attempt to do it (see Section 9.4).

Note that the information theory concepts, which do not significantly depend on the number of parts (compartments) constituting the system and the type of interaction between them, could be applied to such a sort of systems (see also Chapter 4). One of the most popular information concepts is "organisation"; therefore we may assume that

- The better (in a certain sense) the ecosystem is organised the more effectively it uses the free energy of the environment.

Such a principle is consistent with basic principles of non-equilibrium thermodynamics. Different values, defined on the set of (independent of time) flows and storages, for instance *MacArthur's diversity index*, *trophic diversity*, *ascendancy*, *exergy*, *indirect effects* (Patten, 1995), etc., were suggested as measures of organisation. We shall consider in more or less detail only the first four measures; to get more familiar with the latter concept we recommend the works by Patten (1995) and Fath and Patten (1998). Nevertheless, in Section 9.4 we shall consider an example of how to use a classic thermodynamic approach.

9.4. Ecosystem as a metabolic cycle

There is a very close analogy between an ecosystem and metabolic cycle in an organism. Indeed, the metabolic cycle is an open system of step-by-step reactions, which transform input substrates into some "useful" products in such a way that the original products are regenerated. At each intermediate stage the dissipation of energy contained in molecules of substrates occurs, and also the products of transformation are secreted into the environment. At the same time a new portion of substrates can be input from the environment. It is obvious that the cycle is a "chemical machine" performing the work of transformation of some sorts of matter and energy into another (Fig. 9.5). It occurs, and this is very important, with full or partial regeneration of input substrates and dissipation of internal energy. All this is very similar to the description of an ecosystem. Perhaps there is only one distinction: the autotrophic level of any ecosystem uses directly the energy of solar photons, but this distinction is not significant in this context.

Analysis has shown that entropy increase caused by internal processes in this machine is determined by the change of the total Gibbs potential of the system and its environment

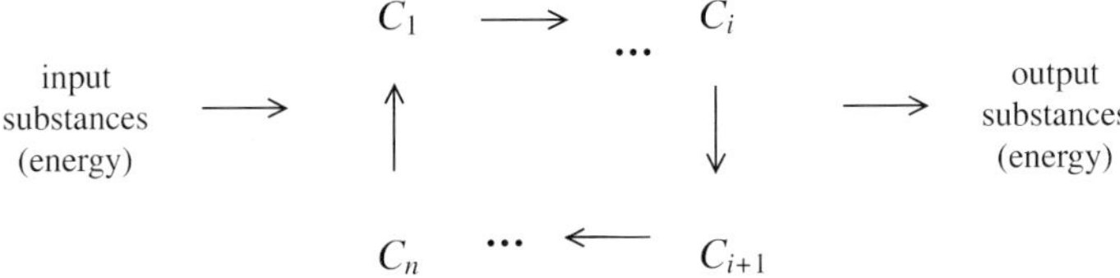

Fig. 9.5. Scheme of abstract metabolic cycle. $C_1, \ldots, C_n$ are intermediate components of a cycle.

taken with opposite sign:

$$Td_iS = -d(G + G_{env}) \tag{4.1}$$

where G and G_{env} are the Gibbs potentials of the system and the environment, the flows of energy and matter provide for the functioning of the machine. Incoming matter passes through the ecosystem along a corresponding trophic chain being stored in compartments and in partial trophic cycles within the system. The turnover time of the general cycle for the whole ecosystem is $\tau = \sum_k \tau_k$, where τ_k is the characteristic time of kth stage. For instance, the residence time of carbon in the kth compartment can be used as τ_k. If there is a "bottle neck" as the slowest stage in the local biogeochemical cycle realised by the ecosystem ($\tau_s \gg \tau_k, k = 1, 2, \ldots; s \neq k$) then $\tau = \tau_s$.

The amount of energy dissipated in the course of a single turnover is defined as

$$\Delta E = T\int_t^{t+\tau} \mathrm{d_i}S = -\int_t^{t+\tau} \mathrm{d}(G + G_{env})$$

$$= -[G(t+\tau) - G(t)] - [G_{env}(t+\tau) - G_{env}(t)]. \tag{4.2}$$

Since after the turnover the ecosystem comes back to the initial state (cyclic process) then $G(t+\tau) = G(t)$, and

$$\Delta E = -[G_{env}(t+\tau) - G_{env}(t)] = -\Delta G_{env}. \tag{4.3}$$

If A is useful work of our cycle and $\Delta_e G_m$ is the change of Gibbs potential of the system caused by the matter exchange between the system and the environment then, in accordance with the First Law, $\Delta E + A = \Delta_e G_m$. From Eqs. (4.2) and (4.3) it follows that after the system has concluded its turnover and returned to the initial state, having done some useful work, the dissipated energy is equal to the decrease of thermodynamic potential (or free energy) of the environment. If the turnover time τ is rather small then the *rate* of energy dissipation, i.e. the function of dissipation will be:

$$\mathrm{Diss} = -\frac{\Delta G_{env}}{\tau} = \frac{\Delta_e G_m}{\tau} - \frac{A}{\tau}. \tag{4.4}$$

Here, A/τ can be interpreted as the *working power* of the system and $\Delta_e G_m/\tau$ as the *power of flows of matter exchange between the system and its environment*; the ratio $\eta = A/\Delta_e G_m$ can be interpreted as the *efficiency coefficient*. The work A can be interpreted as the *exergy* of the system.

These expressions allow us to compare different ecosystems in respect of their energetic efficiency. Indeed, assume that there are two ecosystems with the same values of ΔG_{env} and A: $(\Delta G_{env})_1 = (\Delta G_{env})_2$, $A_1 = A_2$. It is obvious that if $\tau_1 < \tau_2$ then, as follows from Eq. (4.4), $\mathrm{Diss}_1 > \mathrm{Diss}_2$, i.e. the rate of energy dissipation in the first ecosystem is higher than in the second one under the same amount of performed work. Immediately a very attractive teleological evolutionary concept springs to mind (see also Chapter 12):

The evolution of living systems (ecosystems being one sort of them) moves in the direction of decreasing the rate of energy dissipation so that the more perfect systems have lower rates of dissipation.

In the course of evolution the ecosystem structures have become more and more complex: trophic chains became longer, trophic cycles are formed, competition communities become more complex, while the main biochemical mechanisms did not change. As a result the total time τ, which is necessary for the entire completion of the local biogeochemical cycles, or, in other words, the residence time of living biomass within the ecosystem, has to increase. The conclusion is evident: since for the same values of useful work of the total biogeochemical cycle in the ecosystem (the same exergy) and the same exchange flows of matter (between the ecosystem and its environment), so that more complex ecosystems have lesser rates of energy dissipation, they have to be more perfect.

Note that there is another way to decrease the rate of dissipation: to increase the ecosystem exergy conserving the time τ.

However, if we formally apply this concept to real ecosystems then grass is less perfect than forest, since the residence time of the first ecosystem is less than that of the second. Moreover, since *Homo sapiens* possesses the highest exergy then the most perfect ecosystem is an ecosystem with its prevalence. This is nonsense, since both grass and forest and other ecosystems are coexisting on our planet. The point is that we consider the concept to be universal, though it is not necessary that the high rate of dissipation means imperfection of the system in general. The evolutionary way is a movement to some target, but the target is attained under different constraints, so that the evolutionary perfection is most likely to be a compromise between different tendencies and does not need to match the absolute extremum of some universal criterion. In addition note that, of course, only stable systems exist, but there is no direct correspondence between stability and complexity (see Section 6.12, Chapter 6).

9.5. MacArthur's diversity index, trophic diversity and ascendancy as measures of organisation

By analogy with the species diversity index $D = -\sum_{i=1}^{n} p_i \ln p_i$ where p_i are the frequencies of species, MacArthur (1955) has suggested using the so-called "flows diversity index"

$$D_{\mathrm{F}} = -\sum_{s} p(s) \ln p(s) \tag{5.1}$$

where $p(s)$ is a probability of energy transfer along some sth path inside a system; the summation is effected over the complete set of paths, so that $\sum_s p(s) = 1$. The simplest way to calculate these probabilities is the following. First, bearing in mind that flows of matter and energy are often expressed in the same energy units, we can consider matter flows as energy ones. Secondly, as follows from Eq. (3.3)

$$f = \sum_{i=1}^{n} \sum_{j=1, j \neq i}^{n} f_{ij} = \sum_{i=1}^{n} (q_i - y_i).$$

Then $p(s) = p_{ij} = f_{ij}/f$, where s runs over all the indices i and j, and

$$D_F = -\sum_{i=1}^{n} \sum_{j=1, j \neq i}^{n} p_{ij} \ln p_{ij}. \tag{5.2}$$

It is obvious that the measure D_F is maximal, max $D_F = 2 \ln n$ when there is no hierarchy inside the network: all $f_{ij} = f/n^2$, i.e. the total through-flow is homogenously "spread" over the network. We can say that in this case the network is "fully disorganised".

Let us consider the "partially organised" network when flows are directed from compartments with lesser numbers towards compartments with larger ones: $f_{ij} \equiv 0$ if $i \geq j$ (Fig. 9.6a). We also assume that the total through-flow is also homogenously spread over all non-zero internal flows. Then $f_{ij}(i < j) = 2f/n(n-1)$ and max $D'_F = \ln n + \ln(n-1) - \ln 2$. It is obvious that max $D'_F <$ max D_F $(n \geq 2)$. The limit case of such type of organisation is a trophic chain (Fig. 9.6b). It is easy to see that the number of internal flows is $(n-1)$, and max $D''_F = \ln(n-1) <$ max $D'_F <$ max D_F.

One can see that the MacArthur diversity index decreases with increase of ordering (organisation) of the network.

Another generalisation of the concept of diversity is *trophic diversity* or *trophic diversity index* (Svirezhev and Logofet, 1977). Suppose we have a community with its trophic structure represented by a trophic graph. If we take into account the input of the external resource to the system (for instance, the solar energy for the producing species), we obtain additional links between the resource and these species. Graphs of this kind, much simplified as compared with the graphs of real communities, but still keeping the principal features of these structures, are exemplified in Fig. 9.7. The graphs represent the networks shown in Fig. 9.6.

In each graph one may find linear structures, known as trophic chains. In a chain every species is connected with only two other species (the preceding and the succeeding ones), or only with the preceding species, if the chain terminates in the given species. In a sufficiently complex community one and the same species may belong to several chains. For instance, the community illustrated in Fig. 9.7a displays three chains: $R \to 1 \to 4$, $R \to 1 \to 2 \to 4$ and $R \to 1 \to 2 \to 3$. Species 2 and 4 belong to two chains.

Define the number I_i^j—the *trophic index* of the ith species in the jth chain—as the overall number of links between all the vertices in the chain, preceding the ith one (beginning from vertex R up to vertex i). For instance, the index of species 4 in the chain $R \to 1 \to 2 \to 4$ equals 3, whereas the index of species 1 in the same chain is 1

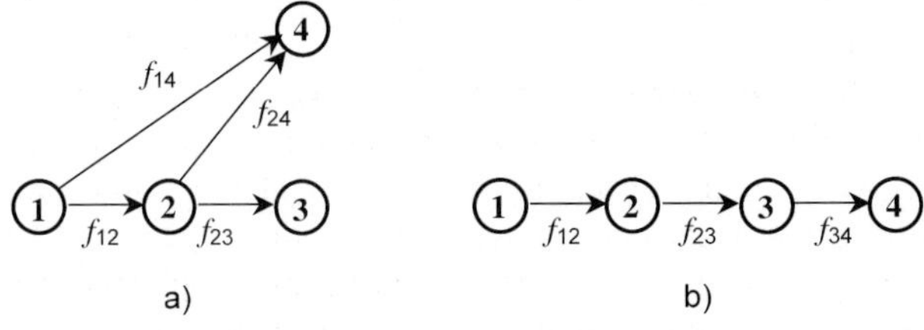

Fig. 9.6. Partially ordered trophic networks.

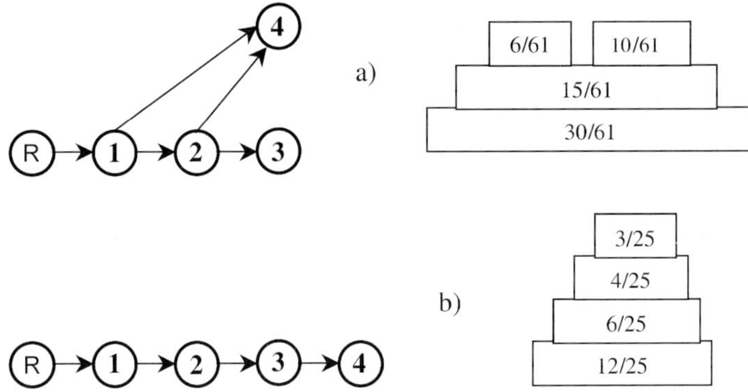

Fig. 9.7. Trophic graphs and the appropriate trophic pyramids maximising the trophic diversity.

(cf. Fig. 9.7a). (The vertices preceding the given one are those which may be reached when moving along the trophic chain in the direction opposite to that of the links.) If a certain ith species belongs to several trophic chains, then the general *trophic index* for this species is $I_i = \sum_j I_i^j$, where the sum is taken over all the chains involving this ith species. For instance, in Fig. 9.7a $I_4 = 5$, while $I_3 = 3$.

If now the species population in the network is N_i, and its frequency is $p_i = N_i / \sum_i N_i$, then the *complexity* of the network may be defined as $I = \sum_i p_i I_i$. The notion of the *trophic frequency* of ith species is defined in terms of complexity and the trophic index:

$$\varphi_i = \frac{p_i I_i}{I}, \quad \sum_i \varphi_i = 1, \quad \varphi_i \geq 0. \tag{5.3}$$

This notion takes into account not only the abundance of a given species, but also its role in the trophic structure and in the community hierarchy. Finally, by analogy with the classic definition for the species diversity, we shall define the *trophic diversity* as

$$D_{\mathrm{T}} = -\sum_{i=1}^{n} \varphi_i \ln \varphi_i. \tag{5.4}$$

We may consider this value as a measure of organisation of the ecosystem. If we now postulate that the trophic diversity is maximal at the equilibrium state, what are the equilibrium compositions we have to obtain? Examples of graphs from Fig. 9.7a,b are used to illustrate this. Clearly, the $\max_{\varphi_i} D_{\mathrm{T}}$ is attained for $\varphi_1 = \varphi_2 = \varphi_3 = \varphi_4 = 1/4$. Calculate the trophic indices for all the four species in the two cases: (a) $I_1 = 1$, $I_2 = 2$, $I_3 = 3$, $I_4 = 5$; (b) $I_1 = 1$, $I_2 = 2$, $I_3 = 3$, $I_4 = 4$. Next, by formula (5.3) one may easily find the optimal values of p_i^*. They are depicted as diagrams in Fig. 9.7 where the area of rectangles is in proportion to p_i^*. We see that the resulting figures resemble the pyramids of populations or biomasses really observed in most natural ecosystems. Hence, it is hoped that this description also reflects some of the really existing regularities, and that it may be useful for prime analysis of ecosystem properties such as complexity and diversity. But it is obviously insufficient for stability analysis, just as the community diversity alone clearly is not sufficient to characterise such a dynamic ecosystem property as stability.

Apparently, from the point of view of information theory, *ascendancy* is one of the most justified measures (Ulanowicz, 1986). The *ascendancy* concept considers an ecosystem as some complex transformer, which transforms the information associated with inflows into the information leaving the system with outflow. The measure of organisation in this case is the mean mutual information of two random values:

1. The hit of some hypothetical particle of through-flowing matter into ith compartment.
2. The outcome of particle out of jth compartment.

Assume that the probability of the first event is proportional to the fraction of incoming flows (including the exchange with environment) with respect to the total through-flows in the system. If $\{\xi\}$ is the set of these events then $\Pr(\xi_i) = Q_i^{\text{in}}/Q$. Analogously, if $\{\eta\}$ is the set of the second events then $\Pr(\eta_j) = Q_j^{\text{out}}/Q$. The probability for events from the sets $\{\xi\}$ and $\{\eta\}$ to occur simultaneously is naturally defined by the fraction of f_{ij}—flow in respect to the total through-flow, $\Pr(\xi_i, \eta_j) = f_{ij}/Q$. Then the entropy of a two-dimensional distribution of this probability is naturally defined as a *measure of organisation* (*ascendancy*):

$$\text{As} = \frac{1}{Q}\sum_{i=0}^{n}\sum_{j=0}^{n} f_{ij} \ln \frac{f_{ij}Q}{Q_i^{\text{in}} Q_j^{\text{out}}}. \tag{5.5}$$

The value of As characterises the degree of complexity of a "flows–storages" diagram, since it has a minimal value As = 0 if and only if every compartment connects with others and the environment by equal bilateral flows (the system is maximally disordered), and the ascendancy is maximal when the diagram is a closed ring of sequentially linked compartments. The latter is considered as an example of maximal order (Ulanowicz, 1995). Comparing Ulanowich's measure, As, with MacArthur diversity index, D_F, we see that despite their formal similarity they fundamentally differ from each other: if the ascendancy increases with the growth of organisation then, vice versa, the MacArthur index decreases.

Ascendancy was estimated for many systems, in particular, for the Cone Spring ecosystem (Ulanowicz, 1995), but since these diagrams resulted from single measurements it is difficult to say something about the degree of organisation of these systems (except a trivial confirmation of their diagrammatic tangle). Nevertheless, Aoki (1995, 1998) has shown that there is a positive correlation between the value of ascendancy and the degree of eutrophication for several lakes in Japan. So, we can say today that the extremity of ascendancy is not yet confirmed by reliable observations, although it is implicitly implied in many ecological works. Developing his method, Ulanowicz (1998) assumed that the increase of ascendancy in the course of evolution is a *tendency* which could be masked and even not realised, since the function of ascendancy is not a strictly maximised goal function.

Note that definition (5.5) has two significant defects:

1. it does not take into account storages,
2. it is not dynamic, i.e. it operates only with a static picture.

Although there are a lot of attempts to overcome these defects, the concept of dynamic organisation is still not developed. Ulanowicz and Abarca-Arenas (1997) have attempted to introduce the storages x_i into the definition of ascendancy. They defined the probabilities $\Pr(\xi_i)$, $\Pr(\eta_j)$ and $\Pr(\xi_i, \eta_j)$ as $\Pr(\xi_i) = x_i/x$, $\Pr(\eta_j) = x_j/x$ and $\Pr(\xi_i, \eta_j) = f_{ij}/Q$ where $x = \sum_{i=0}^{n} x_i$ is the total biomass of all compartments including the environment. Then

$$\mathrm{As} = \frac{1}{Q} \sum_{i=0}^{n} \sum_{j=0}^{n} f_{ij} \ln \frac{f_{ij} x^2}{x_i x_j Q}. \tag{5.6}$$

Unfortunately, this definition contradicts the general properties of probability, since the average probability $\Pr(\xi_i) = \sum_{j=0}^{n} \Pr(\xi_i, \eta_j) = Q_i^{\mathrm{in}} \neq x_i/x$; analogously for $\Pr(\eta_j)$. In addition, we have to introduce some fictive biomass of environment; for this we have to assume (without any reasonable arguments) that the ratio of the total exchange flow to the environment storage is equal to the same ratio of the total internal flow to the total storage of the system.

However, the main problem is how to formulate functional dependences between the change of storages and flows. Of course, we could do it with the help of differential equations, but immediately the problem arises: how to define the ascendancy in this case, and how to define the corresponding probabilities? This is not trivial, since on the one hand, there is the discrete set of states, and on the other hand, the continuous time. This problem is simpler for a finite time interval (Pahl-Wostl, 1995). If we divide the whole interval by r subintervals and denote the flow from ith to jth compartment at the kth subinterval as f_{ij}^k, then the measure of temporal organisation (the integral ascendancy) can be represented as

$$\widehat{\mathrm{As}} = \frac{1}{\hat{Q}} \sum_{k=1}^{r} \sum_{i=0}^{n} \sum_{j=0}^{n} f_{ij}^k \ln \frac{(f_{ij}^k)^2 \hat{Q}}{f_{ij}^{\cdot} f_{i\cdot}^k f_{\cdot j}^k} \tag{5.7}$$

where the point $\cdot$ denotes the summation over the corresponding index, and $\hat{Q} = \sum_{k=1}^{r} \sum_{i=0}^{n} \sum_{j=0}^{n} f_{ij}^k$ is the integral (over all subintervals) total through-flow.

Note that despite a visible diversity of these definitions they are all based on the basic definition of information contained in two- and three-letter words (see Section 4.3, Chapter 4).

9.6. How exergy helps to organise the ecosystem

There are also a lot of attempts to use the exergy as a measure of organisation. The standard definition of exergy in our case is $\mathrm{Ex} = RT_0[xK + x\ln(x/x^{\mathrm{eq}}) - (x - x^{\mathrm{eq}})]$ where $x = \sum_{i=1}^{n} x_i$, x_i are the storages, $K = \sum_{i=1}^{n} p_i \ln(p_i/p_i^{\mathrm{eq}})$ is Kullback's measure and $p_i = x_i/x$, $p_i^{\mathrm{eq}} = x_i^{\mathrm{eq}}/x^{\mathrm{eq}}$. It is postulated that the ecosystem evolves striving to increase its own exergy, which attains its maximum in a steady state. The main problem when we want to test the statement is how to estimate the equilibrium values x_i^{eq}: in fact, we observe the state, which is far from thermodynamic equilibrium. For the particular case (lakes in Japan), Aoki (1993) has suggested a witty idea based on the ergodic paradigm: he has replaced the temporal evolution by the spatial one. He used the values of compartmental storages in the least eutrophic lake as a reference state. Calculating the values of exergy for other lakes, he has got the increase of exergy with growth of eutrophication. The degree of eutrophication was estimated by experts. Note, however, that there is an underwater stone here: by giving the origin milestone (oligotrophic lake) we implicitly give the way.

Nevertheless, we think that the exergy concept (especially in its "genetic definition") plays a very important role in the ecosystem evolution organising such a system's property as stability. In order to clarify our point of view we have to keep in mind a very important result in the theory of Lotka–Volterra equations (Svirezhev and Logofet, 1978).

If we assume that the dynamics of the system are described by Lotka–Volterra equations then

$$\frac{dx_i}{dt} = \varepsilon_i x_i - \sum_{j=1}^{n} \gamma_{ij} x_i x_j, \quad i = 1, \dots, n. \tag{6.1}$$

In studying systems of type (6.1), V. Volterra considered two special classes, conservative and dissipative systems. System (6.1) is said to be *conservative* (in the Volterra sense) if there exists a set of positive numbers $\alpha_1, \dots, \alpha_n$ such that

$$F(x_1, \dots, x_n) = \sum_{i=1}^{n} \sum_{j=1}^{n} \alpha_i \gamma_{ij} x_i x_j \equiv 0. \tag{6.2}$$

If the quadratic form $F(x_1, \dots, x_n)$ is positive definite and the system is called *dissipative*. If the quantity $\text{Ex} = \sum_{i=1}^{n} \alpha_i x_i$ is interpreted as the total weighted biomass of the system then the meaning of the above-given definitions is that interactions between compartments, given by the matrix of γ_{ij}, do not affect the course of the total weighted biomass in the conservative case, but do hamper biomass increase in the dissipative case. Using the Lyapunov method, we can prove (Svirezhev and Logofet, 1978) that the equilibrium $\mathbf{N}^*$ is locally Lyapunov stable (but not asymptotically) in a conservative system, and globally (in the entire positive orthant) asymptotically stable in a dissipative system (see Chapter 6). But if one or several co-ordinates of the model's equilibrium become negative, the corresponding species disappear from the system. Thus, the condition of existence and Lyapunov stability of a feasible equilibrium $\mathbf{N}^*$ (the equilibrium where all the co-ordinates are positive) is not only a sufficient condition, but also a necessary one for *ecological* stability in both conservative and dissipative systems. Ecological (or Lagrange) stability is interpreted as a boundedness of solutions from below and above.

However, the definition of dissipative (in Volterra's sense) systems does not give us their *characterisation*, i.e. some testing criterion, and a search of such $\alpha_1, \dots, \alpha_n$ is a serious problem. Redheffer (1985a) named it as the *problem of Volterra's multipliers*. The problem arises in different fields, when a sign-definite quadratic form as Lyapunov's function is sought. Redheffer (1985b) has proved the theorem about the reduction of an n-dimensional case to a three-dimensional one, which already has a well-known characterisation. He has suggested a computational algorithm for such a kind of reduction.

Our experience in an ecological modelling (in particular, a modelling of multispecies fish pond ecosystems; Svirezhev et al., 1984) has shown that very often, when we used Lotka–Volterra equations and, after calibration (as a result we obtained an estimation of the coefficients γ_{ij}), we tried to use the function of dissipation Diss $= \sum_{i=1}^{n} \sum_{j=1} \gamma_{ij} N_i N_j$ as Lyapunov's function, or to apply Prigogine's theorem using the expression Diss as the rate of internal entropy production, then our attempts were often unsuccessful. This quadratic form was not sign-definite. However, a crazy idea comes to mind: what if we consider the specific "genetic" exergy $\text{ex}_1, \dots, \text{ex}_n$ as Volterra's multipliers $\alpha_1, \dots, \alpha_n$? We have tested the idea in the two particular models mentioned above and have got positive results. Nevertheless, this is only a speculative hypothesis for now, which needs further testing. If the hypothesis would be confirmed then we can say that Volterra's multipliers succeed to obtain from profound "genetic" reasons. This means that the conceptual construction of the model, in general, responds to the paradigm of evolution towards a global stable equilibrium.

So, we see that the different extreme principles and measures of organisation have their own place in the analysis of static compartmental schemes, but all attempts to do a step towards dynamic models and dynamic organisation are at present "not very successive". But the situation is not desperate; for one partial but rather wide class of compartmental models we know how to construct dynamic models by using only static compartmental diagrams (Logofet, 1997; Svirezhev, 1997c; Zavalishin and Logofet, 1997).

9.7. Some dynamic properties of trophic networks

We assume that flows between two compartments depend only on their storages, $f_{ki} = f_{ki}(x_k, x_i)$ and $f_{ik} = f_{ik}(x_i, x_k)$; inflows q_i are either constants or explicitly dependent on time; outflows are proportional to storages, so that $y_i = m_i x_i$. We also assume that every intercompartmental flow belongs to one of these three types: (1) *donor* type, $f_{ki} = \alpha_{ki} x_k$; (2) *recipient* type, $f_{ki} = \beta_{ki} x_k$; (3) *Volterra's "prey–predator"* type, $f_{ki} = \gamma_{ki} x_k x_i$. The latter assumption allows us to express the values of these coefficients as functions of flows and storages at the equilibrium: $\alpha_{ki} = f_{ki}^*/x_k^*$, $\beta_{ki} = f_{ki}^*/x_i^*$, $\gamma_{ki} = f_{ki}^*/x_k^* x_i^*$ and $m_i = y_i^*/x_i^*$. Note that we can consider the expression for input flow in more general form, $q_i' = q_i + \delta_i x_i$. However, we can include the term $\delta_i x_i$ into the expression for output flow, $y_i' = (m_i - \delta_i) x_i$, and by the same token reduce the problem to the preliminary formulation. A dynamic system corresponding to a given diagram is written as

$$\frac{d\mathbf{x}}{dt} = \mathbf{q} + \mathbf{B}\mathbf{x} + \text{diag}\{\mathbf{x}\}\mathbf{C}\mathbf{x} \tag{7.1}$$

where matrices **B** and **C** are the matrices with elements

$$\|B_{ij}\| = \begin{cases} \alpha_{ji} - \beta_{ij} & \text{for } i \neq j, \\ -m_i + \sum_{j \neq i} (\beta_{ji} - \alpha_{ij}) & \text{for } i = j. \end{cases} \qquad \|C_{ij}\| = \begin{cases} \gamma_{ji} - \gamma_{ij} & \text{for } i \neq j, \\ 0 & \text{for } i = j \end{cases} \tag{7.2}$$

and diag{**x**} is the diagonal matrix with elements $x_1, \dots, x_n$. Of course, there is a certain degree of arbitrariness in the construction of these equations: types of intercompartmental flows are given proceeding from ecological expert arguments, but, on the other hand, this arbitrariness brings a certain flexibility into the constructive algorithm. Note that if the total inflow is non-negative, and there is at least a single strictly positive coefficient among non-negative coefficients m_i then all the solutions in Eq. (7.1) will be bounded, i.e. they can always be interpreted from the ecological point of view.

System (7.1) linearised in the vicinity of equilibrium $\mathbf{x}^*$ is written as $\mathrm{d}\Delta\mathbf{x}/\mathrm{d}t = \mathbf{A}\Delta\mathbf{x}$ where $\Delta\mathbf{x} = \mathbf{x} - \mathbf{x}^*$ and $\mathbf{A}$ is the so-called Jacobi's matrix. Knowing it, we can judge about stability of the system, its sensitivity in relation to variations of inflows and, generally speaking, about some general dynamic properties of trophic networks.

An elementary "particle" of any network is the pair of compartments, ith and jth, and its "sort" is defined by characteristics of the flows linking them. It is obvious that the ith compartment may be a donor in relation to the jth one, and then the flow $\vec{ij}$ is a donor type, whereas the jth compartment may be a recipient in relation to the jth one, hence the flow $\vec{ji}$ belongs to a recipient type, etc. In this case, a sort of $\{i,j\}$-pair is defined as a "donor–recipient" (DR). Analogously, other sorts of $\{i,j\}$-pair are defined as the following: "donor–donor" (DD), "donor–Volterra's type" (DV), "recipient–donor" (RD), "recipient–recipient" (RR), "recipient–Volterra's type" (RV), "Volterra's type–donor" (VD), "Volterra's type–recipient" (VR) and "Volterra's type–Volterra's type" (VV). Naturally, the elements of Jacobi's matrix depend on the "sort" of selected pair.

Let $i \neq j$. Then the values of non-diagonal elements are determined as:

$$
\begin{aligned}
&(\mathrm{DD}): A_{ij} = \frac{f_{ji}}{x_j^*};\ (\mathrm{DR}): A_{ij} = 0;\ (\mathrm{DV}): A_{ij} = \frac{f_{ji}}{x_j^*};\\
&(\mathrm{RD}): A_{ij} = \alpha_{ji} - \beta_{ij} = \frac{f_{ji} - f_{ij}}{x_j^*};\ (\mathrm{RR}): A_{ij} = -\beta_{ij} = -\frac{f_{ij}}{x_j^*};\\
&(\mathrm{RV}): A_{ij} = -\beta_{ij} + \gamma_{ji}x_i^* = \frac{f_{ji} - f_{ij}}{x_j^*};\\
&(\mathrm{VD}): A_{ij} = \alpha_{ji} - \gamma_{ij}x_i^* = \frac{f_{ji} - f_{ij}}{x_j^*};\ (\mathrm{VR}): A_{ij} = -\gamma_{ij}x_i^* = \frac{-f_{ij}}{x_j^*};\\
&(\mathrm{VV}): A_{ij} = x_i^*(\gamma_{ji} - \gamma_{ij}) = \frac{f_{ji} - f_{ij}}{x_j^*}.
\end{aligned}
\tag{7.3a}
$$

Let $i = j$. Then $A_{ii} = -m_i + \sum_{k\neq i}(\beta_{ki} - \alpha_{ik}) + \sum_{k\neq i}(\gamma_{ki} - \gamma_{ik})x_k^* = -m_i + \sum_{k\neq i} P_{ik}$ where the items P_{ik} are determined as:

$$
\begin{aligned}
&(\mathrm{DD}): P_{ik} = -\frac{f_{ik}}{x_i^*};\ (\mathrm{DR}): P_{ik} = \frac{f_{ki} - f_{ik}}{x_i^*};\ (\mathrm{DV}): P_{ik} = \frac{f_{ki} - f_{ik}}{x_i^*}.\\
&(\mathrm{RD}): P_{ik} = 0;\ (\mathrm{RR}): P_{ik} = \frac{f_{ki}}{x_i^*};\ (\mathrm{RV}): P_{ik} = \frac{f_{ki}}{x_i^*};\\
&(\mathrm{VD}): P_{ik} = \frac{-f_{ik}}{x_i^*};\ (\mathrm{VR}): P_{ik} = \frac{f_{ki} - f_{ik}}{x_i^*};\ (\mathrm{VV}): P_{ik} = \frac{f_{ki} - f_{ik}}{x_i^*}.
\end{aligned}
\tag{7.3b}
$$

Generally speaking, the Jacobi matrix can be represented as $\mathbf{A} = (\mathbf{M} - \mathbf{Q})\mathrm{diag}\{1/\mathbf{x}^*\}$ where $\mathbf{M}$ is a matrix depending on the equilibrium flow matrix $\mathbf{F}^* = \|f_{ij}^*\|$, $\mathbf{Q}$ and $\mathrm{diag}\{1/\mathbf{x}^*\}$ are diagonal matrices with elements $q_1, \ldots, q_n$ and $1/x_1^*, \ldots, 1/x_n^*$.

Each non-diagonal element in $\mathbf{M}$ can accept one of the values: 0, f_{ji}^*, $-f_{ij}^*$, $f_{ji}^* - f_{ji}^*$, and the diagonal elements are combinations (rather complex) of different flows. In particular, if all intercompartmental flows belong to Volterra-type then $\mathbf{M} = (\mathbf{F}^*)^{\mathrm{T}} - \mathbf{F}^*$ is an anti-symmetric matrix and the Jacobi matrix is represented as

$$\|A_{ij}\| = \begin{cases} \dfrac{f_{ji}^* - f_{ij}^*}{x_j^*} & \text{for } i \neq j, \\ -q_i/x_i^* & \text{for } i = j. \end{cases} \tag{7.4}$$

It is known that in this case the matrix $\mathbf{A}$ is semi-stable (in the sense $\mathrm{Re}\lambda(\mathbf{A}) \leq 0$), but it can always be made stable by a small "movement" of parameters. In fact, we have to assume that any observed equilibrium is stable since it exists. On the other hand, calculating the Jacobi matrix by our observations, we can obtain that some $\mathrm{Re}\lambda(\mathbf{A})$ will be positive, i.e. the corresponding equilibrium will be unstable. Thus, there is a contradiction between the observability of this system and its instability (or more correctly, the instability of its equilibrium). How to resolve this contradiction? What kind of incorrectness could generate it? Firstly, there is some non-coincidence between real and observed values of storages and flows. Secondly, we assume that the system is in equilibrium, while it is not true. And thirdly, we have wrongly defined the system structure, i.e. the number of variables, type of flows, etc. Each of these reasons separately and all these together can provoke this contradiction. Unfortunately, we cannot say what the causal reasons for this contradiction are if we have only the information about storages and flows. At least, we can formulate the following statement: *if the matrix* $\mathbf{A}$ *is unstable then the observed equilibrium cannot exist.* The latter gives us a good method for the selection of observed data.

How to estimate the system response in relation to change in the inflow? We assume that the vector $\mathbf{q}$ changes slowly, quasi-stationary, so that the system is remaining at the equilibrium manifold $\mathbf{x}^* = \mathbf{x}^*(\mathbf{q})$. Then, by applying to the equality (3.3) the implicit function theorem, we obtain

$$\delta\mathbf{q} + \left\|\left(\frac{\partial \mathbf{f}}{\partial \mathbf{x}}\right)^*\right\|\delta\mathbf{x}^* = \left\|\left(\frac{\partial \mathbf{y}}{\partial \mathbf{x}}\right)^*\right\|\delta\mathbf{x}^* \tag{7.5}$$

where $\delta\mathbf{x}^*$ and $\delta\mathbf{q}$ are variations of equilibrium and inflows. By assuming that the inflows do not depend on state variables and taking into account that $\|(\partial\mathbf{f}/\partial\mathbf{x})^*\| - \|(\partial\mathbf{y}/\partial\mathbf{x})^*\|$ is nothing more or less than the Jacobi matrix $\mathbf{A}$, we can rewrite Eq. (7.5) as $\delta\mathbf{q} + \mathbf{A}\delta\mathbf{x}^* = 0$, whence (if the reciprocal matrix $\mathbf{A}^{-1}$ exists)

$$\delta\mathbf{x}^* = -\mathbf{A}^{-1}\delta\mathbf{q}. \tag{7.6}$$

On the other hand, since $\mathbf{x}^* = \mathbf{x}^*(\mathbf{q})$ then $\delta\mathbf{x}^* = \|\delta\mathbf{x}^*/\delta\mathbf{q}\|\delta\mathbf{q}$. In accordance with the standard definition the matrix $\mathbf{S} = \|\delta\mathbf{x}^*/\delta\mathbf{q}\|$ is a *sensitivity* matrix. Comparing it to Eq. (7.6) we see that $\mathbf{S} = -\mathbf{A}^{-1}$ and $\delta\mathbf{x}^* = \mathbf{S}\delta\mathbf{q}$.

Generally speaking, we can define arbitrary norms for the vectors $\delta\mathbf{x}^*$ and $\delta\mathbf{q}$, and also for the matrix $\mathbf{S}$: $|\delta\mathbf{x}^*|$, $|\delta\mathbf{q}|$ and $|\mathbf{S}|$, then $|\delta\mathbf{x}^*| = |\mathbf{S}\delta\mathbf{q}| \leq |\mathbf{S}|\cdot|\delta\mathbf{q}|$. If we assume $|\mathbf{S}| < 1$, i.e. all eigenvalues of $|\mathbf{S}|$, μ_i, must be situated inside a unit circle $|\mu| = 1$, then

$$|\delta\mathbf{x}^*| < |\delta\mathbf{q}|. \tag{7.7}$$

The norms $|\delta\mathbf{q}|$ and $|\delta\mathbf{x}^*|$ can be considered as some measures of external perturbations and the system response in relation to these perturbations; and inequality (7.7) as the formulation of *Le Chatelier's principle*: *the systems, organised by the principle, attempt and, very importantly, are capable, to relieve external perturbations*. In our case, if $|S| < 1$ then input perturbations are always relieved.

Let λ_i be the eigenvalues of Jacobi matrix, then $\mu_i = -(1/\lambda_i)$ and the inequality $|S| < 1$ holds if $|\lambda_i| > 1$. On the other hand, the equilibrium must be stable, i.e. $\mathrm{Re}\lambda_i < 0$; then an intersection of these conditions gives the new general condition for fulfilment of the Le Chatelier principle: $\mathrm{Re}\lambda_i < -1$, which is stricter than the previous ones.

It is obvious that the norm of some vector is identified only with its magnitude, but not with its direction. We cannot say the vector is either positive or negative, it is "good" or "bad" until we shall define ourselves what is "good" and what is "bad", but we may connect the definition of either "good" or "bad" vector $\mathbf{z}$ with the sign of scalar product $(\mathbf{z}, \mathbf{e})$ where $\mathbf{e}$ is a unit vector. For instance, if $(\mathbf{z}, \mathbf{e}) > 0$ then the vector $\mathbf{z}$ is "good", and vice versa, if $(\mathbf{z}, \mathbf{e}) < 0$ then the vector $\mathbf{z}$ is "bad". This is a definition *in average*. In fact, the $(\delta\mathbf{q}, \mathbf{e}) = \sum_{i=1}^n \delta q_i = \delta q$ and $(\delta\mathbf{x}^*, \mathbf{e}) = \sum_{i=1}^n \delta x_i^* = \delta x^*$ are the variation of the total inflow and the total shift in equilibrium, caused by the variation. These are the following possible combinations:

1. Increase of the total inflow $\delta q > 0$ causes an increase of the total equilibrium storage $\delta x^* > 0$.
2. Decrease of the total inflow $\delta q < 0$ causes a decrease of the total equilibrium storage $\delta x^* < 0$.
3. Increase of the total inflow $\delta q > 0$ causes a decrease of the total equilibrium storage $\delta x^* < 0$.
4. Decrease of the total inflow $\delta q < 0$ causes an increase of the total equilibrium storage $\delta x^* > 0$.

It is intuitively clear that if the first two reactions could be called *normal*, then reactions 3 and 4 are *paradoxical*.

Since $\delta x_i^* = \sum_{j=1}^n s_{ij}\delta q_j$ where s_{ij} are elements of the matrix $\mathbf{S}$ then the condition $\delta x^* > 0$ is equivalent to $\sum_{i=1}^n s_i\delta q_i > 0$ where $s_i = \sum_{j=1}^n s_{ji}$. It is clear that the planes $\sum_{i=1}^n \delta q_i = 0$ and $\sum_{i=1}^n s_i\delta q_i = 0$, passing through the origin of the co-ordinates, divide the space of the co-ordinates $\{\delta q_1, \ldots, \delta q_n\}$ on two domains (see Fig. 9.8): in the first domain the system has a normal reaction (in this domain the product $\delta q\delta x^*$ is positive), in the second domain the system has a paradoxical reaction (the product $\delta q\delta x^*$ is negative).

9.8. Stability and reactions of a bog in the temperate zone

Let us return to the diagram shown in Fig. 9.3, which is a compartmental scheme of the carbon local cycle of a bog in the temperate zone.

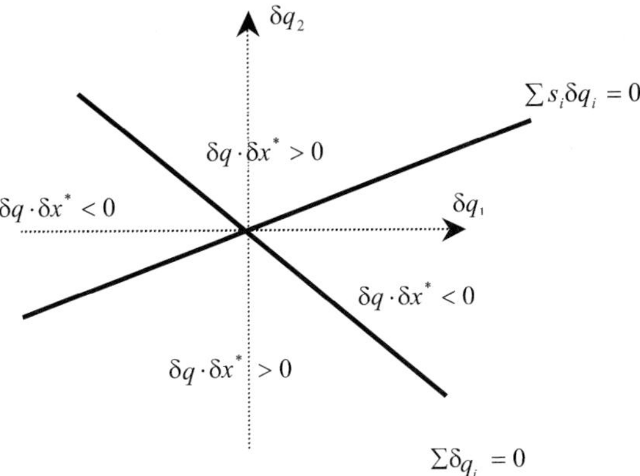

Fig. 9.8. Domains of normal and paradoxical reactions.

The choice of flows type is based on an expert opinion: the rate of the uptake of plant's carbon by animals depends on both storages, so that $f_{12} = \gamma_{12}x_1x_2$; analogously, the flows f_{24} and f_{34} have the donor-type, the flows f_{42} and f_{43} have the Volterra-type. An application of the above-described algorithm results in the non-linear system of differential equations describing the dynamics of the ecosystem (Zavalishin and Logofet, 2001):

$$
\begin{aligned}
\frac{dx_1}{dt} &= q_1 - m_1x_1 - \alpha_{14}x_1 - \gamma_{12}x_1x_2, \\
\frac{dx_2}{dt} &= -m_2x_2 - \alpha_{24}x_2 + (\gamma_{42}x_4 + \gamma_{12}x_1)x_2, \\
\frac{dx_3}{dt} &= -m_3x_3 - \alpha_{34}x_3 + \gamma_{43}x_4x_3, \\
\frac{dx_4}{dt} &= q_4 - m_4x_4 + \alpha_{14}x_1 + \alpha_{34}x_3 + \alpha_{24}x_2 - x_4(\gamma_{43}x_3 + \gamma_{42}x_2).
\end{aligned}
\tag{8.1}
$$

The Jacobi matrix of Eq. (8.1) in equilibrium will be

$$
\mathbf{A} = \begin{Vmatrix}
-\dfrac{q_1}{x_1^*} & \dfrac{-f_{12}^*}{x_2^*} & 0 & 0 \\[2ex]
\dfrac{f_{12}^*}{x_1^*} & 0 & 0 & \dfrac{f_{42}^*}{x_4^*} \\[2ex]
0 & 0 & 0 & \dfrac{f_{43}^*}{x_4^*} \\[2ex]
\dfrac{f_{14}^*}{x_1^*} & \dfrac{f_{24}^* - f_{42}^*}{x_2^*} & \dfrac{f_{34}^* - f_{43}^*}{x_3^*} & -\dfrac{y_4^*}{x_4^*} - \dfrac{f_{43}^* + f_{42}^*}{x_4^*}
\end{Vmatrix},
\tag{8.2}
$$

or, if substituting into Eq. (8.2) the concrete values of flows and storages,

$$\mathbf{A} = \left\| \begin{array}{cccc} -0.1163 & -30.54 & 0.000 & 0.000 \\ 0.0045 & 0.000 & 0.000 & 0.0042 \\ 0.000 & 0.000 & 0.000 & 0.0662 \\ 0.0397 & 17.10 & -7.997 & -0.0834 \end{array} \right\|. \tag{8.3}$$

Its eigenvalues are equal $\lambda_{1,2} = -0.0699 \pm 0.4064\mathrm{i}$, $\lambda_{3,4} = -0.03 \pm 0.6521\mathrm{i}$. Since $\mathrm{Re}\{\lambda_i\} < 0$, then we can say that the observed equilibrium is stable. Note that the knowledge of eigenvalues allows us to evaluate some dynamic characteristics of the system. For instance, the system could be oscillated with the periods of 10 and 16 years, but they are quickly damped (the decrements are equal to 33 and 14 years, correspondingly). Nevertheless, under certain conditions we may expect the appearance of resonance.

The sensitivity matrix $\mathbf{S} = -\mathbf{A}^{-1}$:

$$\mathbf{S} = \left\| \begin{array}{cccc} 0.000 & -223.1 & 14.08 & 0.000 \\ 0.0327 & 0.8499 & -0.0536 & 0.000 \\ 0.070 & 0.7084 & 0.1128 & 0.1250 \\ 0.000 & 0.000 & -15.11 & 0.000 \end{array} \right\|. \tag{8.4}$$

Since all eigenvalues of $\mathbf{S}$ are located outside of the unit circle ($|\lambda_{1,2}|^{-1} \approx 2.4$, $|\lambda_{3,4}|^{-1} \approx 1.53$, whence $|\lambda_i|^{-1} > 1$), then the operator $\mathbf{S}$ is not contracting, and this ecosystem does not satisfy the generalised Le Chatelier principle given by inequality (7.4). This result can be interpreted in the sense that in spite of the fact that the carbon cycle in a bog is stable, and the equilibrium exists and is stable, at the same time no fluctuation of carbon inflows is attenuated and can even be amplified. This shows, for instance, that additional carbon, which has entered the system as a result of inflow fluctuation, is not dissipated by outflows, but it is accumulated inside the system. It leads to the increase of equilibrium storages. It means that the equilibrium state of a bog in the temperate zone is very sensitive in relation to the change of carbon input from external sources. The system does not try to keep its internal state, attenuating the external perturbation, but it changes its state in the direction of perturbation, adapting itself to it. The behaviour of the system may be very complex; resonance, non-linear oscillations and even chaos may appear under certain conditions (Zavalishin and Logofet, 2001).

The considered ecosystem has two real inputs with corresponding inflow q_1 and q_4 (see Fig. 9.3); therefore $\delta q = \delta q_1 + \delta q_4$ and $\delta x^* = s_1 \delta q_1 + s_4 \delta q_4$ where $s_1 = s_{11} + s_{21} + s_{31} + s_{41} \approx 0.1$ and $s_4 = s_{14} + s_{24} + s_{34} + s_{44} \approx 0.13$. In Fig. 9.9 the domains corresponding to normal and paradoxical reactions of a bog in relation to variations of inflows are shown.

One can see that the ecosystem of a bog reacts normally to almost any variations of inflows. Paradoxical reaction is possible if δq_1 and δq_4 have opposite signs.

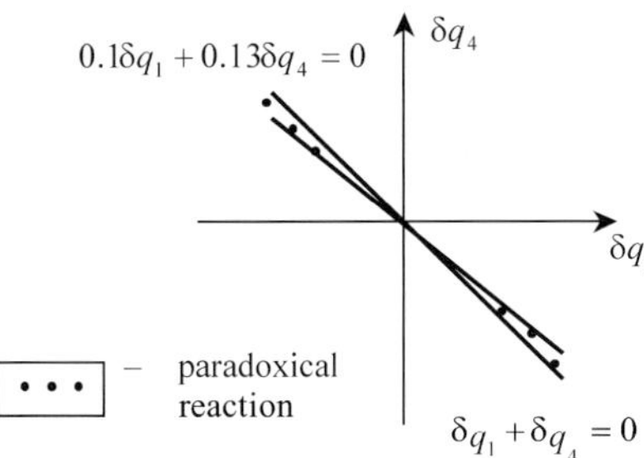

Fig. 9.9. Domains of normal and paradoxical reactions of the ecosystem on variations of inflows.

9.9. Summary of the ecological important issues

A complex ecosystem can be described as an ecological network, indicating how mass, energy and information are transferred from one component to another within the ecosystem. It is interesting that Patten (1991) has shown that the possible interactions between two components (symbiosis, prey–predator, competition, commensalism, amensalism and indifference) all become more beneficial for both components due to the presence of the network that ensures constant cycling of matter, energy and information.

Various measures of the network organisation have been proposed: MacArthur's diversity index, the trophic diversity, ascendancy, exergy storage and indirect effect relative to direct effect. The four latter have been examined in this context and it was demonstrated that they offer different interpretations of network organisation. It is demonstrated that ascendancy is a good measure of the organisation of the network, but the concept has two shortcomings: it is not dynamic and it does not take into account storages. Exergy has the advantage as the measure of ecosystem organisation that it is able to account for storages and genetic information when exergy is calculated by the method presented in Chapter 5. Furthermore it has been shown that ecological networks are able to deal with almost any realistic input in one way or another.

A thermodynamic analysis of bogs has revealed that they show equilibrium storages. Additional carbon entering the system is not dissipated by outflows but accumulates.

Chapter 10
Thermodynamics of vegetation

A greening land, A shining ether...
F. Tyutchev (1865)

On the far planet Venus, Where the gold of Sun is more flaming, Trees have blue leaves...
N. Gumilev

10.1. Introduction. Energetics of photosynthesis

Organic matter in the biosphere has appeared as a result of the activity of autotrophic organisms. They represent a unique group that is able to synthesise organic matter from inorganic. While creating organic matter, they use either the energy of solar radiation (photosynthesis) or chemical energy (chemosynthesis). Photosynthesis, which is the prevailing process, produces the living biomass of vegetation, constituting 99% of the entire biomass of the biosphere. Chemosynthesis plays an important role in the nitrogen cycle and some other processes, but produces very small amounts of organic matter. Therefore, vegetation is a basic component of the biosphere and the main agent of the global biogeochemical cycles, i.e. the functioning of *Biosphera machina*. At the same time, photosynthesis of green plants is the main reason for the existence of all superior forms of life on our planet (and we assign ourselves to this form), since the presence of oxygen in the Earth's atmosphere is a result of photosynthesis.

The basic equation of photosynthesis is

$$CO_2 + H_2O + h\nu \rightarrow (CH_2O) + O_2 + 470\ \mathrm{kJ/mol},$$

where $h\nu$ is a photon and (CH_2O) is a fragment of carbohydrate molecule. As a result of the photoreaction, 470 kJ/mol of energy is released, i.e. the change of enthalpy is $\Delta H = 470\ \mathrm{kJ/mol}$. The change of free energy is equal to $\Delta G = 504\ \mathrm{kJ/mol}$. Since $\Delta G = \Delta H - T\Delta S$, the change of entropy ΔS is equal to $(470 - 504)/293 = -116$ J/K mol (at 20°C). It is necessary to spend eight photons with total energy of about 1470 kJ/mol for the formation of one molecule of O_2. Thus, the coefficient of the solar energy tapping (maximal efficiency coefficient of photosynthesis) is equal to $\eta_{\max} = 504/1470 \approx 0.34$.

Towards a Thermodynamic Theory for Ecological Systems, pp. 243–269

We define the efficiency of photosynthesis in relation to the exergy, i.e. to the value of *useful* work, which can be performed by a photosynthetic system. As a result, we get a significantly lower value for the efficiency coefficient. In this case, exergy is equal to: $\mathrm{Ex} = T|\Delta S|$, where $|\Delta S|$ is the module of entropy change, $|\Delta S| = 116\,\mathrm{J/K\,mol}$, $T = 293$ K and $\mathrm{Ex} = 34\,\mathrm{kJ/mol}$. If we define the "exergetic" efficiency coefficient as $\eta_{\mathrm{ex}} = \mathrm{Ex}/E_{\mathrm{ph}}$, where $E_{\mathrm{ph}} = 1470\,\mathrm{kJ/mol}$ is the energy of photons that are necessary for photoreaction (see above), then $\eta_{\mathrm{ex}} = 0.116 \times 293/1470 \approx 0.023 = 2.3\%$. It is interesting that this value is almost two times lower than "the efficiency coefficient of photosynthesis for the most favourable condition", 5% (Monteith and Unsworth, 1990) and is higher than the efficiency of a "green leaf" machine, 1.68% (Section 5.3). At the same time, the efficiency coefficient for natural vegetation, defined as "the mean ratio of energy expended by the photosynthesis of natural vegetation cover to the incoming solar radiation", usually has an order of magnitude of about 0.5%. What is the reason for the differences in these estimations? This will be explained in Section 10.2.

10.2. Thermodynamic model of a vegetation layer. Fluxes of heat, water vapour and other gases

The leaves of a plant are the organs by means of which the plant assimilates carbon dioxide from the atmosphere. It, in turn, is a "case" from closed cuticle tissue pierced by multiple small holes, so-called "stomata", the size of which can be adapted to the physiological demands of the plant. A very large surface of chloroplasts with grains of chlorophyll is contained in this case. The chloroplast surface communicates with the atmosphere through stomata and intercellular space. Since carbon dioxide can only be assimilated in the soluble form, the chloroplast surface has to be humid. Hence, the humidity in the intercellular space must be very high, and it is usually much higher than the humidity of the atmosphere. Therefore, the CO_2 diffusion from the atmosphere into the leaf with open stomata has to be accompanied by the diffusion of oxygen and water vapour from the leaf into the atmosphere. The latter process, which is called transpiration, is also one of the most important processes that remove metabolic heat. Another important process transporting water vapour into the atmosphere is the evaporation of water from the soil. The sum of transpiration and evaporation is called evapotranspiration. Certainly, both processes are connected with each other: evaporation influences air humidity, and the transpiration rate depends on it. Evapotranspiration maintains the difference of water potential between the atmosphere and soil that is a moving force of the flux of water from roots through leaves into the atmosphere. So, although the transpiration and evaporation are connected, but since only the transpiration affects the production process, we subdivide the system with evapotranspiration as the leading process onto the system (the transpiration and production) and its environment (the evaporation and the humidity of surrounding air).

The flux, whose intensity equals the amount of transpired water per time unit, also transports nitrogen, phosphorus, calcium, magnesium, etc. (in a soluble form), which are necessary (although in small quantities) for the formation of new biomass. This is the best illustration of one evolution principle: *any organ of a living organism is multifunctional.*

In consequence, the functions of vascular plants are far from the optimal. In fact, the expenditure of water resources by vegetation is very wasteful. For instance, the ratio of the increment of dry matter to the amount of evaporated water varies from 1/200 to 1/1000. Although it seems that, at first sight, such a high value of transpiration does not correspond to the physiological demands of the plant, more detailed consideration shows that there are always good reasons for such transpiration: either cooling or CO_2 uptake or transport of water through the plant for the purpose of mineral uptake.

Let us consider one unit of Earth's terrestrial surface with the area A_s, which is covered by a certain type of natural vegetation (e.g. meadow, steppe, forest, etc.). If the vegetation cover is rather dense, then we can consider it as a single leaf (or a continuous photosynthesising media) with area A. The leaf is "packed" along the whole height of vegetation cover d_v, so that the inequality $A > A_s$ is fully possible. Immediately such a concept as the *leaf area index* $L = A/A_s$ arises, so that $A = LA_s$.

Let d_l be the thickness of a leaf. Then the volume V_b occupied by the "wet" biomass of vegetation is equal to $V_b^w = d_l A = d_l L A_s$, and the total "wet" biomass $B_w = \rho_b^w V_b^w = \rho_b^w d_l L A_s$ where ρ_b^w is the density of "wet" biomass.

We assume that the transpiration is realised by some diffusion mechanism consisting of two stages. At the first stage the water vapour diffuses from the wet internal side of a leaf (the chloroplast surface) towards its external side through the stomata and intercellular space. The moving force is the gradient of concentrations of water vapour between the internal side, where the concentration C_i^w is close to the concentration of saturating water vapour C_s^w, when the leaf's temperature is T_l, and the external side with concentration C_e^w. (For instance, if the temperature of leaves is equal to 25°C then $C_s^w \approx 1.8 \times 10^{-2}$ g H_2O/g dry air.) Therefore, the water vapour flux $q_w^{(1)}$ has to be proportional to the gradient of concentration, $q_w^{(1)} \sim (C_s^w - C_e^w)/l_1$, where l_1 is the length of the path, along which the water vapour is moving. It is natural to assume that the length and the leaf thickness have the same order of magnitude. Without loss of generality we can assume that $l_1 \approx d_l$. Since the transport mechanism of water vapour is diffusion, the final expression for the flux will be

$$q_w^{(1)} = \rho \frac{D_w^{(1)}}{l_1}(C_s^w - C_e^w), \tag{2.1}$$

where ρ is the density of dry air and $D_w^{(1)}$ the integral coefficient of diffusion, which naturally depends on the leaf morphology, number and size of stomata, cuticle density, etc.

The second stage of transport is a path from the external surface of the leaf into the atmosphere outside the vegetation layer, where the concentration of water vapour is C^w. The corresponding flux is described as

$$q_w^{(2)} = \rho \frac{D_w^{(2)}}{l_2}(C_e^w - C^w), \tag{2.2}$$

where $D_w^{(2)}$ is the integral diffusion coefficient, and l_2 a characteristic length of the path at the second stage. At this stage diffusion can almost be considered as a turbulent one, and its coefficient depends mainly on the intensity of turbulent exchange. The mean height of vegetation layer, d_v, can be considered as the characteristic length l_2, so that $l_2 = d_v$.

Since the fluxes at the first and second stages flow through different areas (LA_s at the first, A_s at the second stage) and the continuity condition has to be fulfilled, $q_w^{(1)}LA_s = q_w^{(2)}A_s$, or $q_w^{(2)} = Lq_w^{(1)} = q_w$ (L is the leaf area index, for instance, $L \approx 4-5$ for oak forest). The latter is a flux of water vapour through the boundary of vegetation layer with single area ($A_s = 1$). At the first stage of our study we assume that the exchange of heat, water vapour and gases is performed only through the boundary between the vegetation layer and the atmosphere; the exchange between vegetation and soil is neglected (although it exists). The continuity equation allows us to exclude the concentration C_e^w from Eqs. (2.1) and (2.2). As a result we get

$$q_w = \rho \frac{D_w}{l_2}(C_s^w - C^w), \tag{2.3}$$

where the generalised coefficient of diffusion is equal to

$$D_w = \frac{D_w^{(1)} D_w^{(2)}}{D_w^{(1)} + \lambda D_w^{(2)}}, \tag{2.4}$$

where $\lambda = l_1/Ll_2 = d_1/Ld_v$.

If we consider two extreme cases, then formula (2.4) is represented in two different forms:

$$D_w = \frac{D_w^{(1)}}{\lambda} \frac{1}{1+a_1}, \qquad \text{where } a_1 = D_w^{(1)}/\lambda D_w^{(2)} \tag{2.5a}$$

and

$$D_w = D_w^{(2)} \frac{1}{1+a_2}, \qquad \text{where } a_2 = \lambda D_w^{(2)}/D_w^{(1)}. \tag{2.5b}$$

The first case is when $a_1 << 1$, i.e. $\lambda >> D_w^{(1)}/D_w^{(2)}$. Then $D_w/l_2 \approx D_1^{(1)}/l_1$. Therefore, as follows from Eq. (2.3)

$$q_w \approx \rho \frac{D_w^{(1)}}{l_1}(C_s^w - C^w),$$

and if humidity at the external side of the leaf is close to the air humidity, then $q_w \approx q_w^{(1)}$. This means that the total transpiration is mainly determined by the diffusive transport of water vapour within leaves, i.e. the leaf is a "bottleneck" for transpiration. Since the internal diffusion is determined in general by molecular mechanisms, we can set that $D_w^{(1)}$ is the coefficient of the molecular diffusion of water vapour in air, $D_w^{(1)} \approx 0.25\ \text{cm}^2/\text{s}$. The effective area of diffusion depends on the total area of open stomata, which in turn is a function of stomata control. The latter can significantly modify the value of the transpiration rate.

$D_w^{(2)}$ is the coefficient of turbulent diffusion in the surface layer, usually estimated (by the order of magnitude) as $D_w^{(2)} \approx 10^4\ \text{cm}^2/\text{s}$. If there is no wind or the vegetation layer is weakly ventilated, this value is less than $10^4\ \text{cm}^2/\text{s}$ by 1–2 orders of magnitude.

We see that the coefficient of turbulent diffusion is higher by four orders of magnitude than the coefficient of molecular diffusion. Since at standard conditions

$D_w^{(1)}/D_w^{(2)} \approx 0.25 \times 10^{-4}$, the necessary condition of the first extreme case is the realisation of the inequality $l_1 >> 0.25 \times 10^{-4} L l_2 \approx 10^{-4} l_2$. In other words, the vegetation layer has to be thin and consists of plants, which have thick pulpy leaves with small number of stomata. For instance, if $l_2 \approx 100$ cm then the thickness of a leaf $l_1 >> 10^{-2}$ cm.

The necessary condition of the second extreme case is $a_2 << 1$, i.e. $\lambda << D_w^{(1)}/D_w^{(2)}$. In this case $D_w \approx D_1^{(2)}$, i.e. the turbulent diffusion plays the main role in the process of the subsequent transportation of water from the external side of the leaf into the surrounding air. The case could be realised if $l_1 << 10^{-4} l_2$, i.e. vegetation consists of high trees with thin leaves, which have a lot of stomata. Note that the mean thickness of an oak leaf is about 2×10^{-2} cm, so that if $l_2 \approx 10$ m $= 10^3$ cm then $l_1/l_2 \sim 10^{-5}$.

The heat transport from the interior of leaves into the atmosphere is realised by means of molecular diffusion at the first stage with coefficient $D_h^{(1)}$ called the thermal diffusivity of air, and the turbulent diffusion in the second stage. Since $D_h^{(1)} \approx 0.23$ cm^2/s at 25°C, it is natural to assume that $D_h^{(1)} \approx D_w^{(1)}$, and also $D_h^{(2)} \approx D_w^{(2)}$. In this case the diffusion coefficient D_h is the same as before; the characteristic lengths are also the same. Then the turbulent heat flux is equal to

$$q_h = \rho c_p D_w \frac{T_i - T}{l_2}, \tag{2.6}$$

where $c_p = 1$ J/g K is the air heat capacity, T_i and T are the temperatures of the leaf interior and the atmosphere air, respectively. The flux of water vapour connected with evapotranspiration is often called the "latent heat flux". In this case the value q_w is multiplied by the value of the specific enthalpy of water vapour, so that the latent heat flux is expressed in J/cm^2 s. The turbulent heat flux is also called the "sensible heat flux".

The entire transport path of carbon dioxide from the atmosphere to the vegetation layer, then inside the leaf until it reaches chlorophyll cells, can also be divided by two stages, as we were already doing above with water vapour. By analogy, the flux of CO_2 from the atmosphere into the leaf is equal to

$$q_c = \rho \frac{D_c}{l_2}(C^c - C_i^c), \tag{2.7}$$

where C^c and C_i^c are the concentrations of CO_2 (measured in grams of CO_2 per gram of dry air) in the atmosphere outside the vegetation and within the leaf at the level of parenchyma. The generalised coefficient of diffusion is

$$D_c = \frac{D_c^{(1)} D_c^{(2)}}{D_c^{(1)} + \lambda D_c^{(2)}}, \tag{2.8}$$

where $D_c^{(1)}$ is the effective coefficient of CO_2 diffusion through stomata and intercellular space, when the gas moves from the external side of a leaf to the parenchyma, $D_c^{(2)}$ is the coefficient of turbulent diffusion, and $\lambda = l_1/L l_2$. If with respect to $D_c^{(2)}$ we can set $D_c^{(2)} = D_w^{(2)}$, then for $D_c^{(1)}$ we again assume that the transport of CO_2 is determined by its molecular diffusion in air along this path. Therefore, $D_c^{(1)} \approx 0.61 D_w^{(1)} \approx 0.15$ cm^2/s.

The second gaseous flux is a flux of oxygen from the interior of the leaf to the exterior of the vegetation layer. By analogy with the case of carbon dioxide we have

$$q_{\mathrm{ox}} = \rho \frac{D_{\mathrm{ox}}}{l_2}(C_{\mathrm{i}}^{\mathrm{ox}} - C^{\mathrm{ox}}), \tag{2.9}$$

where C^{ox} and $C_{\mathrm{i}}^{\mathrm{ox}}$ are the concentrations of O_2 (measured in grams of O_2 per gram of dry air) in the atmosphere outside the vegetation and within the leaf at the level of parenchyma. The generalised coefficient of diffusion

$$D_{\mathrm{ox}} = \frac{D_{\mathrm{ox}}^{(1)} D_{\mathrm{ox}}^{(2)}}{D_{\mathrm{ox}}^{(1)} + \lambda D_{\mathrm{ox}}^{(2)}}, \tag{2.10}$$

where $D_{\mathrm{ox}}^{(1)}$ is the effective coefficient of oxygen diffusion through stomata and intercellular space when the gas moves from the external side of a leaf to the parenchyma, $D_{\mathrm{ox}}^{(2)}$ is the coefficient of turbulent diffusion, and $\lambda = l_1/Ll_2$. Naturally then $D_{\mathrm{ox}}^{(2)} = D_{\mathrm{w}}^{(2)}$. Since the coefficient of molecular diffusion of oxygen in air $D_{\mathrm{ox}}^{(1)} \approx 0.8D_{\mathrm{w}}^{(1)} \approx 0.2\ \mathrm{cm}^2/\mathrm{s}$,

$$D_{\mathrm{ox}} = \frac{0.8 D_{\mathrm{w}}^{(1)} D_{\mathrm{w}}^{(2)}}{0.8 D_{\mathrm{w}}^{(1)} + \lambda D_{\mathrm{w}}^{(2)}} = \frac{D_{\mathrm{w}}}{0.8 + \lambda D_{\mathrm{w}}}. \tag{2.11}$$

It would be interesting to estimate the relations between these fluxes for vegetation in reality. These estimations help us to answer many questions. For instance, how many grams of water have to evaporate for a plant to assimilate 1 g of carbon from the atmosphere? Or, another question is: how many joules of heat are released directly in the process of assimilation of the atmosphere carbon?

Let us estimate the ratio $q_{\mathrm{w}}/q_{\mathrm{c}}$, which is expressed as grams of water per gram of carbon dioxide. Using Eqs. (2.3), (2.4), (2.7) and (2.8) we get

$$\frac{q_{\mathrm{w}}}{q_{\mathrm{c}}} = \frac{D_{\mathrm{w}}}{D_{\mathrm{c}}} \frac{C_{\mathrm{s}}^{\mathrm{w}} - C^{\mathrm{w}}}{C^{\mathrm{c}} - C_{\mathrm{i}}^{\mathrm{c}}}. \tag{2.12}$$

The ratio of diffusion coefficient can be represented as

$$\frac{D_{\mathrm{w}}}{D_{\mathrm{c}}} = \frac{D_{\mathrm{w}}^{(1)} D_{\mathrm{w}}^{(2)} (D_{\mathrm{c}}^{(1)} + \lambda D_{\mathrm{c}}^{(2)})}{D_{\mathrm{c}}^{(1)} D_{\mathrm{c}}^{(2)} (D_{\mathrm{w}}^{(1)} + \lambda D_{\mathrm{w}}^{(2)})} \approx \frac{1 + 1.64\lambda k}{1 + \lambda k}, \tag{2.13}$$

where $k = D_{\mathrm{w}}^{(2)}/D_{\mathrm{w}}^{(1)} \approx 4 \times 10^{-4}$. Interestingly, even this ratio is contained within the narrow enough interval of 1–1.64, whereas the value of k and λ can vary within rather broad limits. If assuming $l_1 \sim 10^{-2}$ cm, $l_2 \sim 1$ m and $L = 4$, then $\lambda k \sim 1$ and $D_{\mathrm{w}}/D_{\mathrm{c}} \sim 1.32$. Now this estimation is quite satisfactory for the given ratio.

It is known (Budyko, 1977) that on average the difference $C^{\mathrm{c}} - C_{\mathrm{i}}^{\mathrm{c}}$ constitutes about 10% of the CO_2 concentration in the atmosphere, so that $C^{\mathrm{c}} - C_{\mathrm{i}}^{\mathrm{c}} \approx 0.1C^{\mathrm{c}}$, i.e. plants are normally using only a small part of the potential carbon dioxide flux. If we now assume that at temperate latitudes in summer the mean daily temperature is 20°C and the relative humidity is 50%, then $C_{\mathrm{s}}^{\mathrm{w}} - C^{\mathrm{w}} \approx 1.8 \times 10^{-2} - 0.6 \times 10^{-2} = 1.2 \times 10^{-2}$ g H_2O/g dry air. By taking into account $C^{\mathrm{c}} = 0.66 \times 10^{-3}$ g CO_2/g air (it corresponds to 330 ppm) substituting all these values into Eq. (2.12) we finally get $q_{\mathrm{w}}/q_{\mathrm{c}} \approx 240$ g H_2O/g CO_2.

It follows from this that a plant has to evaporate 240 l of water in order to assimilate 1 kg of carbon dioxide, or in order to assimilate 1 kg of carbon it is necessary to evaporate 879 l of water.

In order to answer the second question we have to calculate the rate q_h/q_c. Using formulas (2.6)–(2.8) we get

$$\frac{q_h}{q_c} = c_p \frac{T_l - T}{C^c - C_i^c}. \tag{2.14}$$

Under the same conditions as in the previous case we have $C^c - C_i^c \approx 0.1C^c$, $T_l - T = 5$ K (we assume that the temperature of the leaf interior at the surface of parenchyma is close to the optimal one for photosynthesis: 25°C), $c_p = 1$ J/g K. Then from Eq. (2.14) it follows that $q_h/q_c \approx 76$ kJ/g CO_2. From this it follows that assimilation of 1 g of CO_2 is accompanied by direct dissipation of 76 kJ of heat. If we measure the assimilation in carbon units, then the previous statement could be reformulated as: in order to assimilate 1 g of carbon it is necessary to dissipate 278 kJ of heat.

10.3. Energy balance of a vegetation layer and the energy efficiency coefficient

We have assumed above that the vegetation cover was rather dense so that the transpiration prevails on the evaporation from a bare soil. In this case the equation of energy balance for the vegetation layer is written as

$$R = \gamma_w q_w + q_h + \gamma_c q_c, \tag{3.1}$$

where R is the radiation balance, and q_w, q_h, q_c are the fluxes of water vapour, heat and CO_2, respectively, as already mentioned in Section 10.2. The coefficient γ_w is the specific enthalpy of water vapour $-\gamma_w = 2450$ J/g H_2O. The coefficient γ_c is the energy expense for the assimilation of one unit of CO_2. The latter can be estimated from the thermal effect of photosynthetic reaction, 470 kJ/mol CO_2. This value is the molar enthalpy of CO_2 in the process of its assimilation. By recalculating it to 1 g of CO_2 we get $\gamma_c = 10.7$ kJ/g CO_2.

The flux of assimilated carbon dioxide q_c determines the gross primary production of plants. Therefore, the ratio of two values can be considered as some efficiency coefficient of vegetation: the first is the solar energy expended for the maintenance of photosynthesis and the second is the part of incoming radiation absorbed by the underlying surface, i.e. the radiation balance:

$$\eta_R = \frac{\gamma_c q_c}{R}. \tag{3.2}$$

By substituting Eq. (3.1) into Eq. (3.2) and after simple transformation we get

$$\eta_R = \frac{1}{1 + \frac{1}{\gamma_c}\frac{q_h}{q_c} + \frac{\gamma_w}{\gamma_c}\frac{q_w}{q_c}}. \tag{3.3}$$

In Section 10.2 we calculated the ratios q_w/q_c and q_h/q_c (see Eqs. (2.12) and (2.14)). By substituting these expressions into Eq. (3.3) we have

$$\eta_R = \frac{1}{1 + \dfrac{c_p}{\gamma_c}\dfrac{T_l - T}{C^c - C_i^c} + \dfrac{\gamma_w D_w}{\gamma_c D_c}\dfrac{C_s^w - C^w}{C^c - C_i^c}}. \tag{3.4}$$

If we substitute into Eq. (3.3) the numerical values of $q_w/q_c \approx 240$ g H_2O/g CO_2 and $q_h/q_c \approx 76$ kJ/g CO_2, which were calculated under the climatic condition of temperate latitudes, and set the value 10.7 kJ/g CO_2 as γ_c then $\eta_R = 1/(1 + 7.1 + 55) \approx 1.6\%$ (compare with Section 5.3). If we now keep in mind that for most of Eastern Europe the summer radiation balance is equal to 55–60% of the total solar radiation of the summer season (Budyko, 1963) then the efficiency coefficient, which was recalculated for this value of energy input, is equal to $\eta \approx 0.9\%$.

We have assumed earlier that plants are normally using only a small part of the potential diffusion flux of carbon dioxide. If now we presuppose that all the potential flux is being used, i.e. $C^c - C_i^c \approx C^s$, then the efficiency coefficient η in this theoretical case will be equal to 7%.

10.4. Thermodynamic model of vegetation: internal entropy production

Let us represent the vegetation layer as a certain volume filled by leaves (not necessarily everywhere compactly). It has a unique basement (for instance, 1 cm^2) and a thickness (or height) equal to l_2. It is an open system, and the change of its entropy, caused by the production processes, is represented as

$$\frac{dS}{dt} = \frac{d_i S}{dt} + \frac{d_e S}{dt}, \tag{4.1}$$

where the term $d_i S/dt$ corresponds to dissipation processes (in particular, evapotranspiration and thermal diffusion), while the term $d_e S/dt$ corresponds to an external flow of energy. It can be either the full radiation balance or its share. Then the term dS/dt will correspond to the total increment of biomass, i.e. the gross primary production (per some time unit).

The main assumption is: "all heat produced within the vegetation layer by all metabolic processes, like transpiration and respiration, is exported into the environment (atmosphere). The rate equals the rate of entropy production (or, simply, the entropy production) $d_i S/dt$."

We shall consider the following two processes: (1) transpiration of water by leaves and subsequent transportation of water into the atmosphere by turbulent diffusion, and (2) thermal diffusion of air between interior and exterior of leaves and its subsequent turbulent transportation. In this case the total function of dissipation (see Section 2.5) is

$$\Psi = \Psi_h + (\Psi_w)_{p,T} + \Psi_{hw}, \tag{4.2}$$

where Ψ_h and $(\Psi_w)_{p,T}$ are the items describing the dissipation of energy as a result of the heat conduction and the diffusion of water vapour, correspondingly. The interaction of these two irreversible processes, namely heat conduction and diffusion, leads to the appearance of an additional source of energy dissipation corresponding to the item Ψ_{hw}.

The first item of sum (4.2) is represented as

$$\Psi_{\rm h} = -q_{\rm h}\frac{\Delta T}{l_2}\frac{1}{T}, \tag{4.3}$$

where the flux $q_{\rm h}^{(1)}$ defined by Eq. (2.4) is the specific (per unit area) heat flux, the factor $\Delta T/l_1 = (T - T_{\rm i})/l_1$ is the temperature gradient, $T_{\rm i}$ and T are the temperatures of the leaf interior and the environment (atmosphere). The second item is

$$(\Psi_{\rm w})_{p,T} = -\gamma_{\rm w}q_{\rm w}\frac{\Delta C^{\rm w}}{C^{\rm w}}\frac{1}{l_2}, \tag{4.4}$$

where the specific flux $q_{\rm w}$ is defined by Eq. (2.3), and $\gamma_{\rm w}$ is the specific enthalpy of water vapour; the factor $\Delta C^{\rm w}/C^{\rm w}l_2 = [C^{\rm w} - C_{\rm s}^{\rm w}(T)]/C^{\rm w}l_2$ is the gradient of its potential, $\Delta\mu_{\rm w}/l_2$, at temperature T. The third item

$$\Psi_{hv} = -\gamma_{\rm w}q_{\rm w}\frac{\mu_{\rm w}}{T}\frac{\Delta T}{l_2}, \tag{4.5}$$

where $\mu_{\rm w} = \ln[C_{\rm s}^{\rm w}(T)/C^{\rm w}]$ is the water vapour potential, $C^{\rm w}$ the concentration of water vapour in the atmosphere, and $C^{\rm w} = C_{\rm s}^{\rm w}(T)$ the concentration corresponding to its saturation value at the environmental temperature.

By summing Eqs. (4.3)–(4.5) and substituting the corresponding expressions for the fluxes $q_{\rm h}$ and $q_{\rm w}$ from Section 10.2 we obtain the general expression for the function of dissipation:

$$\Psi = \frac{\rho D_{\rm w}}{l_2^2}\left[c_pT\left(\frac{\Delta T}{T}\right)^2 + \gamma_{\rm w}C^{\rm w}\left(\frac{\Delta C^{\rm w}}{C^{\rm w}}\right)^2 + \left(\gamma_{\rm w}C^{\rm w}\ln\frac{C_{\rm s}^{\rm w}(T)}{C^{\rm w}}\right)\frac{\Delta T}{T}\frac{\Delta C^{\rm w}}{C^{\rm w}}\right], \tag{4.6}$$

where $D_{\rm w} = (D_{\rm w}^{(1)}D_{\rm w}^{(2)})/(D_{\rm w}^{(1)} + \lambda D_{\rm w}^{(2)})$, $\lambda = l_1/Ll_2$. Then the rate of energy dissipation by the vegetation layer of thickness l_2 and unique area will be equal to

$$T\frac{{\rm d_i}S}{{\rm d}t} = \Psi l_2. \tag{4.7}$$

In order to illustrate these formulas we shall use a case study, namely a beech forest near Kiel in the North Germany (Steinborn, 2000). All data are for the year 1991. The first group of data contains information about climatic conditions of the given location: the mean atmospheric temperature during vegetation period is $T =$ 16 °C, the temperature of leaves is $T_{\rm i} = 18$ °C, the number of so-called "shining hours" is equal to $1646\ {\rm h} \approx 6\times10^6$ s, the concentrations of water vapour $C_{\rm s}^{\rm w}(T) = 1.4\times10^{-2}$ g H_2O/g air and $C^{\rm w} = 1.16\times10^{-2}$ g H_2O/g air (relative humidity is 83%). The next group of data provides information about the architecture of the vegetation layer: its thickness is $l_2 = 10\ {\rm m} = 10^3$ cm, the thickness of leaf is 10^{-2} cm and the leaf area index is $L = 4$. Finally, the general information follows: the coefficients of molecular and turbulent diffusion are $D_{\rm w}^{(1)} = 0.25\ {\rm cm^2/s}$ and

$D_W^{(2)} = 10^4\ cm^2/s$, the density of dry air at 20°C is $\rho = 1.2 \times 10^{-3}$ g air/cm^3 and the heat capacity $c_p = 1$ J/K g air.

By substituting all these values into Eq. (4.6) we get $\Psi \approx 0.12 \times 10^{-4}\ W/cm^3$, and, in accordance with Eq. (4.7), $T(d_iS/dt) = 0.12 \times 10^{-4} \times 10^3 = 0.012\ W/cm^2$. Bearing in mind that the mean radiation balance for summer months is equal to 0.0126 W/cm^2, the coincidence is fantastic! We do not know the reason for such a coincidence: either casual or, maybe, causal. At any rate, it is necessary to test the relation using other experimental data.

In fact, all energy absorbed by the vegetation layer of a beech forest is spent on its maintenance, including the production of new biomass, where the share of spent energy is so small that it is impossible to extract it from the total expenditure. This is a value of the next order of magnitude in comparison with the radiation balance. Nevertheless, it becomes possible when using results from Section 10.2.

Indeed, it was shown that the production of 1 kg C of biomass is accompanied by transpiration of 879 l of water. Since this biomass is equivalent to 4.2×10^7 J, the production of 1 J is accompanied by transpiration of 0.021 g H_2O. The flux of evaporated water was estimated in the formula (2.3):

$$q_W = \frac{\rho D_W}{l_2}(C_W^S - C_W),$$

where $D_W = 0.9 \times 10^4\ cm^2/s$. By substituting the corresponding values into the expression we get $q_W = 0.257 \times 10^{-4}$ g H_2O/cm^2 s. Since this "pump" is working in the course of "shining time", the total amount of evaporated water is equal to $(0.257 \times 10^{-4}$ g H_2O/cm^2 s$)(6 \times 10^6$ s$) = 1.542 \times 10^2$ g H_2O/cm^2 that corresponds to 7.34×10^3 J/cm^2. If we compare this value with the gross annual production of our beech forest in 1990, 7.064 kJ/cm^2, we again obtain a very curious coincidence.

Finally, if we assume that all these coincidences are not *casual* but rather reflect some deep *causal* links then the equality "radiation balance ≈ energy dissipation" can be used for estimation of the annual production of vegetation. Indeed, the formula for dissipation can be represented as

$$\text{Diss} \approx \frac{1}{l_2}\gamma_W q_W \frac{\Delta C_W}{C_W} = \frac{1}{l_2}\gamma_W q_W \frac{1-\eta}{\eta}, \tag{4.8}$$

where η is the relative humidity. Here we omit the terms describing the heat flows—they are small in comparison with the latent heat flow, which is determined by the transpiration and subsequent transportation of water vapour into the atmosphere. Since the annual gross production

$$\text{GPP} = \frac{q_W}{0.021}\text{ST}\ \ (\text{J/cm}^2), \tag{4.9}$$

where ST is the "shining time" in seconds; then, by combining Eqs. (4.8) and (4.9), and taking into account that Diss ≈ R in W/cm^2 we get

$$\text{GPP} \approx 2 \times 10^{-2} l_2 R \frac{\eta}{1-\eta}\text{ST}\ \ (\text{J/cm}^2). \tag{4.10}$$

10.5. Vegetation as an active surface: the solar energy degradation and the entropy of solar energy

In the process of any detailed experimental study of the ecosystem several different measurements are carried out; in particular, the data about incoming and outgoing radiations are collected. Naturally, the following idea arises: to attempt to use such values as energy, entropy and exergy as macroscopic characteristics of the ecosystem and its most important component as vegetation. The problem is how to calculate them using the observed data. Several authors recently developed different approaches to the problem (Ulanowicz and Hannon, 1987; Aoki, 1987; Schneider and Kay, 1994; Svirezhev and Steinborn, 2001).

Vegetation is regarded as an active surface interacting with solar radiation and transforming it. Absorption, reflection and emission of radiation are consequences of this interaction, resulting in a new composition of spectrum of outgoing radiation. From a macroscopic point of view this is equivalent to the change of energy and information. It is natural to assume that the difference between these macroscopic variables calculated for incoming and outgoing radiation is defined not only by the properties of these radiation fluxes but also by the properties (state) of the active surface (vegetation), and, as a consequence, by the properties of an ecosystem. A natural idea comes to mind: to use the data of a continuous measurement of the components of radiation balance for the calculation of energy, entropy and exergy. A continuous measurement of the components of the radiation balance for different types of vegetation has become an almost standard procedure giving us information about the transformation of incoming solar radiation by vegetation. These values are used to calculate the dynamics of the radiation balance, i.e. dynamics of the integrated value, which deals with the energy conservation law. However, for a complete description of vegetation as a "thermodynamic machine", the principles responsible for the entropy concept should be implemented in addition to the First Law, namely the Second Law, the Prigogine theorem and the exergy concept (Jørgensen, 1992c).

If we look at the first two principles we see that they can be called "classic" ones, because they use the classic concept of entropy. But as soon as we try to consider the vegetation as some thermodynamic machine and to estimate its efficiency we immediately encounter the following problems: What is a working body of the machine? What is the temperature of the working body? For instance, Ulanowicz and Hannon are considering the Bose–Einstein gas as the working body. What is the temperature of this gas? You can answer that the temperature of incoming photons is equal to the temperature of the Sun (5700 K), but let us remember Gibbs has said that all material processes on our planet are taking place at its temperature. Thus, in all thermodynamic calculations we have to set the Earth's temperature. The first point of view is represented in the work of Essex (1984) and Peixoto and Oort (1992). The second point of view was formulated in the classic Gibbs works; nevertheless, the discussion is continuing.

Another approach differing from the classic one was suggested by Schneider and Kay (1994). According to them a complex structure develops spontaneously if a system is maintained far from thermodynamic equilibrium by a gradient of energy (like solar radiation). The more complex is the structure of a system, the more effectively it "dissipates the gradient". This idea—defined as the "non-equilibrium principle"—is a modern account of Schrödinger's (1944) "order from disorder" premise. Comparing ecosystems that receive the same amount of solar radiation and have the same soil properties and water balance, the most highly developed ecosystem would reradiate its energy on the lowest exergy level, i.e. the ecosystem would display the coldest black body temperature (Kutsch et al., 2001).

This can be explained with the radiation balance, $R = E^{\text{in}} - E^{\text{out}}$, which is the net radiation flux, or in other words, the net radiation transformed into non-radiative energy at the surface (soil, water, vegetation, etc.). If the totals of incoming and outgoing radiation are represented as sums of long- and short-wave components, then $E^{\text{in}} = E_{\text{l}}^{\text{in}} + E_{\text{s}}^{\text{in}}$ and $E^{\text{out}} = E_{\text{l}}^{\text{out}} + E_{\text{s}}^{\text{out}}$ where E_{l}^{in}, $E_{\text{l}}^{\text{out}}$, E_{s}^{in} and $E_{\text{s}}^{\text{out}}$ are these components, respectively. Usually it is assumed that $E^{\text{in}} \approx E_{\text{s}}^{\text{in}}$ and $E^{\text{out}} \approx E_{\text{l}}^{\text{out}} + \alpha_{\text{s}} E^{\text{in}}$ where α_{s} $(0 \leq \alpha_{\text{s}} \leq 1)$ is the so-called *short-wave albedo*. Then $R \approx (1 - \alpha_{\text{s}})E^{\text{in}} - E_{\text{l}}^{\text{out}}$.

Schneider and Kay (1994) propose the ratio of radiation balance and net incoming solar (short-wave) radiation $R_{\text{n}}/K^* = 1 - (L^*/K^*)$ (here the notations used are: $R_{\text{n}} = R = K^* - L^*$, $L^* = E_{\text{l}}^{\text{out}}$ and $K^* = (1 - \alpha_{\text{s}})E^{\text{in}}$) as a functional measure of the system's ability to dissipate the incoming energy. The higher this ratio, and the lower the surface temperature of the system, the higher is its ability to dissipate the radiative gradient. As an argument they give the following results of observations for different surfaces. The surface temperature T is determined as $T = (L^*/\varepsilon\sigma)^{1/4}$ where ε is the emissivity, and σ the Stefan–Boltzmann constant.

If, as Schneider and Kay suggested, we consider ecosystems as energy degraders, then of course all these data are very interesting: they show that the quarry degrades 62% of the net incoming radiation, K^*, into energy in the form of molecular motion, R_{n}, while the 400-year-old forest degrades 91% (Table 10.1). If the more developed ecosystem degrades more energy, then both the ecosystem temperature and the ratio R_{n}/K^* are good indicators of ecosystem integrity. But there are "underwater rocks" in this concept: it is not clear what the term "degradation of energy" means in this case. Formally speaking, if the quarry degrades a lower quantity of energy then it is more organised. On the other hand, if we consider the surface as a heat machine then the expression R_{n}/K^* may be interpreted as its efficiency. Then what is the meaning of the 91% efficiency of the mature fire forest? If, as Schneider and Kay suggested, the ratio R_{n}/K^* points to the degree of exergy degradation,

Table 10.1
Radiative estimates from a thermal infrared multispectral scanner for different ecosystem types in the H.J. Andrews Experimental Forest, Oregon (USA)

	Quarry	Clear-cut	Douglas fir plantation	Natural forest	400-year-old Douglas fir forest
K^* (W/cm^2)	0.0718	0.0799	0.0854	0.0895	0.1005
L^* (W/cm^2)	0.0273	0.0281	0.0124	0.0124	0.0095
R_{n} (W/cm^2)	0.0445	0.0517	0.0730	0.0771	0.0910
R_{n}/K^*	0.62	0.65	0.85	0.86	0.91
T (°C)	50.7	51.8	29.9	29.4	24.7

Data from Luvall and Holbo (1989).

then the interpretation of this value is not quite understandable. By introducing the different information content of incoming and outgoing radiation into our calculation we shall suggest below another concept of exergy which seems more appropriate. Moreover, we can investigate not only the amount of absorbed energy but also its quality, as done in Section 5.6.

Ulanowicz and Hannon (1987) agree with this point of view assuming that "living systems generate more entropy because they are more effective in utilising energy than are the ambient physical systems". Within the framework of classic thermodynamics they suggest a quantitative measure for the entropy calculation of different wavelength intervals. Similar efforts have been made by Press (1976) and Aoki (1987, 1995, 1998), who stress the principal difference between the entropy of scattered and direct solar radiation.

10.6. Vegetation as an active surface: exergy of solar radiation

In this section we apply the method which was described in Section 5.6 and allows us to calculate the exergy of solar radiation interacting with such an active surface as the vegetation (see also Svirezhev and Steinborn, 2001; Svirezhev et al., 2003). But before this, we shall say a few words about standard integral characteristics, describing the interaction widely used in actinometric measurements. Naturally we might be partially repetitive.

Incoming solar radiation is usually described by its energy spectrum, or the density distribution of energy, $E^{\mathrm{in}}(\nu; x, y, t)$ where $\nu \in [\nu_{\min}, \nu_{\max}]$ is the spectral frequency of radiation, the interval $\Omega : [\nu_{\min}, \nu_{\max}]$ includes all possible values of spectral frequencies, t is time, and (x, y) are the coordinates of some point on the globe. At this point the incoming radiation interacts with a surface, and the result of this interaction is the transformation F of incoming radiation into the outgoing radiation $E^{\mathrm{out}}(\nu; x, y)$, so that at the given point and time $E^{\mathrm{in}}(\nu) \overset{F(x,y,t)}{\Rightarrow} E^{\mathrm{out}}(\nu)$. It is obvious that the properties of the transformation F are connected with the characteristics of the reflecting and transforming surface, which could be an ocean surface, vegetation cover, a surface of ice, desert, etc. Since the inflow $E^{\mathrm{in}}(\nu)$ and the outflow $E^{\mathrm{out}}(\nu)$ are known and the transformation F is unknown, the surface transforming incoming solar radiation can be considered as a "black box". By analysing different relations between incoming and outgoing radiation we get some information about the properties of the "black box". Note that the outflow $E^{\mathrm{out}}(\nu)$ contains two components: the reflected radiation and radiation which is passed over the surface, so that $E^{\mathrm{out}}(\nu) = E_{\mathrm{r}}^{\mathrm{out}}(\nu) + E_{\mathrm{p}}^{\mathrm{out}}(\nu)$. In reality, the second component is not registered and, as some estimations show, its value is significantly lower than the first one. That is why we shall assume that $E^{\mathrm{out}}(\nu) \approx E_{\mathrm{r}}^{\mathrm{out}}(\nu)$.

The simplest form of transformation operator is a shift operator $R(\nu)$:

$$E^{\mathrm{out}}(\nu) = E^{\mathrm{in}}(\nu) - R(\nu). \tag{6.1}$$

Here and later on we shall omit the notations x, y, and t by assuming that all the operations are performed at a given point (x, y) and time t. The convolution $R = \int_{\Omega} R(\nu)\mathrm{d}\nu$

plays a very important role in meteorology, and is called the *radiation* (or *energy*) *balance* (see above). Continuous measurement of the radiation balance for the Earth's surface and its components, such as incoming and outgoing short- and long-wave solar radiation, constitute a standard meteorological procedure. Since short-term variations of these values can be considered to be merely a noise, it is possible to apply averaging for monthly and annual intervals. In this way, maps of the annual and monthly radiation balance were obtained (see, for instance, Budyko, 1963).

The second simple form of transformation operator is a contracting operator $\alpha(\nu)$

$$E^{\mathrm{out}}(\nu) = \alpha(\nu)E^{\mathrm{in}}(\nu), \tag{6.2}$$

where $\alpha(\nu)$ is the so-called *spectral albedo*. Its convolution $\hat{\alpha} = (1/\Omega)\int_{\Omega}[E^{\mathrm{out}}(\nu)/E^{\mathrm{in}}(\nu)]\mathrm{d}\nu$ is the *mean* (or *integral*) *albedo*. This value is also a very important parameter in meteorology. Note that the value $\hat{\alpha}$ as defined above is mathematically correct, but unfortunately, in meteorology a different definition of albedo is used (we shall call it the *standard* albedo): $\bar{\alpha} = E^{\mathrm{out}}/E^{\mathrm{in}}$, where $E^{\mathrm{in}} = \int_{\Omega}E^{\mathrm{in}}(\nu)\mathrm{d}\nu$ and $E^{\mathrm{out}} = \int_{\Omega}E^{\mathrm{out}}(\nu)\mathrm{d}\nu$ are the total energies of incoming and outgoing radiation, respectively. This definition is also certainly correct, but the point is that it is yet another definition of the mean. In fact, by applying to the convolution of Eq. (6.2) the mean value theorem we get $\int_{\Omega}E^{\mathrm{out}}(\nu)\mathrm{d}\nu = \int_{\Omega}\alpha(\nu)E^{\mathrm{in}}(\nu)\mathrm{d}\nu = \bar{\alpha}\int_{\Omega}E^{\mathrm{in}}(\nu)\mathrm{d}\nu$, where $\bar{\alpha}$ is a "mean" value $\alpha(\theta)$ for $\theta \in \Omega$. It is often assumed that the interval Ω contains only short-wave radiation (see Section 10.5). Thus, these values are usually shown as the albedo of different surfaces. In the standard BATS scheme (Matthews, 1983, 1984) the interval Ω is divided into two sub-intervals, Ω_{s} and Ω_{l} ($\Omega_{\mathrm{s}} \cup \Omega_{\mathrm{l}} = \Omega$, $\Omega_{\mathrm{s}} \cap \Omega_{\mathrm{l}} = \varnothing$) for short- and long-wave radiation, respectively. The border of division is 0.7 μm. As a result, the two standard albedo, $\alpha_{\mathrm{s}} = E_{\mathrm{s}}^{\mathrm{out}}/E_{\mathrm{s}}^{\mathrm{in}}$ and $\alpha_{\mathrm{l}} = E_{\mathrm{l}}^{\mathrm{out}}/E_{\mathrm{l}}^{\mathrm{in}}$, are derived, where the energy spectra are integrated with respect to corresponding intervals. In standard actinometric measurements the entire spectral interval is divided into the following sub-intervals: Ω_{s}: 0.3–2.8 μm and Ω_{l}: 2.8–100 μm. It is natural that with the higher number of spectral intervals the accuracy of our calculation will increase.

Both types of above-considered operators are linear, and other linear operators do not exist. However, we think that the real transformation operator is more complex and non-linear. We suggest using the *exergy of solar radiation* as such an operator (Section 5.6), which is a Kullback measure of the information increment. We recall the basic definition of the exergy of solar radiation in the context of measurements of the components of the radiation balance.

The spectral exergy (or the exergy flow, but as a rule we shall omit the word "flow") is equal to

$$\mathrm{Ex}(\nu) = E^{\mathrm{out}}(\nu)\ln\frac{E^{\mathrm{out}}(\nu)}{E^{\mathrm{in}}(\nu)} + R(\nu) = E^{\mathrm{in}}(\nu)\alpha(\nu)\ln[\alpha(\nu)] + R(\nu). \tag{6.3}$$

The total exergy, Ex, is a convolution of the spectral exergy: $\mathrm{Ex} = \int_{\Omega}\mathrm{Ex}(\nu)\mathrm{d}\nu$. If the standard Kullback measure of the increment of information is equal to $K = \int_{\Omega}p^{\mathrm{out}}(\nu)\ln[p^{\mathrm{out}}(\nu)/p^{\mathrm{in}}(\nu)]\mathrm{d}\nu$, where $p^{\mathrm{in}}(\nu) = E^{\mathrm{in}}(\nu)/E^{\mathrm{in}}$ and $p^{\mathrm{out}}(\nu) = E^{\mathrm{out}}(\nu)/E^{\mathrm{out}}$ is used, then the expression for the total exergy is rewritten as (compare with Eq. (6.3) of Section 5.6):

$$\mathrm{Ex} = \int_{\Omega} E^{\mathrm{out}}(\nu)\ln\frac{E^{\mathrm{out}}(\nu)}{E^{\mathrm{in}}(\nu)}\mathrm{d}\nu + R = E^{\mathrm{out}}K + E^{\mathrm{out}}\ \ln(E^{\mathrm{out}}/E^{\mathrm{in}}) + R$$

$$= E^{\mathrm{in}}[1 + \bar{\alpha}(K + \ln\bar{\alpha} - 1)] = R\frac{1 + \bar{\alpha}(K + \ln\bar{\alpha} - 1)}{1 - \bar{\alpha}}. \quad (6.4)$$

The value $\eta_{\mathrm{Ex}} = \mathrm{Ex}/E^{\mathrm{in}} = 1 + \bar{\alpha}(K + \ln\bar{\alpha} - 1)$ can be considered as a specific increment of information, ex, per unit of incoming energy. It is interesting that the total exergy is expressed by a non-linear combination of such typical meteorological parameters as the standard albedo $\bar{\alpha} = E^{\mathrm{out}}/E^{\mathrm{in}}$ and the radiation balance R, and also the new variable K (Kullback's information measure).

Since only two spectral intervals are used, the value of K in Eq. (6.4) is

$$K = \frac{1 - \bar{\alpha}}{\bar{\alpha}R}\left[\frac{\alpha_{\mathrm{s}}R_{\mathrm{s}}}{1 - \alpha_{\mathrm{s}}}\ln\frac{\alpha_{\mathrm{s}}}{\bar{\alpha}} + \frac{\alpha_{\mathrm{l}}R_{\mathrm{l}}}{1 - \alpha_{\mathrm{l}}}\ln\frac{\alpha_{\mathrm{l}}}{\bar{\alpha}}\right]. \quad (6.5)$$

Also, in view of the fact that instead of the integral with respect to the full spectrum we have only the integral data for two spectral intervals, the formula for exergy calculation will be approximate. If one looks carefully at the formula, it can be seen that the error appears only when we replace the value of K, or more correctly the factor:

$$F = \int_{\Omega}\left[\frac{\alpha(\nu)R(\nu)}{1 - \alpha(\nu)}\ln\frac{\alpha(\nu)}{\bar{\alpha}}\right]\mathrm{d}\nu,$$

by its approximate value

$$F_{\mathrm{app}} = \left[\frac{\alpha_{\mathrm{s}}R_{\mathrm{s}}}{1 - \alpha_{\mathrm{s}}}\ln\frac{\alpha_{\mathrm{s}}}{\bar{\alpha}} + \frac{\alpha_{\mathrm{l}}R_{\mathrm{l}}}{1 - \alpha_{\mathrm{l}}}\ln\frac{\alpha_{\mathrm{l}}}{\bar{\alpha}}\right].$$

It is easy to show that the factor F is represented as $F = \int_{\Omega} E^{\mathrm{in}}(\nu)\alpha(\nu)\ln\alpha(\nu)\mathrm{d}\nu - \hat{E}^{\mathrm{in}}\bar{\alpha}\ln\bar{\alpha}$. Using the so-called Jensen inequality (Beckenbach and Bellman, 1961) we can write:

$$\int_{\Omega_{\mathrm{s}}} E^{\mathrm{in}}(\nu)\alpha(\nu)\ln\alpha(\nu)\mathrm{d}\nu \geq \hat{E}^{\mathrm{in}}_{\mathrm{s}}\alpha_{\mathrm{s}}\ln\alpha_{\mathrm{s}} \quad \text{and}$$

$$\int_{\Omega_{\mathrm{l}}} E^{\mathrm{in}}(\nu)\alpha(\nu)\ln\alpha(\nu)\mathrm{d}\nu \geq \hat{E}^{\mathrm{in}}_{\mathrm{l}}\alpha_{\mathrm{l}}\ln\alpha_{\mathrm{l}}.$$

Whence follows: $F \geq F_{\mathrm{app}}$, where equality occurs only if $\alpha(\nu)$ are equal to constants within the intervals Ω_{s} and Ω_{l}. In other words, the less the spectral albedo changes within the used spectral intervals, the more exact our approximation will be. However, the approximate value of exergy is always lower than its exact value.

Unfortunately, we were not able to achieve a more exact estimation of the error. In one example of the direct calculation of the error, in which the necessary spectral characteristics were measured for a surface covered by grass (Wilmers, 1984), it was shown that the error is maximal when a day is sunny and clear, and decreases when a day is cloudy. However, even in the worst case it is not higher than 12%.

These data were collected in several different research sites:

1. 100-year-old beech forest in North Germany (Kiel);
2. crop rotation field in North Germany (Kiel);
3. grass community in North Germany (Hamburg);
4. pine forest in Central Germany;
5. wheat field in Oklahoma, USA;
6. moderately managed pine forest in South Germany.

Let us consider the seasonal dynamics of the following observed and calculated parameters as the radiation (energy) balance, exergy, exergy and radiation efficiency coefficients, and Kullback's measure. We take the Kiel sites as an example (see Fig. 10.1).

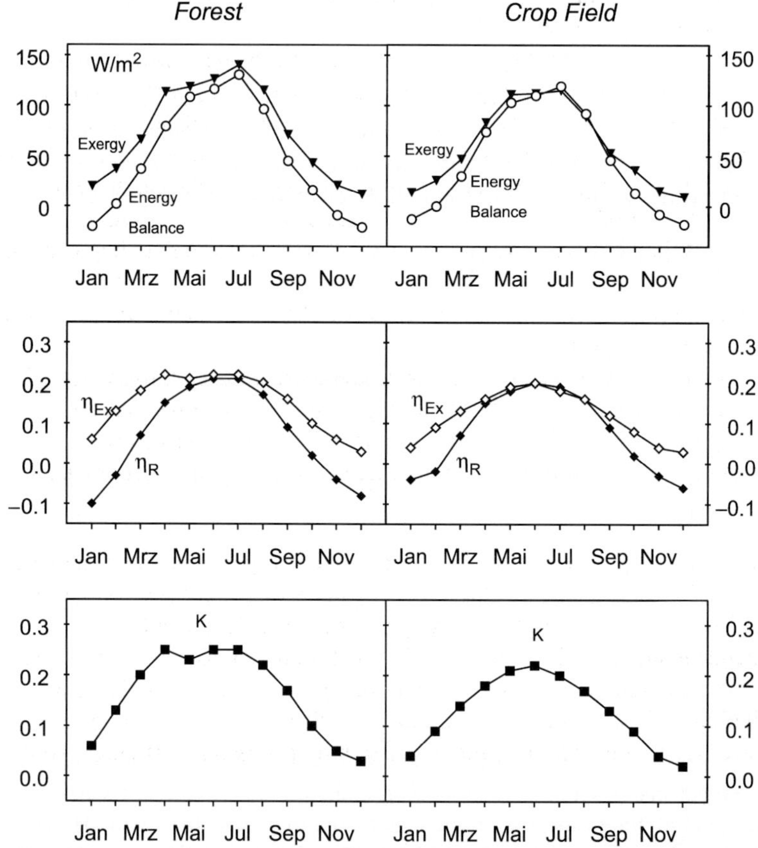

Fig. 10.1. Seasonal dynamics of exergy, radiation (energy) balance, exergy and radiation efficiency coefficients (η_{Ex} and η_R) and Kullback's measure (K) in the research area near Kiel. "Forest" is a 100-year-old beech forest. "Crop field" is a crop rotation field. The results shown are averaged values for the years 1991–1996.

In principle, on both sites all parameters follow the annual dynamics of radiation balance. But there are differences, which will be used below to characterise the specific behaviour of different sites. Usually, exergy is higher than the radiation balance, but during summer, when productivity of plants is the highest, this difference becomes smaller. Correspondingly, exergy efficiency, $\eta_{Ex} = \mathrm{Ex}/E^{in}$, is higher than energy efficiency, $\eta_R = R/E^{in}$, but during summer their values almost coincide. Note that the absolute values of exergy are the result of our calculation with only two spectral frequency intervals. A calculation with more intervals could result in slightly different values.

For the crop field the values of all parameters are slightly lower than for the forest. Looking at exergy and energy balance it can be seen that the difference between the two parameters is larger in the forest. This means that the forest not only absorbs more energy from solar radiation, but it also uses this energy, more efficiently to perform work. During summer this trend becomes clearer. Lower values of the crop field efficiency (see middle part of Fig. 10.1) may be due to higher emission of long wave radiation, resulting from a higher temperature. A thin layer of vegetation usually evaporates less and therefore has a lower cooling ability (Herbst, 1997).

Consequently, incoming energy cannot be dissipated to the same extent as in the forest. Another reason for the lower exergy on the field is that the albedo of crops is higher, which results in a lower amount of available energy, i.e. energy balance. Therefore it can be stated that forests can use the incoming solar radiation more efficiently and more effectively: exergy is higher per unit of absorbed energy, whereas the total amount of absorbed energy is also higher.

All these results allow us to formulate the following teleological hypotheses:

1. Vegetation works as an information machine ($\mathrm{Ex} > R$ and $\eta_{Ex} > \eta_R$ in the course of almost the whole year).
2. Exergy (Ex), energy balance (R) and the increment of information (K) achieve maxima when the productivity of vegetation is also maximal (in the Northern Hemisphere it is usually maximal in June–July). The same statement is also true for the exergy and radiation efficiency coefficients (η_{Ex} and η_R). In this case $\mathrm{Ex} \cong R$ and $\eta_{Ex} \cong \eta_R$.

Before formulating the third hypothesis, it is necessary to keep in mind that the exergy efficiency coefficient η_{Ex} is a function of two independent variables, η_R and K. The total increase of η_{Ex} with growing productivity means that it is a result of the summation of two independent and opposite processes. The first is an increase of η_{Ex} with increasing K, the second is a *decrease* of η_{Ex} with increasing η_R since $\eta_{Ex} \to \eta_R$ (see Fig. 5.4). Using the terminology of the game theory we can say that in order to maximise productivity (under corresponding constraints) the vegetation uses a *minimax* strategy, so that

$$\eta_{Ex} \to \min_{\eta_R} \max_{K} (\eta_{Ex}). \tag{6.6}$$

Note that if there is the natural constraint for η_R: $\eta_R \leq 1$, then the constraint for K is defined by arguments lying outside our formalism. The reasons for this may be related to the structure of vegetation cover. For instance, K-values (1-year averaging, see Table 10.2) for grassland and crops are lower than for forests.

Table 10.2
Mean annual values of exergy and energy balance of different sites

	Kiel forest	Kiel crop	Hamburg grassland	Central Germany forest	South Germany forest	Oklahoma wheat
Ex (W/m^2)	73	59.7	50	86	80	77
R (W/m^2)	50	47	40	77	75	87
η_{Ex}	0.16	0.13	0.12	0.19	0.17	0.15
η_R	0.11	0.11	0.09	0.17	0.16	0.17
K	0.18	0.14	0.12	0.21	0.19	0.16

The Kiel values were averaged over the years 1991–1996. The Hamburg grassland values were averaged over 1973–1993. The Central and South Germany values were averaged over 1974–1997 and 1998–1999. The Oklahoma values are from 1996.

Finally, the third hypothesis can be formulated as:

3. In the process of productivity maximisation, vegetation uses the minimax strategy, minimising the exergy efficiency coefficient with respect to the radiation efficiency coefficient and maximising it with respect to the increment of information.

We think that the statement can be considered as a generalisation of the Jørgensen maximum exergy principle (1997), although applied to one concrete case, i.e. to the interaction between solar radiation and vegetation. If the third hypothesis is a *minimax exergy principle* then we can get a relationship connecting such originally unconnected values as the Kullback measure K and the radiation coefficient η_R. It is natural to assume that these values are independent, if our "active surface" is passive in relation to the incoming radiation. However, if the "active surface" is really "active", i.e. it actively interacts with incoming radiation partly using it and transforming its spectrum (like a vegetation cover), then we have the right to expect that these values will be connected. If we postulate that the *minimax exergy principle* is true, then the productivity of vegetation is maximal when

$$\eta_R = 1 - e^{-K} \quad (\text{or } R = E^{in}(1 - K)). \tag{6.7}$$

Although "*comparaison n'est pas raison*" it would be interesting to compare the annual means of exergy, radiation balance, the corresponding efficiency coefficients and the increment of information (Kullback's measure) for the different sites. The results are shown in Table 10.2. Looking at these sites the differences between more natural and anthropogenic ecosystems are apparent. Besides, in the colder climate of North Germany the values are lower than in South Germany or the US.

To outline the principal difference between forest and agricultural ecosystems we present these values in Fig. 10.2 on the two superposed planes $\{K, \eta_R\}$ and $\{\eta_{Ex}, \eta_R\}$. One can see that the forest ecosystems seem to have higher potential in using exergy regarding both absolute values and efficiency coefficients than crops and grassland. For example, the beech forest in Kiel has higher values for all parameters than the Kiel crop field and the Hamburg grassland. Moreover, it has higher exergy efficiency than the wheat field in

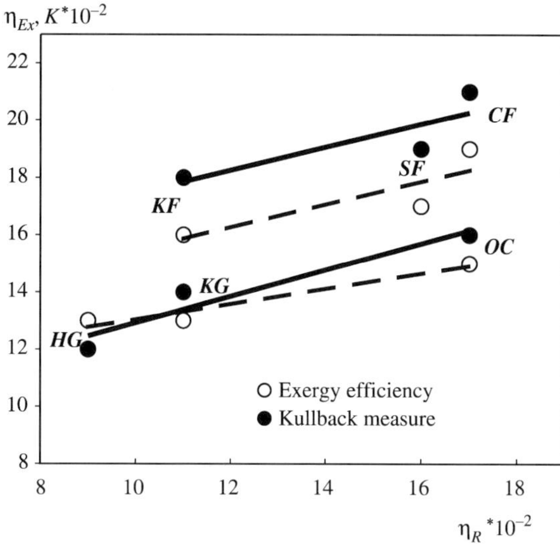

Fig. 10.2. Annual mean values of the exergy efficiency coefficient and Kullback measure in dependence of the radiation efficiency coefficient for different sites. KF, Kiel beech forest; SF, South Germany pine forest; CF, Central Germany pine forest; HG, Hamburg grassland; KC, Kiel crop field; OC, Oklahoma crop field. Dotted lines, linear regression for Kullback's measure; solid lines, linear regressions for the exergy efficiency coefficient.

Oklahoma although the latter has a much higher input of energy. The forests in Central and South Germany also show very high values of exergy and exergy efficiency in relation to the sites with higher anthropogenic pressure.

Certainly, these values are insufficient for any satisfactory statistics. However, the tendencies are as follows (see Fig. 10.2). Firstly, all the points, both $(\eta_{\mathrm{Ex}}, \eta_{\mathrm{R}})$ and (K, η_{R}), can be arranged by lines of regression $\eta_{\mathrm{Ex}} = a_{\mathrm{Ex}} + b_{\mathrm{Ex}}\eta_{\mathrm{R}}$ and $K = a_K + b_K\eta_{\mathrm{R}}$ into two groups corresponding to forest and grass–crop ecosystems.

The regression coefficients are equal to: $a^{\mathrm{f}}_{\mathrm{Ex}} = 0.114$; $a^{\mathrm{f}}_K = 0.134$; $b^{\mathrm{f}}_{\mathrm{Ex}} = b^{\mathrm{f}}_K = 0.4$ for forests and $a^{\mathrm{g}}_{\mathrm{Ex}} = 0.088$; $a^{\mathrm{g}}_K = 0.083$; $b^{\mathrm{g}}_{\mathrm{Ex}} = 0.38$; $b^{\mathrm{g}}_K = 0.45$ for grass and crops. Lines with almost the same gradients indicate that although both the exergy and the information increment depend on the radiation balance (the latter differs in different sites), the dependence is identical for different types of vegetation. This indicates the possible existence of some regulating mechanism that is common for all types of vegetation.

10.7. Simplified energy and entropy balances in the ecosystem

From the viewpoint of thermodynamics, any ecosystem is an open thermodynamic system. An ecosystem being in a "climax" state corresponds to a dynamic equilibrium when the entropy production within the system is balanced by the entropy outflow to its environment.

Let us consider one unit of the Earth's surface, which is occupied by some natural ecosystem (e.g. meadow, steppe, forest, etc.) and is maintained in a climax state. Since the main component of any terrestrial ecosystem is vegetation, we assume that the area unit is covered by a layer, including dense enough vegetation and the upper layer of soil with litter, where dead organic matter is decomposed. The natural periodicity in such a system is equal to 1 year, so that all processes are averaged by a 1-year interval.

Since the energy and matter exchanges between the system and its environment are almost completely determined by the first autotrophic level, i.e. vegetation, the vegetation layer is considered as the system, whereas its environment is the atmosphere and soil. We assume that all exchange flows of matter, and also energy and entropy, are vertical, i.e. we neglect all horizontal flows and exchange between ecosystems located at different spatial points.

In Section 10.3 we have already used the simplified equation of the annual energy balance for a vegetation layer (equality (3.1)): $R = \gamma_w q_w + q_h + \gamma_c q_c$, where the item $\gamma_w q_w$ is the latent heat flux, the item $\gamma_c q_c = \text{GPP}$, and the item q_h is a sensible heat flux, i.e. a turbulent flux transporting heat from the layer surface into the atmosphere. Since in our case there is an additional income of heat from biomass oxidation (respiration and decomposition of dead organic matter), the left side of the balance has to be represented as $R + q_{ox}$ (instead of R) where $q_{ox} = Q_{met} + Q_{dec}$. Here Q_{met} is a metabolic heat, and Q_{dec} is heat released in the process of decomposition. Finally, the equation of energy balance is written as

$$[R - \gamma_w q_w - q_h] + [Q_{met} + Q_{dec} - \text{GPP}] = 0. \tag{7.1}$$

In such a form of representation, all items of the energy balance are arranged in two groups (in square brackets) and members of these groups differ from each other by the orders of magnitude. For instance, the characteristic energies of the processes of a new biomass formation and its decomposition (the second brackets) are lower by a few orders of magnitude than the radiation balance and the energies, which are typical for evapotranspiration and turbulent transfer (the first brackets). As a rule, the terms contained within the second bracket are usually omitted in the standard expression for energy balance: $R = \gamma_w q_w + q_h$ (see, for instance, Budyko, 1977). This suggests that in this case a so-called "asymptotic splitting" could be used. Note that such a kind of method is widely used in the theory of climate: this is a so-called "quasi-geostrophic approximation" (Pedlosky, 1979). Thus, if we follow the logic of asymptotic splitting then instead of exact balance (7.1) we get the two asymptotic equalities:

$$R - \gamma_w q_w - q_h \approx 0 \text{ and } Q_{met} + Q_{dec} - \text{GPP} \approx 0. \tag{7.2}$$

We assume that the fulfilment of the first equality provides the existence of some "thermostat", which should be called the "environment". The thermostat maintains the constancy of temperature and pressure within the environment. Then the fulfilment of the second equality is determined by a consistency of the processes of production on the one hand, and metabolism of vegetation and also decomposition of dead organic matter in litter and soil on the other hand.

In accordance with Glansdorff and Prigogine (1971) the entropy production within the system is equal to $\mathrm{d_i}S/\mathrm{d}t \approx q_{ox}/T$, where T is the temperature and q_{ox} is the heat production of the system. As shown above, the total heat production is a result of two processes: metabolism or respiration ($Q_{\rm met}$) and decomposition of dead organic matter ($Q_{\rm dec}$). Since these processes can be considered as a burning of a corresponding amount of organic matter, the values of $Q_{\rm met}$ and $Q_{\rm dec}$ can be also expressed in enthalpy units. Thus,

$$\frac{{\rm d_i}S}{{\rm d}t} = \frac{1}{T}(Q_{\rm met} + Q_{\rm dec}). \tag{7.3}$$

The value of T is the mean air temperature in the given area averaged along either the entire year, or the vegetation period and the period when decomposition is possible. In the latter case, all other values also have to be averaged along this period. Since the second equality of Eq. (7.2) has to hold, then

$$\frac{{\rm d_i}S}{{\rm d}t} = \frac{\rm GPP}{T}. \tag{7.4}$$

At the dynamic equilibrium the internal entropy production has to be compensated by the entropy export from the system, so that

$$\frac{{\rm d_i}S}{{\rm d}t} = \left|\frac{{\rm d_e}S}{{\rm d}t}\right| = \frac{\rm GPP}{T}. \tag{7.5}$$

The last equality is a quantitative formulation of the so-called "entropy pump" concept (Svirezhev, 1990, 1998b, 2000; Svirezhev and Svirejeva-Hopkins, 1998; Steinborn and Svirezhev, 2000). We assume that the entropy pump "sucks" entropy produced by the ecosystem. As a result, natural ecosystems do not accumulate entropy, and as a result it can exist during a sufficiently long time period. The power of the entropy pump, $|{\rm d_e}S/{\rm d}t|$, depends on the GPP of the natural ecosystem located in situ and the local temperature (see Eq. (7.5)). We assume that the local climatic, hydrological, soil and other environmental conditions are adjusted in such a way that only a natural ecosystem, which is typical of these local conditions, can exist and be in a steady state at this site. We assume that *the local climatic, hydrological, soil and other environmental conditions are adjusted in such a way that only a natural ecosystem corresponding specifically to these local conditions can exist at this site and be here in a steady state (dynamic equilibrium).*

All these statements constitute the main content of the "entropy pump" concept.

Hence, there is no "overproduction of entropy" in natural ecosystems, because the "entropy pump" sucks the entire entropy out of ecosystems, by the same token preparing them for a new 1-year period. Maybe, the picture will be clearer if we consider the process of ecosystem functioning as a cyclic process with 1-year natural periodicity. At the initial point of the cycle the ecosystem is in thermodynamic equilibrium with its environment. Then, as a result of the work done by the environment on the system, it performs a *forced* transition to a new *dynamic* equilibrium. The transition is accompanied by the creation of new biomass, and the ecosystem entropy decreases. After this the reversible *spontaneous* process is started, and the system moves to the initial point producing the entropy in the course of this path. The processes that accompany the transition are metabolism of vegetation and the decomposition of dead organic matter in litter and soil. If the cyclic process is reversible, i.e. the cycle can be repeated infinitely, the total production of entropy by the system has to be equal to its decrease at the first stage. Substantively both these transitions take place simultaneously, so that the entropy is produced within the system, and at the same time it is "sucked out" by the environment. At the equilibrium state, the annual amounts of the entropy produced by the system and the decrease of entropy caused by the work of the environment on the system are equivalent.

Let us imagine that this balance was disturbed (this is a typical situation for agro-ecosystems). Under the impact of new energy and matter inflows, the system moves towards a new state, which differs from dynamic equilibrium of the natural ecosystem. As a rule, the entropy produced by the system along the reversible path to the initial point cannot be compensated by its decrease at the first stage of the cycle. We obtain a typical situation of *entropy overproduction*. The further fate of this overproduced entropy could be different. The entropy can be accumulated within the system. It degrades as a result, and after a while dies (the first fate). The second fate is that the entropy can be "sucked out" by the environment, and the equilibrium will be re-established. In turn, this can be realised in two ways: either by import of an additional low-entropy energy, which can be used for the system restoration, or, unfortunately, by environmental degradation.

10.8. Entropy overproduction as a criterion of the degradation of natural ecosystems under anthropogenic pressure

When we talk about a "reference ecosystem" we take into account a completely natural ecosystem, without any anthropogenic load impacts. To find such an ecosystem today in industrialised countries is almost impossible (except, maybe, in the territories of natural parks). Really all so-called natural ecosystems today are under anthropogenic pressure (stress, impact, pollution, etc.). All these stresses began to act relatively recently (in about the last 100–150 years) in comparison with characteristic relaxation times of the biosphere, so that we can assume with rather high probability that the mechanisms, responsible for the functioning of the "entropy pump", have not yet adapted to the new situation. On the other hand, plants, which are main components of the natural ecosystem, react to anthropogenic stress very quickly, as a rule, by reducing their productivity.

Let us assume that the considered area is influenced by anthropogenic pressure, i.e. the inflow of artificial energy (W) to the system takes place. We include in this notion ("the inflow of artificial energy") both the direct energy inflow (fossil fuels, electricity, etc.) and the inflow of chemical substances (pollution, fertilisers, etc.). The anthropogenic pressure can be described by the vector of direct energy inflow $\mathbf{W}_\mathrm{f} = \{W_1^\mathrm{f}, W_2^\mathrm{f}, \ldots\}$ and the vector of anthropogenic chemical inflows $\mathbf{q} = \{q_1, q_2, \ldots\}$ to the ecosystem. A state of the "anthopogenic" ecosystem can be described by the vector of concentrations of chemical substances $\mathbf{C} = \{C_1, C_2, \ldots\}$, and such a macroscopic variable as the mean gross primary production (GPP) of the ecosystem. The undisturbed state of the corresponding natural ecosystem in the absence of anthropogenic pressure is considered as a *reference state* and is denoted by $\mathbf{C}_0 = \{C_1^0, C_2^0, \ldots\}$ and GPP_0.

We suppose that the first inflow is dissipated inside the system when transformed directly into heat and, moreover, modifies the plant productivity. The second inflow, changing the chemical state of the environment, also modifies the plant productivity. In other words, there is a link between the input variables $\mathbf{W}_f$ and the state variables $\mathbf{C}$ on the one hand, and the macroscopic variable GPP on the other hand. It is given by the function GPP = GPP($\mathbf{C}, \mathbf{W}_f$). Obviously, if we deal with contamination that inhibits plants' productivity, then this function must be monotonously decreasing with respect to its arguments. On the contrary, if the anthropogenic inflows stimulate plants (as with fertilisers) then the function increases. The typical "dose–effect" curves belong to such a functional class.

By formalising the previous arguments we can represent the entropy production within this "disturbed" ecosystem as

$$\frac{d_i S}{dt} = \frac{1}{T}[W + \mathrm{GPP}(\mathbf{C}, \mathbf{W}_f)], \tag{8.1}$$

where the scalar W is a convolution of the inflows $\mathbf{W}_f$ and $\mathbf{q}$, i.e. the total anthropogenic inflow. Note that the convolution may also depend on $\mathbf{C}$, since these internal concentrations are maintained by the inflows $\mathbf{q}$. We shall demonstrate later how this is calculated.

In accordance with the "entropy pump" concept a certain part of the entropy produced is released at this point by the "entropy pump" with power $|d_e S/dt| = \mathrm{GPP}_0/T$, so that the total entropy balance is

$$\frac{dS}{dt} = \sigma = \frac{1}{T}[W + \mathrm{GPP}(\mathbf{C}, \mathbf{W}_f) - \mathrm{GPP}_0]. \tag{8.2}$$

We assume here that despite anthropogenic perturbation the disturbed ecosystem is tending to a steady state again. If we accept it, we must also assume that the transition from natural to anthropogenic ecosystem is performed sufficiently fast so that the "tuning" of the entropy pump does not change. Therefore, it is unique that the residual part of the entropy production given by Eq. (8.2) is compensated by the outflow of entropy to the environment. *This compensation can occur only at the expense of environmental degradation* ($\sigma > 0$) resulting, for instance, from heat and chemical pollution, and a mechanical impact on the system.

We also assume that the relation of succession connects the "natural" and "anthropogenic" ecosystems. We would like to include a few words about the relation of succession. Let us assume that the anthropogenic pressure has been removed, and the succession from the anthropogenic ecosystem towards a natural one has been started. The next stage of this succession, according to our concept, would be a "natural" ecosystem. Really, if the anthropogenic pressure is weak, the "natural" ecosystem (as we understand it) will be of a "wild" ecosystem type, existing at this site.

It is necessary to note, if the anthropogenic ecosystem is an agro-ecosystem, surrounded by forest, a grass–shrubs ecosystem (not a forest) will be successionally close to its "natural" ecosystem. Finally, we can define a *successionally close ecosystem* (i.e. an ecosystem successionally close to an ecosystem under anthropogenic stress) as the first stage of succession of an "anthropogenic" ecosystem where the anthropogenic stress is removed.

What is the "dynamic" sense of "successional closeness", and why do we need the concept? The point is that in this approach we can compare only close steady states, their vicinities must be intersected significantly, and the time-scale of a quasi-stationary transition from a natural to anthropogenic ecosystem and vice versa must be small (in comparison to the temporal scale of succession).

Let us consider the non-local dynamics of the system. We stop the flux of artificial energy (i.e. the anthropogenic energy and chemical fluxes) into the ecosystem. As a result, if the ecosystem is not degraded, a succession will take place at the site, which tends towards the natural ecosystem type specific for the territory (grassland, steppe, etc.). This is a typical reversible situation. Under severe degradation a succession would also take place, but towards another type of ecosystem. This is quite natural, since the environmental conditions have been strongly perturbed (for instance, as a result of soil degradation). This is an irreversible situation. So, the "successional closeness" concept means that we remain in a framework of "reversible" thermodynamics. And if there is no input of artificial energy, the steady-state for a given site (locality) will be presented by the natural ecosystem, as the local characteristics of the "entropy pump" correspond exactly to the natural type of ecosystem.

Nevertheless, there is a small incorrectness. When we discussed a successionally closed system above, we assumed implicitly that at any stage of the succession the system is in a dynamic equilibrium. Since a succession is a transition process between two steady states, this statement is incorrect. However, so far as we can suggest that the temporal scale of ecological succession is much greater than the same for anthropogenic processes, we can consider succession as a thermodynamically quasi-stationary process (simultaneously, we remain inside the model of equilibrium thermodynamics). Nevertheless, if we want to construct a thermodynamic model of succession, we should drop the hypothesis of quasi-stationary transition.

Bearing in mind the meaning of σ, it is obvious that the σ value can be used as the criterion for environmental degradation or as the "entropy fee" which has to be paid by society (actually suffering from the degradation of environment) for modern industrial technologies.

Of course, there is another way to balance the entropy production within the system. For instance, we can introduce an artificial energy and soil reclamation, pollution control (or, generally, ecological technologies). Using the entropy calculation we can estimate the necessary investments (in energy units).

10.9. Energy and chemical loads or how to convolute the vector data

Let us bear in mind that formula (8.2) describing the "degradative" part of the entropy in a disturbed ecosystem also represents the convolution W as a sum of two items: $W = W_{\mathrm{f}} + W_{\mathrm{ch}}$. If the first item W_{f}, which is a convolution of the vector $\mathbf{W}_{\mathrm{f}}$, can be associated with the direct inflows of such a type of artificial energy as electricity, fossil fuels, etc. (energy load) then the second item W_{ch} is associated with the inflows of chemical elements that maintain molar concentrations $\mathbf{C}$ within the system (chemical load).

Since all direct inflows are measured in common energy units, the convolution of $\mathbf{W}_{\mathrm{f}}$ is defined simply: $W_{\mathrm{f}} = \sum_i W_i^{\mathrm{f}}$. The problem is how to calculate W_{ch} (Svirezhev, 1998b).

Let the system, which is described only by the chemical concentrations as state variables, move from an initial state $\mathbf{C}_0$ towards a state with $\mathbf{C}$ that is maintained by the pressure of the chemical load. We assume that the basic concentrations $\mathbf{C}_0$ correspond to a natural ("wild") ecosystem. We assume as well that if the chemical inflows $\mathbf{q}$ were stopped, then the system would evolve to its natural state. In this case the maintenance of concentration $\mathbf{C}$ within the system means that the work against chemical potentials is performed. As a result the entropy is produced inside the system with the rate

$$\frac{\mathrm{d_i}S}{\mathrm{d}t} = R\sum_i A_{i,i_0} q_i = R\sum_i \ln(C_i/C_i^0) q_i, \tag{9.1}$$

where A_{i,i_0} are the affinities. Then

$$W_{\mathrm{ch}} = RT\sum_i \ln(C_i/C_i^0) q_i. \tag{9.2}$$

Strictly speaking, the above-obtained formula is applicable to sufficiently smooth continuous processes, whereas the real process is saw-tooth with 1-year periodicity. Therefore, in order to study slow dynamics of the entropy overproduction, we have to know how to average the function $\sigma(t)$ within a 1-year interval.

Let an anthropogenic pressure begin to act on the natural ecosystem at an initial moment t_0 when $\mathbf{C}(t_0) = \mathbf{C}_0$. If only the entropy pump "sucks out" entropy from the system then the rate of the entropy production will be

$$\frac{\mathrm{d}S}{\mathrm{d}t} = \frac{W_{\mathrm{f}}(t)}{T(t)} + R\sum_i \ln\left(\frac{C_i(t)}{C_i^0}\right)\frac{\mathrm{d}C_i}{\mathrm{d}t} + \frac{r(t)\mathrm{GPP}(t) + D(t)}{T(t)} - \frac{\mathrm{GPP}_0(t)}{T(t)}, \tag{9.3}$$

where $r(t)$ is the respiration coefficient and $D(t)$ is the rate of dead organics' decomposition within a 1-year interval. By integrating with respect to time and applying the Mean Value Theorem we obtain

$$\begin{aligned} &S(t_0+\tau) - S(t_0) = \\ &\frac{1}{T}\left[\int_{t_0}^{t_0+\tau} W_{\mathrm{f}}(t)\mathrm{d}t + \int_{t_0}^{t_0+\tau}[r(t)\mathrm{GPP}(t) + D(t)]\mathrm{d}t - \int_{t_0}^{t_0+\tau}\mathrm{GPP}_0(t)\mathrm{d}t\right] \\ &+R\sum_{i=1}^{n}\left[C_i(t_0+\tau)\ln\left[\frac{C_i(t_0+\tau)}{C_i^0}\right] - (C_i(t_0+\tau) - C_i^0)\right]. \end{aligned} \tag{9.4}$$

We assume that the temperature T remains constant in the course of the entire period of an anthropogenic impact and is equal to τ. We consider a quasi-stationary process in the sense that $[1 - r(t)]\mathrm{GPP}(t) \approx D(t)$, i.e. there is a dynamic equilibrium between the formation of the new biomass and the decomposition of the dead organic matter. Then the total overproduction of entropy in the course of the entire period of an anthropogenic impact will be equal to

$$\sigma_\tau = S(t_0+\tau) - S(t_0) = \frac{1}{T}[\hat{W}_{\mathrm{f}} + \hat{W}_{\mathrm{ch}} + \langle\mathrm{GPP}\rangle\tau - \mathrm{GPP}_0\tau], \tag{9.5}$$

where $\hat{W}_{\mathrm{f}} = \int_0^{\tau} W_{\mathrm{f}}(t+t_0)\mathrm{d}t$ is the total energy load and $\langle \mathrm{GPP} \rangle = 1/\tau \int_0^{\tau} \mathrm{GPP}(t+t_0)\mathrm{d}\tau$ is the mean gross primary production averaged over the interval of anthropogenic impact. The total chemical load can be represented as

$$\hat{W}_{\mathrm{ch}} = RT\sum_i \left[(C_i^0 + \delta C_i)\ln\left(1 + \frac{\delta C_i}{C_i^0}\right) - \delta C_i \right], \tag{9.6}$$

where δC_i are the change of concentrations at the end of anthropogenic impact in comparison with some successionally close natural ecosystem.

Let us consider the following thermodynamic models.

Model 1. We consider an "average" equilibrium. In other words, we assume that the equilibrium condition $\mathrm{d}S/\mathrm{d}t = 0$ holds in some average sense, i.e. $o_{\tau} = 0$. We also assume that the unique way to provide the last equality is to change the productivity so that

$$\langle \mathrm{GPP} \rangle = \mathrm{CPP}_0 - \frac{1}{\tau}(\hat{W}_{\mathrm{f}} + \hat{W}_{\mathrm{ch}}). \tag{9.7}$$

Thus, we obtained the thermodynamic model for the reduction of the ecosystem productivity under anthropogenic pressure.

Model 2. Sometimes we have an experimental (observed) dependence of productivity GPP on W_{f} and **C**. By calculating the average $\langle \mathrm{GPP} \rangle$ we can check the equality (9.7). If $\langle \mathrm{GPP} \rangle < \mathrm{CPP}_0 - (1/\tau)(\hat{W}_{\mathrm{f}} + \hat{W}_{\mathrm{ch}})$ then we assume that there is another, in addition to the reduction in productivity, mechanism which sucks the entropy excess. Such mechanism would be, in general, the environmental degradation. The entropy measure of the degradation is equal to

$$\sigma_{\tau}T = \hat{W}_{\mathrm{f}} + \hat{W}_{\mathrm{ch}} + [\langle \mathrm{GPP} \rangle - \mathrm{GPP}_0]\tau. \tag{9.8}$$

Model 3. Let the dynamics of the anthropogenic system (in particular with respect to the chemical load) be "impulsive", i.e. at the end of each year the system is spontaneously returned to the initial ("natural") state so that at the beginning of every next year the system is starting from the initial state. This situation is typical for agro-ecosystems, which are not too far from the natural ones. This spontaneous transition is accompanied by the entropy production described by Eq. (9.5) for $\tau = 1$ year. Since the averaging interval is equal to 1 year we can consider the average values as the annual ones. If we also assume that from year to year the system is, on average, in the equilibrium state, then we can introduce the following entropy measure for an environmental degradation:

$$\sigma_{\tau}T = W_{\mathrm{f}} + W_{\mathrm{ch}} + \langle \mathrm{GPP} \rangle - \mathrm{GPP}_0, \tag{9.9}$$

where

$$W_{\mathrm{ch}} = RT\sum_i [C_i \ln(C_i/C_i^0) - (C_i - C_i^0)].$$

It is very interesting that the value W_{ch} formally coincides with the "chemical" exergy (see Chapter 5).

10.10. Summary of the ecological important issues

This chapter analyses the photosynthesis thermodynamically. Schneider and Kay (1994) have interpreted the energy balance of an active surface of vegetation: the more complex the structure of a vegetation system, the more effectively it dissipates the gradients and transfers the exergy to an exergy in the form of heat by the temperature of the environment, which seems in contrast to Prigogine's theorem and the interpretation by exergy (Jørgensen, 1997).

First of all, when the total biomass of the system stops increasing, the amount of exergy captured has reached the maximum ($\approx 80\%$ of the solar radiation), but we know that ecosystems can continue the development beyond this limit. Ecosystems have three growth forms: growth by increase of the biomass, growth by increasing the network and growth by increase of the information content. Only the first growth form is associated with more exergy destruction, because more biomass requires more exergy for maintenance, but the exergy destruction per unit of biomass is the same. A more complex network, on the other hand, is able to utilise the exergy of the solar radiation better, and increased information means increased exergy but not increased exergy consumption for maintenance.

The thermodynamic examination of photosynthesis in this chapter has introduced two important efficiency coefficients: the radiation efficiency is defined as the ratio of the amount of energy used for the photosynthesis to the radiation, and the exergy efficiency is the increment of information exergy relative to the total amount of energy coming from the solar radiation. Three teleological hypotheses are formulated as a result of the thermodynamic analysis of the photosynthesis:

(1) Vegetation works as an information machine—the exergy efficiency $>$ the radiation efficiency.

(2) The exergy balance, the energy balance and the increment of information (Kullback's measure of information) have all maxima when the productivity of the vegetation is also at the maximum.

(3) The exergy efficiency coefficient is minimised with respect to the radiation efficiency (approaches the same value of the exergy efficiency as the value of the radiation efficiency) and maximised with respect to increment of information (Kullback's measure of information, K). In other words, the photosynthesis attempts to move the system as far away as possible under the given constraints.

Maintenance of the biological structure requires a major exergy input, including the coverage of evapotranspiration. It implies a high production of entropy, but the entropy is transferred continuously from the ecosystem to the environment, whereby the temperature is maintained within the range suitable for life processes. As discussed in Chapter 3, life not only requires an energy source, but also a transfer of the produced heat (entropy) to the environment. Expressed differently, ecosystems have an effective entropy pump sucking the entire entropy out of the system.

Chapter 11

Thermodynamics of the biosphere

In nature all is harmony,
A consonance fore'er agreed on...
F. Tyutchev, 1865

...It is living matter—the Earth's sum total of living organism—that transforms the radiant energy of the Sun into the active chemical energy of the biosphere. Living matter creates innumerable new chemical compounds and extends the biosphere at incredible speed as a thick layer of new molecular systems. These compounds are rich in free energy in the thermodynamic field of the biosphere. Many of the compounds, however, are unstable, and are continuously converted to more stable form...
...Mechanisms, created by this way—*living organisms*—are principally distinct from other atomic, ionic, or molecular systems in the Earth's crust, both within and outside the biosphere...
V. Vernadsky "Biosphere", 1926

11.1. Introduction

At the beginning of the XIXth century, J.-B. Lamarque had introduced the term "Biosphere". He considered it as the "Scope of Life" and some sort of external cover for Earth. In 1875, the same term was introduced in geology by E. Süss, who distinguished the biosphere as one of Earth's covers. But V. Vernadsky was the person who first created the modern concept of the biosphere. This concept was stated in two lectures published in 1926. Later, it was further developed by Vernadsky himself and by Kostitzin (1935), Timofeev-Resovsky (1961a), Sukhachev (1967) and other Russian scientists. All this work put together allows us to speak of the Russian classical school (Svirezhev, 1974).

According to Vernadsky, the biosphere is an external Earth cover, the Scope of Life (let us recall Lamarque). But he notes also that this definition (as for the Scope of Life) is not complete. The Vernadsky biosphere includes:

- "living matter",
- "bio-generic matter", i.e. organic and mineral substances created by living matter (for instance coal, peat, litter, humus, etc.),
- "bio-inert matter", created by living organisms together with inorganic Nature (water, atmosphere, sediment rocks).

Towards a Thermodynamic Theory for Ecological Systems, pp. 271–299

From the point of view of thermodynamics the biosphere of our planet is a typical open system that exchanges energy and matter with Space and Earth's mantle. Certainly, both the meteoric flux and the volcanism are matter fluxes into the system, and the escaping gases and plates tectonics are typical matter fluxes out of it, but the influence of all these fluxes is negligibly small (at least, for considered periods of time). On the other hand, if we would like to consider the long-term evolution of the biosphere, all the above-mentioned slow fluxes have to be taken into account.

The biosphere is a typical "large" system with an abundance of interacting components. In addition to the methodology of large systems in the study of biosphere processes, we may also use the following two paradigms: the "atomistic" paradigm of bio-geocoenoses (BGC) and the paradigm of invariant structure of biogeochemical cycles.

(a) *Bg-paradigm.* While the Vernadsky concept may be considered as maximally aggregated (it is like a view of the biosphere from the outside), the concept of the BGC suggested by Sukhachev (1967) and developed by Timofeev-Resovsky (1961b) relates to the elementary units of the biosphere, a concept that is basically atomistic in nature. In accordance with the Timofeev-Resovsky definition, the BGC is a part (area) of the biosphere, which has no any essential ecological, geo-morphologic, hydrological, micro-climatic or any other borders within itself. The entire biosphere is divided by these borders on elementary systems, naturally separated from one another. Due to the reality of existence of these boundaries, the BGCs can be considered as half-isolated subsystems, the averaging inside them is quite natural. So, the BGC dynamics can be described by a comparatively small number of variables.

The BGC is also an elementary unit of biogeochemical work in the biosphere. Indeed, all the local nitrogen and phosphorus turnovers are practically closed inside the BGC (excluding denitrification); therefore, in respect to these elements, the BGC may be considered as an almost closed system. If we consider the bio-geocenotic local carbon cycle then we can see that the number of different internal paths of carbon is much higher than the number of carbon inflows and outflows connecting the given BGC with others and with the atmosphere. In other words, the internal structure of BGC is more complex than the structure of its connections with the environment—although all BGCs are connected to the single entity, the biosphere, by the atmosphere and hydrosphere. It is very important that all BGCs have the same structure of the local biogeochemical cycles; therefore, they are dynamically similar, differing from each other only by the parameters. It implies that if we describe the BGC dynamics as the dynamics of local biogeochemical cycles, then the differences between the BGCs are the differences in parameters of the same dynamic systems.

Finally, if the model of the biosphere is primarily the model of global biogeochemical cycles, then the whole biosphere can be considered as a system of loosely interacting elementary subsystems, subjected to the same dynamic laws and relations. Elementary units (subsystems) differ from each other only by different values of parameters. So, we can consider the biosphere as the statistical ensemble.

According to N. Basilevich (personal communication), there are about 50,000 BGCs on the Earth's surface. In principle, we would not have any problems if we had complete information on all the parameters of all BGCs (first of all, their spatial distribution on Earth). However, this is far from being the case. Moreover, our knowledge is unsatisfactory for the most part of the Earth's surface. The only solution is to apply methods of spatial interpolation and extrapolation of the available information (under the assumption of continuity).

(b) *Hypothesis of the invariant structure of biogeochemical cycles.* The second paradigm is based on the hypothesis of the invariant structure of biogeochemical cycles. In other words, the structure of interrelations, along which biogeochemical flows, water and energy are circulated, does not depend on the size of the system itself, or on its geographical location; only the values of flows change. Therefore, in accordance with this concept, we can consider the whole terrestrial vegetation pattern as some united ecosystem described by some average parameters, not differing structurally, for instance, from a forest ecosystem.

However, these two approaches should lead to the same result, as the difference between them is the difference between thermodynamics and statistical mechanics, i.e. the difference in the methods of averaging.

Continuing further study of the biosphere problem, we suggest using the following three basic axioms (see also Svirezhev, 1998a, 1999):

1. The contemporary state of the biosphere is stationary.
2. This state is stable in Lyapunov's sense.
3. This state is structurally stable.

In the process of the study we implement two scientific paradigms: Vernadsky's concept of the biosphere and the evolutionary (Darwinian) paradigm. The evolutionary paradigm gives us the possibility to assert that the current state of the biosphere is the result of the evolutionary process, in that natural selection chooses this state from multiple virtual variants. Therefore, any biosphere model should possess some properties of selectivity, i.e. it should be the non-linear model with multiple equilibriums.

It is obvious that the model has also to comply with the different laws of conservation (energy, matter and momentum), since the biosphere is practically a closed system (with respect to matter) and a "through-flow" system (with respect to energy). The latter means that perhaps the biosphere is a typical dissipative structure.

The model should not be contradictory from the thermodynamic point of view, since the biosphere is an open system far from thermodynamic equilibrium.

An application of these "selective" criteria allows us to assume that such a model would adequately describe the dynamics of the biosphere.

11.2. Comparative analysis of the energetics of the biosphere and technosphere

The biosphere as an open thermodynamic system exists due to a permanent flow of solar energy. Earth receives 3.5×10^{24} J of solar energy annually that maintains the work of the climatic machine. The function of the "green cover" results in 5.5×10^{21} J/year of new biomass. Thus, although vegetation is the main concentrator and transformer of solar energy in the biosphere, it uses only 0.16% of solar energy for the creation of a new biomass. The rest is spent on the process of transpiration by leaves, providing water and nutrient transport, etc. Approximately 60% of it is immediately used for respiration and the remaining 40% is the *annual global production*, which is equal to 2.3×10^{21} J/year (Svirezhev and Svirejeva-Hopkins, 1998).

The energetic characteristics of the biosphere have not significantly changed since the beginning of the era of vascular plants (about one billion years ago). The efficiency coefficient of the autotrophic component, the functioning of which provides the energetic basis for evolution of animals, is equal to $\eta = 2.3 \times 10^{21}/3.5 \times 10^{24} \approx 0.66\%$.

Stability of the biosphere is maintained by the permanent dissipation of energy. In other words, the biosphere is a typical *dissipative system*. This energy flow provides a steady state for 1.84×10^{18} g of living biomass (or 3.5×10^{22} J), and animal biomass constitutes only 0.8% of it, i.e. 1.46×10^{16} g. Animals consume only 3% of the NPP (7.35×10^{19} J/year) (Smil, 1991) that maintains both the metabolism of living matter and its diversity. The latter is the *information basis of evolution*.

At the present time, Earth's technosphere (our technological civilisation) spends about 3×10^{20} J/year to provide its functioning and evolution (Svirezhev and Svirejeva-Hopkins, 1998). This is mainly the energy of fossil fuels and nuclear energy. A share of "pure" biosphere energy (hydropower station and firewood) in this balance is small ($\sim 5\%$).

It is obvious that *Homo sapiens* is a component of both the biosphere and technosphere. If we consider humans as animals, then all human energetic requirements are satisfied through food, and the annual energy food demand per individual is 4×10^9 J. For the current population size of *Homo sapiens* ($\approx 6 \times 10^9$ individuals) annual energy food demand is equal to 2.4×10^{19} J/year. By comparing these values we see that *the energy demand of mankind (as a biological species) is currently equal to one-third of the total biological energy of the biosphere which is accessible to animals* (7.35×10^{19} J/year). Fig. 11.1 represents the dynamics of food energy demand for mankind, using the reconstruction of human population growth from the Neolithic era. Until the Neolithic revolution, when man changed his behaviour from gathering to producing food, he was part of the biosphere, no different from other animals.

At the time of the Neolithic revolution the human population included 4×10^6 individuals, and required the energy supply of 1.6×10^{16} J/year, which was 0.022% of the total energy flow for all animals.

According to the physical theory of fluctuations (Landau and Lifshitz, 1995) the probability of fluctuation which could cause the elimination of *Homo sapiens* is equal to:

$$\mathrm{Pr} = \exp\left[-\frac{\text{energy demand for human population}}{\text{energy supply for all animals}}\right] = \exp\left[-\frac{1.6 \times 10^{16}}{7.35 \times 10^{19}}\right] \approx 99.98\%.$$

From Neolithic times until the beginning of industrial revolution (at the edge of the XVIIIth and XIXth centuries) with its own source of energy (fossil fuels), Man had been only a part of the biosphere. Humans were competing with other species, and had increased their energy demand up to 4×10^{18} J/year, so that the probability of their elimination decreased until $\mathrm{Pr}' = \exp[-4 \times 10^{18}/7.35 \times 10^{19}] \approx 94.8\%$. Looking at these

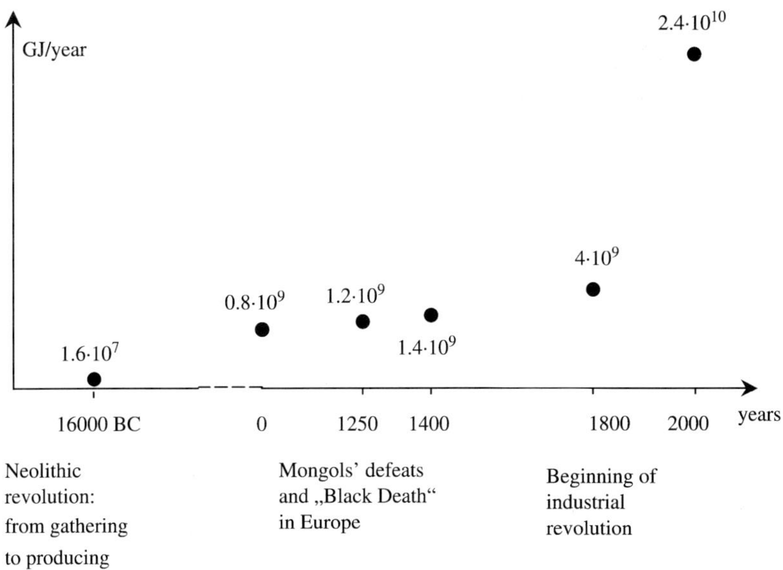

Fig. 11.1. Energy food demand for mankind, 1 GJ = 10^9 J. From Svirezhev and Svirejeva-Hopkins (1998) with changes.

numbers one can say that *Homo sapiens* as a biological species was very fortunate that it had not been eliminated before the technosphere arose.

The NPP of the biosphere, 2.3×10^{21} J/year, is the energy flow, which maintains the diversity of biota. Even now, the energy flow used by the technosphere, 3×10^{19} J/year, is only about 13% of the biosphere NPP. At the present moment, the biosphere and technosphere are in a state of strong competition for common resources, such as land area and fresh water. Contamination of the environment and reduction of the biota diversity are the consequences of the competition.

Since the biosphere (considered as an open thermodynamic system) is at the dynamic equilibrium, then all entropy flows have to be balanced too. Therefore, the entropy excess, which is created by the technosphere, has to be compensated by means of two processes: (1) degradation of the biosphere, and (2) change in the work of Earth's climate machine (in particular, the increase of Earth's average temperature).

Let us assume that all energy consumed by the technosphere is transformed into heat Q. Then the annual entropy produced by the technosphere is equal to $S_{\text{tech}} = Q/T = 3 \times 10^{20}\ \text{J}/287\ \text{K year} \approx 10^{18}\ \text{J/K year}$ at the mean annual temperature of Earth equal to 14°C. The full destruction of biota, which, we assume, is equivalent to its full combustion, gives us the following value of entropy: $S_{\text{d}} = 3.5 \times 10^{22}\ \text{J}/287\ \text{K} \approx 1.210^{20}\ \text{J/K}$. If we assume that the energy consumption of the technosphere would not be increased, then this "anti-entropy storage" of biota would be enough to compensate the entropy produced by the technosphere in the course of the next 120 years. If this techno-generic entropy would be compensated by soil destruction then the agony would be continued in the course of 300–400 years, since the organic matter storage in soil is 3–4 times larger than in biota.

11.3. Myth of sustainable development

Let us talk about the concept of sustainability. Despite its widespread use, there is no rigorous, mathematically correct definition of sustainability. (Note, the same situation exists with the term "stability" except there is a mathematical theory of stability with the rigorous Lyapunov definition, but the "sustainability" concept was still not formalised).

As a result of the Brundtland Commission book "Our common Future. From one Earth to one World" (Oxford University Press, Oxford–New York, 1987), the concept of sustainability is well known today. Nevertheless, it is necessary to note that the concept has a longer time history. Let us recall that V. Vernadsky has introduced a new global system called "Noosphere", which has to be a final stage of the co-evolution of the biosphere and human technological civilisation. Unfortunately, this very attractive idea of *sustainable development* runs counter to the basic laws of physics (the Second Law of Thermodynamics). What arguments can be used to prove this last thesis (see also Svirezhev and Svirejeva-Hopkins, 1998)?

Let us consider the entropy balance of a single area of the Earth's surface occupied by some natural ecosystem (for details see Chapter 10). At the dynamic equilibrium the entropy production within the system is balanced with the entropy flow from the system to the environment. This work is done by the "entropy pump".

Suppose that the considered area is influenced by anthropogenic pressure represented by the direct inflows of artificial energy (energy load) and the inflows of chemical substances (chemical load). This is a typical impact of industry (or, in a broader sense, technological civilisation) on the environment.

If we consider the main characteristic features of technological civilisation, we can see that it creates the energy and chemical loads. These features are:

(a) the use of the non-biosphere sources of energy (fossil fuels, which are the traces of past biospheres, not replenishable by the current biosphere, and nuclear energy);
(b) technological processes increase concentrations of chemical elements in the biosphere (metallurgy, chemical industry, etc.);
(c) dispersion of chemical elements in comparison with their "biotic" concentrations.

All these above-mentioned processes produce entropy that cannot be "sucked" out by the biosphere's "entropy pump". But since the ecosystem should remain in a dynamic equilibrium with its environment, the entropy production (overproduction) of the ecosystem should be compensated by the outflow of entropy to the environment. This compensation can occur only at the expense of environmental degradation in this or, maybe, another location, heat and chemical pollution or caused by mechanical impact on the system. The value of the overproduction (as was shown in Chapter 10) can be used as a criterion for the environmental degradation or as an "entropy fee", which has to be paid by society (really suffering from the degradation of environment) for the modern industrial technologies. Thus, degradation of the environment is the only way to compensate for the overproduction of entropy. The process of overproduction can be non-homogenous in space and then there is the spatial transportation of entropy. This transportation can be either natural or artificial. The natural process of entropy transportation is realised as the wide spreading of different pollutants by natural agents (wind, rivers, etc.). The artificial process is either a purposeful export of industrial waste to other regions, or the import of low-entropy matters (for example, fossil fuels) from other regions. So, the main conclusion is:

Sustainable development is possible only locally, and only at the expense of creating "entropy dumps" elsewhere.

11.4. Thermodynamics model of the biosphere. 1. Entropy balance

From the point of view of thermodynamics, our planet (Earth, *E*) is a closed system, which exchanges with its environment (Space, *S*) only energy. Let us assume that the total mass of Earth's matter is not changing in the course of the considered time interval. Therefore, we can neglect the "slow" input and output flows within "geological" characteristic time and consider only the solar radiation and Earth's heat irradiation as a unique type of exchange between Earth's system and its environment. Moreover, we assume that the planetary radiation balance is also constant. As such a kind of interval we can assume 10^3 years.

Let d_iS_E be the annual production of entropy by Earth and d_eS_E be the annual export of entropy into Space. If the joint system "Earth + Space" is in a dynamic equilibrium, then $d_iS_E = -d_eS_E$, where

$$d_eS_E = \frac{4}{3}dE_{SE}\left(\frac{1}{T_S} - \frac{1}{T_E}\right) \approx -1.8 \times 10^{22} \text{ J/K year,} \tag{4.1}$$

$dE_{SE} \approx 3.6 \times 10^{24}$ J/year is the annual solar energy assimilated by Earth, $T_S = 5770$ K and $T_E = 257$ K are radiation temperatures of the Sun and Earth, and the factor $4/3$ is the Planck form-factor (Ebeling et al., 1990).

Assume that our system (Earth) can be represented as a sum of the following subsystems: atmosphere (A), biota (B), pedosphere (P), hydrosphere (H) and lithosphere (L). Note that the mass of biota can be identified with Earth's phytomass; only water (or salt dilutions) is included into the hydrosphere, polar ices and glaciers are included in the lithosphere, which, in turn, contains Earth's core, mantle and kernel. Each of these subsystems exchanges with each other and with Space by energy and matter. In particular, the matter exchange is realised by means of the global biogeochemical cycles. There is no exchange between biota and lithosphere.

Let $d_iS_j \geq 0$ be the entropy production by *j*th ($j =$ A, B, P, H, L) subsystem, dS_{jk} the export of entropy from *j*th subsystem to *k*th one, dS_{Sk} the export of entropy from Space (environment) to *k*th subsystem. By writing the equations of entropy balance for all the subsystems, then summarising them and taking into account that $dS_{jk} = -dS_{kj}$ for any pair of subsystems we get:

$$\mathrm{d}S_{\mathrm{E}} = \sum_j \mathrm{d}S_j = \sum_j \mathrm{d}S_{\mathrm{S}j} + \sum_j \mathrm{d}_{\mathrm{i}}S_j \tag{4.2}$$

where $\mathrm{d}S_{\mathrm{E}}$ and $\mathrm{d}S_j$ are the annual increments of entropy for Earth (in total) and for each of its subsystems, the sum $\sum_j \mathrm{d}S_{\mathrm{S}j}$ is the annual export of entropy from Space, and the sum $\sum_j \mathrm{d}_{\mathrm{i}}S_j$ is the total annual production of entropy by subsystems. Then for the system "Earth" we have:

$$\sum_j S_j = \text{const}, \quad \sum_j \mathrm{d}S_j = \sum_j \mathrm{d}S_{\mathrm{S}j} + \sum_j \mathrm{d}_{\mathrm{i}}S_j = 0, \tag{4.3}$$

i.e. as one might follow,

$$\mathrm{d}_{\mathrm{i}}S_{\mathrm{E}} = \sum_j \mathrm{d}_{\mathrm{i}}S_j; \quad \mathrm{d}_{\mathrm{e}}S_{\mathrm{E}} = \mathrm{d}S_{\mathrm{SE}} = \sum_j \mathrm{d}S_{\mathrm{S}j}. \tag{4.4}$$

In accordance with Eq. (4.1) the upper limit of the annual entropy production within all subsystems of the system "Earth" is equal to $\sum_j \mathrm{d}_{\mathrm{i}}S_j = 1.8 \times 10^{22}$ J/K.

Note that in spite of the validity of a condition of the Prigogine theorem about the minimum entropy production for the whole system, its application to each subsystem is not correct. Each of them is the non-linear thermodynamic system, in which the dependence of the exchange flows of energy and matter from each other may also be non-linear.

We shall calculate the entropy contained in each subsystem and entropy flows between the subsystems (Venevsky, 1991; Svirezhev, 2001a). For that we shall also use the ideas and methods from Morowitz's book (1968). These calculations are quite cumbersome and, therefore, we shall omit the details. But first let us note that the values of S_j in Eq. (4.2) have different orders of magnitude: $S_{\mathrm{L}} >> S_{\mathrm{A}}, S_{\mathrm{B}}, S_{\mathrm{P}}, S_{\mathrm{H}}$, since the mass of lithosphere is more than six orders of magnitude higher than that of other subsystems.

So, for the subsystems we have:

Atmosphere. We use a polytropic model of the "static" atmosphere (Khrgian, 1983). Such a "static" atmosphere is a non-equilibrium thermodynamic system consisting of the isothermal layers with their temperatures (the temperature of the ground layer is 14°C). In accordance with Landau and Lifshitz (1995) the entropy of such a system is equal to the sum of entropies of each layer, which is the mixture of ideal gases N_2, O_2, CO_2, argon and H_2O vapour. Their specific entropies at temperature 25°C and a pressure of 1 atm are equal to $s_{N_2} = 192$, $s_{O_2} = 205$, $s_{CO_2} = 214$, $s_{Ar} = 155$, $s_{H_2O} = 154$, correspondingly. They are expressed in J/K mole. The calculation of entropy for the mixture of gases, if their molar concentrations and specific entropy for each component are known, is a standard operation in thermodynamics (see, for instance, Landau and Lifshitz, 1995). Then the total entropy of the polytropic atmosphere (under the assumption of the constancy of its composition) is equal to $S_{\mathrm{A}} \approx 3.5 \times 10^{22}$ J/K. Since carbon dioxide is one of the "life-forming" gases, we can calculate the corresponding entropy separately: $S_{\mathrm{A}}^{CO_2} \approx 1 \times 10^{19}$ J/K biota. We assume that the subsystem "biota" is identified with the terrestrial phytomass and submitted into a thermostat with the temperature $T = 14$°C, i.e. with the temperature equal to the mean annual temperature of our planet. The standard (averaging) composition of biomass is the following: liquid H_2O—44%, fixed H_2O—6%, cellulose—37.5%, proteins—8.4%, carbo-hydrates, lipids, etc.—4.1%. The total mass of the dry matter is 1.84×10^{18} g. Since the specific molar entropies for each fraction are known, then the total entropy is easily calculated: $S_B \approx 9.1 \times 10^{18}$ J/K. By excluding the water entropy we immediately get the entropy of dry biomass equal to $S_B \approx 2.8 \times 10^{18}$ J/K. Estimating the value of entropy by another method, Morowitz (1968) has $S_B \approx 2.7 \times 10^{18}$ J/K, i.e. generally speaking, a very similar value.

The main uncertainty here is in the estimation of the fraction of H_2O in a living biomass. The estimates vary in a very wide interval from 40 up to 90% for different plant species. Apparently, the estimation in 50% is understated and, therefore, the total entropy of biota is about 10^{19} J/K.

It is also necessary to take into account that water in biota can be in two states: liquid and ice-like, for instance in crystal lattices of cellulose. Absolute molar entropy of water in the latter state at 14°C is less than the same entropy for liquid water by ≈3 J/K mole. Therefore, increase of the percent of fixed water can decrease the total entropy of biota, but its order of magnitude is not higher than 10^{17}.

So, the entropy of biota (biosphere) is less by three orders of magnitude than the atmosphere entropy and has the same order as the CO_2 entropy in the atmosphere.

11.5. Thermodynamics model of the biosphere. 2. Annual increment of entropy in the biosphere

In order to estimate the annual increment of entropy in the biosphere (biota) we need information about in- and outflows of energy and matter. A lot of such data for 1970 was published by Costanza and Neil (1982), and we shall use them. Again we assume that the subsystem of biota is a thermostat with $T = 14$°C.

The annual increment of entropy

$$dS_B = d_iS_B + dS_{SB} + dS_{PB} + dS_{HB} + dS_{AB}. \tag{5.1}$$

Let us consider each term of Eq. (5.1) separately.

Calculating dS_{SB} we assume that vegetation uses only direct solar energy. Costanza and Neil estimate this flow as $E_{SB} = 1.04 \times 10^{24}$ J/year; then

$$dS_{SB} = \frac{4}{3} dE_{SB} \frac{1}{T_S} \approx 2.403 \times 10^{20} \text{ J/K year}. \tag{5.2}$$

The value of dS_{PB} is determined by flows of mineral elements from soil into plants and flows of dead organic matter from biota into soil. If the first flows are neglected (it is possible since the mass percent of mineral elements in the living biomass is relatively low, as well as their specific entropies), then $dS_{PB} = -s_{bio}^{287} q_{DOM}$, where the annual flow of dead organic matter $q_{DOM} = 3.34 \times 10^{17}$ g/year (living biomass contains about 50% water), and $s_{bio}^{287} \approx 2.47$ J/K g is the specific entropy of living biomass at 287 K. The first value is taken from Costanza and Neil, and the second value was obtained using the results of Section 11.4. In fact, $S_B \approx 9.1 \times 10^{18}$ J/K, and the total mass of the dry matter is

1.84×10^{18} g. Then $s_{\text{bio}}^{287} = 9.1 \times 10^{18}/2 \times 1.84 \times 10^{18} \approx 2.47$ J/K g and

$$\mathrm{d}S_{\text{PB}} = -s_{\text{bio}}^{287} q_{\text{DOM}} \approx -8.25 \times 10^{17} \text{J/K year.} \tag{5.3}$$

The value of $\mathrm{d}S_{\text{HB}} = s_{\text{H}_2\text{O}}^{287} q_{\text{H}_2\text{O}}$, where $q_{\text{H}_2\text{O}} \approx 6.67 \times 10^{19}$ g H_2O/year is the mass of liquid water consumed annually by biota (Costanza and Neil, 1982) and $s_{\text{H}_2\text{O}}^{287} \approx 3.89 \times$ J/K g (Morowitz, 1968). Then

$$\mathrm{d}S_{\text{HB}} = s_{\text{H}_2\text{O}}^{287} q_{\text{H}_2\text{O}} \approx 2.595 \times 10^{20} \text{ J/K year.} \tag{5.4}$$

The value of $\mathrm{d}S_{\text{AB}}$ consists of the following exchange flows of entropy:

1. Entropy flow caused by diffusion of CO_2 through stomata into leaf, $\mathrm{d}S_{\text{AB}}^{\text{CO}_2}$;
2. Entropy flow caused by diffusion of O_2 through stomata into the atmosphere, $\mathrm{d}S_{\text{AB}}^{\text{O}_2}$;
3. Entropy flow caused by evapotranspiration, $\mathrm{d}S_{\text{AB}}^{\text{H}_2\text{O}}$.

Let us consider each of these flows separately. Analogously with $\mathrm{d}S_{\text{HB}}$ we have: $\mathrm{d}S_{\text{AB}}^{\text{CO}_2} = s_{\text{CO}_2}^{287} q_{\text{CO}_2}$ and $\mathrm{d}S_{\text{AB}}^{\text{O}_2} = -s_{\text{O}_2}^{287} q_{\text{O}_2}$, where q_{CO_2} and q_{O_2} are the annual amounts of carbon dioxide and oxygen consumed and released by vegetation in the process of photosynthesis, correspondingly: $q_{\text{CO}_2} = 2.255 \times 10^{17}$ g CO_2/year and $q_{\text{O}_2} = 1.64 \times 10^{17}$ g O_2/year. Accordingly with Morowitz, the values of corresponding specific entropies are equal to $s_{\text{CO}_2}^{287} \approx 4.86$ J/g K and $s_{\text{O}_2}^{287} \approx 6.41$ J/g K. Then

$$\begin{aligned} \mathrm{d}S_{\text{AB}}^{\text{CO}_2} &= s_{\text{CO}_2}^{287} q_{\text{CO}_2} \approx 1.096 \times 10^{18} \text{ J/K year,} \\ \mathrm{d}S_{\text{AB}}^{\text{O}_2} &= -s_{\text{O}_2}^{287} q_{\text{O}_2} \approx -1.051 \times 10^{18} \text{ J/K year.} \end{aligned} \tag{5.5}$$

The entropy flow $\mathrm{d}S_{\text{AB}}^{\text{H}_2\text{O}}$ is a jump of entropy caused by the phase transition "liquid water → water vapour": $\mathrm{d}S_{\text{AB}}^{\text{H}_2\text{O}} = q_{\text{H}_2\text{O}}^{\text{BA}}(\mathrm{d}Q/T_{\text{B}})$, where $q_{\text{H}_2\text{O}}^{\text{BA}} = 5.67 \times 10^{19}$ g H_2O/year is the annual total mass of evapotranspirated water, $\mathrm{d}Q = -2514$ J/g H_2O is energy of the phase transition, and $T_{\text{B}} = 287$ K is the temperature of biota. Then

$$\mathrm{d}S_{\text{AB}}^{\text{H}_2\text{O}} = q_{\text{H}_2\text{O}}^{\text{BA}}(\mathrm{d}Q/T_{\text{B}}) \approx -4.97 \times 10^{20} \text{ J/K.} \tag{5.6}$$

We shall summarise all these flows in accordance with their orders of magnitude. So, the sum of flows, having 20 decimal exponents, is

$$\mathrm{d}S_{20} = \mathrm{d}S_{\text{SB}} + \mathrm{d}S_{\text{HB}} + \mathrm{d}S_{\text{AB}}^{\text{H}_2\text{O}} = (2.403 + 2.595 - 4.97) \times 10^{20} \approx 2.8 \times 10^{18}, \tag{5.7}$$

i.e. the summation of the exchange flows of entropy related to the global energy and water cycles leads to the loss of two orders of magnitude. Note that the value of $\mathrm{d}S_{20}$ will be lower if a jump of entropy caused by the phase transition from fixed water to liquid in the process of evapotranspiration is taken into account, i.e. *two* phase transitions are considered. The entropy jump is equal to

$$\Delta S^{287} = p q_{\text{H}_2\text{O}}^{\text{BA}} (s_{\text{fix,H}_2\text{O}}^{287} - s_{\text{H}_2\text{O}}^{287}), \tag{5.8}$$

where p is the fraction of fixed water in the total evapotranspiration, and $(s_{\text{fix,H}_2\text{O}}^{287} - s_{\text{H}_2\text{O}}^{287}) = -0.168$ J/K g H_2O is the difference of corresponding entropies. The value of p is usually estimated as 12%; then $\Delta S^{287} = -1.14 \times 10^{18}$ J/K. Assuming that $p = 30\%$, we get

$\Delta S^{287} = -2.85 \times 10^{18}$ J/K. Therefore, with a very high probability we obtain that $dS_{20} \approx 1 \times 10^{18}$ J/K. Moreover, since the percent of water in biota is just a little understated, we can state that the value of dS_{20} has the order of magnitude lower than 18.

The summation of flows with 18 decimal exponents gives the value $dS_{18} = dS_{AB}^{CO_2} + dS_{AB}^{O_2} = (1.096 - 1.051) \times 10^{18} \approx 4.5 \times 10^{16}$ J/K year, i.e. the exchange flows of entropy related to CO_2 and O_2 are almost balanced, so that their sum is also reduced by two orders of magnitude.

In order to obtain the total entropy balance we have to estimate the internal production of entropy, $d_i S_B$, by biota, which is mainly connected with chemical reactions of the formation of structural molecules of plant and animal biomass (the formation of proteins and cellulose). Organic compounds containing phosphorus take an active part in such a type of reaction, therefore $dS_{AB}^{CO_2}, dS_{AB}^{O_2} >> d_i S_B$ by virtue of relation of the orders of magnitude in the phosphorus and carbon biochemical cycles (Morowitz, 1968).

Finally, since the exchange entropy flow between the pedosphere and biota is $dS_{PB} = -8.25 \times 10^{17}$ J/K year, the annual entropy balance for the biosphere is equal to

$$\begin{aligned} dS_B &\approx dS_{20} + dS_{18} + dS_{PB} = 1 \times 10^{18} + 4.5 \times 10^{16} - 0.825 \times 10^{18} \\ &\approx 10^{17} \text{ J/K year}. \end{aligned} \tag{5.9}$$

So, with the accuracy of two orders of magnitude (entropy of the biosphere is about 10^{19} J/K) we can state that the entropy balance for the biosphere in 1970 was equal to zero, i.e. *in spite of the anthropogenic impact the biosphere was in dynamic equilibrium*. The balance is shown in Fig. 11.2.

The result is very important, since the hypothesis about a quasi-stationary state of the contemporary biosphere plays one of the main roles in globalistics (Svirezhev, 1997a, 1998a). Should we test this statement somehow? Generally speaking, numerous estimations show that zero energy balance of the biosphere is fulfilled with sufficient accuracy that testifies to the advantage of the stationary hypothesis, but this is not entirely sufficient. In order to test whether the thermodynamic condition is sufficient, the balance between the internal entropy production and its export into the environment has to be estimated. The estimation has shown that it is almost equal to zero. Therefore, the system is in dynamic equilibrium with its environment, i.e. the contemporary biosphere is in a quasi-stationary state.

11.6. Exergy of solar radiation: global scale

In Section 10.6 of Chapter 10, we calculated the intraseasonal exergy balance for several types of vegetation localised in different geographical sites (two forests, grass and agriculture). We shall consider here a global scale of the problem (see also Svirezhev et al., 2003). In particular, it is interesting to see what new characteristics are added to the overall portrait of our planet by the global map of total exergy.

In order to calculate the global spatial distribution of exergy we used the surface radiation budget (SRB) data set collected and processed by Rossow and Schiffer (1991)

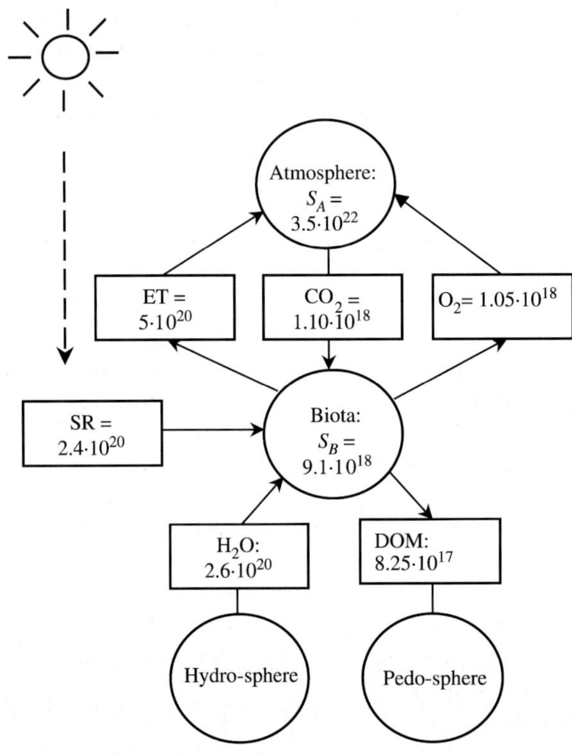

Fig. 11.2. Annual entropy balance for the biosphere (biota) in the 1970s. SR—solar radiation; CO_2—carbon dioxide (net); H_2O—liquid phase of water; DOM—dead organic matter. All storages are in J/K, all flows are in J/K year.

and Darnell et al. (1992, 1996). This data set was developed on the $1° \times 1°$ grid (360×180 points). It consists of monthly means covering the period July 1983–June 1991. It covers the spectral range of 0.2–50 μm, and is divided into two regions: the short-wave (SW, 0.2–5.0) and the thermal long-wave (LW, 5.0–50) micrometers. Both the downward and net radiation at the surface are given.

Among others the following parameters are given: all-sky downward short-wave flux (SWDWN, insolation), total sky net short-wave flux (SWNET, absorbed), all-sky downward long-wave flux (LWDWN), total sky net long-wave flux (LWNET; all in W/m^2). In our notations $E^{\text{in}} = \text{SWDWN} + \text{LWDWN}$, $R = \text{SWNET} + \text{LWNET}$, $\bar{\alpha} = 1 - (\text{SWNET} + \text{LWNET})/(\text{SWDWN} + \text{LWDWN})$. As above, we again deal with two spectral intervals, Ω_s and Ω_l for short- and long-wave radiation, respectively, with the border of division at 5 μm, and then each spectrum (incoming and outgoing radiation) is represented only by two numbers, which are the integrals in respect to the corresponding interval. Other values used in the exergy calculations are defined as $R_s = \int_{\Omega_s} R(\nu)\mathrm{d}\nu = \text{SWNET}$, $R_l = \int_{\Omega_l} R(\nu)\mathrm{d}\nu = \text{LWNET}$, $\alpha_s = E_s^{\text{out}}/E_s^{\text{in}} = 1 - (\text{SWNET}/\text{SWDWN})$ and $\alpha_l = E_l^{\text{out}}/E_l^{\text{in}} = 1 - (\text{LWNET}/\text{LWDWN})$.

Determination of the SRB requires information concerning the surface conditions (temperature, reflectivity, emissivity), the overlying atmosphere (composition, transmissibility, temperature, etc.), and the top-of-the-atmosphere insolation. Extensive work has been done to validate the methods and results. Recent descriptions are given in Darnell et al. (1992), Gupta et al. (1992, 1993) and Whitlock et al. (1995). These studies include comparison with more detailed radiative transfer models, with the results of other SRB algorithms and with surface measurements.

We reduced the data to the $2° \times 2°$ grid for easier handling. For our exergy analysis we used the data for 1990. The reason is that 1990 is often referred to in the climate change debate because it was decided to relate the dynamic of future CO_2 emissions to 1990.

For calculation of the exergy spatial distribution we use formulas (6.4) and (6.5) (Chapter 10), and also data about the components of the radiation balance. The results are represented as a map of the annual global exergy (see Fig. 11.3a), where the data are averaged by the annual time interval. Note that since we have only the integral data for two spectral intervals, the exergy estimation gives the value rather lower than exact one.

In addition, a map of global vegetation (biomes map) was taken from the site of NASA's Earth observatory: http://earthobservatory.nasa.gov/Library/LandCover/land_cover_3.html.

This map shows a breakdown of general vegetation types over large areas of Earth, and contains 14 categories (see Fig. 11.4b). To determine what type of vegetation covers a given region, NDVI data were analysed and periodically compared with ground test measurements of the mapped areas.

The spatial distribution of the annual total exergy is shown in Fig. 11.3a. One can see that the exergy reaches its maximal values in the "red–orange" domains 1–4, where the highest degree of transformation of incoming radiation occurs. What kind of specific properties do these domains possess? It is known that these are so-called "oceanic gyres", i.e. the regions of the World Ocean with maximal circulation (Peixoto and Oort, 1992). This circulation allows for a very efficient transformation of the incoming radiation. As radiation is highest in these areas, we see the global maximum of exergy here. A map of the energy balance would show similar results, but the exergy map manifests these domains much more distinctly. Regions 1 and 2 are the well-known "equatorial upwellings" (famous El Niño events happen in region 1), region 3 is the "Arab–Somali upwelling", region 4 is the "Darwin upwelling". In other words, exergy can be a good indicator for crucial regions of the ocean.

One of the possible interpretations of exergy is the maximal useful work which can be performed by the system in the course of movement towards thermodynamic equilibrium with the environment (see Chapter 5). The First Law of Thermodynamics states that $\delta Q = \mathrm{d}U + \delta A$, i.e. heat δQ, absorbed by the system from its environment, is spent for the increase of internal energy of the system, $\mathrm{d}U$, and the performance of work against external forces. In our case, the system is an active surface transforming incoming radiation, and $\delta Q = E^{\mathrm{in}} - E^{\mathrm{out}} = R$, $\delta A = \mathrm{Ex}$, so that $R = \mathrm{d}U + \mathrm{Ex}$, or

$$\mathrm{d}U = R - \mathrm{Ex}, \tag{6.1}$$

where $\mathrm{d}U$ is the full differential of internal energy of the system. Therefore, U is a state

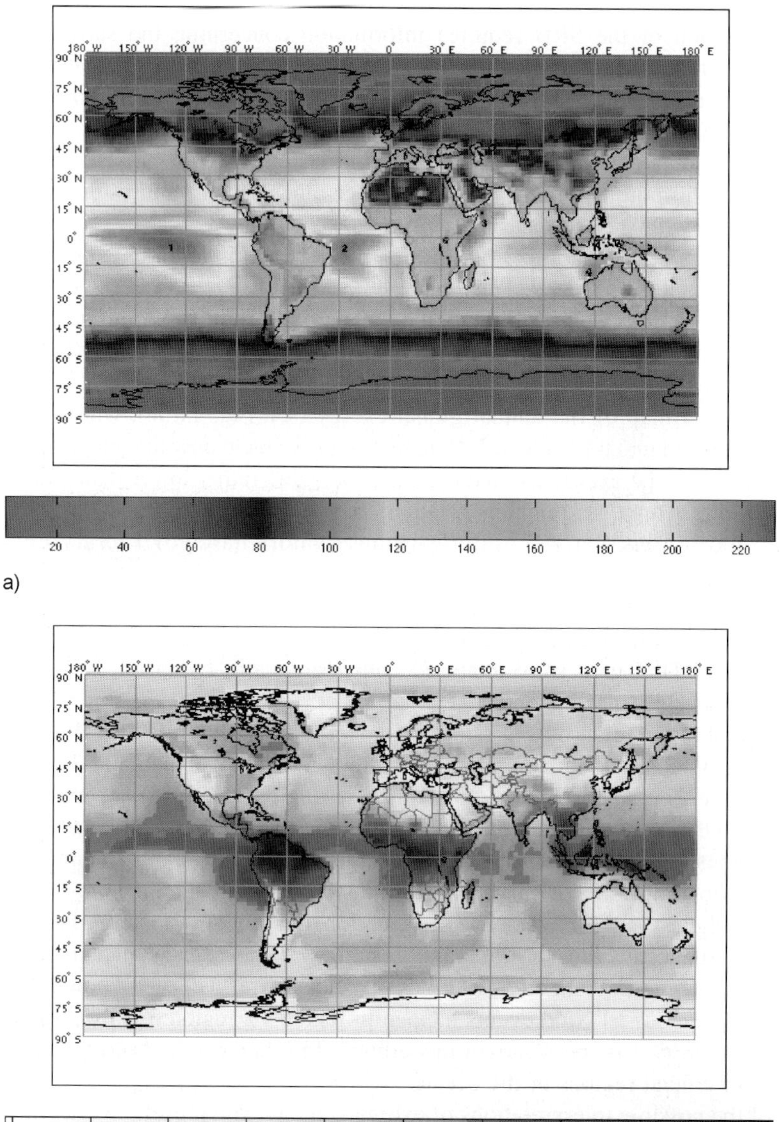

Fig. 11.3. Global maps (annual means, W/m^2) of exergy (a) and internal energy increment $\Delta\hat{U}$ (b).

variable. Let us consider a 1-year interval. Then the values $R = \hat{R}$ and $\text{Ex} = \hat{\text{Ex}}$ have to be annual means of their intraseasonal values, but the annual change of internal energy is defined as $\Delta\hat{U} = U_{\text{end}} - U_{\text{beg}}$ where U_{beg} and U_{end} are the values of U at the beginning and the end of the year. Since, in our case, all these processes are periodical with a 1-year period

Colour version of Fig. 11.3. on page 284.

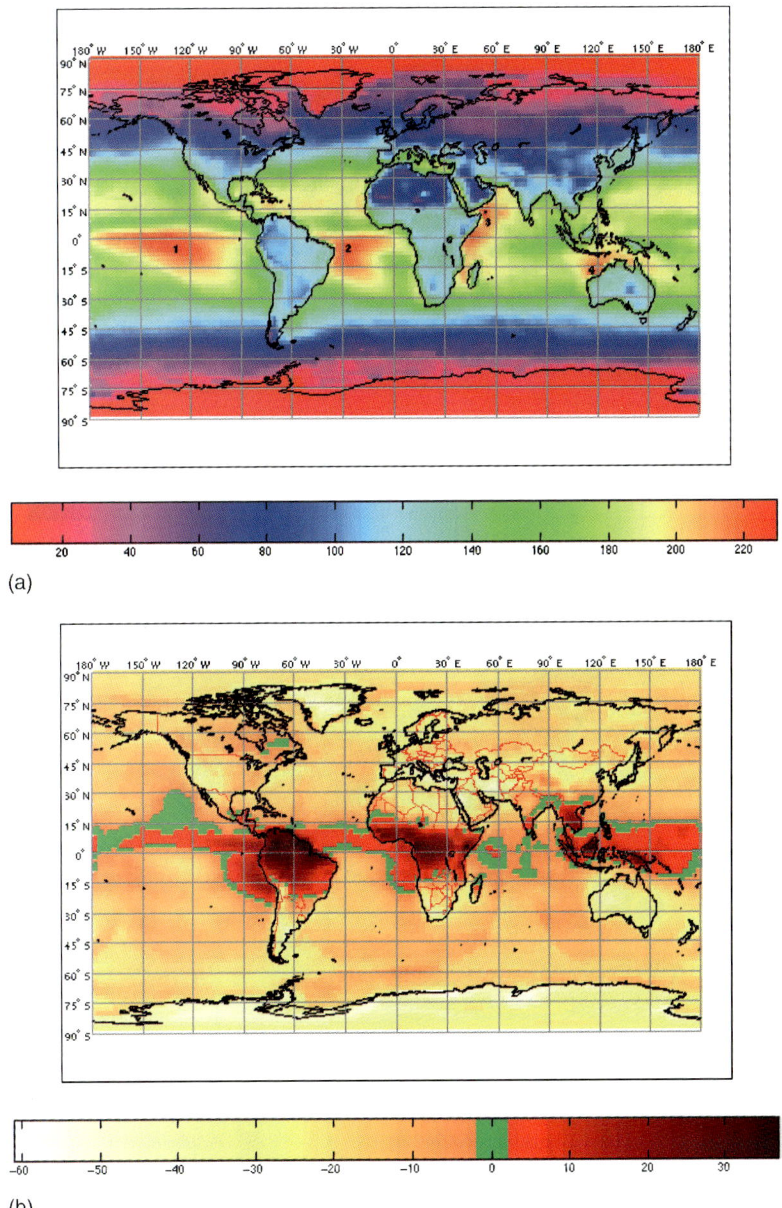

Fig. 11.3. Global maps (annual means, W/m^2) of exergy (a) and internal energy increment $\Delta\hat{U}$ (b).

Colour version of Fig. 11.4. on page 285.

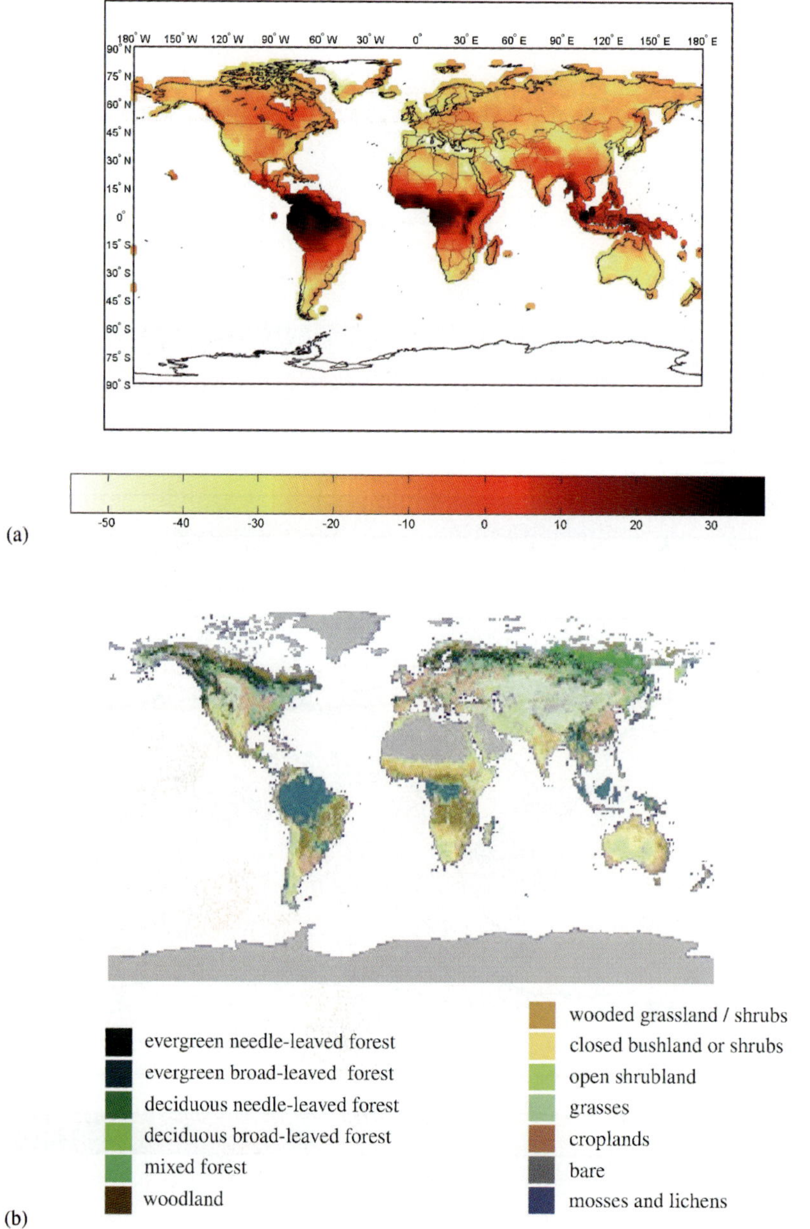

Fig. 11.4. Continental maps of internal energy increment $\Delta\hat{U}$ (annual mean, W/m^2) (a) and vegetation (b).

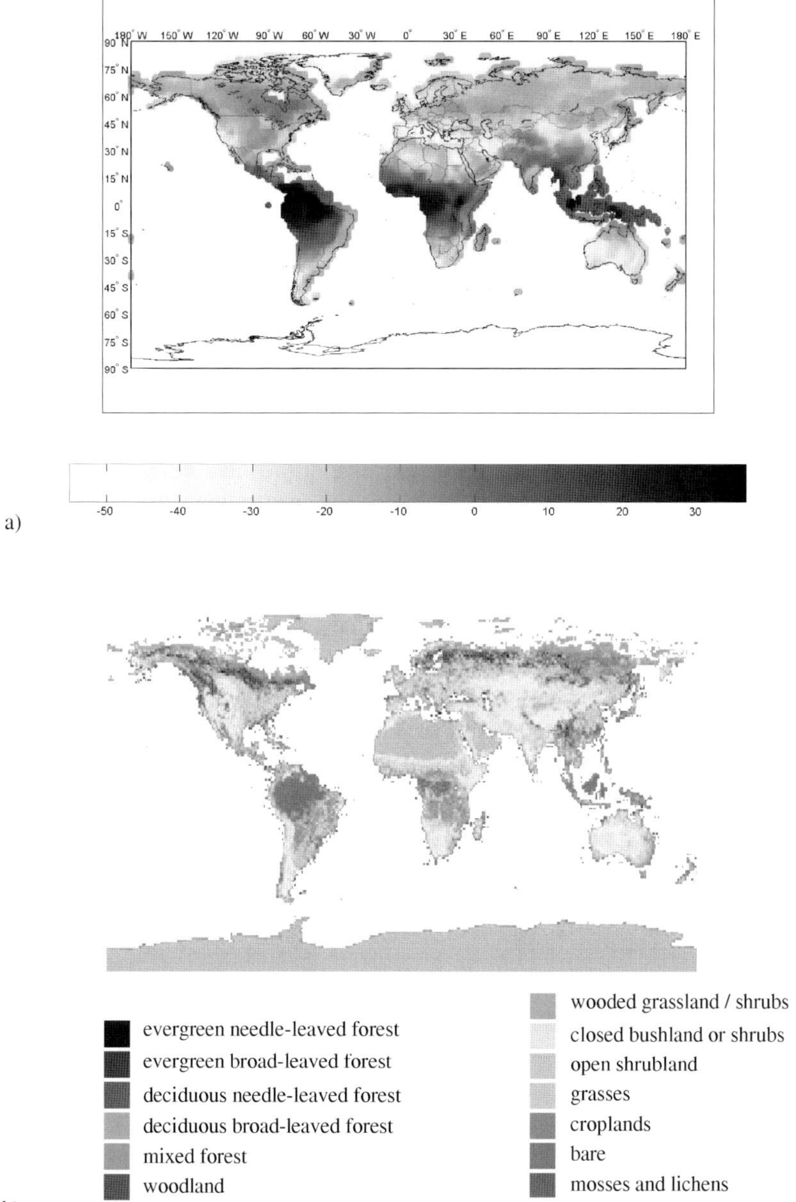

Fig. 11.4. Continental maps of internal energy increment $\Delta\hat{U}$ (annual mean, W/m^2) (a) and vegetation (b).

then the values of the state variables at the beginning and the end of the year have to coincide, $U_{\text{end}} = U_{\text{beg}}$. Therefore, $\Delta\hat{U} = \hat{R} - \hat{\text{E}}\text{x} = 0$.

However, if we construct a global map of the annual increment of internal energy, $\Delta\hat{U}$, we can see that the condition $\Delta\hat{U} = 0$ is far from being fulfilled everywhere. There are regions where the annual increment of internal energy is positive and very large (these are equatorial regions of the oceans and continents). There are also regions with large negative values of increment (polar and sub-polar regions). Formally, this means that in all these regions the energy conservation law is not fulfilled! However, this contradiction is quickly resolved if one recalls that only the global value of $\Delta\hat{U}$, $\Delta\hat{U}_{Gl} = \int_{\Sigma} \Delta\hat{U}(x, y)dxdy$, where Σ is the whole surface of the globe, has to equal zero. The real calculations prove this fact.

It is natural that in the process of integration the regions with positive and negative $\Delta\hat{U}$ will compensate each other. A physical mechanism of the compensation is a transition of some part of internal energy (for instance, in the form of mass and energy transport by aerial and oceanic currents) from domains with positive $\Delta\hat{U}$ into domains where $\Delta\hat{U} < 0$. Then, from the point of view of thermodynamics, the isoline $\Delta\hat{U}(x, y) = 0$ and its neighbourhood have to be a relatively quiet zone.

The corresponding map is shown in Fig. 11.3b, in which the isoline $\Delta\hat{U}(x, y) = 0$ and its vicinity are marked in green. These "green domains" separate the "red" and "yellow" regions from each other. Expert opinion assumes that the "red" regions in the ocean are characterised by a very high intensity of kinetic movements, upwellings and a very active biological life. Carbon dioxide is emitted from the ocean into the atmosphere. The continental "red" regions correspond to domains with the maximal biological production. Maybe (but this is a hypothesis) these regions are also CO_2 sources. The "yellow" regions are also very active, but the activity is of a different sort. In the ocean these regions absorb carbon dioxide from the atmosphere and are the regions of circular currents and downwellings. The continental "yellow" regions are domains with low biological production. It may be that these regions are CO_2 sinks.

There is not enough data to either validate or disprove this hypothesis. The reasoning (not proof) is as follows. It is known that the tropical ocean releases CO_2, while the polar ocean absorbs it. The reason is that CO_2 solubility is inversely related to the temperature. The internal energy increment $\Delta\hat{U}$, as distinguished from the radiation balance, is a continuous function: there are no gaps on the boundary "earth–ocean". Therefore, it may be suggested that we have the same situation in terrestrial systems: tropical ecosystems could possibly release CO_2, while polar ones absorb it. This fact is also confirmed by vast reserves of humus in taiga and broad-leaved forests, and the absence of humus in the tropics (cf. Ulanowicz and Hannon, 1987).

When we look at the map of continental distribution of $\Delta\hat{U}$ (Fig. 11.4a), we get the visual impression that the regions with positive and high values for the increment of internal energy correspond to areas with a highly developed vegetation cover (like tropical rain forests). For confirmation, it is sufficient to compare Fig. 2 with the vegetation map represented in Fig. 11.4b. Continuing the visual comparison between other vegetation types and the corresponding gradations of the increment of internal energy, we see that the global pattern formed by the spatial distribution of $\Delta\hat{U}$ (Fig. 11.4a) and the global vegetation pattern (Fig. 11.4b) are very similar, at least when we compare these maps simply as integral picturesque objects. All these are in good agreement with the table of "exergy utilisation and storage in a comparative set of ecosystems" (Jørgensen et al., 2000), according to which deserts utilise only 2% of exergy, while the contribution of tropical and old-growth deciduous forests is 70–72%.

Unfortunately, we do not know any reliable statistical method which would allow us to estimate quantitatively the degree of similarity between the two colour patterns. For this reason, at this stage we restrict the analysis to the visual comparison.

Since the NPP of different types of vegetation (biomes) decreases in an almost monotonous way from the equator to the poles, and the character of decrease is similar for $\Delta\hat{U}(x, y)$, then we can assume that there is a close correlation between these values, and the value of $\Delta\hat{U}(x, y)$ would be a measure of NPP. In order to prove this, additional investigations are needed.

11.7. Exergy of the biosphere

It seems that Vinogradov (1959) was the first to compare the chemical compositions of living and non-living matter (in Earth's biota and crust, correspondingly). Later on many authors repeated these estimations, but their corrections were insignificant. Therefore, for the estimation of the chemical exergy of the biosphere we have used (Svirezhev, 1997b) the original Vinogradov data. Some of them are shown in Table 11.1.

If we assume that the Earth's crust (non-living matter) is a system in thermodynamic equilibrium, we can calculate the exergy of living matter, i.e. the exergy of the biosphere, where the non-living matter of the Earth's crust is considered as some reference state. In other words, we consider the biosphere as some chemical system (for instance, an "active membrane"), which either concentrates or disperses chemical elements in comparison with their basic concentrations in the Earth's crust.

Let s_i be the content of ith element in the biosphere (biota) in percent in respect to mass and s_i^0 be the same in the crust, which is considered as a residence state. If m_i is the atomic weight of ith element then its molar concentration c_i will be equal to $c_i = (s_i/m_i)M$, where M is the total biomass of the biosphere. In accordance with the standard definition of chemical exergy (see formula (4.6), Chapter 5), the exergy of the biosphere will be

$$\begin{aligned}\text{Ex} &= RT_0 \sum_{i=1}^{n} [c_i \ln(c_i/c_i^0) - (c_i - c_i^0)] \\ &= RT_0 \sum_{i=1}^{n} \frac{1}{m_i} [s_i M \ln(s_i M/s_i^0 M_0) - (s_i M - s_i^0 M_0)] \qquad (7.1)\end{aligned}$$

where M and M_0 are the total mass of the biosphere and the Earth's crust, correspondingly. Strictly speaking, we take into account only that part of the crust which interacts with

Table 11.1
Chemical composition of living (biota) and non-living (crust) matter (in % to weight)

Element	H	C	Si	Al	O	N	P
Biota	1.05×10^1	1.8×10^1	2.0×10^{-1}	5.0×10^{-3}	7.0×10^1	3.0×10^{-1}	7.0×10^{-2}
Crust	1.00	3.56×10^{-1}	2.6×10^1	7.45	4.9×10^1	4×10^{-2}	1.2×10^{-1}

the living matter of the biosphere (Vernadsky includes this part into the biosphere). The temperature T_0 is the temperature of this part. Since we assume that it is in thermal equilibrium with biota then, without loss of generality, we can say that $T_0 = T = 287$ K. A specific exergy of one weight unit of living matter is $\text{ex} = \text{Ex}/B$.

It is very important that the specific exergy depends not only on the relative element composition of both the living matter and the crust (i.e. the percentage content of different chemical elements in these systems, in the biosphere and in the crust), but, to a significant degree, its value depends on the ratio $r = M/M_0$, i.e. on the relative value of crust matter involved in the fast (in comparison with geological times) global biogeochemical cycles. It will become clear if we rewrite Eq. (7.1) as

$$\text{ex} = K_W + \sigma(\ln r - 1) + \frac{\sigma^*}{r} \tag{7.2}$$

where

$$K_W = RT\sum_{i=1}^{n} \frac{s_i}{m_i} \ln\left(\frac{s_i}{s_i^0}\right), \quad \sigma = RT\sum_{i=1}^{n} \frac{s_i}{m_i}, \quad \sigma^* = RT\sum_{i=1}^{n} \frac{s_i^0}{m_i}.$$

Taking into account that all concentrations in Vinogradov's table are evaluated in the relative weight units (%) and using formula (7.2), we can calculate the values of K, σ and σ^*: $K_W = 777$, $\sigma = 395$, $\sigma^0 = 130$. All these values are measured in J/g of living matter.

Let us calculate the specific exergy for different values of r. The first simplest hypothesis is the assumption that the total amount of matter does not change in the course of transition from the non-living state to the living one, i.e. the peculiar conservation law of matter has been realised and $M = M_0$, i.e. $r = 1$. In other words, there is dynamic equilibrium between biota and crust, i.e. between the living and non-living matter of the biosphere. The latter is understood in Vernadsky's wide sense, where the whole matter of crust has been involved into the "Big Living Cycle". According to Vernadsky the Earth's crust is a result of the biosphere's activity, the trace of the past biospheres.

By setting $r = 1$ in Eq. (7.2) we obtain immediately that specific exergy of 1 g of living matter is equal to 512 J/g.

The specific exergy can be presented as the sum $\text{ex} = \sum_{i=1}^{n} (\text{ex})^i$, where $(\text{ex})^i$ is the contribution of partial exergy, corresponding to ith element, into the total exergy. Let us compare these contributions (see Table 11.2).

In fact, the main contribution belongs to hydrogen (~71%) (i.e. water, carbo-hydrates, etc.). The second place is occupied by carbon (~21%). Then come silicon, aluminium

Table 11.2
Partial specific exergy for different chemical elements (in J/g of living matter)

Element	H	C	Si	Al	O	Na	Mg	Fe
$(\text{ex})^i$	365	106	21.3	6.59	5.88	2.39	2.06	1.76

and oxygen (4.15%, 1.3% and 1.12%), respectively. The contribution of others is negligibly small.

Since the summary contribution of hydrogen, carbon and silicon is equal to 96.2%, we can speak about our biosphere as the "hydrogenous–carbonate–silicon" biosphere. If we compare "carbon exergy" of 1 g of living matter (106 J) and the so-called "carbon equivalent": ~8–9 kJ/g of raw biomass, we can conclude that the part of "structural", "creative" exergy is about 1.2%, i.e. it is very small in comparison with "heat" enthalpy. The latter is equal to the number of calories obtained in the process of biomass burning.

In Chapter 5, a new measure of exergy based on the genetic complexity of different organisms was suggested. In accordance with this measure, if the "exergy cost" of detritus is equal to 1, then the "exergy cost" of most plants and trees is about 30. Note that global vegetation is the leading actor of our biosphere. And if the free energy of 1 g of detritus is equal to ~18.4 kJ/g, then the specific exergy of living matter of the biosphere must be equal to 550 kJ/g. There is a contradiction, is there not? How can we resolve this?

The traces of the past biospheres are contained in the "stratosphere" (a part of the crust) that is created in the process of sedimentation. It has accumulated sediments during the long time of evolution of the biosphere. Ronov (1980) estimates the total mass of organic carbon in the stratosphere as the value of $\sim 1.2 \times 10^{22}$ g, and the total mass of carbon in the contemporary biosphere is $\sim 3 \times 10^{18}$ g (Svirezhev et al., 1985). Note that we include in the biosphere not only the carbon of living matter, but also carbon of the atmosphere and pedosphere, since this carbon also participates in the "Small Carbon Cycle" connecting the atmosphere, biota and pedosphere into a single entity. Then the contemporary value of $r = 2.5 \times 10^{-4}$. Substituting the value into Eq. (7.2) we get that the specific exergy will be equal to ~520 kJ/g of living matter. Comparing this value with the "genetic" exergy we can see that they are very close.

Let us imagine the young biosphere, which is developed on the thin and young crust. This young biosphere is very aggressive and all the crust matter is involved into processes of chemical interaction and exchange with the biosphere. Then, we get the case with $r = 1$, which was considered above, whereas the contemporary value of $r = 3.6 \times 10^{-6}$. This coincidence seems to be very interesting. Note that the main contribution into the specific exergy gives the term σ^0/r, i.e. the term corresponding to the processes working against the increase of entropy, and that separate a thin film of living matter from the immense mass of the crust. The latter, in turn, is the entropy storage of the past biospheres. It becomes clear that the thinner this film, the better the ability of the living matter to do this type of work. In other words, the smaller the value of r, the larger the value of specific exergy. But this situation is typical for the old biosphere when it was in equilibrium a long time ago.

We see that the main role in the formation of comparatively low exergy is played by the term K_W (see Eq. (7.2)) determined by the chemical composition of living matter. In other words, at the first stages of the biosphere formation the exergy of living matter is determined mainly by its chemical composition and, as a consequence, by the type of chemical processes used by life forms for the formation of their matter.

11.8. Exergy and the evolution

The evolutionary paradigm is one of the main paradigms in ecology. It was already used in the previous section, when we estimated the exergy of living matter. Now let us consider how to apply the exergy concept to some evolutionary problem.

It is trivial that every point on the surface of our planet is different from any other point and therefore offering different conditions for the various life forms. This enormous heterogeneity explains why there are so many species on Earth (see also Section 4.7). There is, so to say, an ecological niche for "everyone", and "everyone" may be able to find a niche where the organism is best fitted to utilise the resources.

Ecotones, the transition zones between two ecosystems, offer a particular variability in life conditions, which often results in a particular richness of species diversity. Studies of ecotones have recently drawn much attention from ecologists, because ecotones have pronounced gradients in the external and internal variables giving a clearer picture of the relations between them.

Margalef (1991) claims that ecosystems are anisotropic, meaning that they exhibit properties with different values when measured along axes in different directions. It means that the ecosystem is not homogeneous in relation to properties concerning matter, energy and information, and that the entire dynamics of the ecosystem works toward increasing the differences.

These variations in time and space make it particularly difficult to model ecosystems and to capture the essential features of ecosystems. However, the hierarchy theory (Pattee, 1973) applies these variations to develop a natural hierarchy as a framework for ecosystem description and theory. The strength of the hierarchy theory is that it facilitates studies and modelling of ecosystems.

Darwin's theory describes the competition among species and states that the species best fitted to the prevailing conditions in the ecosystem will survive. Darwin's theory can, in other words, describe the changes in ecological structure and the species composition, but cannot be directly applied quantitatively, e.g. in ecological modelling (see Chapters 7–9, 12 and 13).

All the species in an ecosystem are confronted with the question: how is it possible to survive or even grow under the prevailing conditions? The prevailing conditions are considered as *all* factors influencing the species, i.e. all external and internal factors including those originating from other species. This explains the co-evolution, as now any change in the properties of one species will influence the evolution of the other species.

Species are generally more sensitive to stress than functional properties of ecosystems. Schindler (1988) observed in experimental acidifications of lakes that functional properties such as primary production, respiration and grazing were relatively insensitive to the effects of a continued exposure to acidification, while early signs of warning could be detected at the level of species composition and morphologies. This underlines the importance of the development of structurally dynamic models able to predict the change in focal properties of the species, which would correspond to a shift in species composition (see Chapter 13).

All natural external and internal factors of ecosystems are dynamic—the conditions are steadily changing, and there are always many species waiting in the wings, ready to take over, if they are better fitted to the emerging conditions than the species dominating under the present conditions. There is a wide spectrum of species representing different combinations of properties available for the ecosystem. The question is, which of these species are best able to survive and grow under the present conditions, and which species are best able to survive and grow under the conditions one time step further and two time steps further and so on? The necessity in Monod's sense is given by the prevailing conditions—the species must have genes or maybe rather phenotypes (meaning properties) which match these conditions to be able to survive. But the natural external factors and the genetic pool available for the test may change randomly or by "chance".

Steadily, new mutations (misprints are produced accidentally) and sexual recombinations (the genes are mixed and shuffled) emerge and steadily give new material to be tested against the question: which species are best fitted under the conditions prevailing just now?

These ideas are illustrated in Fig. 11.5. The external factors are steadily changing and some of them even relatively fast, partly randomly (e.g. the meteorological or climatic factors). The species of the system are selected among the species available and represented by the genetic pool, which again is slowly, but surely changing, randomly or by "chance". What is named ecological development is the changes over time in nature caused by the dynamics of the external factors, giving the system sufficient time for the

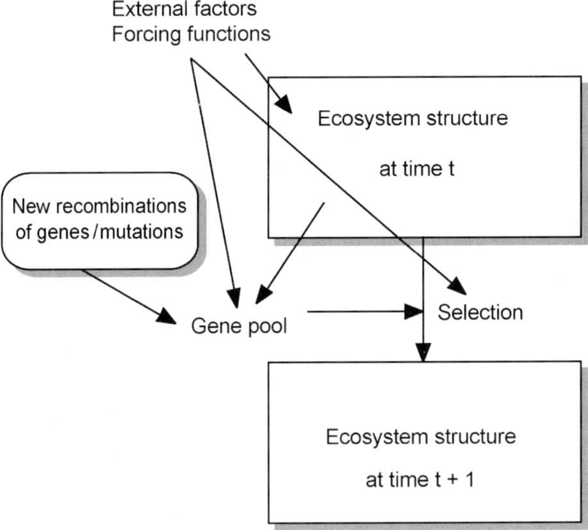

Fig. 11.5. Conceptualisation of how the external factors steadily change the species composition. The possible shifts in species composition are determined by the gene pool, which is steadily changing due to mutations and new sexual recombinations of genes. The development is, however, more complex. This is indicated by (1) arrows from "structure" to "external factors" and "selection" to account for the possibility that the species are able to modify their own environment (see below) and thereby their own selection pressure; (2) an arrow from "structure" to "gene pool" to account for the possibilities that the species can to a certain extent change their own gene pool.

reactions, including an organisation of the network. There is, however, one change that seems to keep ongoing continuously and to have a directional component. This is the change referred to as evolution that is defined as the change in the properties of populations of organisms over time. The population is the unit of evolution, although genes, individuals and species also play a role.

Evolution is closely related to the genetic pool. It is the result of the relations between the dynamics of the external factors and the dynamics of the genetic pool. The external factors steadily change the conditions for survival, and the genetic pool steadily comes up with new solutions to the problem of survival.

Darwin's theory assumes that populations consist of individuals that

1. On average produce more offspring than is needed to replace them upon their death—this is the property of high reproduction. An enormous amount of genetic variation is produced in every generation, but only a few individuals of the generation will survive to produce the next generation.
2. Have offspring which resemble their parents more than they resemble randomly chosen individuals in the population—this is the property of inheritance.
3. Vary in heritable traits influencing reproduction and survival (i.e. fitness)—this is the property of variation. All the individuals of a population differ genetically from each other.

Only a (minor) part of the individuals forming a population survives and reproduces. These survivors are not a random sample of the population, but they are characterised by the possession of certain properties/attributes that favour survival under the prevailing environmental conditions. The survivors have the properties that make them particularly well adapted for the environment. The more survivors, the more exergy the system has—the survival is measured by the distance from thermodynamic equilibrium. There are, of course, no agents involved in the selection, but the individuals without the best fitness to the prevailing environmental conditions will be eliminated. The organisms with the right combination of properties adapted to the prevailing environmental conditions are the organisms that will be able to contribute most to the exergy of the system through survival.

All the above-mentioned three properties are parts of the presentation in Fig. 11.5. High reproduction is needed to get a change in the species composition caused by changes in external factors. Variability is represented in the short- and long-term changes in the genetic pool, and inheritance is needed to see an effect of the fitness test in the long run.

Without inheritance every new generation would start from the same point and it would not be possible to maintain the result of the fitness test. Evolution is able to continue from the already obtained results.

The species are continuously tested against the prevailing conditions (external as well as internal factors) and the better they are fitted, the better they are able to maintain and even increase their biomass. The specific rate of population growth may even be used as a measure for the fitness (see e.g. Stenseth, 1986). But the property of fitness must of course be inheritable to have an effect on the species composition and the ecological structure of the ecosystem in the long run. Natural selection has been criticised for being a tautology: fitness is measured by survival, and survival of the fittest therefore means the survival of the survivors. However, the entire Darwinian theory including the above-mentioned three assumptions cannot be conceived as a tautology, but may be interpreted as follows: the species offer different solutions to survival under given prevailing conditions, and the species that have the best combinations of properties to match the conditions have also the highest probability of survival and growth. The formulation by Ulanowicz (1986) may also be applied: those populations are the fittest that best enhance the autocatalytic behaviour of the matter–energy loops in which they participate.

Anthropogenic changes in external factors (i.e. anthropogenic pollution) have created new problems, because new genes fitted to these changes do not develop overnight, while most natural changes have occurred many times previously and the genetic pool is therefore prepared and fitted to meet the natural changes. The spectrum of genes is sooner or later able to meet most natural changes, but not all of the man-made changes, because they are new and untested in the ecosystem.

Evolution moves toward increasing complexity in the long run; see Fig. 11.6. The fossil records have shown a steady increase of species diversity. There may be destructive forces, for instance anthropogenic pollution or natural catastrophes (for a shorter time), but the probability that

1. new and better genes are developed;
2. new ecological niches are utilised

will increase with time. The probability will even (again excluding the short time perspective) increase faster and faster, as the probability is roughly proportional to

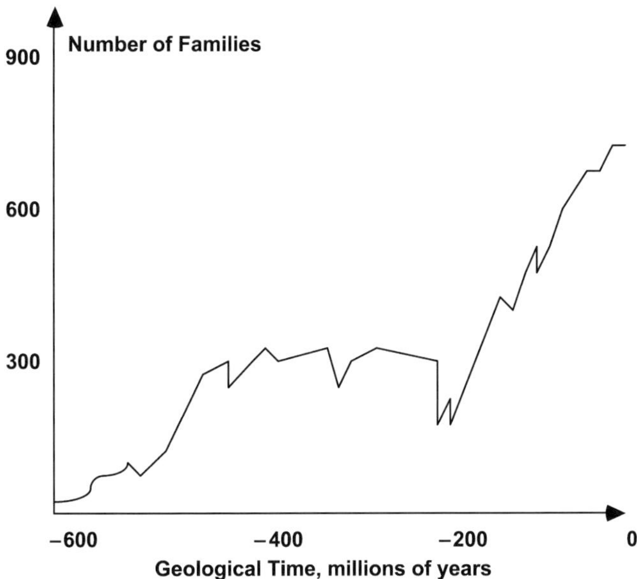

Fig. 11.6. Changes in species diversity over geological time. Redrawn from Raup and Sepkowski (1982).

the amount of genetic material where mutations and new sexual recombinations can be developed.

It is equally important to note that a biological structure is more than an active non-linear system. In the course of its evolution, the biological structure is continuously changed in such a way that its structural map itself is modified.

The overall structure thus becomes a representation of all the information received. Biological structure represents through its complexity a synthesis of the information with which it has been in communication (Schoffeniels, 1976). Evolution is maybe the most discussed topic in biology and ecology and millions of pages have been written about evolution and its ecological implications.

Today the basic facts of evolution are taken for granted and interest has shifted to more subtle classes of fitness/selection, i.e. towards an understanding of the complexity of the evolutionary processes. One of these classes concerns traits that influence not only the fitness of the individuals possessing them, but also the entire population. These traits overtly include social behaviours, such as aggression or cooperation, and activities that through some modification of the biotic and abiotic environment feed back to affect the population at large, for instance pollution and resource depletion.

The more species are on Earth, the more possibilities the ecosphere offers to utilise the available resources and ecological niches and the more possibilities the ecosphere has to adapt to new, expected or unexpected, conditions. The number of species should therefore be an indirect measure of the utilisation of the available resources to move away from thermodynamic equilibrium, or expressed differently, as the width of ecological information. The same considerations are applied when the survival of species is discussed. Endangered species often have very little diversity in the gene pool (Lewin, 1994).

The depth of ecological information may be expressed by using the weighting or conversion factors in Table 5.1 in Chapter 5 for the most developed organism at a given time. We do not know the exact number of species today: a rough estimation is 10^6–10^7, and we do not have good estimations of the number of species during evolution at all. It is therefore proposed to use the number of marine families (Raup and Sepkowski, 1982; see Fig. 11.6) as a proper relative measure of the ecological information width. We do know approximately from fossil records when each species emerged. Consequently, we could get a first relative estimation of an evolution index by multiplication of these possible expressions for the width and the depth of biological information that should be closely related to the relative evolution of the exergy of the biosphere (Jørgensen, 2000b).

Fig. 11.7 shows the development over geological time from 550 Ma to today of the proposed above *evolution index* = (*number of marine families*) × (*weighting factor for the most developed organism at a given time*). The applied weighting factors are shown in Table 11.3. Note in Figs. 11.6 and 11.7 the decrease in evolution index about 220 Ma ago and 65 Ma ago due to extinction of a high number of species (the dinosaurs—65 Ma ago, probably by a catastrophic event). The overall trend is (in spite of some fluctuations due to catastrophic events) toward an increase of the evolution index. When the index is decreasing it is due to a sudden and/or major change of the environmental conditions, because the forcing functions of the ecosphere are changed stochastically. The index will,

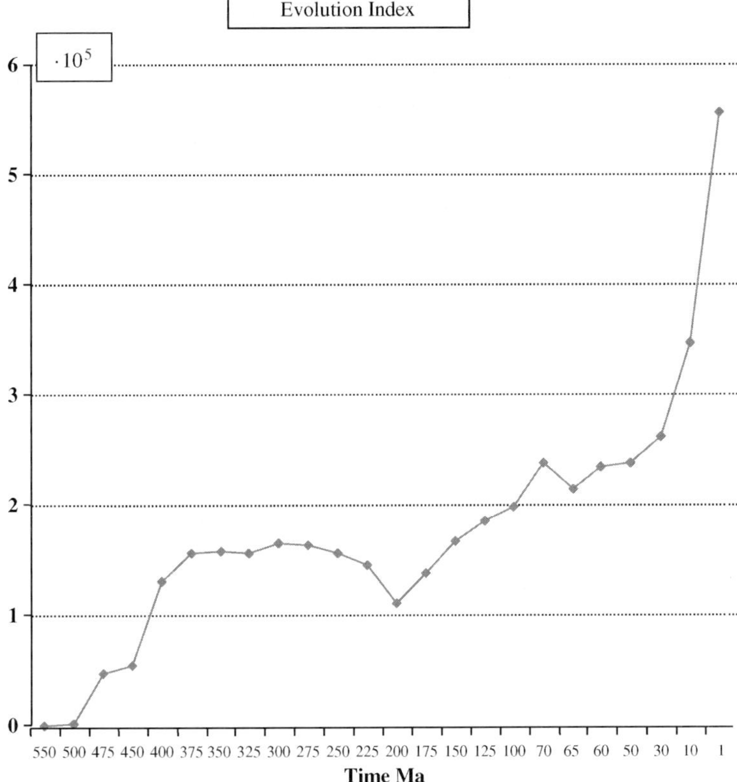

Fig. 11.7. The introduced evolution index is shown versus time. The evolution index is a rough estimation of evolution based on the number of families and the emergence of more and more developed species. Notice the Cambrian explosion, i.e. the fast increase in the evolution index around 500 Ma ago. It is also possible to see the decrease in the evolution index around 200 and 65 Ma ago due to massive extinction of many species, probably due to a catastrophic event.

Table 11.3
Weighting factors applied to calculate the evolution index shown in Fig. 11.7

Ma	Applied weighting factor
550	30
475	250 (primitive fish)
400	330
375	350 (amphibians)
300	370 (reptiles)
200	400 (mammals)
30	440 (monkeys)
10	500 (apes)
1	716 (human)

however, recover again when evolution has had sufficient time to find a new solution to the suddenly changed and emerged life conditions.

The plot illustrates, of course relatively and very roughly, how the ecosphere is moving more and more away from thermodynamic equilibrium by increasing useful knowledge about the ways to utilise the resources and adapt to the currently changing conditions. The amount of biomass has not necessarily changed much over this period of time, except when the terrestrial ecosystems started to be exploited about 400 Ma ago. The amount of inorganic material on Earth available to form complicated and coordinated living matter (this is about 20 elements) has not changed. The changes in the exergy of the biosphere are therefore almost solely due to the increased width and depth of information. The plot in Fig. 11.7 may therefore also be considered as a plot of the relative change in the exergy of the biosphere. The figure should, however, be considered only as a first rough attempt to quantify the evolution.

The following terms are used to cover the various forms of selections (Wilson, 1978):

1. *Individual selection*. The component of natural selection that operates on the differential fitness of individuals within local and homogeneous populations.
2. *Group selection*. The component of natural selection that operates on the differential productivity of local populations within the more global population.
3. *Egoism*. All traits promoted by individual selection $d > r$ in linear selection modelling, where d is the effect on the fitness of the individual itself and r is the effect on every other member of the local population.
4. *Weak altruism*. All non-egoistic traits selections, where $0 < d < r$ in linear selection models.
5. *Strong altruism*. All non-egoistic traits selections, where $0 > d$, when r is sufficiently great in linear selection models.

It can be shown that all these types of selections actually take place in nature, and that many observations support the various selection models that are based on these types of selections. Kin selection has been observed with bees, wasps and ants (Wilson, 1978). Prairie dogs endanger themselves (altruism) by conspicuously barking to warn fellow dogs of an approaching enemy (Wilson, 1978), and a parallel behaviour is observed for a number of species.

Co-evolution explains the interactive processes among species. It is difficult to observe co-evolution, but it is easy to understand that it plays a major role in the entire evolution process. The co-evolution of herbivorous animals and plants is a very illustrative example. The plants will develop toward a better spreading of seeds and a better defence towards herbivorous animals. This will in the latter case create a selection of herbivorous animals that are able to cope with the defence. Therefore, the plants and herbivorous animals will co-evolve.

Co-evolution means that the evolution process cannot be described as reductionistic, but that the entire system is evolving. A holistic description of the evolution of the system is needed.

The Darwinian and Neo-Darwinian theories have been criticised from many sides. It has for instance been questioned whether the selection of the fittest can explain the relatively high rate of evolution. Fitness may be measured here by the ability to grow and reproduce under the prevailing conditions. It implies that the question raised according to the Darwinian theories (see the discussion above) is: "which species have the properties that give the highest ability for growth and reproduction?" We shall not go into the discussion in this context, it is another very comprehensive theme, but just mention that the complexity of the evolution processes is often overlooked in this debate. Many interacting processes in evolution may be able to explain the relatively high rate of evolution that is observed.

Seven examples below are used to illustrate that many processes (a) interact, (b) accelerate the rate of evolution and (c) increase the complexity of the evolutionary processes.

1. A mother tiger is an excellent hunter and therefore she is able to feed many offspring and bring her good "hunting genes" further in the evolution. Her tiger kittens have a great probability to survive because they get sufficient food. But in addition she can teach them her hunting strategy and will have more time to care for them in general, because of her successful hunting. So, the kittens not only survive (i.e. the genes survive), but also a better nursing and hunting strategy survives from one tiger generation to the next. We can say in our "computer age" that not only the hardware (the genes) but also the software (the know how) survives.

2. McClintock has observed by working with maize that genes on chromosomes actually move around or transpose themselves; they even appear to change in relation to environmental stress factors. He proposes the idea that the genetic program is not necessarily fixed in each individual. Other geneticists have found what have been dubbed "jumping genes" and to a certain extent confirm this idea. Jumping genes are often named *transposons* and many workers have labelled them "selfish DNA" (Dawkins, 1989). These discoveries may form the basis for a revolution in biological thinking: the reductionist image of a genetic blueprint may be false.

3. Cairns et al. (1988) showed that when bacteria lacking an enzyme for metabolising lactose were grown in a lactose medium, some of them underwent a mutation that subsequently enabled them to produce the enzyme. This mutation violated the long-held central dogma of molecular biology, which asserts that information flows only one way in the cell—from genes to RNA, to protein and enzyme. Here, the information was obviously going in the reverse direction. An enzyme coded for by a particular gene was feeding back to change that gene itself.

4. A problem of mutations with large effects on development is that they are usually selectively disadvantageous. However, Augros and Stanciu (1987) claim that a subsidiary peak occurs through a different and novel mechanism, which may be explained by a mutation of the D-genes—the genes that control the development of the organism.

5. Symbiosis is generally very well developed in nature. Poly-cellular organisms are a result of symbiotic relationships among many unicellular organisms according to Lynn Margulis, as can be recognised from the endo-symbiosis in all organisms. It may explain the jumps in the evolution: two or more "properties" are suddenly united and create a symbiotic effect (see Mann, 1991).

6. Fischer and Hinde (1949) describe how the habit of opening milk bottles has spread among blue and great tits. Milk bottles were left on the doorsteps of households and were

raided by these songbirds, which open them by tearing off their foil caps. The birds then drink the cream from the top of the bottles. The habit has probably spread through some type of social learning or social enhancement. A novel and learned behaviour appears to have modified these birds' environments in ways that have subsequently changed the selection pressures that act back on the bird themselves (Sherry and Galef, 1984). None has shown any genetic response to these altered selection pressures.

This example illustrates what Odling-Smee and Patten (1992) call "ecological inheritance", which they assert works parallel to the genetic inheritance. The ecological inheritance is a result of the species' ability to change their environment and thereby to a certain extent modify the selection pressure on themselves.

Nobody dealing with evolution would deny these possibilities of the species to modify their own environment, but the influence of this ability on the evolution process has most probably been underestimated. Odling-Smee and Patten attempt to emphasise the importance by introduction of the concept of "envirotype" as a supplement to genotype and phenotype.

A total image of the evolution will require a holistic approach to account for many simultaneously interacting processes. Evolution is a result of many simultaneous processes that are interacting in a very complex way. This may enable us to explain the relatively high rate of evolution.

7. A further complication is the so-called morpho-genes or D-genes. The developmental processes, as mentioned in the fourth example above, are obviously extremely important for the evolution processes, but it would not be possible to go into more detail in this context. Further information can be found in Dawkins (1982, 1989) and Augros and Stanciu (1987).

11.9. Summary of the ecological important issues

The concept of sustainable development is from a thermodynamical point of view unrealistic, if we consider the entire ecosphere and technosphere. Only locally is sustainable development possible and only as a result of an entropy dump elsewhere.

It is possible to set up an exergy balance for the entire biosphere and on a global scale.

Figs. 11.3 and 11.4 give a clear global image of the exergy balance. The exergy map indicates the upwelling areas in the oceans very clearly. The tropical forest areas opposite the deserts have a high utilisation of the exergy in the solar radiation. The net primary production is almost monotonously declining from the equator to the poles—so the exergy pattern and vegetation pattern follow, not surprisingly, the radiation pattern very closely.

The exergy of the biosphere has been calculated based upon the difference in chemical composition between biota and the crust. The crust represents the thermodynamic equilibrium and the composition of the biosphere represents the free energy of the biosphere—the chemical energy needed to provide the characteristic composition of the biosphere. An expression can also easily be found for specific exergy: see Eq. (7.2); r, the ratio of living matter to the total weight of the biosphere, is found to be 2.5×10^{-4}. When this value is applied, the specific exergy is found to be 520 kJ/g, which can be compared with the expected value, namely 30×18.4 kJ/g $= 550$ kJ/g, where 30 is the

β value of average plants/vegetation; see Table 5.1. The two values are very close, which may be considered as a support for the weighting factors based on the genetic information.

The external factors steadily change the species composition. The introduction of exergy calculations in ecological models makes it possible to describe this shift in the species composition, as will be presented later. Exergy is, so to say, a measure of the survival, as it accounts for biomass and information. Which properties give the best survival can be determined by testing which properties (parameters in the models) give the highest exergy.

The exergy weighting factors have also been used in an attempt to express the evolution quantitatively by multiplication of the number of families and the weighting factors of the most developed species. The number of families expresses the possibilities the ecosphere offers to utilise the available resources and the ecological niches—we could call it the width of the biological information—and the weighting factor expresses the amount of feedbacks and regulation mechanisms the most advanced species have—we could call it the depth of the biological information. The multiplication of the two would, therefore, be a relative measure of the overall increase of the biological information. Fig. 11.7 shows this quantification of the evolution.

Chapter 12

Teleology and extreme principles: a tentative Fourth Law of Thermodynamics

All is for the best in the best of all possible worlds.
Voltaire "Candide"

Teleology is a lady no biologist can live without, but whose company seems shameful in society.
G. von Brükke

12.1. Introduction

It does not seem that we are living in an ordered world, and nature is not an exception to this, but we *wish* to live in such a world, which is close to the Hellenistic one. Its Aristotelian philosophy of life was teleological and deterministic in principle (a different form of philosophy was scarcely thinkable in the well-ordered Graeco-Roman society in its prime). On the other hand, our mentality had also been formed under the influence of Eastern philosophy, which considered Chance to be the motive (driving) force of development and evolution: "Again I saw that under the sun the race is not to the swift, nor the battle to the men of skill; but time and chance happen to them all" (*The Bible*. Ecclesiastes, 9). The fundamental duality of our "scientific" approach to the problems of the development and evolution of natural systems, which manifests itself in the form of a dual classification of dynamical systems (into deterministic and stochastic ones), is a result of the influence of these two basic paradigms. For instance, the modern synthetic theory of evolution is a typical compromise between these opposite paradigms. On the one hand, there is the Aristotelian paradigm, which was developed by Lucretius: "Survival of the fittest". He illustrated this principle by the following verse:

Multaque tum interisse animantum saecla necesses
Nec potuisse propagando procudere prolem.
Nam quaecumque vides vesci vitalibus auris,
Aut dolus, aut vitrus, aut denique mobilitas est
Ex ineunte aevo genus id tutata reservans

Lucretius *De Rerum Natura*. Liber 5, 855.

Towards a Thermodynamic Theory for Ecological Systems, pp. 301–323

On the other hand, when the mutation process is considered as a driving force of evolution, this is an acceptance of the Ecclesiastesian "Chance" paradigm.

Nevertheless, determinism is more familiar for us. In the 1960s, in the USSR, a book under the title "The Foreseen Future" was published. Certainly, in 2003, many of its prognoses look very naïve, but one very impressive statement it made was that "Mankind cannot live under conditions of an unpredictable future". We feel more comfortable if we believe that all living systems have a certain aim, and their being is meaningful.

Note that if we want to look for aims, goal functions, etc. in ecology, then our *credo* must be the *Panglossism*. Pangloss, Candide's teacher, believed that "All is for the best in the best of all possible worlds" (Voltaire's "Candide"). We *believe* that a certain aim exists, and a system attempts to maximise (or minimise) some goal function. Unfortunately, it seems to us, the number of such functions is a continuum. It is not a problem to construct an arbitrary number of similar functions; the problem is rather how to select and interpret them.

Although a trend to apply the various extreme (teleological) principles in sciences with sufficiently long histories (mechanics, physics) appeared a long time ago, in ecology it has started only recently. Note that ecology itself is a young science. A trend to formulate the basic regularities of population and ecosystem dynamics in the form of some extreme principles is a consequence of the perennial teleleologicity of our thinking on the one hand and, not least, of the quest for more concise and elegant description on the other. Though it is hardly probable that there is a goal that the population, community or ecosystem tries to attain, nevertheless our hypothesis, which is purely auxiliary in nature, is that such a goal exists and may put all the variety of ecological mechanisms in a proper perspective. We think that the first attempt in this line was probably the hypothesis of Lotka (1922), i.e. that in the process of its own evolution the ecosystem tends to increase the energy throughflow, reaching a maximum at equilibrium (of course with regard to different constraints). Lotka went so far as to suggest calling this statement "the Fourth Law of Thermodynamics", so that S.-E. Jørgensen was not the first to use this term when he has also named his exergy principle the Fourth Law.

It is interesting that thermodynamics with its one-directional time, with monotonous increasing entropy and with the monotonous decreasing function of internal entropy production, is a very favourable field for various teleological concepts and interpretations, where ecologists can meet each other without fear of being embarrassed by their teleological Ladies.

12.2. The maximum power principle

The balance of energy flows for ecosystems can be written as

$$\frac{\mathrm{d}E}{\mathrm{d}t} = q_{\text{in}}^{\text{e}} - q_{\text{out}}^{\text{e}}, \tag{2.1}$$

where $E(t)$ is the energy storage in the form of biomass, different structures, etc., or captured energy, q_{in}^{e} and $q_{\text{out}}^{\text{e}}$ are energy in- and outflow. It is obvious that the derivative $\mathrm{d}E/\mathrm{d}t$ is a power, the energy flow *through* the ecosystem is equal to $q_{\text{out}}^{\text{e}}$, so that for growing systems, for which $(\mathrm{d}E/\mathrm{d}t) > 0$, $q^{\mathrm{e}}_{\mathrm{out}} < q^{\mathrm{e}}_{\mathrm{in}}$. Thus, in equilibrium $q^{\mathrm{e}}_{\mathrm{out}}$ reaches a maximum. We see that Lotka's principle holds, but its statement is trivial.

Let us rewrite the Gibbs equation in a form that is typical for "irreversible" thermodynamics:

$$\frac{\mathrm{d}E}{\mathrm{d}t} = \frac{\mathrm{d}'A_{\mathrm{us}}}{\mathrm{d}t} + \frac{\mathrm{d}'Q}{\mathrm{d}t} + \frac{\mathrm{d}A_{\mathrm{irrev}}}{\mathrm{d}t}. \tag{2.2}$$

Here A_{us} is the so-called "useful" work performed within the system, Q is the energy (heat) imported into the system from the environment and A_{irrev} is the irreversible loss of energy (note that the expressions $\mathrm{d}'A_{\mathrm{us}}$, $\mathrm{d}'Q$, and $\mathrm{d}'A_{\mathrm{irrev}}$ are not full differentials). Even if we neglect this term in Eq. (2.2), then the subdivision of $\mathrm{d}E/\mathrm{d}t$ into two items is very arbitrary and depends on what we imply by the concept of "usefulness". The import of energy and the export of heat in ecosystems are non-spontaneous processes, which are realised by means of some special "pumps". Such active systems as ecosystems have their own internal pumps; therefore, they have to possess a high degree of internal organisation, i.e. their structure has to be rather complex. We assume, therefore, that some share of the useful work must be used for the creation of this structure and the maintenance of its functioning. Furthermore, a remaining share of the work is used for the growth of biomass and maintenance of metabolic and reproductive process. We shall consider the total work spent for the maintenance of biomass and structure and also the work performed against some generalised "forces of friction" as the total metabolism of the ecosystem. As a result, metabolic heat is produced, and it must be exported from the ecosystem.

As a rule, the characteristic time taken by an ecosystem to evolve is much longer than the mean generation time of species that form the ecosystem; therefore, their life cycles can be considered as fast processes. In other words, the new biomass decomposes and releases heat; the latter must also be exported.

Note that we have not yet used any such thermodynamic concepts as free energy and entropy, although when we talked above about the transformation of work to heat we implicitly used the Second Law.

Summing together everything mentioned above and comparing Eqs. (2.1) and (2.2), we can write

$$q^{\mathrm{e}}_{\mathrm{in}} = \frac{\mathrm{d}'Q}{\mathrm{d}t}, \qquad q^{\mathrm{e}}_{\mathrm{out}} = -\frac{\mathrm{d}'A_{\mathrm{us}}}{\mathrm{d}t} - \frac{\mathrm{d}'A_{\mathrm{irrev}}}{\mathrm{d}t} \tag{2.3}$$

(in accordance with the definition both these works are negative). If we now assume additionally that irreversible processes within the ecosystem cannot, in principle, be controlled, then we can easily see that Lotka's principle, asserting that $q^{\mathrm{e}}_{\mathrm{out}} \to \max$, is equivalent to the following statement: *In the process of its own evolution the ecosystem tends to increase the useful power, reaching a maximum at equilibrium, i.e.*

$$\left|\frac{\mathrm{d}'A_{\mathrm{us}}}{\mathrm{d}t}\right| \to \max. \tag{2.4}$$

Thus, a useful work is *exergy* in accordance with its basic definition, so we can talk about maximisation of the exergy increment. Here the value of Q is considered only as the imported energy (heat), whereas in addition to energy the ecosystem also imports matter. In this case the Gibbs equation is written as (De Groot and Mazur, 1962)

$$\frac{\mathrm{d}E}{\mathrm{d}t} = \frac{\mathrm{d}'A_{\mathrm{us}}}{\mathrm{d}t} + \frac{\mathrm{d}^{*}Q}{\mathrm{d}t} + \sum_{k} h_k \frac{\mathrm{d_e}N_k}{\mathrm{d}t} + \frac{\mathrm{d}A_{\mathrm{irrev}}}{\mathrm{d}t}, \tag{2.5}$$

where a flux of the so-called "adduced" heat is defined as

$$\frac{\mathrm{d}^{*}Q}{\mathrm{d}t} = \frac{\mathrm{d}'Q}{\mathrm{d}t} - T\sum_{k} s_k \frac{\mathrm{d_e}N_k}{\mathrm{d}t}. \tag{2.6}$$

Here s_k and $h_k = \mu_k + Ts_k$ are the specific molar entropy and enthalpy, $(\mathrm{d_e}N_k/\mathrm{d}t)$ are the exchange matter flows between the ecosystem and the environment, which can also be represented as a difference of in- and outflows, $(\mathrm{d_e}N_k/\mathrm{d}t) = (\mathrm{d_e}N_k/\mathrm{d}t)^{\mathrm{in}} - (\mathrm{d_e}N_k/\mathrm{d}t)^{\mathrm{out}}$. The following is obvious.

So, we have almost obtained Odum's formulation of maximum power principle (MPP). H.T. Odum used the principle to explain the structure and processes of ecosystems (Odum and Pinkerton, 1955).

H.T. Odum constructed his argument by analogy with the construction of the function of dissipation (see Sections 2.5 and 3.3):

$$\mathrm{Power} = \sum_{k} X_k J_k, \tag{2.7}$$

where X_k and J_k are generalised thermodynamic forces and fluxes. The kth flux—"ecoflux"—can be defined as $J_k = (\mathrm{d}N_k/\mathrm{d}t)$, where N_k is the biomass (or biomass density) of kth species (component) of an ecosystem (Odum et al., 1960). The organic matter accumulated in the biomass of kth species may be defined as the "ecoforce", X_k, equal to the Gibbs free energy (or "ecopotential") difference "released" by the process, ΔG_k, per unit of biomass, N_k, expressed, for instance, in carbon units. Thus, the driving "ecoforce", $X_k = \Delta G_k/N_k$, is a function of the concentration of biomass and organic matter. Note that $\Delta G_k > 0$ for all forced, non-spontaneous processes, for photosynthesis, for instance. It is such processes that form the living matter. Then, formally, the function of power will be

$$\mathrm{Power} = \sum_{k} \Delta G_k \left(\frac{1}{N_k} \frac{\mathrm{d}N_k}{\mathrm{d}t} \right) = \sum_{k} \Delta g_k \frac{\mathrm{d}N_k}{\mathrm{d}t}, \tag{2.8}$$

where $\Delta g_k = \Delta G_k/N_k$ is the specific change of Gibbs free energy.

Power, as we have seen, is the increase in biomass density per time unit converted to free energy. Notice that the MPP focuses on a rate, in Eq. (2.8) indicated as $(\mathrm{d}N_k/\mathrm{d}t)$, the ecoflow, multiplied with the fraction that is able to do useful work, i.e. $\Delta G_k/N_k = \Delta g_k$. Maximum power thereby becomes equal to the rate of through-flow of useful energy.

Later on Odum (1983a,b) defined the MPP as a maximisation of *useful* power. Following this, Eq. (2.8) is applied to the ecosystem level by summing up all the contributions to the *total* power that are useful. This means that non-useful power is not included in the summation. The difference between the useful and non-useful power will be further discussed below, because the emphasis on *useful* power is perhaps the key to understanding Odum's principle and utilising it to interpret ecosystem properties.

Brown et al. (1993) and Brown (1995) have restated the MPP in more biological terms. According to the restatement, it is the transformation of energy into work (consistent with the term useful power) that determines success and fitness. Many ecologists have incorrectly assumed that natural selection tends to increase efficiency. If this were true endothermic organisms could never have evolved. Endothermic birds and mammals are extremely inefficient compared with reptiles and amphibians. They expend energy at high rates in order to maintain a high, constant body temperature, which, however, gives high levels of activities independent of environmental temperature (Turner, 1970). Brown (1995) defines fitness as reproductive power, the rate at which energy can be transformed into work to produce offspring. This interpretation of the MPP is more consistent with the maximum exergy principle (see Section 12.3) than with Lotka's and Odum's original idea.

In a recent book named "Maximum Power—The Ideas and Applications" by Odum and Hall (1995), a clear interpretation of the MPP has been presented, as has been applied in ecology by H.T. Odum. The principle claims that power or output of useful work is maximised—not the efficiency and the rate, but the trade-off between a high rate and high efficiency yielding "most useful energy = useful work". It is illustrated in Fig. 12.1.

Hall uses an interesting semi-natural experiment by Warren (1970) to illustrate the application of the principle in ecology. Streams were stocked with different levels of predatory cut-throat trout. When predator density was low, there was considerable invertebrate food per predator, and the fish used relatively little maintenance of food searching energy per unit of food obtained. With a higher fish-stocking rate, food became less available per fish, and each fish had to use more energy searching for it. Maximum production occurred at intermediate fish-stocking rates; i.e. at intermediate rates, the fish utilised their food.

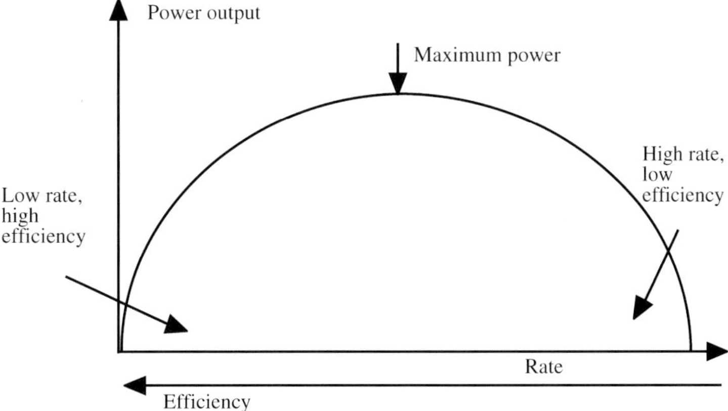

Fig. 12.1. The maximum power principle claims that the development of an ecosystem is a trade-off (a compromise) between the rate and the efficiency, i.e. the maximum power output per unit of time.

Hall (Odum and Hall, 1995) also mentions another example. Deciduous forests in moist and wet climates tend to have a leaf area index (LAI) of about 6. Such an index is predicted from the maximum power hypothesis applied to the net energy derived from photosynthesis. Higher LAI values produce more photosynthate, but do it less efficiently because of the respiration demand by the additional leaf. Lower leaf area indices are more efficient per leaf, but draw less power than the observed intermediate values of roughly 6.

According to Gilliland (1982) and Andresen (1983), the same concept applies for regular fossil fuel power generation. The upper limit of efficiency for any thermal machine such as a turbine is determined by the Carnot efficiency. A steam turbine could run at 80% efficiency, but it would need to operate at a nearly infinitely slow rate. Obviously, we are not interested in a machine that generates revenues infinitely slowly, no matter how efficiently. Actual operating efficiencies for a modern steam powered generator are, therefore, closer to 40%, roughly half the Carnot efficiency.

The examples show that the MPP is embedded in the irreversibility of the World. The highest process efficiency can be obtained by endoreversible conditions, meaning that all irreversibilities are located in the coupling of the system to its surrounding; there are no internal irreversibilities. Such systems will, however, operate very slowly. Power is zero for any endoreversible system. If we want to increase the process rate, it will imply that we also increase the irreversibility and thereby decrease the efficiency. The maximum power is the compromise between endoreversible processes and very fast completely irreversible processes.

12.3. Hypothesis: a thermodynamic law of ecology

In previous chapters, we examined how much of a comprehensive ecosystem theory could be derived from the three laws of thermodynamics. The laws were cast as restrictions to contain growth and development, whose processes of course have to satisfy the conservation principle (the First Law) for applicable parameters, degrade energy (the Second Law) and evacuate effluent heat to surroundings. What we found can be summarised as follows: energy flow through a system, defining it as open or at least non-isolated, is *necessary* for continued existence (partly deduced from the Third Law), and a flow of usable energy is *sufficient* to form an ordered structure, called a dissipative system (Prigogine, 1980). Morowitz (1992) referred to this latter as a Fourth Law of Thermodynamics, but it would seem more appropriate if such a law could be expanded to state *which* ordered structure among possible ones will be selected. A hypothesis about this selection has been introduced more than two decades ago (Jørgensen and Mejer, 1977; Mejer and Jørgensen, 1979; see also Jørgensen, 1982, 1992a, 2002a).

Second-Law dissipation acts to tear down the structures and eliminate the gradients, but it cannot operate unless the gradients are established in the first place. Structure and organization can be expressed in different units, such as the number of state variables, number of connections in an interactive web and *kJ* of exergy that corresponds to distance from thermodynamic equilibrium. Biological systems, especially, have many possibilities for moving away from equilibrium, and it is important to know along which pathways among possible ones a system will develop. This leads to the following hypothesis (Jørgensen et al., 2000):

"If a system receives an input of exergy, it will utilise this exergy to perform work. The work performed is first applied to maintain the system far away from thermodynamic equilibrium whereby exergy is lost by transformation into heat at the temperature of the environment. If more exergy is available, the system is moved further away from thermodynamic equilibrium, as reflected in growth of gradients. If more than one pathway to depart from equilibrium is offered, the one yielding the most work under prevailing conditions, and ultimately moving the system the farthest from thermodynamic equilibrium under the prevailing conditions, providing the most ordered structure, tends to be selected. Or if differently expressed: among the many ways for ecosystems to move away from thermodynamic equilibrium, the one maximising of time derivative of exergy, $\mathrm{d(Ex)}/\mathrm{d}t$, under the prevailing conditions will be selected."

This is a restatement and expansion of Jørgensen and Mejer (1977, 1979). A paradox appears to exist in conflicting criteria, the joint maximization of two diametrically opposed properties—storage, which is build up, and dissipation, which is teardown. This chapter will try to resolve this paradox in an ecological context, and in the process, expose the complexity of the interplay between thermodynamics and the growth of order in ecosystems and the ecosphere.

However, we have to note that all these definitions contain one incorrectness: we assume implicitly that exergy is a function of state and extensive variable (namely, in this case we can talk about exergy storage, captured exergy, dissipated exergy, etc.), whereas exergy is a function of two values: current state and reference (initial) state. Therefore, the change of exergy, $\mathrm{d'Ex}$, is not a full differential. Nevertheless, we can transform $\mathrm{d'Ex}$ to a total differential, i.e. consider exergy as a function of state, if we assume that *different ecological systems have the same reference state.* For instance, if such a reference state is "inorganic soup" or detritus, then any ecosystem evolution (short- or long-time) must start from this initial state. In other words, we define a single "zero" for all ecosystems, as we do in the classic thermodynamics, when we define "zeros" for energy and entropy.

If the exergy is an extensive state variable, then it can be represented as a bilinear form:

$$\mathrm{Ex} = \sum_k \mathrm{ex}_k N_k, \tag{3.1}$$

where N_k is the mass or concentration of kth component and ex_k is its specific exergy. Naturally, the amount of useful work that can be performed in the ecosystem is defined by both the total amount of biomass and the structure. The latter is determined by information coded in genomes of species that are joined in the ecosystem. Therefore, it is logical to consider the specific "genetic" exergy (see Section 5.7) as ex_k in Eq. (3.1).

In accordance with that formulated above, the exergy extreme principle is written as

$$\frac{\mathrm{d(Ex)}}{\mathrm{d}t} = \sum_k N_k \frac{\mathrm{d(ex}_k)}{\mathrm{d}t} + \sum_k \mathrm{ex}_k \frac{\mathrm{d}N_k}{\mathrm{d}t} \to \max. \tag{3.2}$$

In fact, the genome evolves very slowly; therefore, the derivatives $\mathrm{d(ex}_k)/\mathrm{d}t$ are smaller by several orders of magnitude than $\mathrm{d}N_k/\mathrm{d}t$. So we can neglect the first item if we consider microevolutionary dynamics, as well as the second item if macroevolutionary dynamics is interesting.

Unfortunately, today a universally adopted definition of exergy does not exist. We think that it is mainly explained by some indeterminacy in the concept of "useful work". The choice of either concept does very often depend on what kind of process takes place in the system. For instance, if the main processes are chemical, and they are taking place by constant temperature and volume, then the exergy is equivalent to both free energy and thermodynamic potential (Gibbs free energy). Of course, here we can talk about a partial case—the situation is more or less standard. However, different authors in different models define the meaning of "useful work" in different ways, and this choice is, as a rule, subjective. As a result, we have the whole spectrum of different definitions for exergy, and we often cannot establish among them a certain relation of equivalency. The situation can be described very well by a paraphrase of John von Neumann's motto to Chapter 4: "...nobody knows what exergy is in reality, that is why in the debate you will always have an advantage". That is why we have to be very careful when currently known ecological phenomena, conformities and principles are interpreted from the "exergical" point of view.

Just as it is not possible to prove the first three Laws of Thermodynamics by deductive methods, so the above hypothesis could only be "proved" inductively. In Section 12.4 a number of concrete cases, contributing generally to the support of the hypothesis, are presented. Models are used in this context to test the hypothesis. Consistency of the exergy-storage hypothesis with other theories (goal functions, orientors) describing ecosystem development will also be examined. Finally, it is discussed how these theories together form a pattern that can be considered a workable ecosystem theory.

The hypothetical Fourth Law of Thermodynamics is proposed to explain growth observed in ecological systems. *Growth* is defined as an increase in measurable quantity, often in ecology set to be biomass, but an ecosystem can grow in three different ways (Jørgensen et al., 2000):

1. The biomass or biological structure can grow.
2. The complexity of the structure can grow, i.e. the number of components, connections and thereby feedback's number in the trophic network are increasing. It implies that the mass and energy cycle and the goal system through-flow increase. See also Chapter 7 on the trophic chain.
3. The information can grow, which means that the level of organisation including the number of feedback mechanisms increases.

See also the relationship between Kullback's measure of information and exergy presented in Section 4.3.

In general, growth means an increase in a system's size, while development is an increase in organisation independent of a system's size. Growth is measured as mass or energy change per unit of time, for instance kg/day, while storage-specific growth is measured in 1/units of time, for instance 1/24 h. Development may take place without any change (growth) in biomass. In thermodynamic terms, a growing system is one moving away from thermodynamic equilibrium. At equilibrium, the system cannot do any work.

All its components are inorganic, have zero free energy (exergy) and all gradients are eliminated. Everywhere in the Universe there are structures and gradients resulting from growth and developmental processes cutting across all levels of organisation. A gradient is understood as a difference in an intensive thermodynamic variable, such as temperature, pressure or chemical potential.

12.4. Supporting evidence

Several case studies from Jørgensen et al. (2000) and Jørgensen (2002a) are presented below in which alternative energy-use pathways representing probably different gains in stored exergy are compared. More examples can be found in these references. However, Chapter 13 is devoted to showing a network between ecological observations, ecological rules and the pattern of an ecosystem theory, which is presented last in this chapter.

(1) *Size of genomes.* In general, biological evolution has been towards organisms with an increasing number of genes and diversity of cell types (Futuyma, 1986; compare also with Section 4.7). If a direct correspondence between free energy and genome size is assumed, this can be reasonably taken to reflect increasing exergy storage accompanying the increased information content and processing of "supreme" organisms.

(2) *Le Chatelier's Principle.* The exergy-storage hypothesis might be taken as a generalised version of Le Chatelier's Principle. Biomass synthesis can be expressed as a chemical reaction:

$$\text{energy} + \text{nutrients} = \text{molecules with more free energy (exergy) and organisation} \\ + \text{dissipated energy.}$$

According to Le Chatelier's Principle, if energy is put into a reaction system at equilibrium the system will shift its equilibrium composition in a way to counteract the change. This means that more molecules with more free energy and organisation will be formed. If more pathways are offered, those giving the most relief from the disturbance (displacement from equilibrium) by using the most energy, and forming the most molecules with the most free energy, will be the ones followed in restoring equilibrium.

For example, the sequence of organic matter oxidation (e.g. Schlesinger, 1997) takes place in the following order: by oxygen, nitrate, manganese dioxide, iron (III), sulphate and carbon dioxide. This means that oxygen, if present, will always out-compete nitrate which will out-compete manganese dioxide, and so on. The amount of exergy stored as a result of an oxidation process is measured by the available kJ/mol of electrons, which determines the number of adenosine triphosphate (ATP) molecules formed. In this case exergy is the same as free energy. ATP represents an exergy storage of 42 kJ/mol. Usable energy as exergy in ATPs decreases in the same sequence as indicated above. This is as expected, if the exergy-storage hypothesis is valid (Table 12.1). If more oxidising agents are offered to a system, the one giving the highest storage of free energy will be selected. In Table 12.1, the first (aerobic) reaction will always out-compete the others because it gives the highest yield of stored exergy. The last (anaerobic) reaction

Table 12.1
Yields of kJ and ATPs per mole of electrons, corresponding to 0.25 mol of CH_2O oxidised

Reaction	kJ/mol e^-	ATPs/mol e^-
$CH_2O + O_2 \rightarrow CO_2 + H_2O$	125	2.98
$CH_2O + 0.8NO_3^- + 0.8H^+ \rightarrow CO_2 + 0.4N_2 + 1.4H_2O$	119	2.83
$CH_2O + 2MnO_2 + H^+ \rightarrow CO_2 + 2Mn_2^+ + 3H_2O$	85	2.02
$CH_2O + 4FeOOH + 8H^+ \rightarrow CO_2 + 7H_2O + Fe^{2+}$	27	0.64
$CH_2O + 0.5SO_4^{2-} + 0.5H^+ \rightarrow CO_2 + 0.5HS^- + H_2O$	26	0.62
$CH_2O + 0.5CO_2 \rightarrow CO_2 + 0.5CH_4$	23	0.55

The released energy is available to build ATP for various oxidation processes of organic matter at pH 7.0 and 25°C.

produces methane; this is a less complete oxidation than the first because methane has a greater exergy content than water.

Numerous experiments have been performed to imitate the formation of organic matter in the primeval atmosphere of Earth four billion years ago (Morowitz, 1968). Energy from various sources was sent through a gas mixture of carbon dioxide, ammonia and methane (compare also with the discussion in Section 3.6). Analyses have shown that a wide spectrum of compounds, including several amino acids contributing to protein synthesis, is formed under these circumstances. There are obviously many pathways to utilise the energy sent through simple gas mixtures, but mainly those forming compounds with rather large free energies (high exergy storage, released when the compounds are oxidised again to carbon dioxide, ammonia and methane) will form an appreciable part of the mixture (Morowitz, 1968).

(3) *Photosynthesis.* There are three biochemical pathways for photosynthesis: (1) the C_3 or Calvin–Benson cycle, (2) the C_4 pathway and (3) the crassulacean acid metabolism (CAM) pathway. The latter is least efficient in terms of the amount of plant biomass formed per unit of energy received. Plants using the CAM pathway are, however, able to survive in harsh, arid environments that would be inhospitable to C_3 and C_4 plants. CAM photosynthesis will generally switch to C_3 as soon as sufficient water becomes available (Shugart, 1998). The CAM pathways yield the highest biomass production, reflecting exergy storage under arid conditions, while the other two give highest net production (exergy storage) under other conditions. While it is true that a gram of plant biomass produced by the three pathways has different free energies in each case, in a general way improved biomass production by any of the pathways can be taken to be in a direction that is consistent, under the conditions, with the exergy-storage hypothesis.

(4) *Leaf size.* Givnish and Vermelj (1976) observed that leaves optimise their size (thus mass) for the conditions. This may be interpreted as meaning that they maximise their free-energy content. The larger the leaves, the higher their respiration and transpiration, and the more solar radiation they can capture. Deciduous forests in moist climates have a LAI of about 6%. Such an index can be predicted from the hypothesis of highest possible leaf size, resulting from the trade-off between having leaves of a given size versus maintaining leaves of a given size (Givnish and Vermelj, 1976). Size of leaves in a given environment depends on the solar radiation and humidity regime, and while, for example,

sun and shade leaves on the same plant would not have equal exergy contents, in a general way leaf size and LAI relationships are consistent with the hypothesis of maximum exergy storage.

(5) *Biomass packing.* The general relationship between animal body weight, *W*, and population density, *N*, is $N = A/W$, where *A* is a constant (Peters, 1983); see also Eq. (6.6) in Chapter 3. The highest packing of biomass depends only on the aggregate mass, not on the size of individual organisms. This means that it is biomass rather than population size that is maximised in an ecosystem, as density (number per unit area) is inversely proportional to the weight of the organisms. Of course, the relationship is complex. A given mass of mice would not contain the same exergy or number of individuals as an equivalent weight of elephants. Also, genome differences (Example 1) and other factors would figure in. If other proposed goal functions as, for instance, exergy destruction, then biomass packing would follow the relationship $N = A/W^{0.65-0.75}$, because respiration expressing the conversion of exergy to heat is proportional to the weight in the exponent 0.65–0.75 (Peters, 1983). As this is not the case, biomass packing and the free energy associated with this lend general support for the exergy-storage hypothesis.

(6) *Cycling.* If a resource (for instance, a limiting nutrient for plant growth) is abundant, it will typically recycle faster. This is a bit strange, because recycling is not needed when a resource is non-limiting. A modelling study (Jørgensen, 2002a) indicated that free-energy storage increases when an abundant resource recycles faster. Fig. 12.2 shows

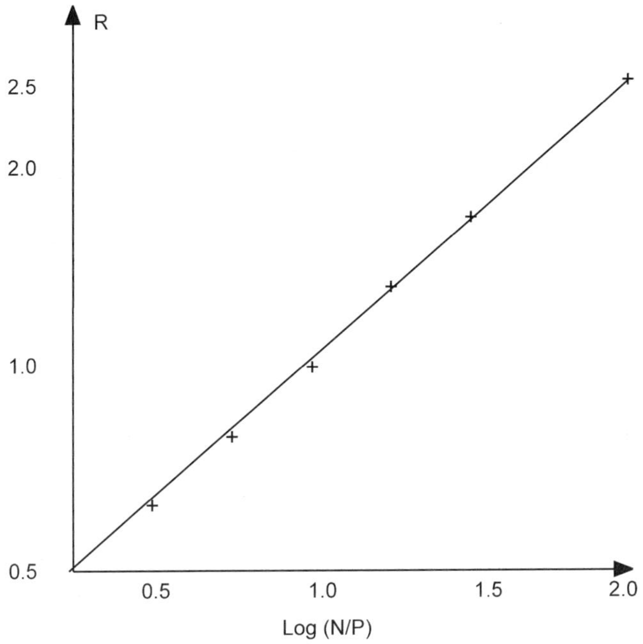

Fig. 12.2. Log–log plot of the ratio of nitrogen to phosphorus turnover rates, *R*, at maximum exergy versus the logarithm of the nitrogen/phosphorus ratio, log(N/P). The plot is consistent with Vollenweider (1975).

such results for a lake eutrophication model. The ratio, R, of nitrogen (N) to phosphorus (P) cycling, which gives the highest exergy, is plotted in a logarithmic scale versus log(N/P). The plot in Fig. 12.2 is also consistent with empirical results (Vollenweider, 1975).

Of course, one cannot "inductively test" anything with a model, but the indications and correspondences with data tend to support in a general way the exergy-storage hypothesis. The cycling ratio giving the highest ascendancy is also correlated similarly to the N/P ratio (R. Ulanowicz, personal communication).

(7) *Fitness.* Brown et al. (1993) and Marquet and Taper (1998) examined patterns of animal body size. They explained frequency distributions for the number of species as functions of body size in terms of fitness optimisation. Fitness can be defined as the rate at which resources in excess of those required for maintenance are used for reproduction (Brown, 1995). This definition suggests channelling of resources to increase the free-energy pool. Fitness, so interpreted, may be taken as consistent with the exergy-storage hypothesis.

(8) *Structurally dynamic modelling.* Dynamic models whose structure changes over time are based on non-stationary or time-varying differential or difference equations. We will refer to these as *structurally dynamic models.* A number of such models, mainly of aquatic systems (Jørgensen, 1986, 1988, 1990, 1992a,b; Nielsen, 1992a,b; Jørgensen and Padisák, 1996; Coffaro et al., 1997; Jørgensen and de Bernardi, 1997, 1998), have been investigated to see how structural changes are reflected in free-energy changes. The latter were computed as exergy indexes (see Section 12.5). Time-varying parameters were selected iteratively to give the highest index values in a given situation at each time step (see Jørgensen and Padisák, 1996). Such informal procedures for system identification are complicated and prone to error. Final results, and whether local versus global optima are realised, etc. are very sensitive to initial choices made. Even so, at the least, it was always observed that maximum exergy index values could not be achieved without changing parameter values, i.e. without structural dynamics. The technicalities of parameter fitting aside, this overall result means that system structure must change if its free-energy storage is to be continually maximised. Changes in parameters, and thus system structure, do not only reflect changes in external boundary conditions, but also mean that such changes are necessary for the ongoing maximisation of exergy. For all the models investigated along these lines, the changes obtained were in accordance with actual observations (see references). These studies therefore affirm, in a general way, that systems adapt structurally to maximise their exergy content.

It is noteworthy that Coffaro et al. (1997), in his structural–dynamic model of the Lagoon of Venice, did not calibrate the model describing the spatial pattern of various macrophyte species such as *Ulva* and *Zostera*, but used exergy-index optimisation to estimate parameters determining the spatial distribution of these species. He found good accordance between observations and model, as was able, using this method *without* calibration, to explain more than 90% of the observed spatial distribution of various species of *Zostera* and *Ulva*. Some examples of structurally dynamic models will be presented in Chapter 13.

(9) *E.P. Odum's attributes.* In summary, phenological progression in ecosystems, when viewed through the lens of exergy relationships, bears unmistakable resemblances to the growth of organisms, succession of communities and evolution of taxa. All these

processes can be seen as proceeding on different space–time scales more or less under the exergy principles of growth as outlined in this volume (see Table 12.2 for a partial list). This list is complete in accordance with E.P. Odum's attributes (see also Table 2.5 in Section 2.4). This implies that all the characteristics of ecosystem development, summarised in E.P. Odum's attributes, comply with the maximum exergy-storage principle.

The conclusion from these examples and other examples presented in Jørgensen et al. (2000) and Jørgensen (2002b) is that they do not constitute a rigorous test of the exergy-storage hypothesis. This is impossible because exergy cannot be measured for ecological systems. They are too complex. However, through modelling and recourse to many examples, a kind of "inductive verification" is possible. That is what this section has tried to show, namely that the hypothesis provides a plausible objective function over a broad selection of actual systems and circumstances. Assistance from modelling depends on

Table 12.2
A partial list of characteristics of developed ecosystems that are in accordance with the exergy principles of this chapter

Characteristics	Explanation
1. High biomass	To utilise available nutrients and water to produce the highest stored exergy
2. High respiration, transpiration and other catabolic processes	To maintain the system far from thermodynamic equilibrium
3. Gradient development	The system moves as far as possible away from thermodynamic equilibrium
4. High information content	To utilise maximally the flow of exergy and resources
5. High level of specialisation and differentiation	To utilise space and time heterogeneity to gain the highest possible level of exergy
6. High level of adaptation and buffer capacity	To meet the challenge of changing forcing functions
7. High levels of network complexity and organisation	A consequence of the first four characteristics
8. Big size of (some) organisms	To minimise specific entropy production and thereby the cost of maintenance when exergy flow becomes limiting
9. Highly developed history	Caused by all developmental processes
10. High indirect/direct effect ratio	A consequence of the complex network
11. Irreversible processes	A consequence of system history
12. Both bottom-up and top-down regulation	To utilise all available avenues to build as much dissipative structure as possible
13. Symbioses developed	Two or more species move simultaneously further from thermodynamic equilibrium
14. Diversity of processes	To utilise all available avenues to build as much dissipative structure as possible

deriving a valid substitute measure for absolute exergy, an index covering the storage of both biomass and information that can be used in modelling studies to give further credential to the hypothesis. In Chapter 5 such an index—the relative exergy index—was developed.

12.5. Other ecosystem theories

Boltzmann (1905) said that the struggle for existence is a struggle for free energy available for work, which is a very close definition to the maximum exergy principle introduced in Chapter 5. Similarly, Schrödinger (1944) pointed out that organisation is maintained by extracting order from the environment. These last two principles may be interpreted as the systems that are able to gain most exergy under the given conditions, i.e. to move the farthest from thermodynamic equilibrium, will prevail. Exergy is defined as the useful or available energy of the system relative to the environment. Such systems will gain most biogeochemical energy available for doing work and therefore have most energy stored to be able to struggle for their existence. There seems to be a certain parallelism, therefore, between the three formulations of principles. However, in spite of external similarity between the MPP and the exergy principle, they are not equivalent (the differences will be discussed later on).

Boltzmann proposed that "life is a struggle for the ability to perform work", which is exergy. Referring to one of the thermodynamic identities, which can be written in verbal form:

$$\text{free energy} = \text{energy} - \text{temperature} \cdot \text{entropy}$$

that can be interpreted as (Straskraba et al., 1997)

$$\text{energy} - \text{disorder} = \text{energy} + \text{order}.$$

The difference between free energy and exergy is the ability with exergy to select a case-dependent reference state.

Ecological (and biological) growth and development have very much to do with the evolution of order in the material of organised matter, and work must be done to create this order out of the background (reference state) of somewhat less order. Teleology is frequently brought into the discussion about the origins of order, in the form of "objective functions", "goal functions", "optimisation criteria", "extreme principles" and "orientors" (e.g. Müller and Leupelt, 1998). This volume's central hypothesis, exergy-storage maximisation, is one such goal function, or in Aristotelian terms, a "final cause". In this section a selection of others is reviewed, all of them being criteria for purposeful ecological growth and development. How the different approaches can be united in an ecosystem theory will be discussed in Section 12.6.

(1) *Maximum biomass.* Biomass is stored energy, some of which can be turned into work. This portion is exergy, the inherent order in which is taken into account through multiplication by N_i's (Eq. (4.3) in Chapter 5). Eqs. (5.1) and (5.2) in Chapter 5 show even more clearly the two contributions—by N the total matter, and by Kullback's measure K of the increment of information. The ability of a species to perform work in an ecosystem, its

exergy or free energy, is thus proportional not only to its information content, but also to its biomass. Margalef (1968), Straskraba (1979, 1980) and Brown (1995) have all proposed the use of biomass as an ecological goal function. As biomass is storage and has exergy, its maximisation would be at least partly consistent with the exergy-storage hypothesis. For entire systems, however, this would require different weighting factors, as shown in Eqs. (7.9) and (7.10) in Chapter 5, to account for the different information (order) inherent in the different biological species.

(2) *Maximum power* (for details, see Section 12.2). The transformation of energy to perform work is correlated with the amount of exergy available (stored or in passage) within the system. The more exergy is stored, the more is available to be drawn on for work at a later stage, which requires conversion from storage to through-flow. In order to achieve storage, however, there must first be boundary flows (inputs) to sequester. Ecosystems must, therefore, contain balanced mixes of diametrically opposed quantities, storages and flows. Through-flow and storage are inversely related (see "ascendancy" in Chapter 11). One can be traded for the other, as determined by the composition of organic "stores" and biotic "storers" and "processors" that in aggregate determine whole-system turnover. Rapid turnover decreases storage and increases through-flow, and vice versa. A nice link between exergy storage and work performance was demonstrated by Salomonsen (1992) for two lakes with significantly different levels of eutrophication. He showed that the exergy/maximum power ratio was approximately the same in both cases.

(3) *Minimum specific entropy*. Mauersberger (1983, 1995) proposed a "minimum entropy principle" consistent with the principle of *least specific dissipation* from far-from-equilibrium thermodynamics (Prigogine, 1947). Johnson (1990, 1995) investigated least specific dissipation over a wide range of ecological case studies. Aoki (1988, 1989, 1993, 1995) compared entropy production, which reflects exergy utilisation, in terms of maintenance versus exergy storage in different lake ecosystems. He found that eutrophic lakes capture and store more exergy, then subsequently use it for maintenance. This is consistent with the general observation (e.g. Jørgensen, 1982; Salomonsen, 1992) that eutrophic lakes have more biomass, thus more stored exergy but, following on from this, also greater through-flow and dissipation, though less specific dissipation, than mesotrophic or oligotrophic lakes. Biomass-specific exergy, in other words, decreases with increasing eutrophication.

(4) *Maximum emergy*. Odum (1983a,b) introduced another goal function, "embodied energy", later contracted to *emergy*. Section 7.9 has presented the idea behind this concept. Embodied energy is expressed in solar energy equivalents; see Section 7.9. Though exergy and emergy are conceptually and computationally very different quantities, and though emergy calculates how much solar energy it costs to build a structure whereas exergy expresses the actual work potential for growth once built, the two measures correlate well when computed for models (Jørgensen, 1994). The difference between the two concepts may be expressed as follows: emergy expresses the costs in solar radiation equivalent, while exergy expresses the result (the stored working capacity). When two species are compared, the ratio between the two may be very different, but when we compare entire ecosystems, it is not surprising that the costs and the results are parallel concepts. Bastianoni and Marchettini (1996) found that a natural lagoon had a high exergy/emergy ratio while a man-made wastewater lagoon had a low value. Nature is, therefore,

apparently better able to utilise the emergy in order to obtain exergy than man-made systems (Bastianoni, 1998).

(5) *Ascendancy*. Another network measure of whole-system contributions to growth and development is *ascendancy* (Ulanowicz, 1986, 1997). According to this theory, in the absence of overwhelming external disturbances, living systems exhibit a propensity to increase in an "ascendant" direction (see also Chapter 9). As ascendancy is well correlated with stored exergy (Jørgensen, 2002b), maximising ascendancy is similar to maximising exergy storage. The relationship is not straightforward, however. Considering Example 6 in Section 12.4, increased cycling at steady state increases both the through-flow and storage that can be derived from boundary inputs (Patten et al., 1997). One is traded for the other, depending on the composition of components, which determines system turnover. Rapid turnover decreases storage and increases through-flow, and vice versa. As ascendancy is dominated by its extensive variable, through-flow, if this is maximised, then storage must be sacrificed accordingly in the steady-state relationship. But, as through-flow and storage are closely coupled, if through-flow is maximised so necessarily is the storage to which this may contribute. Conversely, the greater the storage in a system is, the more is available to be converted to through-flow as circumstances warrant. Maximisation of ascendancy, a measure heavily dominated by through-flow, can thus be taken as generally consistent with the exergy-storage hypothesis.

In conclusion, initially different concepts about how energy and matter are related to one another in complex systems organisation turn out in many cases—but not all—to be merely nuances in expression of the same central phenomena: that basically local negentropy production (exergy storage) opposes but enables entropy production (exergy destruction).

12.6. Towards a consistent ecosystem theory

The properties of ecosystems can only be revealed using a pluralistic view. It is therefore not surprising that there are many different ecosystem theories published in the scientific literature. It is, on the other hand, necessary to try to unite the theories and examine if they are tied up in contradictions or form a pattern that can be used to give a better understanding of the nature of ecosystems and to solve the global environmental problems. The goal is to give a common framework of reference for *further* development of a more profound and comprehensive ecosystem theory than the one we are able to present today. The pattern should serve as a "conceptual diagram", which can be used as a basis for further discussion of how ecosystems behave. We are still in an early stage of an ecosystem—theoretical development—and it may be argued that this attempt is premature, but experience from modelling has taught us that it is better to conclude one's thoughts in a conceptual diagram at an early stage and then be ready to make changes than to let all modelling efforts wait until all details are known, as this will never be the case due to the immense complexity of Nature (Jørgensen, 2001b, 2002b). Moreover, recent development in ecosystem theory has made it possible to conclude that the theories presented here are indeed consistent and supplementary.

The centre of the pattern presented below is the tentative Fourth Law of Thermodynamics, but it cannot be excluded that other tentative laws could be the core of an ecosystem theory. What can we conclude from these (tentative) laws about ecosystem properties? Can we, as is known from physics, formulate a limited number of laws and explain a very large number of observations (Jørgensen, 2001b, 2002b; see also Chapter 1)? This question has been answered with a clear "yes" in this volume. The recent development in system ecology represents a paradigm shift. The paradigm that is now receding has dominated our culture for several hundred years. It views the universe as a mechanical system composed of elementary building blocks. The new paradigm is based on a holistic worldview. The world is seen as an integrated whole and recognises the fundamental interdependence of all phenomena.

The support for the validity of the tentative law in its present formulation (see Section 12.3) is strong and may be summarised in the following three points:

1. It may be considered a translation of Darwin's theory to thermodynamics and is consistent with the basic, thermodynamic laws. The selected organisation is the one that offers most "survival" and may be measured as exergy.
2. The application of the hypothetical law in models gives (many) results that are consistent with ecological observations (see Section 12.3 of this chapter, Jørgensen, 2002b, and Chapter 13).
3. It is possible to validate models that describe the change in properties of the components, the so-called structurally dynamic models, by application of exergy as a goal function (see Section 12.3 and Chapter 13).

We need a number of different complementary approaches to explain ecosystems, which is not surprising as much simpler physical phenomena, light, for instance, need two different descriptions, namely, as waves and as particles. Several ecosystem theories have been presented in the scientific literature during the last 2–3 decades; see Section 12.5. At first glance they look very different and seem to be inconsistent, but a further examination reveals that they are not so different and that it should be possible to unite them in a consistent pattern. It has been accepted by the system scientists since 1998/1999 but, as a result of two meetings in 2000, one in Porto Venere, Italy, in late May and one in Copenhagen, in early June in conjunction with the American Society of Limnology and Oceanography meeting, it can now be concluded that a consistent pattern of ecosystem theories has been formed. Several system ecologists agreed on the pattern presented below as a working basis for further development in system ecology. This is of the utmost importance for progress in system ecology because, with a theory in hand, it will be possible to explain many rules that are published in ecology and applied ecology that again explain many ecological observations, as has been discussed in Sections 12.3–12.5 and will be discussed further in Chapter 13. We should, in other words, be able to attain the same theoretical basis that characterises physics: a few basic laws, which can be used to deduce rules that explain observations. It has, therefore, also been agreed that one of the important goals in system ecology would be to demonstrate (prove) the links between ecological rules and ecological laws.

The first contribution to a clear pattern of the various ecosystem theories came from the network approach used often by Patten. He has shown by a mathematical analysis of networks in steady state (representing, for instance, an average annual situation in an ecosystem with close to balanced inputs and outputs for all components in the network) that the sum of through-flows in a network (which is maximum power) is determined by the input and the cycling within the network. The input (the solar radiation) again is determined by the structure of the system and its biomass (the stored exergy is a function of the biomass distributed over the structure). Furthermore, the more the complexity of the structure, the more maintenance is needed. Therefore, the more exergy dissipated, the greater the inputs are. Cycling, on the other hand, means that the same energy (and correspondingly, exergy) is utilised better in the system, and therefore more biomass distributed over the structure (exergy) can be formed without increase of the inputs. It has been shown previously that more cycling means an increased ratio of indirect to direct effects, while increased input does not change the ratio of indirect to direct effects; see Chapters 13 and 14.

Fath and Patten (2000) used these results to determine the development of various variables used as goal functions (exergy, power, entropy, etc.). An ecosystem is of course not setting goals, but a goal function is used to describe the direction of development an ecosystem will take. Their results can be summarised as follows:

1. Increased inputs (more solar radiation is captured) mean more biomass is distributed over the same structure, more exergy stored, more exergy degraded and, therefore, also means higher entropy dissipation, more through-flow (power), increased ascendancy, but no change in the ratio indirect to direct effect or in the retention time for the energy in the system = total exergy stored in the system/input exergy per unit of time.
2. Increased cycling implies the same biomass is distributed over a more complex structure, more exergy stored, more through-flow, increased ascendancy, increased ratio of indirect to direct effect, increased retention but no change in exergy degradation.

Almost simultaneously Jørgensen et al. (2000) published a paper, which claims that ecosystems manifest three growth forms (see also Section 12.3):

I. Growth of physical structure (biomass), which is able to capture more of the incoming energy in the form of solar radiation, but also requires more energy for maintenance (respiration and evaporation).
II. Growth of the complexity of network, which means more cycling of energy and matter.
III. Growth of information (more developed plants and animals with more genes), from r-strategists to K-strategists, that wastes less energy but also usually carries more information.

These three growth forms may be considered as an integration of E.P. Odum's attributes, describing changes in an ecosystem associated with development from the early to the mature stage; eight of the most applied attributes associated with the three growth forms should be mentioned (for the complete list of attributes see Table 2.5). These changes are the following:

1. Ecosystem biomass (biological structure) increases.
2. More feedback loops (including recycling of energy and matter) are built.
3. Respiration increases.
4. Respiration relative to biomass decreases.
5. Bigger animals and plants (trees) become more dominant.
6. The specific entropy production (relative to biomass) decreases.
7. The total entropy production will first increase and then stabilise on approximately the same level.
8. The amount of information increases (more species, species with more genes, the biochemistry becomes more diverse).

Growth form I covers attributes 1, 3 and 7. The biomass increases according to attribute 1, which implies that the respiration also increases because it costs more exergy to maintain more biomass. This also means that the entropy production will increase.

Growth form II covers 2 and 6. When the network increases, there will be more feedback mechanisms available for regulation of the network. The energy and matter will thereby circle to a higher extent, which means that more biomass can be supported with the same total input and output of the exergy.

Growth form III covers the attributes 4, 5, 7 and 8. Bigger and more developed species will take over according to growth form III. It implies more biomass in relation to respiration and while the total entropy production is not changed the specific entropy production is decreasing.

Holling (1986)—see Fig. 12.3—has suggested how ecosystems progress through the sequential phases of renewal (mainly growth form I), exploitation (mainly growth form II), conservation (dominant growth form III) and creative destruction. The latter phase also fits into the three growth forms but will require further explanation. The creative destruction phase is a result of either external or internal factors. In the first case (for instance, hurricanes and volcanic activity), further explanation is not needed, as an ecosystem has to use the growth forms under the prevailing conditions that are determined by external factors. If the destructive phase is a result of internal factors, the question is "why would a system be self-destructive?"

A possible explanation is: a result of the conservation phase is that almost all nutrients will be contained in organisms, which implies that there are no nutrients available to test new and possibly better solutions to move further away from thermodynamic equilibrium or, if expressed in Darwinian terms, to increase the probability of survival. This is also implicitly indicated by Holling, as he talks about creative destruction.

Therefore, when new solutions are available, it would, in the long run, be beneficial for the ecosystem to decompose the organic nutrients into inorganic components, which can be utilised to test the new solutions. The creative destruction phase can be considered as a method to utilise the three other phases and the three growth forms more effectively in the long run (Jørgensen and Fath, submitted). This is indicated in the figure as "trend of each further cycle" and it is shown that the ecosystem is moving towards a higher specific

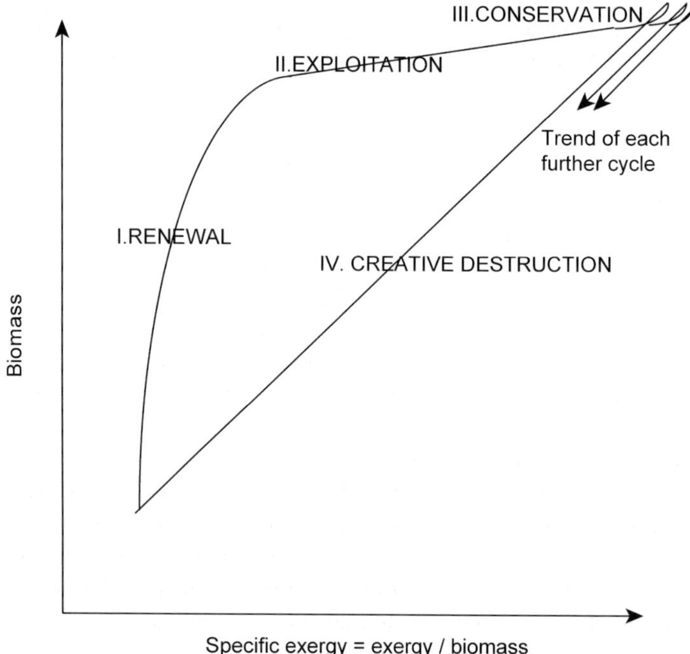

Fig. 12.3. Holling's four phases of ecosystems, described in terms of biomass versus specific exergy. The presentation is inspired by Ulanowicz (1997).

exergy (and maybe biomass), if the inorganic components are available to form more biomass for each cycle.

Five of the presented hypotheses to describe ecosystem growth and development are examined with respect to three growth forms:

A. The entropy production tends to be minimum (this is proposed by Prigogine (1947, 1980) for linear systems at steady non-equilibrium state, not for systems that are far from equilibrium). It is applied by Mauersberger (1983, 1995) to derive expressions for bioprocesses at a stable stationary state. See Section 12.5 and Chapter 3.
B. Natural selection tends to make the energy flux through the system a maximum, so far as compatible with the constraints to which the system is subject (Odum, 1983a,b). This is also called the MPP (see Section 12.2).
C. Ecosystems will organise themselves to maximise the degradation of exergy (Kay, 1984).
D. A system that receives a through-flow of exergy will have a propensity to move away from thermodynamic equilibrium, and if more combinations of components and processes are offered to utilise the exergy flow, the system has the propensity to select the organisation that gives the system as much stored exergy as possible; see Sections 12.1 and 12.3 (Jørgensen and Mejer, 1977, 1979; Mejer and Jørgensen, 1979; Jørgensen, 1982, 1997, 2002b).

E. An ecosystem will have a propensity to develop towards a maximisation of the ascendancy (Ulanowicz, 1986); see Section 12.5 and Chapter 9.

The usual description of ecosystem development illustrated, for instance, by the recovery of Yellow Stone Park after fire, an island born after a volcanic eruption, reclaimed land, etc., is well covered by Odum (1969): at first the biomass increases rapidly which implies that the percentage of captured incoming solar radiation also increases. But since species diversity of plants at the first stage of succession is very low, there are a lot of lacunas in the ability of plants to capture solar radiation in *necessary* spectral intervals. Moreover, vegetation cover at this stage has not developed a storey structure that naturally decreases its ability to capture incoming radiation, in particular diffusive radiation and radiation, which has already passed through upper storeys. In this time the energy needed for maintenance increases with growth of biomass and flux of metabolic heat increases too. But the heat must be dissipated and for this the mechanism of transpiration is insufficient: in addition, the mechanism of turbulence transport is needed. The latter can be provided by the sufficiently complex architecture of vegetation cover. There are not such types of possibilities at this stage; therefore, the possibilities of simple biomass growth (form I) are quickly exhausted. It is obvious that the information component of exergy is very small, and the value of exergy is determined by the enthalpy of biomass, i.e. captured *exergy* is equal to captured *energy*.

Finally, we note that the through-flow (of useful energy), exergy dissipation and the entropy production also increase due to the increased need of energy for maintenance.

Growth form II becomes dominant at the next stage of succession when the ecosystem structure begins to be complicated, although there is an overlap of the two growth forms. Further unstructured growth of biomass does, therefore, not increase the specific exergy of biomass. The complication of structure leads to increase of the information component of exergy and increase of biomass diversity. The role of different species in the ecosystem and their influence on such important general characteristics as, for instance, stability becomes more and more different in comparison with the first stage of proportional simple growth. This means that their biomasses are diversified, and their specific exergies differ more and more from each other. The ecosystem can still improve the ecological network and can still replace r-strategists by K-strategists, small animals and plants with bigger ones, etc. However, growth form II does not require more exergy for maintenance. Exergy degradation is, therefore, not increasing but is maintained on a constant level.

Finally, if growth forms I and II are typical microevolutionary processes (an ecosystem succession belongs to such a type), then growth form III is a macroevolutionary process a when the genotype information is changed as a result of mutations and consequent natural selection, so that the number of non-nonsense genes increases, and less developed species could be more developed. Let us keep in mind the so-called Cope's law: *the later descendent may be increasingly larger than its ancestors.* For instance, the horse is today much bigger than the horse fossils from 20–30 million years ago.

The accordance with the five descriptors + specific entropy production and the three growth forms based on this description of ecosystem development is shown in Table 12.3.

Table 12.3
An accordance between the growth forms and the proposed descriptors

	Growth form I	Growth form II	Growth form III
Exergy storage	Up	Up	Up
Power/through-flow	Up	Up	Up
Ascendancy	Up	Up	Up
Exergy destruction	Up	Equal	Equal
Retention time	Equal	Up	Up
Entropy production	Up	Equal	Equal
Exergy/biomass = specific exergy	Equal	Up	Up
Entropy/biomass = specific entropy production	Equal	Down	Down
Ratio indirect/direct effects	Equal	Up	Up

Based upon the results, it is possible to formulate the following hypothesis, which unites the five hypotheses:

"Ecosystem development in all stages will move away from thermodynamic equilibrium and select the components and the organisation that yields the highest flux of useful energy through the system and the most exergy stored in the system. This corresponds also to the highest ascendancy."

12.7. Some final comments

(1) In general, growth means increase in system size, while development is an increase in organisation independent of system size. Growth is measured as mass or energy change per unit of time, while storage-specific growth is measured in 1/units of time. Development may take place without any change (growth) in biomass. Ulanowicz (1986) uses "growth" and "development" as extensive and intensive aspects of the same process; they may often co-occur. In thermodynamic terms, a growing system is one moving away from thermodynamic equilibrium. At the state of equilibrium the system cannot do any work. All its components are inorganic, have zero free energy (exergy) and all gradients are eliminated. Everywhere in the universe there are structures and gradients resulting from growth and developmental processes cutting across all levels of organisation. A gradient is understood as a difference in an intensive thermodynamic variable, such as temperature, pressure, altitude or chemical potential. Second-law dissipation acts to tear down the structures and eliminate the gradients, but it cannot operate unless the gradients are established in the first place. Structure and organisation can be expressed in different units, such as number of state variables, number of connections in an interactive web and kJ of exergy, which corresponds to distance from thermodynamic equilibrium. Biological systems, especially, have many possibilities for moving away from equilibrium, and it is important to know along which pathways among the possible ones a system will

develop. This leads to the following hypothesis, formulated in accordance with Jørgensen et al. (2000):

"If a system receives an input of exergy, it will utilise this exergy to perform work. The work performed is applied to: (1) maintain the system, dissipating the residue as heat (degraded exergy) to the system's surroundings, (2) move the system further from thermodynamic equilibrium, reflected in growth of gradients, and if more than one pathway to depart from equilibrium is offered, the one yielding the most work, and ultimately moving the system far away from thermodynamic equilibrium under the prevailing conditions, i.e. giving the most ordered structure, tends to be selected."

This is a restatement and expansion of Jørgensen and Mejer (1977). A paradox appears to exist in conflicting criteria, the joint storage maximisation of the two diametrically opposed properties storage that is build-up and dissipation, or teardown. This chapter has tried to resolve this paradox in an ecological context, and in the process expose the complexity of the interplay between thermodynamics and the growth of order in ecosystems and the ecosphere.

(2) Application of intuitively clear concepts, such as "input", "captured", "dissipation" and "degradation", which seems so clear in application to energy, induces certain difficulties in the case of exergy; in spite of that exergy can be considered as extensive a state variable as energy. In order to avoid these difficulties, we would like to come up with some analogue of them.

It is known that the total exergy of ecosystem is described by expression (3.1), $\mathrm{Ex} = \sum_k \mathrm{ex}_k N_k$, where ex_k are specific exergies. We assume the following:

1. Let h_{det} be a specific enthalpy of detritus and $\mathrm{ex}_k^{\mathrm{gen}}$ the specific genetic exergy of kth species biomass. Then in the process of ecosystem microevolution (succession)

$$\underset{t\to\infty}{\mathrm{Ex}(t)} = \sum_k \mathrm{ex}_k^{\mathrm{gen}} N_k(t) \tag{7.1}$$

passing from the first stage of growth (form I), which at $\mathrm{Ex}(t) = h_{\mathrm{det}} \sum_k N_k(t)$ until the dynamic equilibrium, which at its value $\mathrm{Ex}^{\mathrm{eq}} = \sum_k \mathrm{ex}_k^{\mathrm{gen}} N_k^{\mathrm{eq}}$.
2. Exergy in the process of its degradation decreases to the initial value $h_{\mathrm{det}} \sum_k N_k$.

   In the framework of this formalism exergy is produced within the system as entropy; it can be imported and exported using such material carriers as energy and matter.

Chapter 13
Application of exergy as ecological indicator and goal function in ecological modelling

Model every system and if it not modelable make it modelable.
Sven Erik Jørgensen
(changed from Galilei)

A vision without action is just a dream; an action without a vision just passes time; a vision with an action changes the world.
Nelson Mandela

13.1. Introduction

Exergy has been applied in order to improve two important tools for environmental management, namely assessment of ecosystem integrity and ecological models.

About 15 years ago, the environmental managers proposed to find ecological indicators that were able to assess the integrity of ecosystems, or you may say "take the pulse" of the ecosystem. The idea was to be able not only to assess (preferably quantitatively) a few indicators of the ecosystem integrity, but also (if possible) to come up with a diagnosis. If the ecosystem is not sound, what would we name the disease? It was realised that the first step in a process of cure would be to set up a quantitative diagnosis. How bad was the eutrophication or the toxic substance pollution for instance? Exergy and specific exergy have been applied as ecological indicators:

1. by comparison and integrity assessment of eutrophied lakes (Jørgensen, 2002b),
2. by comparison and integrity assessment of coastal zones (Jørgensen, 2002b),
3. by integrity assessment of Mondego Estuary in Portugal (Marques et al., 2002),
4. by integrity assessment of Chinese lakes (Xu et al., 2000),
5. as ecological indicators for coastal lagoons in Europe (ISPRA, 2001),
6. for integrity assessment of different farming systems (Jørgensen, 2002b),
7. for integrity assessment in a situation where toxic contamination of ecosystems has taken place.

Towards a Thermodynamic Theory for Ecological Systems, pp. 325–349

The application of exergy as ecological indicator is presented in Section 13.2 where the relationship between exergy and ecosystem integrity will be discussed, and in Section 13.3 where Example 1 in the above list is used to illustrate the application.

If we follow the general modelling procedure, we will attain a model that describes the processes in the focal ecosystem, but the parameters will represent the properties of the state variables as they are in the ecosystem during the examination period. They are not necessarily valid for another period because we know that an ecosystem can regulate, modify and change them, if a response to the change in the prevailing conditions is needed, determined by the forcing functions and the interrelations between the state variables. Our present models have rigid structures and a fixed set of parameters, reflecting that no changes or replacements of the components are possible. We need, however, to introduce parameters (properties) that can change according to changing forcing functions and general conditions for the state variables (components) to continuously optimise the ability of a system to move away from the thermodynamic equilibrium. So, we may hypothesise that the change of these properties (parameters) can be accounted for in our model by the use of an ecological goal function. The idea currently is to test if a change of the most crucial parameters produces a higher goal function of the system and, if that is the case, to use that set of parameters.

The types of models that can account for the change in species composition as well as for the ability of the species, i.e. the biological components of our models, to change their properties, i.e. to adapt to the prevailing conditions imposed on the species, are sometimes called structurally dynamic models in order to indicate that they are able to capture structural changes. They may also be called the next or fifth generation of ecological models to underline that they are radically different from previous modelling approaches and can do more, namely describe changes in species composition or changes in the properties of the species.

It could be argued that the ability of ecosystems to replace present species with other better fitted species can be considered by the construction of models that encompass *all* actual species for the entire period that the model attempts to cover. This approach has, however, two essential disadvantages. The model becomes first of all very complex, as it will contain many state variables for each trophic level. It also implies, of course, that the model will contain many more parameters that have to be calibrated and validated and this will introduce high uncertainty into the model's results and will render the application of the model very case-specific (Nielsen, 1992a,b). In addition, the model will still be rigid and, having continuously changing parameters, even without changing the species composition will not give the model the property of the ecosystems (Fontaine, 1981). It can be shown to be very important that ecological models reflect the flexibility and adaptability that characterise organisms. If a model includes many rigid state variables (species), there will only be one species that will have a combination of properties that gives the best chance for survival in a given situation. The other species will have a combination of the properties that makes survival and growth more difficult, and they cannot become complete (Nielsen, 1992a,b).

Several goal functions have been proposed, but only very few models that account for change in species composition or for the ability of the species to change their properties within some limits have been developed.

Bossel (1992) applies maximisation of a benefit or satisfaction index based on balancing weighted surplus *orientor* satisfactions on a common satisfaction scale. The approach is used to select the model structure of continuous dynamic systems and is able to account for the ecological structural properties. The approach seems very promising, but has unfortunately not been applied to ecological systems except in three cases.

Straskraba (1979) uses a maximisation of biomass as the governing principle. The model computes the biomass and adjusts one or more selected parameters to achieve the maximum biomass at every instance. The model has a routine which computes the biomass for all possible combinations of parameters within a given realistic range. The combination that gives the maximum biomass is selected for the next time step and so on.

Exergy has been used most widely as a goal function in ecological models. It has been applied up to now in 13 case studies, where significant changes in the species composition or the properties of the species were observed: for four shallow lakes—Søbygård Lake, Denmark (Jørgensen, 2002b), Glumsø Lake, Denmark (Jørgensen, 2002b), Mogan Lake, Turkey (Zhang et al., 2003a,b) and Lake Balaton, Hungary (Jørgensen and Padisák, 1996); two population dynamic models (Jørgensen, 2002b); for Mondego Estuary, Portugal (Jørgensen et al., 2002a); for Lake Annone Italy (Jørgensen and de Bernardi, 1997); for the Lagoons of Venice (Coffaro et al., 1997); to explain the success and failure of biomanipulation (Jørgensen and de Bernardi, 1998); to explain the intermediate disturbance hypothesis (Jørgensen and Padisák, 1996); to explain the change in the properties of Darwin's finches (Jørgensen and Fath, in press) and to explain the hysteresis in the shift from submerged vegetation and to phytoplankton-dominated eutrophication and back again to submerged vegetation by reduction of the nutrient input (Zhang et al., 2002). For all 13 case studies, the models were able to simulate the observed changes with a standard deviation similar to other model studies.

Moreover, it has been found possible to improve the parameter estimation by the use of exergy. If one parameter is not known with sufficient accuracy, it is possible to find this parameter as the value which yields the highest exergy for the model of the ecosystem considered (Jørgensen, 1995b, 2002b). For eutrophication models it has furthermore been attempted to combine a normal calibration of some parameters with a determination of the combination of other parameters that gives the highest exergy (Jørgensen, 2001a).

Finally, it should be mentioned that it is possible to obtain a better calibration of models developed for ecosystems that show seasonal changes of species composition, for instance a eutrophication model, where the phytoplankton and zooplankton species in the spring, summer and fall are often different. The usually applied calibration procedure determines one parameter set covering the entire year, while we, by the use of exergy optimisation, can find the current change of parameters that reflects the change of species composition, the so-called succession. Not surprisingly the application of a current optimisation of the exergy will therefore offer a better conformity between model simulations and observations (Jørgensen et al., 2002c; Zhang et al. 2003a,b). Exergy optimisation is, of course, only used for the parameters of the organisms, while physical–chemical parameters are calibrated according to the normally applied procedure. Only the living components show flexibility and adaptability.

As seen from this overview, the exergy maximisation is a useful tool in modelling and can offer some clear advantages in obtaining good model results. One case study, namely the change in the beak size of Darwin's finches, has been selected to illustrate this structurally dynamic type of model and these applications in modelling. It is presented in Section 13.5.

13.2. Exergy and specific exergy as ecological indicators

More and more environmental managers include ecological considerations in their management strategy, and they have therefore asked ecologists and system ecologists the following question: how can we express that an ecosystem is ecologically sound and measure it? The doctor of medicine attempts to express the health condition of his patient by the use of indicators such as, for instance, the blood pressure, the temperature, the kidney function, etc. The environmental manager is similarly searching for ecological indicators that can assess the ecosystem integrity. As an ecosystem is a very complex system, it is not surprising that it is not an easy task to find good ecological indicators to give the appropriate information on the ecosystem integrity, although many ecologists and system ecologists have been and are working with this problem.

Rapport (1995) even uses the phrases "to take nature's pulse", "the problem of detecting diseases in nature" and "clinical ecology" to stress the parallelism to human pathology.

von Bertalanffy (1942, 1952) characterised the evolution of complex systems in terms of four major attributes:

(1) progressive integration (entails the development of integrative linkages between different species of biota and between biota, habitat and climate),
(2) progressive differentiation (progressive specialisation as systems evolve biotic diversity to take advantage of abilities to partition resources more finely and so forth),
(3) progressive mechanisation (covers the growing number of feedbacks and regulation mechanisms),
(4) progressive centralisation (it probably does not refer to a centralisation in the political meaning, as ecosystems are characterised by short and fast feedbacks and decentralised control, but to the more and more developed cooperation among the organisms (the Gaia effect) and the growing adaptation to all other components in the ecosystem).

Costanza (1992) summarises the concept definition of ecosystem integrity as the following integrity: (1) homeostasis, (2) absence of disease, (3) diversity or complexity, (4) stability or resilience, (5) vigour or scope for growth and (6) balance between system components. He emphasises that it is necessary to consider all or at least most of the definitions simultaneously. Consequently, he proposes an overall system integrity index, $\mathrm{HI} = VOR$, where V is system vigour, O the system organisation index and R the resilience index. With this proposal Costanza probably touches on the most crucial ecosystem properties covering ecosystem integrity.

Measures of integrity should reflect the two aspects of the organisational state of an ecosystem: functional and structural. Function refers to the overall activities of the ecosystem. Structure refers to the interconnection between the components of the system. Measures of function would indicate the amount of energy being captured by the system.

It could be covered by measuring the exergy captured by the system. Measures of structure would indicate the way in which energy is moving through the system. The exergy stored in the ecosystem could be a reasonable indicator of the structure. If an ecosystem is able to maintain its organisation far from thermodynamic equilibrium in spite of the changing environmental condition, the ecosystem is said to have integrity. If an ecosystem is unable to maintain its organisation then it has lost its integrity. Integrity is, therefore, associated with the ability of the system to reach and maintain its optimum operating point.

Table 13.1 summarises a set of ecological indicators (referred from Kay, 1991). As seen, the stressed system has flows that are about 20% less (except export that drops by about 8%, but measured relative to the import, it even increases), the biomass drops about 35% and the ascendancy drops by slightly more than 20%.

Exergy expresses, in accordance with Chapter 5, the biomass of the system and the genetic information, which this biomass is carrying. Only a relative exergy index, however, can be calculated. It measures the relative, approximate distance from thermodynamic equilibrium, but it is only based on the ecological components included in the calculation. It includes the exergy embedded in the ability to make ordering processes, which are carried out by the information stored in the genes. Exergy also expresses the energy needed to decompose the system into inorganic matter (Svirezhev, 1992), and the work the system can perform by a proper use of these decomposition processes.

Table 13.1
Ecosystem indicators for the Crystal River March Gut Ecosystem

Indicator	Control	Stressed ecosystem
Biomass	1,157,000	755,000
Total		
Imports	7400	6000
Total throughput	22,800	18,000
Production	3300	2600
Exports	950	870
Respiration	6400	5100
Living matter		
Production	400	330
Exports	320	250
Respiration	3600	3100
To detritus	5700	4300
Detritus		
Import	0	0
Production	2900	2200
Exports	640	620
Respiration	2800	2100
Food web		
Cycles	142	69
Nexuses	49	36
Finn's cycling index	10.2%	9.3%
Ascendancy	28,500	22,400

See also Kay (1991). Unit: mg/m^2/day.

The relative exergy index may be used as a measure for ecosystem integrity and will at least partly cover (1)–(5) of the total of six points given in the definitions by Costanza (1992) (see above).

(1) The homeostasis is embodied in the information of the organisms on how they will meet certain changes by feedback reactions. This information is determined mainly by the genes, which are considered in the calculation of exergy.
(2) Absence of disease is reflected in the biomass, since any ecosystem disease sooner or later will be accompanied by a drop in biomass. As the higher organisms have more genes, exergy will be particularly sensitive to a drop in biomass of these organisms, which is considered as an advantage for the use of exergy as an ecological indicator.
(3) Living matter has higher complexity compared with the same elements in organic form (detritus), which has higher exergy than the elements in inorganic form. Exergy will thereby be a measure of complexity, but not necessarily of diversity. The exergy will generally increase as the ecological niches are better utilised by an increased biodiversity but there are cases, for instance of eutrophication of aquatic ecosystem, where the exergy increases and the biodiversity decreases; see below for further explanation.
(4) Exergy can be shown by the use of statistics on modelling studies to cover a sum of buffer capacities (see below) and is thereby related to the resistance of the ecosystem.
(5) Growth increases in biomass and the genes contain information on how to utilise the resources for growth. The evolution has steadily opened new pathways to utilise the resources (including the ecological niches) better and better. Exergy, which is used for both information and biomass measures, therefore can be considered as the potential for growth. It is interesting in this context that there is a relationship between the exergy stored in the ecosystem and the ability of the system to capture exergy from the solar radiation.
(6) The balance between the system components and the biodiversity are not covered by the use of exergy as integrity indicator because, for instance, eutrophic systems often have a low biodiversity and a biased distribution of the biomass, but a high exergy.

The conclusion from this comparison of Costanza's ecosystem integrity definition and the concept of exergy is that there is a need for supplementary ecosystem integrity indicators. Exergy does not cover all the aspects of ecosystem integrity presented in Costanza's definition.

Specific exergy, ex = Exergy/Volume or Biomass (see Chapter 5) seems to be opposed to the total exergy, Ex, to be a candidate for a better coverage of points 3 and 6 in the definition of ecosystem integrity given above:

(3′) Many model studies in this volume and ecological studies (Weiderholm, 1980) clearly show that increased biodiversity means that there is a higher probability of a better utilisation of the available resources, i.e. the value of ex increases. A better utilisation of all ecological niches is accompanied by a higher biodiversity. Thereby the value of ex also measures the structural complexity and the ratio of biomass to total mass (biomass + inorganic matter). A development towards a more complex organism (with more genes) will also result in a higher ex.

(6′) A better utilisation of all ecological niches means that there will be more species and thereby a better balance between system components, which again may ensure a better balance between various buffer capacities.

Reactions of ecosystems to perturbations have been widely discussed in relation to the stability concepts. However, this discussion has in most cases not considered the enormous complexity of regulation—and feedback mechanisms.

An ecosystem is a soft system that will *never* return to exactly the same point again. It will attempt to maintain its functions on the highest possible level, but never with exactly the same biological and chemical components in the same concentrations again. The species composition or the food web may or may not have changed, but at least it will not be the same organisms with exactly the same properties. In addition, it is unrealistic to consider that the same combination of forcing functions will occur again. We can observe that an ecosystem has the property of resilience, in the sense that ecosystems have a tendency to recover after stress; but a complete recovery, meaning that exactly the same situation will happen again, will never be realised. The combination of external factors—the impact of the environment on the ecosystem—will never appear again, and even if they would, the internal factors—the components of the ecosystem—would meanwhile have changed and can therefore not react in the same way as the previous internal factors did. Resistance is another widely applied stability concept. It covers the ability of the ecosystem to resist changes when the external factors are changed. An ecosystem will always be changed when the conditions are changed; the question is what exactly is changed and by how much?

It is observed that increased phosphorus loading gives decreased diversity (Ahl and Weiderholm, 1977; Weiderholm, 1980), but eutrophic lakes *are* highly stable. A similar relationship is obtained between the diversity of the benthic fauna and the phosphorus concentration relative to the depth of the lakes (Weiderholm, 1980); see also Chapter 7.

The concept of buffer capacity (Jørgensen, 1988) has a definition, which allows for a numerical formalisation, for instance in modelling, and it is furthermore applicable to real ecosystems as it acknowledges that *some* changes will always take place in the ecosystem in response to changed forcing functions. The question is how large these changes are in relation to changes in the conditions (the external variables or forcing functions).

In accordance with the concept the buffer capacity, β, is defined as

$$\beta = 1/(\partial(\text{state variable})/\partial(\text{forcing function})). \tag{2.1}$$

In a multi-dimensional case, when there are n variables x_i and m forcing functions f_k, i.e. parameters or functions of time describing the impact of the environment on the system, then the buffer capacity is a matrix of $(n \times m)$-dimension:

$$\boldsymbol{\beta} = \|\beta_{ik} = 1/(\partial x_i/\partial f_k\|, \qquad i = 1, \ldots, n; \;\; k = 1, \ldots m, \tag{2.2}$$

and we may consider all combinations of state variables and forcing functions. It implies that even for one type of change there are many buffer capacities corresponding to each of the state variables.

High nutrient concentrations favour, to a certain extent, large phytoplankton species.

It was found by statistical analysis of modelling results with many different models (Jørgensen and Mejer, 1977; Jørgensen, 1992a, 1994), that there is a correlation between exergy and the buffer capacities; see Section 6.4. Some buffer capacities may be reduced even when the exergy increases, as mentioned above for the eutrophication case, but it is more than compensated by the increase of other buffer capacities. These results are consistent with the relation that exergy measures the energy needed to decompose the system to inorganic components (Svirezhev, 1992).

The above observations explain why it has been very difficult to find a relationship between ecosystem stabilities in the broadest sense and species diversity, as already discussed a few times. Stability of ecosystems in its broadest ecological sense should be considered as a multi-dimensional concept, and the relation between species diversity and stability is therefore not simple and can be only revealed by a multi-dimensional relation. If species diversity decreases, the stability (represented by buffer capacities) may decrease in some directions, but will increase in others. It may be formulated as the following: if the system can offer a better survival, i.e. bigger buffer capacities in relation to the changing forcing functions by decreasing the diversity, the system will not hesitate to react accordingly. See also Chapters 6 and 9.

The above-mentioned relation between exergy and buffer capacities indicates that point 4 in the definition of ecosystem integrity is globally covered by the use of exergy as ecological indicator, but because there is almost an infinite number of buffer capacities, and it would therefore be impossible to cover them all, this relationship between exergy and buffer capacities can only be applied semi-quantitatively in practice. As the concepts of ecosystem stability, resilience and ecosystem integrity are multi-dimensional, it will often be necessary to supplement the computations of exergy by relevant and focal buffer capacities. Buffer capacities related to the management situation should be selected. If we are concerned with the influence of toxic substances, the buffer capacities based upon the changes provoked by the input of toxic substances should be selected. If we are concerned with acid rain and its influence on the forest, we should find the buffer capacities relating the pH of rain water to the growth of trees in the forest, and so on. The result will be a limited number of buffer capacities. To keep the ecosystem healthy, we should consider these focal buffer capacities in our environmental management strategies. As long as the buffer capacities can withstand the stress with only minor changes, the ecosystem should be considered healthy.

Costanza (1992) has proposed an overall system integrity index consisting of system vigour, system organisation and system resilience. System vigour and the global system resilience are in accordance with the above presentation described by using exergy. The organisation is better covered by the specific exergy as it is highly dependent on the species diversity and their organisation, and independent of the total biomass. As stability is multi-dimensional it would be an improvement in the assessment of ecosystem integrity to include focal buffer capacities related to actual or possible stress situations.

As pointed out by Costanza (1992) these concepts will require a heavy dose of systems modelling. It would be possible to assess the concentrations of the most important species or classes of species, and then calculate the exergy and the specific exergy, but it would require a dynamic model based upon mass balances to find the buffer capacities, as they relate changes in forcing functions with changes in state variables, unless it can be presumed that the relationships between forcing functions and state variables are linear. This does not imply that a new model has been developed for every new case study, because models have a certain generality and the experience from one modelling study to the next is essential. Furthermore, if the seasonal changes in exergy and specific exergy should be assessed, it will either require many measurements throughout the year or the development of a model that is able to simulate the seasonal changes. Exergy and specific exergy usually vary significantly during the year to reflect the ability of the ecosystem to cope with the changes in temperature, precipitation and other climatic factors. A model will, furthermore, have the advantage that it can answer questions such as: how will the ecosystem integrity change if the forcing functions are changed so and so?

These considerations lead to the following tentative procedure for a practical assessment of ecosystem integrity.

1. Set up relevant questions related to the integrity of a considered ecosystem.
2. Assess the most important mass flows and mass balances related to these questions.
3. Make a conceptual diagram of the ecosystem, containing the components of importance for the mass flows defined under 2.
4. Develop a dynamic model (if the data are not sufficient, a steady-state model should be applied) using the usual procedure (see, for instance, Jørgensen, 1994).
5. Calculate exergy, specific exergy and relevant buffer capacities by the use of the model. If the model is dynamic it will also be possible to find the seasonal changes in exergy, specific exergy and buffer capacities.
6. Assess the ecosystem integrity: high exergy, specific exergy and buffer capacities imply a good ecosystem integrity. If the exergy and specific exergy are high, but one of the focal buffer capacities is low, then the "medicine" is to improve the structure of the ecosystem to assure a higher focal buffer capacity. If the exergy is high, but the specific exergy and some focal buffer capacities are low, we would probably be dealing with a eutrophic system, where the "medicine" should be a reduction of the nutrient loadings. Based upon the values of the three indicators, different measures should be taken to improve the ecosystem integrity.

13.3. Assessment of ecosystem integrity. An example: a lake ecosystem

Given below is an example of the application of exergy, specific exergy and buffer capacity as indicators of ecosystem integrity (Jørgensen, 1994). The same approach has been applied to the models of wetlands, fishponds, streams and agricultural systems with the same general results. The case study should therefore be considered as an illustrative example. A eutrophication model with seven state variables has been applied: nutrients, phytoplankton, zooplankton, planktivorous fish, carnivorous fish, detritus and sediment. The usual equations (see, for instance, Jørgensen, 1976, 1986) are applied, but the following characteristics according to ecological observations have been introduced.

1. For grazing and predation the threshold concentrations are used, below which no grazing and predation take place.

2. The predation by the most carnivorous fish on the planktivorous fish is reduced above a certain concentration of phytoplankton to consider that carnivorous fish are hunting by sight.
3. The growth rate of phytoplankton and zooplankton are reduced step-wise from low nutrient concentrations to high nutrient concentrations in accordance with the observations that bigger species prevail at a higher level of eutrophication as discussed above.
4. Adaptation to changed temperature is used by changing the optimum temperature accordingly.
5. The flow rate relative to the volume is 0.1, which assures a fast reaction to the nutrient concentration in flowing water.
6. The model does not distinguish between phosphorus and nitrogen, but assumes that they are present in the ratio used by phytoplankton, i.e. 1:7. The photosynthesis follows the uptake of nutrient by a factor of 12, corresponding to 11 times as much uptake of carbon, hydrogen, oxygen and other elements than that of nitrogen and phosphorus.

The model computes the exergy and the specific exergy. The buffer capacities for changes in phytoplankton, zooplankton and the two classes of fish, when the inputs of nutrients and the temperature are changed, are found using a sensitivity analysis. A partial buffer capacity, β_{ik}, can be estimated as kth *forcing function*/ith *state variable*. It is observed that a small change in a forcing function implies a change in a selected state variable. The ratio between the changes corresponds to the buffer capacity. A high buffer capacity means that the given component of the ecosystem is resistant against changes in the given forcing function. Changes are more easily controlled. There are, of course, many buffer capacities corresponding to many combinations of forcing function and state variables.

Figs. 13.1–13.3 show some of the results. Exergy, specific (or sometimes also called structural) exergy and buffer capacity of phytoplankton to changed nutrients loading are plotted versus the nutrient concentration in flowing water. As seen the exergy increases with the increased nutrient input due to the resulting higher total biomass concentration. Specific exergy has a maximum at the nutrient concentration about 2 mg/l. With higher nutrient inputs, the specific exergy declines due to an unequal distribution of the biomass. Particularly the phytoplankton and the planktivorous fish increase on behalf of zooplankton and carnivorous fish. A structural change is observed, which is consistent with general observations in lakes; see also the results presented in Figs. 13.1 and 13.3. The buffer capacity of phytoplankton to changed nutrient loading has a minimum at a total nutrient input of about 2. It increases as discussed above with the increased nutrient loading mainly due to the slower growth rate.

Other changes in buffer capacities can be summarised as follows.

1. The buffer capacity for the influence of nutrients on the carnivorous fish is increasing with increasing nutrient loading. After a certain nutrient input the concentrations of carnivorous fish remain low at almost the same level independent of the nutrient concentration.

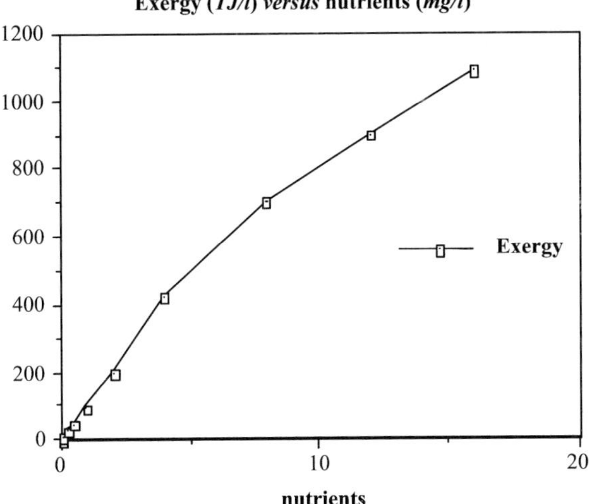

Fig. 13.1. The results obtained by the use of a eutrophication model with seven state variables are shown. The exergy is plotted versus the total inputs of nutrients (nitrogen and phosphorus, mg/l). The curve reflects the increase in biomass, which grows slower than linear with the nutrient concentration.

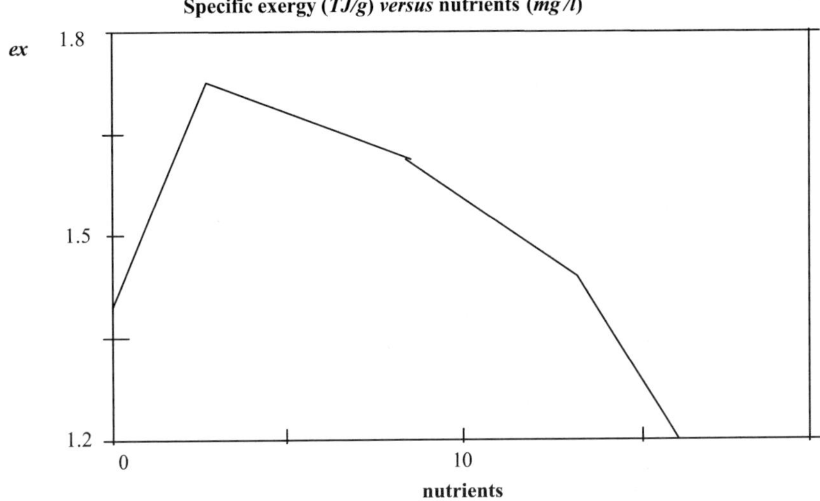

Fig. 13.2. The results obtained from the eutrophication model with seven state variables are shown. The specific exergy is plotted versus the total inputs of nutrients (nitrogen and phosphorus).

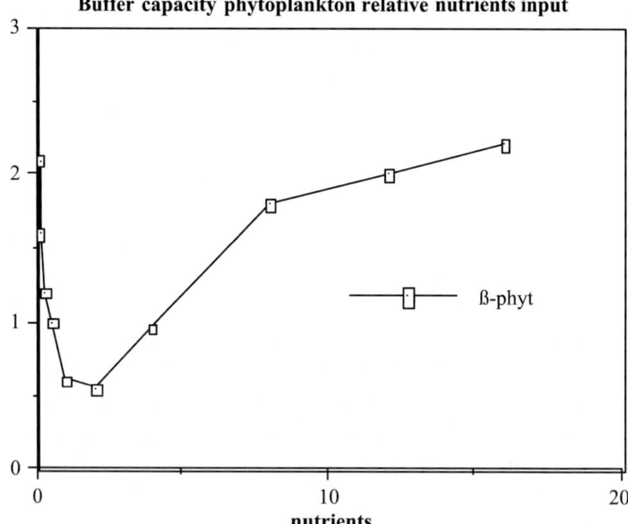

Fig. 13.3. Buffer capacities of phytoplankton to changed nutrients loading (mg/l) are plotted versus the nutrient concentration in flowing water.

2. The buffer capacities for the influence of nutrients on zooplankton and planktivorous fish have a maximum at 1 mg/l and are decreasing above this concentration by increasing nutrients input.

   The results are consistent with the general observations and previous model studies, referred above. The buffer capacities are generally either increasing or constant up to a nutrient level of about 1–4 mg/l, except for the influence of nutrients on phytoplankton. At about the same concentration, specific exergy has its maximum.

   The results may be interpreted as follows: up to a total nutrient concentration of about 2 mg/l, the examined buffer capacities are quite constant or even increasing, except for the influence of nutrients on phytoplankton. Specific exergy, which measures the ability of the system to utilise the resources, is increasing along the same line.

3. The buffer capacities for the influence of temperature are generally decreasing by increasing nutrient input above a nutrient concentration of about 2–4 mg/l except for carnivorous fish, where the buffer capacity is consequently increasing slightly with increasing nutrient input.

In this range of nutrient loading the top-down and the bottom-up controls are working in parallel; see Sommer (1989). A nutrient loading up to this level seems therefore fully acceptable, but if the loading increases above this level, crucial buffer capacities and the specific exergy decrease.

Some buffer capacities measuring the influence of nutrient input on phytoplankton and carnivorous fish increase, and so does the exergy, but it reflects only the ability of the system to meet changes in forcing functions (in the case of the nutrient loading) by such changes in the structure that the direct influence of these changes is reduced (in this case the influence of nutrients on phytoplankton).

The results of an ecosystem integrity analysis, as has been illustrated in this case study, can only be interpreted semi-quantitatively, because the accuracy of the underlying model does not allow very precise quantifications. The case study has, on the other hand, shown that the analysis is very useful, as it enables us to assess approximately the acceptable level of nutrient loadings and to see the consequences in relation to buffer capacities if we increase the loading above this level, i.e. to predict the expected outcome of the structural changes in the system. In this range of nutrient loading the top-down control has collapsed; see Sommer (1989). A nutrient concentration above approximately 2 mg/l should consequently be omitted and measures should be taken to reduce the nutrient loadings accordingly.

Fifteen lake case studies taken from ICLARM's ECOPATH survey on various ecosystems (Christensen and Pauly, 1993) were compared in relation to exergy and specific exergy. The figures are taken from the steady-state average situations based on observations. The exergy increases with increasing eutrophication for the 15 lakes. Fig. 13.4 shows, from these 15 case studies, the specific exergy (= exergy divided by the

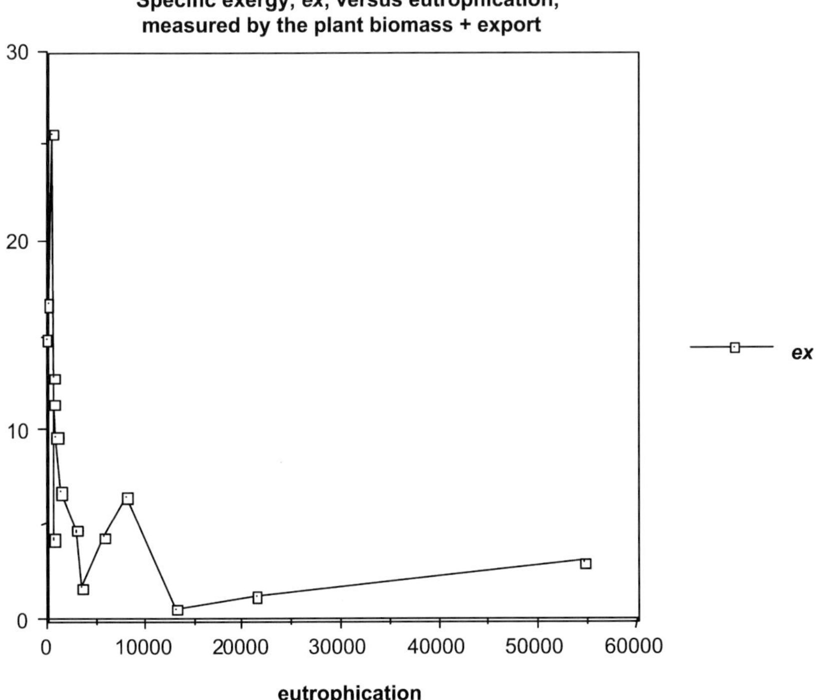

Fig. 13.4. The specific exergy (kJ/g) is plotted versus the eutrophication for 15 lake studies.

total biomass versus the eutrophication) measured by the biomass of phytoplankton and macrophytes including export. The specific exergy has a maximum at a relatively low level of eutrophication, which is to be compared with a medium level at the model exercise (see Fig. 13.2) as most of the 15 examined lakes are eutrophic to hypereutrophic. Fig. 13.4 indicates that the results found by modelling studies (Figs 13.1–13.2) are also valid when lake observations are used.

13.4. Thermodynamics of controlled ecological processes and exergy

A model of an ecosystem described by n state variables N_i can be written as the system of ordinary differential equations

$$\frac{dN_k}{dt} = F_k(N_1, \ldots, N_n; f_1(t), \ldots, f_m(t); \alpha_1, \ldots, \alpha_r), \qquad k = 1, \ldots, n \tag{4.1}$$

where $f_1(t)$ are the so-called *driving functions* describing processes that take place in the environment and do not depend on the ecosystem state such as temperature, solar radiation, etc. Parameters α_s are internal characteristics of the ecosystem, determining its functioning and its interaction with the environment such as, for instance, rates of nutrient uptake, respiration coefficients, maximal intrinsic rates of the ecosystem species, mortality coefficients, etc. Certainly, all these can depend on driving functions. It is natural to assume that values of all these parameters result in a long-term evolution of ecosystems while the maximal degree of adaptation is being attained. A measure of the adaptation is exergy. However, in order to reach this state some of the parameters have to change in the course of evolution, increasing the degree of adaptation, i.e. they are *control* variables. We assume for simplicity that all α_s are the control variables; since α_s vary within some natural borders binding the domain Ω_α, $\alpha_s \epsilon \Omega_{\alpha.}$ Since $f_k(t)$ are given functions, Eq. (4.1) is written as a non-autonomous dynamical system:

$$\frac{dN_k}{dt} = F_k(N_1, \ldots, N_n; \alpha_1, \ldots, \alpha_r; t), \qquad k = 1, \ldots, n. \tag{4.2}$$

Below, we follow the concepts of "Pontryagin's maximum principle" (Pontryagin et al., 1969).

We do not concretise the expression for exergy now; we shall do it later. Let the exergy derivative be

$$\frac{d\text{Ex}}{dt} = -F_0(N_1, \ldots, N_n; \alpha_1, \ldots, \alpha_r; t), \tag{4.3}$$

then the current value of exergy is

$$\text{Ex}(t) = -\int_{t_0}^{t} F_0(N_1, \ldots, N_n; \alpha_1, \ldots, \alpha_r; t)dt. \tag{4.4}$$

If we introduce a new variable $N_0 = -\text{Ex}$ extended system (4.2) is written as

$$\frac{dN_k}{dt} = F_k(N_1, \ldots, N_n; \alpha_1, \ldots, \alpha_r; t), \qquad k = 0, 1, \ldots, n. \tag{4.5}$$

We consider one more system in relation to auxiliary variables ψ_k:

$$\frac{d\psi_k}{dt} = -\sum_{j=0}^{n} \frac{\partial F_j}{\partial N_k} \psi_k, \qquad k = 0, 1, \ldots, n. \tag{4.6}$$

If we take into consideration the Hamiltonian

$$H(\boldsymbol{\psi}, \mathbf{N}, \boldsymbol{\alpha}, t) = \sum_{k=0}^{n} \psi_k F_k, \tag{4.7}$$

then all these equations are written as the following Hamiltonian system:

$$\frac{dN_k}{dt} = \frac{\partial H}{\partial \psi_k}, \qquad \frac{d\psi_k}{dt} = -\frac{\partial H}{\partial N_k}, \qquad k = 0, 1, \ldots, n. \tag{4.8}$$

The system starting at time t_0 from initial point $\mathbf{N}^0$ moves with $\boldsymbol{\alpha}(t) \in \Omega_a$. At time t_1 its exergy is equal to $\mathrm{Ex}_1 = \mathrm{Ex}(N^1, \alpha(t_1), t_1)$. In accordance with one of the Pontryagin theorems about optimality, this value is maximal if such a non-zero vector-function $\boldsymbol{\psi}(t) = (\psi_0, \psi_1, \ldots, \psi_n)$ corresponding to the optimal functions $\boldsymbol{\alpha}(t)$ and $\mathbf{N}(t)$ exists that

1. For any $t \in [t_0, t_1]$ Hamiltonian (4.7) attains its maximum at the optimal point $\alpha(t)$:

$$\sup_{\boldsymbol{\alpha}} [H(\boldsymbol{\psi}(t), \mathbf{N}(t), \boldsymbol{\alpha}(t), t)] = M(\boldsymbol{\psi}(t), \mathbf{N}(t), t), \tag{4.9}$$

2. The following relationships hold:

$$\psi_0(t) = \text{constant} \leq 0, \qquad M(\boldsymbol{\psi}(t), \mathbf{N}(t), t) = -\int_t^{t_1} \sum_{k=1}^{n} \frac{\partial F_k}{\partial t} \psi_k(t) dt. \tag{4.10}$$

Consider the system from the thermodynamic point of view. Since the total time derivative of exergy is

$$\frac{d\mathrm{Ex}}{dt} = \frac{\partial \mathrm{Ex}}{\partial t} + \sum_{k=1}^{n} \frac{\partial \mathrm{Ex}}{\partial N_k} \frac{dN_k}{dt},$$

then

$$\frac{\partial \mathrm{Ex}}{\partial t} = -\left[F_0 + \sum_{k=1}^{n} \frac{\partial \mathrm{Ex}}{\partial N_k} F_k \right]. \tag{4.11}$$

One of the optimality conditions is $\psi_0(t) = \text{constant} \leq 0$; without loss of generality we set $\psi_0(t) = -1$. If in addition we set $(\partial \mathrm{Ex}/\partial N_k) = -\psi_k$, $k = 1, \ldots, n$ then Eq. (4.11) is represented as

$$\frac{\partial \mathrm{Ex}}{\partial t} = H(\boldsymbol{\psi}, \mathbf{N}, \boldsymbol{\alpha}, t). \tag{4.12}$$

However, in accordance with Pontryagin's maximum principle at any $t \in [t_0, t_1]$ along the optimal trajectory leading to the maximal value of exergy at t_1 the Hamiltonian

$H(\boldsymbol{\psi}, \mathbf{N}, t) = \max_{\boldsymbol{\alpha}} H(\boldsymbol{\psi}, \mathbf{N}, \boldsymbol{\alpha}, t)$. Therefore, *everywhere along the optimal trajectory the rate of exergy growth*, $\partial \mathrm{Ex}/\partial t$, *is maximal for the given state and parameters.*

We have obtained a very important result proving the equivalency of local and global forms of the "exergy maximum principle". Of course, the equivalency has been proved for a special class of system. The problem of how large is the class is open until now, but it seems to us that a lot of chemical and ecological systems belong to this class. For instance, in the case of a chemical system, when exergy is equivalent to thermodynamic potential, $G(\mathbf{N}, t) = \sum_{k=1}^{n} \mu_k N_k$, $H(\boldsymbol{\psi}, \mathbf{N}, \boldsymbol{\alpha}, t) = \partial G(\mathbf{N}, t)/\partial t$, $\psi_k = -\mu_k$ (μ_k is the chemical potential), and the equations describing a state of the system are the chemical kinetic ones.

In the case of the stationary environment when the right sides in the model equations do not explicitly depend on time, the Hamiltonian also does not depend on time and the function $M(\boldsymbol{\psi}(t), \mathbf{N}(t), t) \equiv 0$ along the optimal trajectory. Therefore, $\partial \mathrm{Ex}/\partial t \equiv 0$, and $\mathrm{Ex} = \mathrm{Ex}(\mathbf{N})$, i.e. the exergy is only a function of the state. This can be interpreted as "the adaptation is possible only in a changing environment".

Defining the exergy at time t as $\mathrm{Ex} = \sum_{i=1}^{n} \mathrm{ex}_i^g N_i(f_1(t), \ldots, f_m(t); \alpha_1, \ldots, \alpha_r)$ where ex_i^g are the specific "genetic" exergies (see Chapters 5 and 12) and N_i, $i = 1, \ldots, n$ are the solution of Eq. (4.1), we formulate the principle of adaptation as a local extreme principle for exergy.

Any ecosystem which is unable to forecast the future state of environment and does not have a sufficiently long memory, i.e. it is a Markovian system, adjusts its interference with the environment in such a way that the maximal increment of exergy is provided.

Mathematically this principle is expressed as:

Let $\mathrm{dEx}/\mathrm{d}t = \sum_{i=1}^{n} \mathrm{ex}_i^g F_i(\mathbf{N}, \mathbf{f}, \boldsymbol{\alpha})$ where $\mathbf{f} = (f_1, \ldots, f_m)$. For the given variation of $\mathbf{f} : \delta f_k$, $k = 1, \ldots, m$ the parameters $\boldsymbol{\alpha}$ vary in such a manner that $(\mathrm{d}Ex/\mathrm{d}t)$ tends to maximum. Thus

$$\delta(\mathrm{dEx}/\mathrm{d}t) = \sum_{i=1}^{n} \sum_{k=1}^{m} \mathrm{ex}_i^g \frac{\partial F_i}{\partial f_k} \partial f_k + \sum_{i=1}^{n} \sum_{s=1}^{r} \mathrm{ex}_i^g \frac{\partial F_i}{\partial \alpha_s} \partial \alpha_s = 0. \tag{4.13}$$

By denoting $a_s = \sum_{i=1}^{n} \mathrm{ex}_i^g \dfrac{\partial F_i}{\partial \alpha_s}$ and $\varphi_k = \sum_{i=1}^{n} \mathrm{ex}_i^g \dfrac{\partial F_i}{\partial f_k}$ equality (4.13) is represented as

$$\sum_{s=1}^{r} a_s(t) \delta \alpha_s = -\sum_{k=1}^{m} \varphi_k(t) \delta f_k = K(t). \tag{4.14}$$

The variations $\delta \alpha_s$ have to belong to a domain $\delta \Omega_a$ containing the zero point. In addition, they are not to fall outside the domain of admissible α_s: Ω_α.

So, the adaptation principle given by Eq. (4.14) imposes a single constraint on variations $\delta \alpha_s$, therefore the system can adapt to a changing environment by changing its internal parameters. Moreover, if the line $\sum_{s=1}^{r} a_s \delta \alpha_s = K$ and domain $\delta \Omega_a$ are mutually disjointed, this means that the environment is changing so fast that the system is not able to adapt to the change. In this case the optimal strategy is to await further steps.

Certainly, there is a large uncertainty, and to reduce it we may introduce additional relationships between $\delta \alpha_a$: the problem is that we do not know how to do it. However, we could go another way, namely to reduce this uncertainty by increasing the uncertainty in the environment. We assume that there is a set of K: $K_p, p = 1, 2, \ldots, P$ belonging to some probabilistic distribution. For instance, this may be various temperatures typical for this site and season, etc. As a rule, $P >> r$, and we have an overdetermined system of linear algebraic equations

$$\sum_{s=1}^{r} a_s^p \, \delta\alpha_s = K_p, \qquad p = 1, \ldots, P \tag{4.15}$$

where

$$a_s^p(t) = \sum_{i=1}^{n} \mathrm{ex}_i^g \frac{\partial F_i(\mathbf{N}, \mathbf{f}^p, \boldsymbol{\alpha})}{\partial \alpha_s}$$

and $\mathbf{f}^p$ is a value of the driving vector-function for a concrete case. Analogously the value of K_p is calculated. System (4.15) can be solved by the method of least squares.

So, we obtain a single solution (in a certain sense) of Eq. (4.15), $\alpha_s^*(t)$, $s = 1, \ldots, r$, which can be considered as a mean adaptive strategy. Note that this approach (in another formulation) forms the basis of the structurally dynamic modelling (see Section 13.5).

13.5. Modelling the selection of Darwin's finches

As mentioned in Section 13.1, one example is applied in this chapter to illustrate the ideas behind structurally dynamic modelling and the application of exergy maximisation for parameter estimation and improved calibrations. Jørgensen and Fath (in press) compare a model that contains the available information on the selection of the beak size of Darwin's finches, a model with three classes of beak size and the structurally dynamic model which currently changes the beak size in accordance with a maximisation of the exergy. This latter model is presented in detail below.

All three models are based upon the detailed information about Darwin's finches which can be found in Grant's book (1986, 1999). All additional references are also in the book. The models reflect, therefore, the available knowledge which in this case is comprehensive and sufficient to validate even the ability of the third model to describe the changes in the beak size as a result of climatic changes, causing changes in the amount, availability and quality of the seeds that make up the main food item for the finches. The medium ground finches, *Geospiza fortis*, on the island Daphne Major were selected for these modelling cases due to very detailed case-specific information found in Grant (1986, 1999). The conceptual diagram of the model in the STELLA format is shown in Fig. 13.5.

The juvenile finches are promoted to adult finches 120 days after birth, according to growth curves by Boag (1984). The mortality of the adult finches is expressed as a normal mortality rate (Grant, 1986) + an additional mortality rate due to food shortage and also caused by a disagreement between beak depth and the size and hardness of seeds.

The beak depth can vary between 3.5 and 10.3 mm (Grant, 1986, 1999). Abott et al. (1977) have published an accordance (see Fig. 13.6) between beak size and a special parameter $\sqrt{DH}$, where D is the seed size and H the seed hardness which both are dependent on the precipitation, particularly in the months January–April (Grant and Grant, 1980; Grant, 1985). It is possible to determine a handling time for the finches for a given $\sqrt{DH}$ as a function of the beak depth (Grant, 1981), which explains that

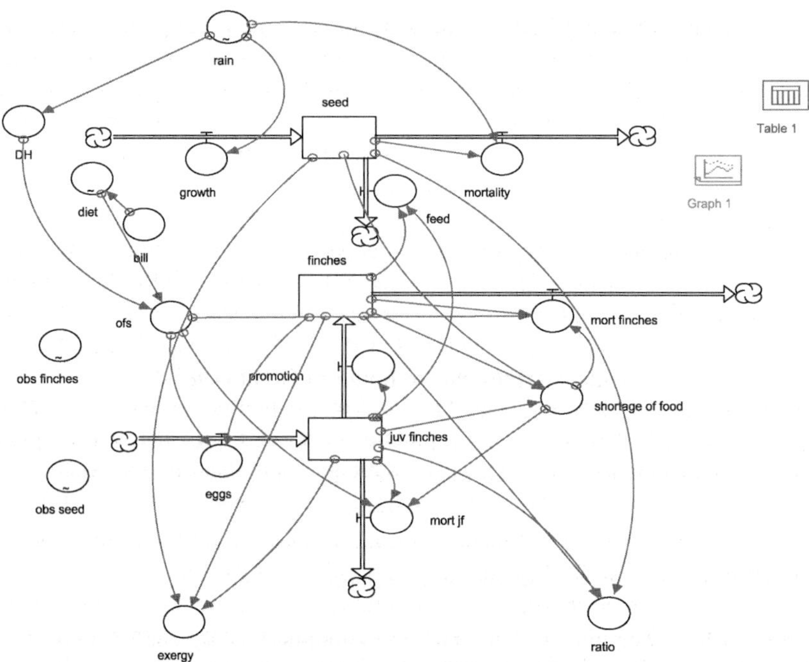

Fig. 13. 5. Conceptual diagram of the presented structurally dynamic model of Darwin's finches. The model has three state variables: juvenile finches, adult finches and seed. The beak size is currently adjusted to give the highest exergy.

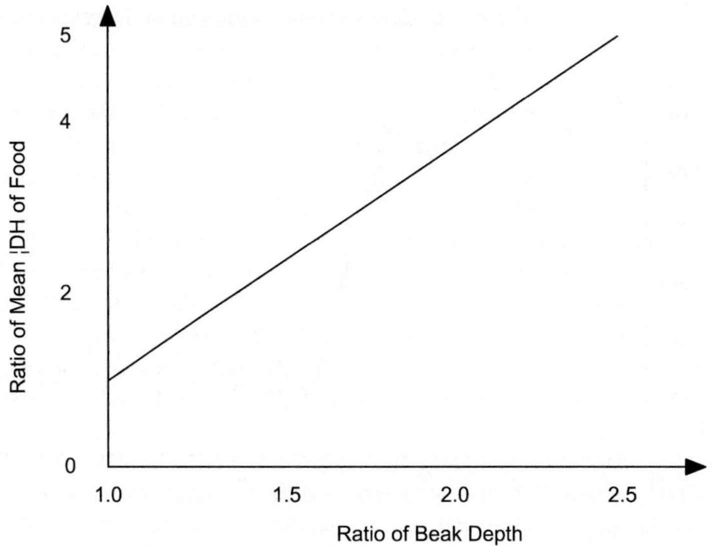

Fig. 13.6. The relative beak size is plotted versus $\sqrt{DH}$, where D is the seed size and H the seed hardness.

the accordance between $\sqrt{DH}$ and the beak depth becomes an important survival factor. Fig. 13.6 (Abott et al., 1977) is used in the model in order to find a function called "diet", which is compared with $\sqrt{DH}$, and to find how well the beak depth fits into the $\sqrt{DH}$ of the seed. This fitness function is based on information given by Grant (1981) about the handling time. It is named *ofs* in the model. It influences, as mentioned above, the mortality of adult finches, but also has an impact on the number of eggs laid and the mortality of the juvenile finches.

The growth rate and mortality of seeds is dependent on the precipitation, which is a forcing function known as a function of time (Grant, 1986, 1999). A function called shortage of food is calculated from the food required by the finches (Grant, 1985) and from the food available (the seed state variable). How the food shortage influences the mortality of juvenile finches and adult finches can be found in Grant (1985).

The seed biomass and the number of *G. fortis* as a function of time from 1975 to 1982 are known (Grant, 1985). These numbers from 1975 to 1976 (compare with Figs. 13.7 and 13.8) have been used to calibrate the coefficients determining:

1. The influence of the fitness function, *ofs*, on (a) the mortality of adult finches, (b) the mortality of juvenile finches, and (c) the number of eggs laid.
2. The influence of food shortage on the mortality of adult and juvenile finches is known (Grant, 1985). The influence is therefore calibrated within a narrow range of values.
3. The influence of precipitation on the seed biomass (growth and mortality).

All other parameters are known from the literature.

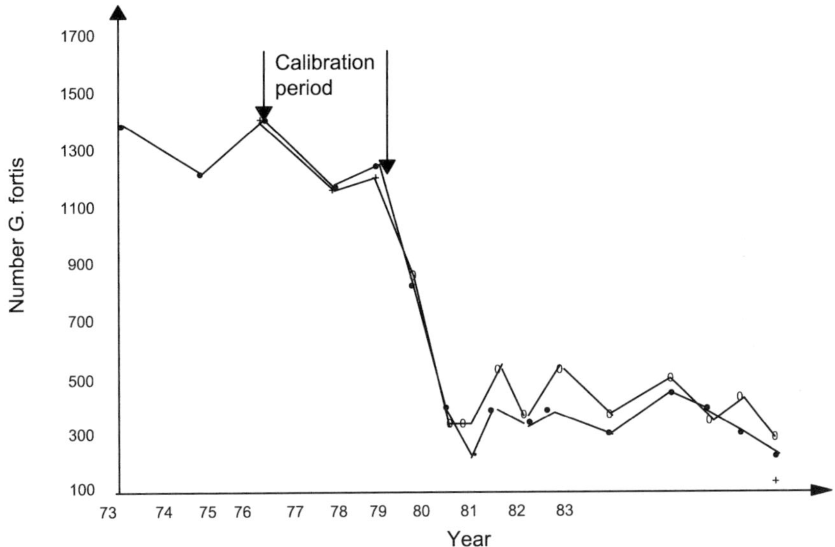

Fig. 13.7. The observed number of finches (•) from 1973 to 1983, compared with the simulated result (○). 1975 and 1976 were used for calibration and 1977/1978 for the validation referred to in Table 13.3.

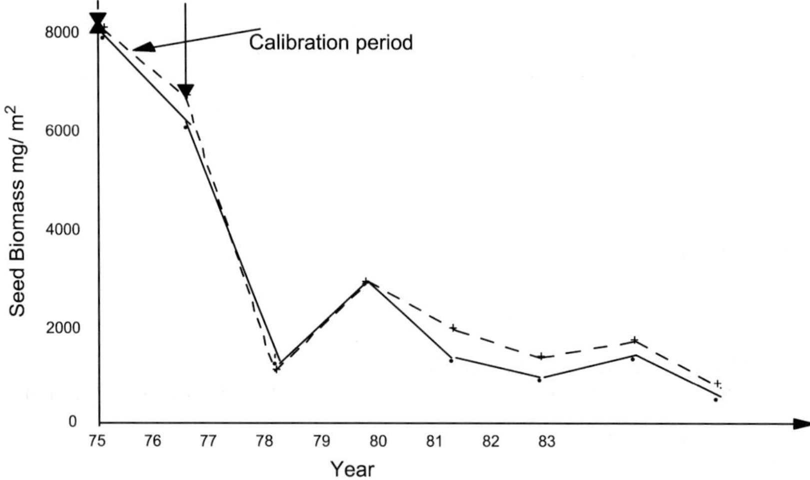

Fig. 13.8. The observed number of seeds (full line) compared with the simulated results (dotted line) from 1975 to 1983. The two first years, 1975/1976 observations were used for calibration, while the next 2 years are used for the validation referred to in Table 13.3.

This model is structurally dynamic, since it uses a current adjustment of the beak depth according to optimisation of a goal function. An exergy index, which measures the distance of the system (or rather the system as described by the model) from thermodynamic equilibrium, is used as a goal function. Exergy is calculated as $\mathrm{ex}_1^g N_1 + \mathrm{ex}_2^g N_2$, where ex_i^g refers to the information carried by the seed ($i = 1$) and by the finches ($i = 2$), and N_i refers to the densities of the seed and the finches. The application of this goal function presumes that the ecosystem tends to move as far as possible away from thermodynamic equilibrium under the prevailing conditions. The model was calibrated by the use of the data from 1975 and 1976. The precipitation and the beak depth (initial value in 1975–1976 4.5 mm) are known as a function of time according to Boag and Grant (1981, 1984). If these functions give the approximately right beak size as a function of time, the right number of seeds and *G. fortis* for the validation period, then the model is validated satisfactory. The discrepancy between predicted and observed beak size, number of seeds and *G. fortis* can be used as a measure of the model validation. The model is validated by the use of the data from 1977/1978; the number of finches, the biomass of seeds and the beak depth are used. These two years were used as it is considered to be most important that the model can simulate the shift in numbers that takes place during this period.

The model equations and the corresponding numerical algorithm are represented in Table 13.2 (in STELLA format).

The validation results can be seen from a comparison between the observations and the simulated results in Figs. 13.7 and 13.8 and Table 13.3. Notice that the number of parameters to be calibrated is only five, while a model that would contain n age classes would have $5n$ parameters.

Table 13.2
The equations for the presented structurally dynamic model of Darwin's finches

```
finches(t)/dt = (promotion-mort_finches)
INIT finches = 1170

promotion = IF(TIME > 120)THEN(juv_finches)ELSE(0)

mort_finches = (0.00125 + 0.0022 * (1-ofs) + 0.0026 * shortage_of_food) * finches

juv_finches(t)/dt = (eggs-promotion-mort_jf)
INIT juv_finches = 0

eggs = IF(TIME < 35)THEN(finches * (0.039-(1-ofs) * 0.05))ELSE(0)

promotion = IF(TIME > 120)THEN(juv_finches)ELSE(0)
mort_jf = IF(juv_finches > 50) THEN(0.0155 * juv_finches + 0.007 * (1-0.006 * shortage_of_food)ELSE(0)

seed(t)/dt = (growth-mortality-feed)
INIT seed = 8000

growth = 7.9 * rain

mortality = 0.0028 * seed-14 * (rain)^2
feed = finches * 0.024 * 0.055 + juv_finches * 0.015 * 0.1
beak = 12
DH = 50-12 * rain
exergy = seed * 120 * 20 + (finches * 16 + juv_finches * 9) * 370
ofs = 1-((DH)^0.5-diet) * 0.48
ratio = seed * 120/(16 * (finches + juv_finches))
shortage_of_food = IF(finches * 5 + juv_finches * 2 > seed)THEN(1)ELSE(0)
diet = GRAPH(beak)
(7.00, 4.00), (7.56, 4.25), (8.11, 4.50), (8.67, 4.75), (9.22, 5.00), (9.78, 5.26), (10.3, 5.55), (10.9, 5.80),
(11.4, 6.10), (12.0, 6.50)
obs_finches = GRAPH(TIME)
(0.00, 200), (30.4, 230), (60.8, 257), (91.2, 288), (122, 317), (152, 346), (182, 380), (213, 360), (243, 335),
(274, 320), (304, 305), (335, 290), (365, 272)
obs_seed = GRAPH(TIME)
(0.00, 3000), (30.4, 3100), (60.8, 3250), (91.2, 3400), (122, 3550), (152, 3800), (182, 4100), (213, 3820),
(243, 3630), (274, 3480), (304, 3350), (335, 3180), (365, 3000)
rain = GRAPH(TIME)
(0.00, 2.40), (33.2, 2.00), (66.4, 2.90), (99.5, 2.30), (133, 2.20), (166, 1.80), (199, 0.92), (232, 0.43), (265, 0.45),
(299, 0.5), (332, 0.4), (365, 1.70)
```

Table 13.3
Validation of the model

Percent average deviation between observed and simulated value	11.6
Linear regression value for observed vs. simulated finches (R^2)	0.971
Linear regression value for observed vs. simulated seed (R^2)	0.811
Simulated beak size at end of 1977 (observed value = 9.96 mm)	9.96
Simulated beak size at mid-1978 (observed value = 10.04 mm)	10.04
Number of parameters	5

13.6. Exergy of the global carbon cycle: how to estimate its potential useful work

Now we come back from geological times to the current state of the biosphere in order to estimate its capability to maintain the biological evolution. But at the beginning we would like to bear in mind several estimations connected with capabilities of the climate machine (see, for instance, Section 5.3).

Lorenz (1955) has introduced the concept of "available potential energy of the atmosphere", A, which maintains its general circulation. He defined it as the difference between the total potential energy, P, and the minimum total potential energy, which could result from any adiabatic redistribution of mass. This value shows how much of the mechanical work can be performed by the atmosphere in an ideal case. In accordance with Lorenz, the ratio $A/P = 1.75\%$. If we take into account the internal energy of the biosphere, U, then $A/(P + U) = 0.5\%$. In reality, the kinetic energy will still be less: about 10% of A. This means that in spite of the fact that the potential energy of the atmosphere is enormous, only its contemptible part is able to perform mechanical work, in particular, to "turn on" the machine of global circulation.

This result has stimulated us to muse upon: what percentage of the biosphere's NPP should be engaged in performing the useful work? Under this value we may imply, for instance, the maintenance of the animal kingdom (including humans) or the measure of ability for developing and reconstructing the biosphere.

How to estimate this work? Before that we shall talk about some data on the global carbon cycle. A studied system is the biota; its state is described by the total amount of carbon contained in the living matter of biota. Such compartments as the atmosphere and pedosphere, as well as the ocean containing *inorganic* carbon are considered as the environment of the system—the biota. A sum of carbon contained in the terrestrial and oceanic biota (living matter) is equal to $C_b^0 = 610$ Gt, in vegetation + 3 Gt, in marine biota = 613 Gt (den Elzen et al., 1995). The atmosphere and the biota are connected by inflow, which is equal to the annual net primary production NPP = 60 Gt for terrestrial + 6 Gt for oceanic biota. We see that the contribution of the oceanic part to the total amount of carbon contained in the biota, and to the exchange of carbon between the atmosphere and the biota compartment, is relatively small; in spite of it the net production of phytoplankton is equal to approximately 35–40 Gt. Actually the biological turnover of carbon within the surface ocean is practically closed. Since zooplankton grazes up to 80% of phytoplankton and its metabolism is very high, it plays a main role (with bacteria) in the process of decomposition of organic matter. Chemical (not biological) mechanisms play a main role in the carbon exchange and establishing equilibrium between the atmosphere and the ocean. So, when we consider the whole biogeochemical cycle we could be confined by the consideration of its terrestrial part, so that NPP = 60 Gt and $C_b^0 = 610$ Gt.

The global biogeochemical cycle of carbon is considered as a closed thermodynamic system. In favour of this hypothesis we should say: (a) anthropogenic emission of carbon dioxide today is sufficiently small in comparison to biological production and (b) if the biosphere is at the equilibrium then the amount of carbon, which is equal to the annual net production, must be released into the atmosphere as a result of decomposition of the dead organic matter contained in the litter and humus. The latter process is *spontaneous*.

The characteristic time of this process is 1 year. Certainly, when a new equilibrium is established, the amount of carbon in humus is also changed but this process is much slower.

We assume that the biosphere operates in a periodical regime (with a 1-year period). At the beginning the vegetation accumulating green biomass, by the same token, disturbs the equilibrium between the biota (biosphere) and the atmosphere and accumulates the exergy. Then the new biomass goes to the litter and soil and approximately the same amount of old biomass is decomposed with the emission of carbon dioxide into the atmosphere. As a result the accumulated exergy dissipates (but within another "decomposing" system), and the biosphere comes back to the primary equilibrium state. An annual cycle is completed. Certainly, two such stages, separated from each other, represent an ideal scheme; in reality they are overlapping.

The exergy accumulated during an annual cycle is equal to (see Section 5.4) $\mathrm{Ex} = C_b^1 \ln(C_b^1/C_b^0) - (C_b^1 - C_b^0)$ where $C_b^1 = C_b^0 + \mathrm{NPP}$. Assuming $\mathrm{NPP} = \Delta C \ll C_b^0$, the expression for exergy is re-written as

$$\mathrm{Ex} \approx \frac{(\Delta C)^2}{2C_b^0}. \tag{6.1}$$

Substituting the corresponding numerical values into Eq. (6.1) we get the estimation of exergy to be equal to 2.95 Gt C. Note that in this model the exergy is estimated in carbon units, like biomass.

It is necessary to say a few words about the energy cost of carbon contained in biomass. If we take into account only the "free" carbon or the carbon dioxide then the energy cost (contents) of 1 g C is equal to $RT/m_c = 8.4 \times 287/12 \approx 200$ J. On the other hand, the enthalpy estimation of the energy cost of 1 g C of biomass is 42 kJ. This example shows that the factor RT in the expression for chemical exergy ought to be used with care. For instance, when we calculate the "genetic" exergy, it is estimated in relation to detritus energy contents (detritus enthalpy), $\sim$18.5 kJ/g. If we take into account that carbon content in detritus is about 46% (Bazilevich, 1993) then the energy contents of 1 g C in detritus is equal to $\sim$40 kJ; that is very close to the standard value for living biomass.

Keeping in mind the interpretation of exergy as a quantity of mechanical (or useful) work, which can be performed by the system, the quantity Ex = 2.95 Gt C can be considered as an upper limit of the work performed by the terrestrial biosphere. This is only 4.9% of the whole energy flow of carbon cycle contained in the NPP; the rest is spent to support the turnover of carbon.

How can this work be interpreted? We think that this is maintenance of metabolism of the terrestrial superior animals (including humans), although the terrestrial cycle can be closed without terrestrial animals, since the main decomposers here are soil protists and microorganisms. It is interesting that the value 4.9% practically coincides with the estimation of the fraction of energy, which passes from autotrophic to the next trophic level in terrestrial ecosystems, and is approximately 5–10% (Odum, 1971a,b). There is also another hypothesis: this work can be used for the development and reconstruction of the biosphere.

Considering the total amount of carbon in the system, 610 Gt, as an analogue of its potential energy, we see that only $\mathrm{Ex}/C_b^0 = 2.95/610 \approx 0.48\%$ of it can be considered as an "available potential energy". It is interesting that this value practically coincides with Lorenz's estimation of such type of value for the atmosphere, 0.5%.

Calculating the exergy residual (see Section 5.4)—a measure of the system's ability to evolve—we get

$$\delta\mathrm{Ex} \approx \frac{1}{6(C_b^0)^2}(\Delta C)^3 \approx 0.097\ \mathrm{Gt}. \tag{6.2}$$

It is interesting that the ratio $R_{real} = \mathrm{Ex}/\delta\mathrm{Ex} \approx 30.4$, i.e. it is equal to the lower estimation of weighting factor in the specific "genetic" exergy for plants (see Table 5.1). If we take into account that almost all biomass of the biosphere is the biomass of plants, then we obtain a very surprising coincidence!

Calculations by the VECODE vegetation model under doubling CO_2-scenario (carbon amount in the atmosphere increases by two times, $((C_a^0)_1 = 1500$ Gt) gives an increase of carbon in the vegetation of 30% $((C_b^0)_1 = 793$ Gt), and an increase of production of 25% $(P_1 = 75$ Gt) (Cramer et al., 2001). By substituting these values into Eqs. (6.1) and (6.2) we get $\mathrm{Ex}_1 \approx 3.5$ Gt C and $\delta\mathrm{Ex}_1 \approx 0.11$ Gt C. The ratio $R_{model}(\mathrm{Ex}/\delta\mathrm{Ex}) \approx 31.8$ is almost unchanged in comparison with the real value, 30.4. Repeating the same calculations for the Moscow Biosphere Model (Svirezhev et al., 1985)—$P_2 = 70.6$ Gt and $(C_b)_2 = 678$ Gt—we get $\mathrm{Ex}_2 \approx 3.67$ Gt C, $\delta\mathrm{Ex}_2 \approx 0.127$ Gt C and $(\mathrm{Ex}/\delta\mathrm{Ex})_2 \approx 28.9$. We see that the ratio $\mathrm{Ex}/\delta\mathrm{Ex}$ is practically the same for both models and the real cycle. It is natural to expect this if we assume that this ratio reflects the genetic nature of plants, which of course has not changed with climate change.

The ratio R can be used as a criterion of *thermodynamic consistency* for different "mechanistic" models of the global carbon cycle. In an "ideal" model, if a modelled biosphere does not undergo some kind of catastrophe, the ratio value, which is calculated using model data, must be approximately equal to its current value, 30.4. If the model ratio differs from the current one then this can indicate the possible thermodynamic inconsistency of the model. For instance, the degree of the thermodynamic inconsistency defined as $TI = |R_{model} - R_{real}|/R_{real}$ is the same for both the Moscow Biosphere Model, 4.9%, and for VECODE, 4.6%.

When this value is applied, the specific exergy is found to be 520 kJ/g, which can be compared with the expected value, namely 30×18.4 kJ/g $= 550$ kJ/g, where 30 is the β-value for some average plant; see Table 5.1. The two values are very close, which may be considered in favour of calculating the weighting factors based on the genetic information.

The external factors steadily change species composition. The introduction of exergy calculations in ecological models makes it possible to describe this shift in the species composition, as will be presented later. Exergy is, so to speak, a measure of the survival as it accounts for biomass and information. Properties that give the best survival can be determined by testing those values of models' parameters that provide the highest exergy.

The exergy weighting factors have also been used in an attempt to express the evolution quantitatively by multiplication of the number of families and the weighting factors of the most developed species. The number of families expresses the possibilities that the ecosphere offers to utilise the available resources and ecological niches—we could call it the width of the biological information—and the weighting factor expresses the amount of feedback and regulation mechanisms the most advanced species have—we could call it the depth of the biological information. The product of the two would, therefore, be a relative measure of the overall increase of the biological information. Fig. 11.7 demonstrates this quantification of the evolution.

Postscriptum

The reader of this volume will hopefully agree with the authors that we have developed an ecosystem theory. The theory has been presented with the concept of exergy as the central issue. However, as pointed out many times in the volume, there are other approaches that are equally valid and which in some context may offer a better understanding than the approach presented here. These other approaches are, however, completely consistent with the theory presented in this volume. The various approaches form a pattern of ecological theories. It is not surprising that complex systems such as the ecosystem require more than one explanation to give a complete description. Even light requires two descriptions to be consistent with the observations; how many descriptions do we need for the ecosystem that is much more complex than light? So, the message is hopefully that various theories form a pattern and that we have a theory which can be applied much more widely than has been the case up to today. The pattern of theories has emerged in system ecology quite recently. The theory is, therefore, not without flaws. On the contrary, during the coming years it is necessary to improve the theory. Some of the formulation will be reformulated and expressed more clearly. Other basic laws may be added. Some detail of the present fundamental laws will perhaps be shown to be wrong and will therefore require more radical changes of the basic laws. It is not surprising, because this is how science develops.

The ecosystem theory presented in the previous 13 chapters *can* be reformulated in a compressive form as the following eight laws:

1A. Mass, for instance accounted as the elements, is conserved in ecosystems.
1B. Energy is conserved in ecosystems.
2. All processes in ecosystems are irreversible (this is probably the most useful way to express the Second Law of Thermodynamics in ecology). Another useful formulation, where the Second Law is applied to an ecosystem, is:
   Ecosystems are driven by an input of low entropy energy, which is transmitted, after its usage for maintenance of the ecosystem, to the environment as high entropy energy.
3. At 0 K, neither disorder nor order (structure) can be created. At increasing temperature, the processes creating order (structure) happen faster, but also the cost of maintaining the structure in the form of disordering processes gets higher. Carbon-based life is, therefore, found at the temperatures where there is a good balance between the rates of two opposite processes, i.e. at about 250–350 K.
4. When a system receives a through-flow of exergy, the system will, after covering the maintenance energy, use the exergy to move away from thermodynamic equilibrium. If more possibilities are offered, the one which most moves the system farther away (indicated as growth) from thermodynamic equilibrium will be selected.

Towards a Thermodynamic Theory for Ecological Systems, pp. 351–354

5. The growth of an ecosystem is possible by increase in the physical structure (biomass), by increase in the network (more cycling) and by increase of information embodied in the system. All three growth forms imply that the system is moving away from thermodynamic equilibrium and are associated with an increase of (1) the exergy stored in the ecosystem, (2) the through-flow (power) and (3) the ascendancy. The ecosystem receiving solar radiation will attempt to get maximum exergy storage, maximum power and maximum ascendancy.
6. The carbon-based life on Earth has a characteristic basic biochemistry, which all organisms share. This implies that many biochemical compounds can be found in all living organisms. They have, therefore, almost the same elementary composition and the composition of all organisms can be represented in a relatively narrow range of about 25 elements in total.
7. Biological systems are organized hierarchically (O'Neill et al., 1986). The variables describing the state at any level are determined by the processes at their immediate lower levels. Thereby, each level gets emerging properties.

These laws can also be used in ecology to explain ecological rules and observations. It is intuitively clear that the time has come to turn ecology into a theoretical science in line with physics. However, the basic ecological theory is young and, only a few years ago, it was acknowledged that the various ecological theories form a consistent pattern that can be used as a theoretical foundation of ecology.

Hopefully, all ecologists will use the opportunity to explain their observations and rules theoretically, but unfortunately the interactions between general ecologists and system ecologists (more theoretically inclined ecologists) are very limited. It is the hope of the authors of this volume that the presented pattern of ecological theory will encourage general ecologists to apply the ecological theories more widely. Let us propose that the big international and national meetings in ecology are used to encourage interactions between the two ecologies for the mutual benefit. A weak point in theoretical ecology can only be revealed by an application of the theory to explain observations made by the general ecologists and the ecological theory can be applied by general ecologists to support ecological observations. All scientific disciplines require not only observations and theory, but also a close cooperation between the two. Rapid and further progress in ecology is only possible through many and frequent interactions between observations and theory.

We would, however, propose to work with seven basic laws of system ecology to explain observations and rules in ecology and to understand reactions of ecosystems. Only by working with the laws can we find the corrections, improvements and additions to these laws that are needed. Only by trying to make ecology a theoretical science will we be able to build a solid theory of ecosystems in ecology.

Note that, at present, in environmental sciences, and particularly in ecology, we have a curious state-of-the-arts. On the one hand, a lot of different empirical information (experiments and observations in vitro and in vivo, etc.) on biological objects is collected, sometimes to such a fantastic degree of detail that it even becomes unclear how to use this information for further investigations. On the other hand, there are large lacunas in our positive knowledge about the mechanisms of ecosystem functioning, the character of interaction between an ecosystem's components, etc. In order to fill these lacunas we have to perform such an amount of work that it is not serious even to speak about it. As a result, these lacunas are filled by different hypotheses that are quite reasonable sometimes, but sometimes amount to no more than just being witty. Hence, it seems to us that the application of thermodynamic concepts gives a maximal guarantee of success on the way towards the construction of a general ecological theory.

Such properties of "information" provision in ecology naturally form two classes of ecological models. For the first class, the process of "natural selection" on the set of models, where the information is reasonably averaged and maximally aggregated in order to construct a suitable "thermostat" (in the thermodynamic sense of the word), is typical. The latter implies the possibility of representation of the modelled object as some unique system and its environment. Note that such a type of representation is a foundation stone of the exergy concept. The final result of selection is "analytical" models (for instance, a "prey–predator" model) that are outwardly simple, but their dynamics are complex. The Modelsauria that form the second class cannot survive here. Nevertheless, these models that try to use all the present information do exist (these are so-called simulation or portrait models). The algorithm of their construction is very simple: this is the Irish ragout recipe described by Jerome K. Jerome in "Three men in a boat": to put everybody into one pot, to mix it up, and then to boil. Of course, in this case we avoid the complex of "to splash out a tot with water"; that is inherent in models of the first class, but then we can never distinguish the simulation of reality from the artefact of modelling.

How can we get over the scarcity of information in ecological modelling? Firstly, we can use the temporal hierarchy that is usual in the ecosystem. Such an approach (well known in thermodynamics) allows distinguishing between "fast" and "slow" variables. In this case, we can always assume that the system reaches its equilibrium quickly, and then it slowly evolves in a quasi-stationary regime. The latter is a typical dynamics of the system in classic thermodynamics. In any case, the application of different time scales allows significant simplification of the original problem.

The second method is the use of different optimal principles for the completeness of models. In those cases where we do not have sufficient information about some "intimate" control mechanisms in the living systems, we exploit the perpetual teleologicity of the human mentality. It seems to us that the teleological formulations are more "scientific". The latter is very important when it is necessary to get over the subconscious suspicion of the representatives of descriptive natural sciences, of mathematical and physical ("simplifying the real complexity of Nature") models. Perhaps the teleological formulation of several biological rules and laws is a manifestation of this subconscious reaction(?). In the end, we are simply very lucky that the descriptive language of thermodynamics is teleological.

So, at the stage of construction of the primary model, we use:

1. Conservation laws, later on, are set as an invariable part of the model structure.
2. Some general information about the system's temporal and spatial hierarchies.
3. Some general information about the structure of matter, energy and information flows.
4. Extreme (optimal) teleological principles.

Note that any thermodynamic theory is constructed in accordance with this scheme (including even the classic thermodynamics).

Finally, concluding our book, we would like to stress once again that a thermodynamic theory for ecological systems may by no means be associated with the solved or nearly solved problems. We are still greatly restricted by the burden of ideas and concepts from classic thermodynamics, and therefore any new ideas, concepts and methods are only to be welcomed. We have introduced a few ideas in this volume, but many more are still needed. Perhaps we can say that such a thermodynamic theory is only in the formative stage, therefore...

(*to be continued*)

References

Abott, I., Abott, L.K., Grant, P.R., 1977. Comparative ecology of Galapagos Ground Finches: evaluation of the importance of floristic diversity and interspecific competition. Ecological Monographs 47, 151–184.

Ahl, T., Weiderholm, T., 1977. Svenska vattenkvalitetskriterier. Eurofierande ämnen. SNV PM (Swed), 918.

Alexeev, V.V., 1975. About applicability of the statistical mechanics methods to the description of biocoenoses. Biophysics 20(6), 1133–1136.

Andresen, 1983. Optimization of Exergy. Dr. Scient Thesis at Copenhagen University, 212pp.

Aoki, I., 1987. Entropy balance in Lake Biva. Ecol. Model. 37, 235–248.

Aoki, I., 1988. Entropy laws in ecological networks at steady state. Ecol. Model. 42, 289–303.

Aoki, I., 1989. Ecological study of lakes from an entropy viewpoint—Lake Mendota. Ecol. Model. 49, 81–87.

Aoki, I., 1992. Exergy analysis of network systems at steady state. Ecol. Model. 62, 183–193.

Aoki, I., 1993. Inclusive Kullback index—a macroscopic measure in ecological systems. Ecol. Model. 66, 289–300.

Aoki, I., 1995. Entropy production in living systems: from organisms to ecosystems. Thermochim. Acta 250, 359–370.

Aoki, I., 1998. Entropy and exergy in the development of living systems: a case study of lake-ecosystems. J. Phys. Soc. Jpn. 67, 2132–2139.

Augros, R., Stanciu, G., 1987. The New Biology. Discovering the Wisdom of Nature. Shambhala, Boston.

Bak, P., 1996. How Nature Works. Springer, New York, 212pp.

Barbashin, E.A., 1967. Introduction to Stability Theory. Nauka, Moscow.

Bastianoni, S., 1998. A definition of 'pollution' based on thermodynamic goal functions. Ecol. Model. 113, 163–166.

Bastianoni, S., Marchettini, N., 1996. Emergy/exergy ratio as a measure of the level of organization of systems. Ecol. Model. 99, 33–40.

Bauer, E.C., 1935. Theoretical Biology (in Russian). VIAM, Moscow.

Bazilevich, N.I., 1973. Bigeochemistry of the main types of global vegetation. Proceedings of the Fifth Meeting of the USSR Botanical Society, Kiev, pp. 239–244.

Bazilevich, N.I., 1993. Biological Productivity of Ecosystems of Northern Eurasia. Nauka, Moscow, 293pp.

Beckenbach, E.F., Bellman, R., 1961. Inequalities. Springer, Berlin.

Blumenfeld, L.A., 1977a. Problems of Biological Physics. Nauka, Moscow, 336pp.

Blumenfeld, L.A., 1977b. Probleme der molekularen Biophysik. Akademie-Verlag, Berlin.

Boag, P.T., 1984. Growth and allometry of external morphology in Darwin's Ground Finches on Isle Daphne Major, Galapagos. Journal of Zoology, London 204, 413–441.

Boag, P.T., Grant, P.R., 1981. Intense natural selection in a population of Darwin's Finches. In the Galapagos. Science 214, 82–85.

Boag, P.T., Grant, P.R., 1984. Darwin's Finches on Isla Daphne Major, Galapagos: breeding and feeding ecology in climatically variable environment. Ecological Monographs 54, 463–489.

Boltzmann, L., 1905. The Second Law of Thermodynamics (Populare Schriften. Essay No. 3 (Address to Imperial Academy of Science in 1886)). D. Riedel, Dordrecht, Reprinted in English in: Theoretical Physics and Philosophical Problems, Selected Writings of L. Boltzmann.

Bonner, J.T., 1965. Size an Cycle. An Essay on the Structure of Biology. Princeton University Press, New Jersey, 219pp.

Bossel, H., 1992. Real structure process description as the basis of understanding ecosystem. Ecol. Model. 63, 261–276.

Brillouin, L., 1956. Science and Information Theory. Academic Press, New York, 320pp.

Brown, J.H., 1995. Macroecology. The University of Chicago Press, Chicago, IL.
Brown, J.H., Marquet, P.A., Taper, M.L., 1993. Evolution of body size: consequences of an energetic definition of fitness. Am. Nat. 142, 573–584.
Brown, M.T., Mc Clanahan, T.R., 1992. Energy Systems. Overview of Thailand. Center for Wetlands, University of Florida, Gainesville, FL, 310pp.
Budyko, M.I. (Ed.), 1963. Atlas of the Earth's Thermal Balance. Hydrometeoizdat, Leningrad.
Budyko, M.I., 1977. Global Ecology. Mysl', Moscow, 328pp.
Burger, R., 2000. The Mathematical Theory of Selection, Recombination, and Mutation. Wiley Interscience, New York.
Cairns, J., Overbaugh, J., Miller, S., 1988. The origin of mutants. Nature 355, 142–145.
Cavalier-Smith, T., 1985. The Evolution of Genome Size. Wiley, Chichester, 480pp.
Cerbe, G., Hoffmann, H.J., 1996. Einführung in the Thermodynamik, 11th edn. Carl Hanser Verlag, München/Wien, 470pp.
Chernavsky, D.S., Chernavskaya, N.S., 1984. Problem of the new information in evolution. In: Zotin, A.I. (Ed.), Thermodynamics and Control of Biological Processes. Nauka, Moscow, pp. 247–254.
Chetaev, N.G., 1955. The Stability of Motion. GTI, Moscow, English version: Pergamon Press, New York, 200pp.
Christensen, V., Pauly, D., 1993. Trophic Models of Aquatic Ecosystems. ICLARM/Danida, 390pp.
Coffaro, G., Bocci, M., Bendoricchio, G., 1997. Structural dynamic application to space variability of primary producers in shallow marine water. Ecol. Model. 102, 97–114.
Costanza, R., 1992. Toward an operational definition of ecosystem health. In: Costanza, R., Norton, B.G., and Haskell, B.D. (Eds.), Ecosystem Health. Island Press, Washington, DC, pp. 239–256.
Costanza, R., Neil, C., 1982. The energy embodied in the products of the biosphere. In: Mitsch, W., Bosserman, R., and Klopathek, A. (Eds.), Energy and Ecological Modelling. Elsevier, Amsterdam, pp. 743–755.
Costanza, R., Sklar, F.H., 1985. Articulation, accuracy and effectiveness of mathematical models: a review of freshwater wetland applications. Ecol. Model. 27, 45–69.
Cramer, W., Bondeau, A., Woodward, F.I., Prentice, I.C., Betts, R.A., Brovkin, V., Cox, P.M., Fisher, V., Foley, J., Friend, A.D., Kucharik, C., Lomas, M.R., Ramankutty, N., Sitch, S., Smith, B., White, A., Young-Molling, C., 2001. Dynamic responses of global terrestrial ecosystems to changes in CO_2 and climate. Global Change Biol. 7, 357–373.
Darnell, W.L., Staylor, W.F., Gupta, S.K., Ritchey, N.A., Wilber, A.C., 1992. Seasonal variation of surface radiation budget derived from International Satellite Cloud Climatology Project C1 data. J. Geophys. Res. 97, 15741–15760.
Darnell, W.L., Staylor, W.F., Gupta, S.K., Ritchey, N.A., Wilber, A.C., 1996. Surface Radiation Budget: A Long-Term Global Dataset of Shortwave and Longwave Fluxes. EOS Transactions, Electronic Supplement.
Dawkins, R.D., 1982. The Extended Phenotype. Freeman, Oxford.
Dawkins, R.D., 1989. The Selfish Gene, 2nd edn. Oxford University Press, Oxford, 465pp.
de Groot, S.R., Mazur, P., 1962. Non-Equilibrium Thermodynamics. North-Holland, Amsterdam.
den Elzen, M., Beusen, A., Rotmans, J., 1995. Modelling global biogeochemical cycles: an integrated assessment approach, GLOBO Report Series No. 7. RIVM, Bilthoven, 104pp.
Ebeling, W., Engel, A., Feistel, R., 1990. Physik der Evolutionsprozesse. Akademie Verlag, Berlin, 374pp.
Eigen, M., 1981. Darwin und die Molekularbiologie. Angew. Chem. 93, 221.
Essex, C., 1984. Radiation and the irreversible thermodynamics of climate. J. Atmos. Sci. 41, 1985–1991.
Fath, B., Patten, B.C., 1998. Network orientors: a utility goal function based on network synergism. In: Müller, F. and Leupelt, M. (Eds.), Eco Targets, Goal Functions and Orientors. Springer, Berlin, pp. 161–176.
Fath, B., Patten, B.C., 2000. Goal Functions and Network Theory. Presented in Porto Venere May 2000 at the Second Conference on Energy.
Feistel, R., Ebeling, W., 1981. On the thermodynamics of irreversible processes in ecosystems. Studia Biophys. 86, 237.
Fenchel, T., 1974. Intrinsic rate of natural increase: the relationship with body size. Oecologia 14, 317–326.
Fischer, J., Hinde, R.A., 1949. The opening of milk bottles by birds. Br. Birds 42, 347–357.
Fonseca, J.C., Marques, J.C., Paiva, A.A., Freitas, A.M., Madeira, V.M.C., Jørgensen, S.E., 2000. Nuclear DNA in the determination of weighting factors to estimate exergy from organisms biomass. Ecol. Model. 126, 179–190.

Fontaine, T.D., 1981. A self-designing medel for testing hypotheses of ecosystem development. In: Dubois, D. (Ed.), Progress in Ecological Engineering and Management by Mathematical Modelling. Proceedings of the Second International Conference on the State-of-the-Art Ecological Modelling, April 1980, Liège, Belgium, pp. 281–291.

Frautschi, S., 1988. Entropy in an expanding universe. In: Weber, B.H., Depew, D.J., and Smith, J.D. (Eds.), Entropy, Information and Evolution. New Perspectives in Physical and Biological Evolution. MIT Press, Cambridge, MA, 488pp.

Futuyma, D.J., 1986. Evolutionary Biology, 2nd edn. Sinauer Associates Inc, Sunderland, MA, 600pp.

Gibbs, J.W., 1948. Collected Works, vols. 1,2. Yale University Press, New Haven.

Gilliland, M.W., 1982. Embodied Energy Studies of Metal and Fuel Minerals. Report to National Science Foundation.

Givnish, T.J., Vermelj, G.J., 1976. Sizes and shapes of liana leaves. Am. Nat. 110, 743–778.

Glansdorff, P., Prigogine, I., 1971. Thermodynamics of Structure, Stability and Fluctuations. Wiley Interscience, New York.

Gödel, K., 1986. Collected Works, vol. 1. Oxford University Press, New York, 488pp.

Goel, N.S., Maitra, S.C., Montroll, E.W., 1971. On the Volterra and other nonlinear models of interacting populations. Rev. Modern Phys. 43, 231–276.

Grant, P.R., 1981. The feeding of Darwin's Finches on *Tribulus cistoides*. American Scientists 69, 653–663.

Grant, P.R., 1985. Interspecific competition in fluctuating environments In: Diamond, J.M. and Case, T.J. (Eds.), Community Ecology. Harper & Row, New York, pp. 173–191.

Grant, P.R., 1986. Ecology and Evolution of Darwin's Finches. Princeton University Press, New Jersey, 492pp. Reprinted with foreword and afterword in 1999.

Grant, P.R., Grant, B.R., 1980. Annual variation in finch number, foraging and food supply on Sala Daphne Major, Galapagos. Oecologia 46, 55–62.

Gupta, S.K., Darnell, W.L., Wilber, A.C., 1992. A parameterization for longwave surface radiation from satellite data: recent improvements. J. Appl. Meteorol. 31, 1361–1367.

Gupta, S.K., Wilber, A.C., Darnell, W.L., Suttles, J.T., 1993. Longwave surface radiation over the globe from satellite data: an error analysis. Int. J. Remote Sens. 14, 95–114.

Handbook of Mathematical Economy (in four volumes). North-Holland, Amsterdam, 1991.

Hannon, B., 1973. The structure of ecosystems. J. Theor. Biol. 41, 534–546.

Hannon, B., 1979. Total energy cost in ecosystems. J. Theor. Biol. 80, 271–293.

Hannon, B., 1982. Energy discounting. In: Mitsch, W., Ragade, R., Bosserman, R., and Dillon, J. (Eds.), Energetics and Systems. Ann Arbor Science Publishers, Ann Arbor, MI, pp. 73–100.

Harries, J.E., 2000. Physics of the earth's radiative energy balance. Contemp. Phys. 41, 309–322.

Hastie, N., 2001. Perspective, p. 121–128 in Essays in Biochemistry. In: Chapman, K.E. and Higgins, S.J. (Eds.), Portland Press, Eynsham, UK, 131pp.

Herbst, M., 1997. Die Bedeutung der Vegetation für den Wasserhaushalt ausgewählter Ökosysteme. University of Kiel, Kiel, 119pp.

Herendeen, R., 1981. Energy intensity in ecological economic systems. J. Theor. Biol. 91, 6076–6620.

Hirschfelder, J.O., Curtiss, C.F., Bird, R.B., 1954. Molecular Theory of Gases and Liquids. Wiley, New York, 631pp.

Holling, C.S., 1986. The resilience of terrestrial ecosystems: local surprise and global change. In: Clark, W.C. and Munn, R.E. (Eds.), Sustainable Development of the Biosphere. Cambridge University Press, Cambridge, pp. 292–317.

ISPRA, 2001. The Application of Exergy as Ecological Indicator for Coastal Lagoons in Europe, Internal Research Report, ISPRA.

Ivanitzky, G.R., Esipova, N.G., Abagian, R.A., Shnol, S.E., 1985. Blocks perfecting of genetic text as a factor of the acceleration of the biological evolution. Biophysics 30, 418–421.

Jeffries, C., 1988. Mathematical Modeling in Ecology: A Workbook for Students. Birkhauser, Boston.

Johnson, L., 1990. The thermodynamics of ecosystem. In: Hutzinger, O. (Ed.), The Handbook of Environmental Chemistry, The Natural Environmental and The Biogeochemical Cycles, vol. 1. Springer, Heidelberg, pp. 2–46.

Johnson, L., 1995. The far-from-equilibrium ecological hinterlands. In: Patten, B.C., Jørgensen, S.E., and Auerbach, S.I. (Eds.), Complex Ecology, The Part-Whole Relation in Ecosystems. Prentice-Hall PTR, Englewood Cliffs, NJ, pp. 51–104.

Johnson, L., 2002. Imperfect Symmetry, Thermodynamics in Ecology and Evolution. Torgoth Publishing, Sidney, Canada, 220pp.

Jørgensen, S.E., 1976. A eutrophication model for a lake. J. Ecol. Model. 2, 147–165.

Jørgensen, S.E., 1982. A holistic approach to ecological modelling by application of thermodynamics. In: Mitsch, W., Ragade, R.K., Bossermann, R.W., and Dillon, J. (Eds.), Systems and Energy. Ann Arbor Press, Ann Arbor, MI, 192pp.

Jørgensen, S.E., 1984. Parameter estimation in toxic substance models. Ecol. Model. 22, 1–12.

Jørgensen, S.E., 1986. Structural dynamic model. Ecol. Model. 31, 1–9.

Jørgensen, S.E., 1988. Use of models as experimental tools to show that structural changes are accompanied by increased exergy. Ecol. Model. 41, 117–126.

Jørgensen, S.E., 1990. Ecosystem theory, ecological buffer capacity, uncertainty and complexity. Ecol. Model. 52, 125–133.

Jørgensen, S.E., 1992a. Parameters, ecological constraints and exergy. Ecol. Model. 62, 163–170.

Jørgensen, S.E., 1992b. Development of models able to account for changes in species composition. Ecol. Model. 62, 195–208.

Jørgensen, S.E., 1992c. Integration of Ecosystem Theories: A Pattern. Kluwer, Dordrecht, 383pp.

Jørgensen, S.E., 1994. Fundamentals of Ecological Modelling, 2nd edn., Developments in Environmental Modelling, vol. 19. Elsevier, Amsterdam, 628pp.

Jørgensen, S.E., 1995a. The growth rate of zooplankton at the edge of chaos: ecological models. J. Theor. Biol. 175, 13–21.

Jørgensen, S.E., 1995b. Exergy and ecological buffer capacities as measures of the ecosystem health. Ecosystem Health 1, 150–160.

Jørgensen, S.E., 1997. Integration of Ecosystem Theories: A Pattern, 2nd edn. (revised). Kluwer Academic Publishers, Dordrecht, 388pp.

Jørgensen, S.E., 2000a. Application of exergy and specific exergy as ecological indicators of coastal areas. Aquatic Ecosystem Health and Management 3, 419–430.

Jørgensen, S.E., 2000b. The Principles of Pollution Abatement. Elsevier, Amsterdam, The Netherlands, 520pp.

Jørgensen, S.E., 2001a. Parameter estimation and calibration by use of exergy. Ecol. Model. 146, 299–302.

Jørgensen, S.E., 2001b. Toward a consistent pattern of ecosystem theories. Sci. World 1, 71–75.

Jørgensen, S.E., 2002a. Explanation of ecological rules and observation by application of ecosystem theory and ecological models. Ecol. Model. 158, 241–248.

Jørgensen, S.E., 2002b. Integration of Ecosystem Theories: A Pattern, 3rd edn. (revised). Kluwer Academic Publisher, Dordrecht, 428pp.

Jørgensen, S.E., Bendoricchio, G., 2001. Fundamentals of ecological modelling, 3rd edn. Elsevier, Amsterdam, 628pp.

Jørgensen, S.E., de Bernardi, R., 1997. The application of a model with dynamic structure to simulate the effect of mass fish mortality on zooplankton structure in Lago di Annone. Hydrobiologia 356, 87–96.

Jørgensen, S.E., de Bernardi, R., 1998. The use of structural dynamic models to explain successes and failures of biomanipulation. Hydrobiologia 359, 1–12.

Jørgensen, S.E., Fath, B., Patten, B.C., Straskraba, M. Growth and development. Oikos (submitted for publication).

Jørgensen, S.E., Fath, B. Modelling the Selection of Darwin's Finches. Ecol. Model. (in press).

Jørgensen, S.E., Fath, B. A Thermodynamic Theory for Ecosystems. Ecol. Model. (submitted for publication).

Jørgensen, S.E., Mejer, H.F., 1977. Ecological buffer capacity. Ecol. Model. 3, 39–61.

Jørgensen, S.E., Mejer, H.F., 1979. A holistic approach to ecological modelling. Ecol. Model. 7, 169–189.

Jørgensen, S.E., Padisák, J., 1996. Does the intermediate disturbance hypothesis comply with thermodynamics? Hydrobiologia 323, 9–21.

Jørgensen, S.E., Logofet, D., Svirezhev, Yu., 1995. Exergy principle and exergical systems in ecological modeling In: Patten, B.C. and Jørgensen, S.-E. (Eds.), Complex Ecology: The Part-Whole Relation in Ecosystems. Prentice-Hall PTR, Englewood Cliffs, NJ, pp. 585–608.

Jørgensen, S.E., Marques, J.C., Nielsen, S.N., 2002. Structural changes in an estuary, described by models and using exergy as orientor. Ecol. Model. 158, 233–240.

Jørgensen, S.E., Mejer, H., Nielsen, S.N., 1998. Ecosystem as self-organizing critical systems. Ecol. Model. 111, 261–268.

Jørgensen, S.E., Nielsen, S.N., Mejer, H., 1995. Emergy, environ, exergy and ecological modelling. Ecol. Model. 77, 99–109.

Jørgensen, S.E., Patten, B.C., Straskraba, M., 1999. Ecosystem emerging: 3. Openness. Ecol. Model. 117, 41–64.

Jørgensen, S.-E., Patten, B.C., Straškraba, M., 2000. Ecosystems emerging: 4. Growth. Ecol. Model. 126, 249–284.

Jørgensen, S.E., Ray, S., Berec, L., Straskraba, M., 2002. Improved calibration of a eutrophication model by use of the size variation due to succession. Ecol. Model. 153, 269–278.

Jørgensen, S.E., Verdonschot, P., Lek, S., 2002. Explanation of the observed structure of functional feeding groups of aquatic macro-invertebrates by an ecological model and the maximum exergy principle. Ecol. Model. 158, 223–232.

Kauffman, S.A., 1991. Antichaos and adaption. Sci. Am. 265(2), 64–70.

Kauffman, S.A., 1993. Origins of Order: Self-Organization and Selection in Evolution. Oxford University Press, Oxford.

Kay, J., 1984. Self Organization in Living Systems. Thesis. Systems Design Engineering, University of Waterloo, Ontario, Canada.

Kay, J., 1991. A non-equilibrium thermodynamic framework for discussing integrity. Environ. Mgmt 15, 483–495.

Kay, J., Schneider, E.D., 1992. Thermodynamics and measures of ecological integrity, Proceedings "Ecological Indicators". Elsevier, Amsterdam, pp. 159–182.

Kerner, E.H., 1957. A statistical mechanics of interacting biological species. Bull. Math. Biophys. 19, 121–146.

Kerner, E.H., 1959. Further consideration on the statistical mechanics of biological associations. Bull. Math. Biophys. 21, 217–255.

Kharkevich, A.A., 1963. The cost of information, Problems of Cybernetics, vol. 9. Nauka, Moscow, pp. 71–102.

Khinchin, A.Ya., 1943. Mathematical Foundations in Statistical Mechanics. GTI, Moscow.

Khinchin, A.Ya., 1953. The entropy concept in the theory of probability. Uspekhi mat. Nauk 8(3), 3–20.

Khinchin, A.Ya., 1957. Mathematical Foundations in Information Theory. Dover Publications, New York.

Khlebnikov, V., 1910. Attempt of construction of one concept in Natural Sciences: Symbios and Metabios. News of Students Life, Sanct-Peterburg 1, 11–12.

Khrgian, A.G., 1983. Physics of the Atmosphere (in Russian), vol. I. HMI, Leningrad.

Klix, F. (Ed.), 1974. Organismische Informationsverarbeitung. Akademie-Verlag, Berlin.

Kostitzin, V.A., 1935. Evolution de l'atmosphere: Circulation Organique, Epoques Glaciares. Hermann, Paris, 125pp.

Kostitzin, V.A., 1937. La Biologie Mathematique. Armand Colin, Paris, 223pp.

Krapivin, V.F., Svirezhev, Yu. M., Tarko, A.M., 1983. Mathematical Modelling of the Global Biosphere Processes. Moscow, Nauka, 237pp.

Kullback, S., 1959. Information Theory and Statistics. Wiley, New York.

Kutsch, W.L., Steinborn, W., Herbst, M., Baumann, R., Barkmann, J., Kappen, L., 2001. Environmental indication: a field test of an ecosystem approach to quantify biological self-organisation. Ecosystems 4, 49–66.

Landau, L.D., Lifshitz, E.M., 1995. Statistical Physics (in Russian). Nauka, Moscow, English translation from the second edition: Statistical Physics, Pergamon Press, London, 1958.

La Salle, J., Lefschetz, S., 1961. Stability by Liapunov's Direct Method. Academic Press, New York.

Lewin, B., 1994. Genes V. Oxford University Press, Oxford, 620pp.

Lewontin, R.C., 1969. The Meaning of Stability, Diversity and Stability in Ecological Systems, Brookhaven Symposium in Biology, vol. 22. National Bureau of Standards, US Department of Commerce, Springfield, VA.

Li, W.-H., Grauer, D., 1991. Fundamentals of Molecular Evolution. Sinauer, Sunderland, MA, 430pp.

Lindeman, R.L., 1942. The trophic dynamic aspect of ecology. Ecology 23, 399–418.

Logofet, D.O., 1993. Matrices and Graphs: Stability Problems in Mathematical Ecology. CRC Press, Boca Raton, FL.

Logofet, D.O., 1997. Svicobians of the compartment models and DaD-stability of the Svicobians: aggregating '0-dimensional' models of global biogeochemical cycles. Ecol. Model. 104, 39–49.
Lorenz, E.H., 1955. Available potential energy and the maintenance of the general circulation. Tellus VII(2), 157–167.
Lorenz, E.H., 1963. Deterministic nonperiodic flow. J. Atmos. Sci. 20, 130–141.
Lorenz, E.H., 1967. The nature and theory of the general circulation of the atmosphere. WMO Publication, 218, WMO, Geneva, Switzerland, 161pp.
Lotka, A.J., 1922. Contribution to the energetics of evolution. Proc. Natl Acad. Sci. USA 8, 147–150.
Lotka, A.J., 1925. Elements of Physical Biology. Williams & Wilkins, Baltimore.
Lotka, A.J., 1956. Elements of Mathematical Biology. Dover Publications, New York, 465pp.
Lurie, D., Wagensberg, J., 1979. Entropy balance in biological development and heat dissipation in embryogenesis. J. Nonequilibrium Thermodyn. 4, 127.
Lurie, D., Wagensberg, J., 1984. Extreme principle for the biomass diversity in ecology. In: Zotin, A.I. (Ed.), Thermodynamics and Control of Biological Processes. Nauka, Moscow, pp. 153–163.
Luvall, J.C., Holbo, H.R., 1989. Measurements of short-term responses of coniferous forest canopies using thermal scanner data. Remote Sens. Environ. 27, 1–10.
Lyons, S.K., Willig, M.R., 2000. A hemisphere assessment of scale dependence in latitudinal gradients of species richness. Ecology 80, 2483–2491.
MacArthur, R.H., 1955. Fluctuations of animal populations and a measure of community stability. Ecology 36, 533–536.
Mann, C., 1991. Lynn Margulis: science's unruly earth mother. Science 252, 378–381.
Margalef, R.A., 1951. A practical proposal to stability. Publ. Inst. Biol. Apl., Univ. Barselona 6, 5–19.
Margalef, R.A., 1968. Perspectives in Ecological Theory. Chicago University Press, Chicago, IL.
Margalef, R.A., 1991. Networks in ecology In: Higashi, M. and Burns, T.P. (Eds.), Theoretical Studies of Ecosystems: The Network Perspectives. Cambridge University Press, Cambridge, pp. 41–57.
Margalef, R.A., 1995. Information theory and complex ecology. In: Patten, B.C. and Jørgensen, S.E. (Eds.), Compex Ecology. Prentice-Hall PTR, New Jersey, pp. 40–50.
Marques, J.C., Jørgensen, S.E., 2002. Three selected observations interpreted in terms of a thermodynamic hypothesis. Contribution to a general theoretical framework. Ecol. Model. 158, 213–222.
Marques, J.C., Pardal, M.A., Nielsen, S.N., Jørgensen, S.E., 1997. Analysis of the properties of exergy and biodiversity along an estuarine gradient of eutrophication. Ecol. Model. 102, 155–167.
Marques, J.C., Pardal, M.A., Nielsen, S.N., Jørgensen, S.E., 2002. Application of holistic ecological indicators of ecosystems' integrity: a case study in the Mondego Estuary. p. 551–564 in Aquatic Ecology of the Mondego River Basin—Global Importance of Local Experience. Coimbra University Press, Coimbra, 577pp.
Marquet, P.A., Taper, M.L., 1998. On size and area: patterns of mammalian body size extremes across land masses. Evol. Ecol. 12, 127–139.
Matthews, E., 1983. Global vegetation and land use: new high-resolution data bases for climate studies. J. Clim. Appl. Meteor. 22, 474–487.
Matthews, E., 1984. Prescription of Land-Surface Boundary Conditions in GISS GCM II: A Simple Method Based on High-Resolution Vegetation Data Sets, NASA TM-86096. National Aeronautics and Space Administration, Washington, DC.
Mauersberger, P., 1981. Entropie und freie Enthalpie in aquatische Ökosysteme. Acta Hydrodyn. 26, 67.
Mauersberger, P., 1983. General principles in deterministic water quality modelling. In: Orlob, G.T. (Ed.), Mathematical Modeling of Water Quality: Streams, Lakes and Reservoirs, International Series on Applied Systems Analysis, vol. 12. Wiley, New York, pp. 42–115.
Mauersberger, P., 1995. Entropy control of complex ecological processes. In: Patten, B.C. and Jørgensen, S.E. (Eds.), Complex Ecology: The Part-Whole Relation in Ecosystems. Prentice-Hall, Englewood Clifs, NJ, pp. 130–165.
May, R.M., 1973. Stability and Complexity in Model Ecosystems. Princeton University Press, Princeton, NJ, 265pp.
Mejer, H.F., Jørgensen, S.E., 1979. Energy and ecological buffer capacity. In: Jørgensen, S.E. (Ed.), State-of-the-Art of Ecological Modelling (Environmental Sciences and Applications, 7). Proceedings of a Conference on Ecological Modelling, 28 August–2 September 1978, Copenhagen, International Society for Ecological Modelling, Copenhagen, pp. 829–846.

Monteith, J.L., Unsworth, H., 1990. Principles of Environmental Physics. Edward Arnold, London, 241pp.
Morowitz, H.J., 1968. Energy flow in biology, Biological Organisation as a Problem in Thermal Physics. Academic Press, New York, 179pp. See review by H.T. Odum, Science, 164 (1969) 683–684.
Morowitz, H.J., 1970. Entropy for Biologists: An Introduction to Thermodynamics. Academic Press, New York, 195pp.
Morowitz, H.J., 1978. Foundations of Bioenergetics. Academic Press, New York.
Morowitz, H.J., 1992. Beginnings of Cellular Life. Yale University Press, New Haven.
Müller, F., Leupelt, M. (Eds.), 1998. Eco Targets, Goal Functions and Orientors. Springer, Berlin, 618pp.
Nicolis, G., Prigogine, I., 1977. Self-Organization in Non-Equilibrium Systems: From Dissipative Structure to Order Through Fluctuations. Wiley Interscience, New York, 492pp.
Nicolis, G., Prigogine, I., 1989. Exploring Complexity. An Introduction. W.H. Freeman and Company, New York, 314pp.
Nielsen, J.M., 2003. Proteins. Ingeniøren 13, 12–13.
Nielsen, M., 2001. Proteins. Ingeniøren 32, 12–13.
Nielsen, R., Haugaard, 2001. Hunting the proteins (In Danish Jagten på proteinerne). Ingeniøren. 16, 20(April), 20–21.
Nielsen, R.H., 1999. The Archaean Sea were teeming of life. Ingeniøren. Naturvidenskab. 6(February), 16–17.
Nielsen, S.N., 1992a. Application of Maximum Exergy in Structural Dynamic Models. PhD Thesis. National Environmental Research Institute, Denmark, 51pp.
Nielsen, S.N., 1992b. Strategies for structural–dynamical modelling. Ecol. Model. 63, 91–102.
Odling-Smee, Patten, 1992. Personal communication.
Odum, E.P., 1959. Fundamentals of Ecology, 2nd edn. W.B. Saunders, Philadelphia, PA.
Odum, E.P., 1969. The strategy of ecosystem development. Science 164, 262–270.
Odum, E.P., 1971. Fundamentals of Ecology, 3rd edn. W.B. Saunders, Philadelphia, US, 360pp.
Odum, E.P., 1983. Basic Ecology, vol. 1. Saunders College Publishing, Philadelphia, 320pp.
Odum, H.T., 1956. Primary production in flowing waters. Limnol. Oceanogr. 1, 102–117.
Odum, H.T., 1971. Environment, Power, and Society. Wiley Interscience, New York, 331pp.
Odum, H.T., 1983. System Ecology. Wiley Interscience, New York, 510pp.
Odum, H.T., 1988. Self-organization, transformity, and information. Science 242, 1132–1139.
Odum, H.T., 1996. Environmental Accounting—Emergy and Decision Making. Wiley, New York, 370pp.
Odum, H.T., 2002. Explanation of ecological relationships with energy system concepts. Ecol. Model. 158, 201–212.
Odum, H.T., Hall, A.S., 1995. Maximum Power. The Ideas and Applications of H.T. Odum. University Press of Colorado, Denver, US, 390pp.
Odum, H.T., Pinkerton, R.C., 1955. Time's speed regulator: the optimum efficiency for maximum power output in physical and biological systems. Am. Sci. 43, 331–343.
Odum, H.T., Cantlon, J.E., Kornicker, L.S., 1960. An organizational hierarchy postulate for the interpretation of species–individual distribution: Species entropy and ecosystem evolution and the meaning of a species-variety index. Ecology 41, 395–399.
O'Neill, R.V., DeAngelis, D.L., Waide, J.B., Allen, T.F.H., 1986. A Hierarchical Concept of Ecosystems. Princeton University Press, Princeton, NJ, 253pp.
Onsager, L., 1931. Reciprocal relations in irreversible processes. I. Phys. Rev. 37, 405–426.
Oster, G.F., Perelson, A.S., Katchalsky, A., 1973. Network Thermodynamics. Quart. Rev. Biophys. 6, 1–121.
Ostwald, W., 1931. Gedanken zur Biosphäre. Wiederabdruck. BSB B.G. Teubner Verlagsgesellschaft, Leipzig (reprinted 1978).
Our common future. From one Earth to one World: A Report of the Brundtland Commission, 1987. Oxford University Press, Oxford.
Pahl-Wostl, C., 1995. The Dynamic Nature of Ecosystems: Chaos and Order Entwined. Wiley, New York.
Pattee, H.H. (Ed.), 1973 Hierarchy Theory: The Challenge of Complex System. George Braziller, New York, 156pp.
Patten, B.C., 1968. Mathematical models of plankton production. Int. Revue Ges. Hydrobiol. 53, 357–408.
Patten, B.C., 1982. Environs: relativistic elementary particles for ecology. Am. Nat. 119, 179–219.

Patten, B.C., 1991. Network ecology: indirect determination of the life environment relationship in ecosystems. In: Higashi, M. and Burns, T.P. (Eds.), Theoretical Studies of Ecosystems: The Network Perspective. Cambridge University Press, Cambridge, pp. 288–351.

Patten, B.C., 1995. Network integration of ecological extremal principles: exergy, emergy, power, ascendancy, and indirect effects. Ecol. Model. 79, 75–84.

Patten, B.C., 1997. Synthesis of chaos and sustainability in a nonstationary linear dynamic model of the American black bear (*Ursus americanus Pallas*) in the Adirondack Mountains of New York. Ecol. Model. 100, 11–42.

Patten, B.C., Higashi, M., Burns, T.P., 1990. Trophic dynamics in ecosystem networks: significance of cycles and storage. Ecol. Model. 51, 1–28.

Patten, B.C., Straskraba, M., Jørgensen, S.E., 1997. Ecosystem Emerging: 1. Conservation. Ecol. Model. 96, 221–284.

Pedlosky, J., 1979. Geophysical Fluid Dynamics. Springer, New York, 619pp.

Peixoto, J.P., Oort, A.O., 1992. Physics of Climate. Springer, New York, 520pp.

Peters, R.H., 1983. The Ecological Implications of Body Size. Cambridge University Press, Cambridge, 329pp.

Petoukhov, V.K., 1985. Thermodynamical estimate of an upper limit for the kinetic energy generation rate in a planetary atmosphere. Izvestiya Acad. Science USSR, Atmos. Ocean. Phys. 21(5), 460–465.

Pianka, E.R., 1978. Evolutionary Ecology. Harper and Row, New York.

Pierce, J.R., 1980. An Introduction to Information Theory. Dover Publications, New York.

Polischuk, E.M., 1971. Some Statistical Criteria of Stability in Multi-Species Biocoenoses, The Works of the Agrophysical Institute, vol. 30. HMI, Leningrad, pp. 202–211.

Pontryagin, L.S., Boltyansky, V.G., Gamkrelidze, R.V., Misshenko, E.F., 1969. Mathematical Theory of Optimal Processes. Nauka, Moscow, 384pp.

Press, W.H., 1976. Theoretical maximum for energy from direct and diffuse sunlight. Nature 264, 734–735.

Prigogine, I., 1947. L' Etude Thermodynamique des Processus Irreversibles. Desoer, Liege.

Prigogine, I., 1955. Introduction to Thermodynamics of Irreversivle Processes. Thournes, Chicago, The second edition by Wiley Interscience, New York, 1967.

Prigogine, I., 1980. From Being to Becoming: Time and Complexity in the Physical Sciences. Freeman, San Franscisco, CA, 220pp.

Prigogine, I., Stengers, I., 1979. La nouvelle Alliance. Gallimard, Paris, 288pp.

Prigogine, I., Wiame, J.M., 1946. Biologie et Thermodynamique des Phenomenes Irreversibles. Experientia 2, 451.

Pritz, A.K., 1974. Principle of Stationary States for Open Systems and the Population Dynamics, Kaliningrad, 123pp.

Rapport, D.J., 1995. Preventive ecosystem health care: the time is now. Ecosystem Health 1, 127–128.

Raup, D.M., Sepkowski, J.J., 1982. Mass extinctions in the marine fossil record. Science 215, 1501–1503.

Redheffer, R., 1985. Volterra multipliers I. SIAM J. Alg. Disc. Meth. 6, 570–589.

Redheffer, R., 1985. Volterra multipliers II. SIAM J. Alg. Disc. Meth. 6, 612–623.

Reynolds, C.S., 2002. Ecological pattern and ecosystem theory. Ecol. Model. 158, 181–200.

Ronov, A.B., 1980. The Sediments Sphere of the Earth (Quantitative Conformities of Structure, Composition, and Evolution). Nauka, Moscow.

Rossow, W.B., Schiffer, R.A., 1991. ISCCP cloud data products. Bull. Am. Meteor. Soc. 72, 2–20.

Rouche, N., Habets, P., Laloy, M., 1977. Stability Theory by Liapunov's Direct methods. Springer, New York.

Rubin, A.B., 1999. Biophysics, vol. I. University Books Hous, Moscow, 448pp.

Ruelle, D., Tackens, F., 1971. On the nature of turbulence. Commun. Math. Phys. 20, 167–192.

Russel, L.D., Adebiyi, G.A., 1993. Classical Thermodynamics. Saunders College Publishing, Harcourt Brace Jovanovich College Publishers, Fort Worth, 620pp.

Rutten, M.G., 1971. The origin of Life by Natural Causes. Elsevier, Amsterdam.

Salamolard, M., Butet, A., Leroux, A., Bretagnolle, V., 2000. Responses of an avian predator to variations in prey density at a temperate latitude. Ecology 81, 2428–2441.

Salomonsen, J., 1992. Properties of exergy. Power and ascendancy along a eutrophication gradient. Ecol. Model. 62, 171–182.

Scheffer, M., Carpenter, S., Foley, J.A., Folke, C., Walker, B., 2001. Castrophic change of ecosystems. Nature 413, 591–596.

Schindler, D.W., 1988. Effects of acid rain on freshwater ecosystems. Science 239, 149–157.

Schlesinger, W.H., 1997. Biogeochemistry. An Analysis of Global Change, second ed. Academic Press, San Diego, 680pp.

Schneider, E., Kay, J., 1994. Life as a manifestation of the second law of thermodynamics. Math. Comput. Model. 19(6–8), 25–48.

Schoffeniels, E., 1976. Anti-Chance. Pergamon Press, New York, 188pp.

Schrödinger, E., 1944. What is life? Cambridge University Press, Cambridge, 212pp.

Shannon, C.E., Weaver, W., 1963. The Mathematical Theory of Communication (first published in 1949). University of Illinois Press, Champaign, IL, 388pp.

Sherry, D.F., Galef, B.G., 1984. Cultural transmission without imitation: milk bottle opening by birds. Anim. Behav. 62, 937–938.

Shieh, J.H., Fan, L.T., 1982. Estimation of energy (enthalpy) and energy (availability) contents in structurally complicated materials. Energy Resour. 6, 1–46.

Shnol, S.E., 1989. About the full determination of the biological evolutionary trajectories, or about the limit perfection attained in the course of the process of natural selection at really small intervals of time, Ontogenesis, Evolution and Biosphere: in memoria N.V. Timofeev-Resovsky. Nauka, Moscow, pp. 215–222.

Shugart, H.H., 1998. Terrestrial Ecosystems in Changing Environments. Cambridge University Press, Cambridge, 534pp.

Simon, H.A., 1973. The organisation of complex systems, Pattee, Hierarchy Theory. Braziller, New York, pp. 3–27.

Smil, V., 1991. General Energetics: Energy in the Biosphere and Civilization. Wiley Interscience, New York, 369pp.

Sommer, U., 1989. Toward a Darwinian ecology of plankton. In: Sommer, U. (Ed.), Plankton Ecology: Succession in Plankton Communities. Springer, Berlin, 390pp.

Steinborn, W., 2000. Quantifizierung von Ökosystem-Eigenschaften als Grundlage für die Umwelbewertung. Dissertation. Christian-Albrecht-Universität, Kiel.

Steinborn, W., Svirezhev, Yu., 2000. Entropy as an indicator of sustainability in agro-ecosystems: North Germany case study. Ecol. Model. 133, 247–257.

Stenseth, N.C., 1986. Darwinian evolution in ecosystems: a survey of some ideas and difficulties together with some possible solutions. In: Casti, J.L. and Karlquist, A. (Eds.), Complexity. Language and Life: Mathematical Approaches. Springer, Berlin, pp. 107–123.

Straskraba, M., 1979. Natural control mechanisms in models of aquatic ecosystems. Ecol. Model. 6, 305–322.

Straskraba, M., 1980. The effects of physical variables on freshwater production: analyses based on models. In: Le Cren, E.D. and McConnell, R.H. (Eds.), The Functioning of Freshwater Ecosystems (International Biological Programme 22). Cambridge University Press, Cambridge, pp. 13–31.

Straskraba, M., Gnauk, A.H., 1985. Freshwater Ecosystems: Modeling and Simulation, Developments in Environmental Modelling, vol. 8. Elsevier, Amsterdam.

Straskraba, M., Jørgensen, S.E., Patten, B.C., 1997. Ecosystem Emerging: 2. Dissipation. Ecol. Model. 96, 221–284.

Sukhachev, V.N., 1967. Structure and dynamics of biogeocoenoses, Structure and Types of Matter. Nauka, Moscow, pp. 560–577.

Svirezhev (Svirejev), Yu.M., 1974. Systems Analysis of the Biosphere: Concepts of the Russian Classical School, Towards Our Future, vol. II. Pergamon Press, London, pp. 121–144.

Svirezhev, Yu.M., 1976. Vito Volterra and the modern mathematical ecology. In: Volterra, V. (Ed.), Mathematical Theory of Struggle for Existence. Nauka, Moscow, (the postscript to the Russian translation of this book).

Svirezhev, Yu.M., 1977. Systems Analysis of the Biosphere: Concepts of Russian Classical School, Problems of Cybernetics, vol. 32. Moscow, Nauka, pp. 225–236 (in Russian).

Svirezhev, Yu.M., 1983. Modern Problems in Mathematical Ecology. Invited Lecture. Proceedings of the International Congress of Mathematics, Warsaw, vol. II, pp. 1677–1693.

Svirezhev, Yu.M., 1987. Non-linear Waves, Dissipative Structures, and Catastrophes in Ecology. Nauka, Moscow, 368pp.

Svirezhev, Yu.M., 1989. Coevolution of the Biosphere and Mankind: The Modern Globalistics and Concepts of Russian Classical School, Ontogenesis, Evolution and Biosphere: in memoria N.V. Timofeev-Resovsky. Nauka, Moscow, pp. 254–264.

Svirezhev, Yu.M., 1990. Entropy as a measure of environmental degradation. Proceedings of the International Conference on Contaminated Soils. Karlsruhe, Germany.

Svirezhev, Yu.M., 1991. Phenomenological Thermodynamics of Interacting Populations. J. Gen. Biol. 52(6), 840–854.

Svirezhev, Yu.M., 1992. Thermodynamic Ideas in Ecology, Lecture on the School of Mathematical Ecology. ICTP, Trieste.

Svirezhev, Yu. (Ed.), 1997a. The Brandenburg Biosphere Model: A New Version of the Moscow Biosphere Model. PIK, Potsdam, 835pp.

Svirezhev, Yu.M., 1997b. Exergy of the Biosphere. Ecol. Model. 96, 309–310 (Corrigendum to the article: Ecol. Model. 104, 307–308).

Svirezhev, Yu.M., 1997c. On some general properties of trophic networks. Ecol. Model. 99, 7–17.

Svirezhev, Yu., 1998a. Globalistics: a new synthesis. Philosophy of global modelling. Ecol. Model. 108, 53–65.

Svirezhev, Yu.M., 1998b. Thermodynamic orientors: how to use thermodynamic concepts in ecology? In: Müller, F. and Leupelt, M. (Eds.), Eco Targets, Goal Functions and Orientors. Springer, Berlin, pp. 102–122.

Svirezhev, Yu.M., 1999. Virtual biospheres: complexity versus simplicity. In: Rossi, C., Batianoni, S., Donati, A., and Marchettini, N. (Eds.), Tempos in Science and Nature: Structure, Relations, and Complexity, Annals of the New York Academy of Science, 879, pp. 368–382.

Svirezhev, Yu., 2000. Stability concepts in ecology. In: Jørgensen, S.E. and Müller, F. (Eds.), Handbook of Ecosystem Theories and Management. CRC Press LCC, Boca Raton, FL, pp. 361–383.

Svirezhev, Yu., 2001a. Thermodynanics of the biosphere. In: Jørgensen, S.E. (Ed.), Thermodynamics and Ecological Modelling. CRC Press LLC, Boca Raton, FL, pp. 349–364.

Svirezhev, Yu., 2001b. Thermodynamics and theory of stability. In: Jørgensen, S.E. (Ed.), Thermodynamics and Ecological Modelling. CRC Press LLC, Boca Raton, FL, pp. 117–132.

Svirezhev, Yu.M., 2002. Simple spatially distributed model of the global carbon cycle and its dynamic properties. Ecol. Model. 155, 53–69.

Svirezhev, Yu.M., Logofet, D.O., 1978. Stability of Biological Communities. Nauka, Moscow, 324pp (English version: Stability of Biological Communities, Mir, Moscow, 1983).

Svirezhev, Yu.M., Logofet, D.O., 1995. The Mathematics of Community Stability. In: Patten, B.C. and Jørgensen, S.-E. (Eds.), Complex Ecology: The Part-Whole Relation in Ecosystems. Prentice-Hall PTR, Englewood Cliffs, NJ, pp. 343–371.

Svirezhev, Yu.M., Passekov, V.P., 1982. Fundamentals of Mathematical Population Genetics. Nauka, Moscow, 482pp (English version: Fundamentals of Mathematical Evolutionary Genetics, Kluwer, Dordrecht, 1990, 395pp.).

Svirezhev, Yu.M., Steinborn, W.H., 2001. Exergy of solar radiation: information approach. Ecol. Model. 145, 101–110.

Svirezhev, Yu., Svirejeva-Hopkins, A., 1997. Diversity of the biosphere. Ecol. Model. 97, 145–146.

Svirezhev, Yu.M., Svirejeva-Hopkins, A., 1998. Sustainable biosphere: critical overview of basic concept of sustainability. Ecol. Model. 106, 47–61.

Svirezhev, Yu.M., Krysanova, V.P., Voinov, A.A., 1984. Mathematical modelling of a fish pond ecosystem. Ecol. Model. 21, 315–337.

Svirezhev, Yu.M., Krapivin, V.F., Tarko, A.M., 1985. Modelling of the main biosphere cycles. In: Malone, T.F. and Roederer, J.G. (Eds.), Global Change. Cambridge University Press, Cambridge, pp. 298–313.

Svirezhev, Yu.M., Steinborn, W.H., Pomaz, V., 2003. Exergy of solar radiation: global scale. Ecol. Model. 169, 339–346.

Sweinsdottir, S., 1997. Highest probability of life on the moon of Jupiter, Europa. Ingeniøren 21(May), 10–11.

Szargut, J., 1998. Exergy Analysis of Thermal Processes: Ecological Cost. Presented at a workshop in Porto Venere, May 1998.

Szargut, J., Morris, D.R., Steward, F.R., 1988. Exergy Analysis of Thermal, Chemical and Metallurgical Processes. Hemisphere Publishing Corporation/Springer, New York/Berlin, 312pp.

Taft, R.J., Mattrick, J.S., Andrew, P.S. Genome-wide increases in non-coding DNA positively correlate with increasing biological complexity. Draft paper.

Taylor, P.H., Gaines, S.D., 1999. Can Rapopport's rule be rescued? Modeling causes of the latitudinal gradient in species richness. Ecology 80, 2474–2482.

Thoma, J., 1977. Energy, Entropy and Information. IIASA, Laxenburg. Austria, pp. 77–32.

Tiezzi, E., 2003. The Essence of Time (125pp.) and The End of Time (202pp.). WIT Press, Southampton, UK.

Timofeev-Resovsky, N.V., 1958. Microevolution, elementary phenomena, material and factors of evolutionary process. Bot. J. 43, 317–336.

Timofeev-Resovsky, N.V., 1961a. About some principles of the biohorological units of classification. Proc. Ural Branch USSR Ac. Sci. 27, 290–311.

Timofeev-Resovsky, N.V., 1961b. Miassovo's Lecture.

Timofeev-Resovsky, N.V., 1968. The Biosphere and Mankind. UNESCO Bull. 1, 3–10.

Tripet, F., Richner, H., 1999. Density-dependent processes in the population dynamics of a bird ectoparasite Cedrato-phyllus Gallinae. Ecology 80, 1267–1277.

Turner, F.B., 1970. The ecological efficiency of consumer populations. Ecology 51, 741–742.

Ulanowicz, R.E., 1986. Growth and Development, Ecosystems Phenomenology. Springer, New York, 204pp.

Ulanowicz, R.E., 1995. Network Growth and Development: Ascendency. In: Patten, B.C. and Jørgensen, S.-E. (Eds.), Complex Ecology: The Part-Whole Relation in Ecosystems. Prentice-Hall PTR, Englewood Cliffs, NJ, pp. 643–655.

Ulanowicz, R.E., 1997. Ecology, The Ascendent Perspective. Columbia University Press, New York, 201pp.

Ulanowicz, R.E., 1998. Network orientors: theoretical and philosophical considerations why ecosystems may exhibit a propensity to increase in ascendancy. In: Müller, F. and Leupelt, M. (Eds.), Eco Targets, Goal Functions and Orientors. Springer, Berlin, pp. 177–192.

Ulanowicz, R.E., Abarca-Arenas, L.G., 1997. An informational synthesis of ecosystem structure and function. Ecol. Model. 95, 1–10.

Ulanowicz, R.E., Hannon, B.M., 1987. Life and the production of entropy. Proc. Royal Soc. London 232, 181–192.

Usher, M.B., Williamson, M.H., (Eds.) 1974. Ecological Stability. Chapman & Hall, London.

Venevsky, S.V., 1991. Entropy as a function of the biosphere. J. Gen. Biol 52(6), 900–915.

Vernadsky, V.I., 1926. The Biosphere (in Russian). Nauchtekhizdat, Leningrad, English version: Vernadsky V.I., 1998. The Biosphere (complete annotated edition), Copernicus, New York, 192pp.

Vinogradov, A.P., 1959. Chemical evolution of the Earth. The USSR Academy Scientific Publisher, Moscow.

Volkenstein, M.V., 1988. Biophysics. Nauka, Moscow, 592pp.

Vollenweider, R.A., 1975. Input–output models with special references to the phosphorus loading concept in limnology. Schweiz. Z. Hydrol. 37, 53–84.

von Bertalanffy, L., 1942. Theoretische Biologie. Verlag von Gebrüder Bornträger, Berlin.

von Bertalanffy, L., 1952. Problems of Life. Wiley, New York, 240pp.

von Bertalanffy, L., 1956. Biophysik des Fließgleichgewichts. Vieweg, Braunschweig.

Walter, H., 1964. Die Vegetation der Erde in öko-physiologischer Betrachtung, Die tropischen und subtropischen Zonen, vol. 1. Fischer, Jena, 538pp.

Walter, H., 1968. Die Vegetation der Erde in öko-physiologischer Betrachtung, Die gemässigten und arktischen Zonen, vol. 2. Fischer, Jena, 1001pp.

Warren, C.E., 1970. Biology and Water Pollution Control. W.B. Saunders, Philadelphia, 340pp.

Weiderholm, T., 1980. Use of benthos in lake monitoring. J. Water Pollut. Control Fed. 52, 537.

Whitlock, C.H., Charlock, T.P., Staylor, W.F., Pinker, R.T., Laszlo, I., Ohmura, A., Gilgen, H., Konzelman, T., DiPasquale, R.C., Moats, C.D., LeCroy, S.R., Ritchey, N.A., 1995. First global WCRP shortwave surface radiation budget data set. Bull. Am. Meteor. Soc. 76, 905–992.

Wilmers, F., 1984. Variations of plant Albedo due to latitude and altitude above sea level. In: Lieth, H. (Ed.), Interaction between Climate and Biosphere. Swets & Zeitlinger B.V., Lisse, pp. 75–90.

Wilsey, B.J., Potvin, C., 2000. Biodiversity and ecosystem functioning: importance of species evenness in an old field. Ecology 81, 887–892.

Wilson, D.S., 1978. Prudent predation: a field test involving three species of tiger beetles. Oikos 31, 128–136.

Wolfram, S., 1984a. Cellular automata as models of complexity. Nature 311, 419–424.

Wolfram, S., 1984b. Computer software in science and mathematics. Sci. Am. 251, 140–151.

Xu, F.-L., Jørgensen, S.E., Tao, S., 2000. Ecological indicators for assessing freshwater ecosystem health. Ecol. Model. 116, 77–92.

Yaglom, A., Yaglom, I., 1973. Probability and Information. Nauka, Moscow.

Zavalishin, N.N., Logofet, D.O., 1997. Modelling of ecosystems using given storage-flow diagrams. Math. Model. 9(9), 3–17.

Zavalishin, N.N., Logofet, D.O., 2001. Dynamic compartmental models of the carbon cycle in the ecosystem of the transition bog. Math. Model. 13(4), 3–18.

Zhang, J., Jørgensen, S.E., Tan, C.O., Beklioglu, M., 2003a. A structurally dynamic modelling—Lake Mogan, Turkey as a case study. Ecol. Model. 164, 103–120.

Zhang, J., Jørgensen, S.E., Tan, C.O., Beklioglu, M., 2003b. Hysteresis in vegetation shift—Lake Mogan prognoses. Ecol. Model. 164, 227–238.

Zhang, J., Jørgensen, S.E., Mahler, H. Examination of structurally dynamic eutrophication models. Ecol. Model. (in press).